Advances in Intelligent Systems and Computing

Volume 756

Series editor

Janusz Kacprzyk, Polish Academy of Sciences, Warsaw, Poland
e-mail: kacprzyk@ibspan.waw.pl

The series "Advances in Intelligent Systems and Computing" contains publications on theory, applications, and design methods of Intelligent Systems and Intelligent Computing. Virtually all disciplines such as engineering, natural sciences, computer and information science, ICT, economics, business, e-commerce, environment, healthcare, life science are covered. The list of topics spans all the areas of modern intelligent systems and computing such as: computational intelligence, soft computing including neural networks, fuzzy systems, evolutionary computing and the fusion of these paradigms, social intelligence, ambient intelligence, computational neuroscience, artificial life, virtual worlds and society, cognitive science and systems, Perception and Vision, DNA and immune based systems, self-organizing and adaptive systems, e-Learning and teaching, human-centered and human-centric computing, recommender systems, intelligent control, robotics and mechatronics including human-machine teaming, knowledge-based paradigms, learning paradigms, machine ethics, intelligent data analysis, knowledge management, intelligent agents, intelligent decision making and support, intelligent network security, trust management, interactive entertainment, Web intelligence and multimedia.

The publications within "Advances in Intelligent Systems and Computing" are primarily proceedings of important conferences, symposia and congresses. They cover significant recent developments in the field, both of a foundational and applicable character. An important characteristic feature of the series is the short publication time and world-wide distribution. This permits a rapid and broad dissemination of research results.

More information about this series at http://www.springer.com/series/11156

Jolanta Mizera-Pietraszko
Pit Pichappan · Lahby Mohamed
Editors

Lecture Notes in Real-Time Intelligent Systems

 Springer

Editors
Jolanta Mizera-Pietraszko
Department of Computer Science, Institute
   of Mathematics and Computer Science
Opole University
Opole
Poland

Lahby Mohamed
University of Hassan II
Casablanca
Morocco

Pit Pichappan
Digital Information Research Foundation
Chennai
India

ISSN 2194-5357   ISSN 2194-5365   (electronic)
Advances in Intelligent Systems and Computing
ISBN 978-3-319-91336-0   ISBN 978-3-319-91337-7   (eBook)
https://doi.org/10.1007/978-3-319-91337-7

Library of Congress Control Number: 2018942337

Printed on acid-free paper

This Springer imprint is published by the registered company Springer International Publishing AG
part of Springer Nature
The registered company address is: Gewerbestrasse 11, 6330 Cham, Switzerland

Preface

Nowadays, powerful intelligent systems adopt variety of sophisticated novel approaches aimed to respond to the challenges of the real-world problems. Across almost all the domains, dynamical progress in development of innovative conceptual models, paradigms, and techniques of big data processing reflects in science and profoundly transforms the industrial infrastructure. Simply saying, artificial intelligence supports human intelligence in most areas.

This volume explores the current methodologies based on the state of the art in real-time intelligent computation. It presents the studies that address the real-world challenges and provides a complete synthesis of the research reflections in computer-aided intelligence.

The 46 research works published in this volume have been selected for oral presentations out of a total of 149 submissions received from many countries. They were presented at the Second International Conference on Real-Time Intelligent Systems held at Casablanca in Morocco during October 18–20, 2017. RTIS 2017 conference was organized by Digital Information Research Foundation in India and the UK and sponsored by University of Hassan II in Morocco.

This collection of chapters offers the latest trends in development of real-time intelligent computation and provides the comprehensive solutions to numerous challenging issues.

For that reason, the proceedings of the RTIS 2017 conference are mainly recommended not only to the scientific community but also to IT consultants, software developers, and all those interested in intelligent computing.

The Editors would like to express special thanks to this book series Editor Professor Janusz Kacprzyk and the Springer Editors for their work on publication of this volume. We are very thankful to the RTIS 2017 Chairs, Program Committee Members, and the authors who contributed to the growth of the domain.

Jolanta Mizera-Pietraszko
Pit Pichappan
Mohamed Lahby

Organization of the RTIS 2017 Conference

Honorary General Chairs

Idriss Mansouri	President, Hassan II University, Casablanca, Morocco
Jolanta Mizera-Pietraszko	Opole University, Poland

General Chairs

Abderrahim Sekkaki	University Hassan II, Morocco
Simon Fong	University of Macau, Macau

Program Chairs

Pit Pichappan	Digital Information Research Lab, India
Mohamed Lahby (TPC Chair)	UH2C, Casablanca, Morocco

Organizing Committee

Mohamed Lahby	ENS, University Hassan II, Casablanca, Morocco
Raouyane Brahim	FSAC, University Hassan II, Casablanca, Morocco
Said Jai Andaloussi	FSAC, UH2C, Casablanca, Morocco
Noreddine Gherabi	ENSA of Khouribga, Hassan 1st University, Morocco
Khalid Jebari	Faculty Chouaid Doukkali, Univ. El Jadida, Morocco
Taoufik Rachad	ENSIAS, UM5, Rabat, Morocco

International Program Committee

Abdallah Shami	University of Western Ontario, Canada
Ahmed Ali	COMSATS Institute of Information Technology, Pakistan
Abdelwahab Naji	ENSET Mohammedia, UH2C, Casablanca, Morocco
Ahmed Karmouch	University of Ottawa, Canada
Alain Richard Ndjiongue	University of Johannesburg, South Africa
Ali Shaukat	Simula Research Laboratory, Norway
Amr Ali-Eldin	Leiden University, The Netherlands
Benchaïba Mahfoud	University of Science and Technology Houari Boum, Algeria
Brahim Raouyane	FSAC, UH2C, Casablanca, Morocco
Cherkaoui Leghris	FST Mohammedia, UH2C, Casablanca, Morocco
Essaid Sabir	ENSEM, UH2C, Casablanca, Morocco
Idy Diop Cheikh	Anta Diop University, Senegal
Jerzy Józefczyk	Wroclaw University of Science and Technology, Poland
Jun Liu	Ulster University, UK
Kanae Matsui	Tokyo Denki University, Japan
Katarzyna Gdowska	AGH University of Science and Technology, Poland
Khalid Jebari	Faculty Chouaid Doukkali, Univ. El Jadida, Morocco
Leila Fetjah	FSAC, University Hassan II, Casablanca, Morocco
Mahmoud A. Doughan	Lebanese University, Lebanon
Mansoor Ahmed	COMSATS Institute of Information Technology, Pakistan
Markos Papageorgiou	Technical University of Crete, Greece
Michele Ottomanelli	Technical University of Bari, Italy
Mohamed Al-Sarem	Taibah University, Medina, Saudi Arabia
Mohamed Amine	ERRIAS, FSAC, UH2C, Casablanca, Morocco
Mohamed El Khaili	ENSET Mohammedia, UH2C, Casablanca, Morocco
Mohamed Kissi	FST Mohammedia, UH2C, Casablanca, Morocco
Mohamed Lahby	ENS, UH2C, Casablanca, Morocco
Mohamed Reda Chbihi Louhdi	FSAC, UH2C, Casablanca, Morocco
Mohammed Moujjabir	FST Mohammedia, UH2C, Casablanca, Morocco
Noreddine Gherabi	ENSA of Khouribga, Hassan 1st University, Morocco
Paolo Delle Site	University Niccolò Cusano, Rome, Italy

Pascal Lorenz	University of Haute Alsace, France
Roberta Di Pace	University of Salerno, Italy
Ronald R. Yager	Machine Intelligence Institute, Iona College, USA
Saeed Ghazi Maghrebi	Islamic Azad University, Iran
Said Jai Andaloussi	FSAC, UH2C, Casablanca, Morocco
Saïd Nouh	UH2C, Casablanca, Morocco
Schahram Dustda	TU Wien, Austria
Shaukat Ali	Simula Research Laboratory, Norway
Suat Ozdemir	Gazi University, Turkey
Taoufik Rachad	ENSIAS, UM5, Rabat, Morocco
Tarek Bejaoui	University of Carthage, Tunisia
Youssef Baddi	UCD, El Jadida, Morocco
Zbigniew Banaszak	Warsaw University of Technology, Poland
Madjed Bencheikh Lehocine	Constantine 2 University, Algeria
Bellafkih Mostafa	INPT, Rabat, Morocco
Belmekki Abdelhamid	INPT, Rabat, Morocco
Khoukhi Fadoul	FST Mohammedia, Morocco
Sabbar Wafae	FSJES-Casablanca, Morocco
Saddoune Mohammed	FST Mohammedia, Morocco
Faiq Gmira	FSJESAS Hassan II University, Casablanca, Morocco

Students Committee

Youness Abakarim	ENS, UH2C, Casablanca, Morocco
ElMehdi Belbacha	ENS, UH2C, Casablanca, Morocco
Khalid Elfahssi	FSAC, UH2C, Casablanca, Morocco
Karim Benzidane	FSAC, UH2C, Casablanca, Morocco
Jamal Mawane	FST Mohammedia, UH2C, Casablanca, Morocco

Contents

Application Areas of Real-Time Intelligence

Modelling Intelligent Systems

Modeling, Design and Development of a Multi-agent Decision Support System for the Real-Time Control of the Operating Theaters

Fatima Taif[✉], Abedelwahed Namir, and Mohamed Azouazi

Faculty of Sciences Ben M'Sik, Casablanca, Morocco
taiffatima@gmail.com, Abd.namir@gmail.com, Azouazii@gmail.com

Abstract. The effective management of the hospital system user depends on multiple factors including the anticipation and responsiveness. A system of this exceptional situation is characterized by sudden onset of several elements that disrupt the execution of operatory program underway. Dealing with this kind of situations, the hospital must have a tool to help making decisions on time. In this context, we modeled the process of emergency care in the operating room with an IDSS (Interactive Support System Decision) embedded in a system multi agent (MAS). Specifically, the agents are assisted by a decision support system for planning elective surgery and the allocation of the necessary human and medical resources. Agents express their preferences using the method ELECTRE III to resolve the differences. The negotiation mechanism is based on the CNP (Contract Net Protocol). The protocol developed on JADE (Java Agent Development Framework) provides message exchanges between agents and their proposes predefined behaviors. The approach is tested through simple scenarios

Keywords: Coordination · Decision · Modeling · Multi-agents
Operating theaters · Real time

1 Introduction

The operating theaters are a very costly bottleneck in most hospital systems. All patient flows from the various surgical departments converge towards the operating theaters. In order to manage these flows, an operational programming is established. It is a question of specifying, in an operative program, the list of patients to be operated as well as their order of passage in each operating theaters and possibly in the post-operation care theaters (SSPI). However, this operating program is not often respected because of disruptions (arrival of emergencies) which constitute a reality of the operating theaters, which necessitates taking decisions and carrying out rapid actions.

The work presented in this paper describes a robust approach to guarantee the performance of the operative and flexible program allowing the real-time control of the operating theaters and to help the decision-makers of this block manage the hazards by

© Springer International Publishing AG, part of Springer Nature 2019
J. Mizera-Pietraszko et al. (Eds.): RTIS 2017, AISC 756, pp. 3–16, 2019.
https://doi.org/10.1007/978-3-319-91337-7_1

defining. This approach is based on Integration the multi agents (MAS) in Interactive Decision Support System (IDSS). We define a supervisor agent, and the service agents.

The (IDSS) [1, 2] is an interactive computer-based system intended to help decision makers use communications technologies, data, documents, knowledge and/or models to identify and solve problems, complete decision process tasks, and make decisions.

The remainder of this paper is organized as follows: Some of the health care applications based on agent and, a general description of the problem is illustrated in the Sect. 2, in Sect. 3 we propose a multi-agent architecture based (IDSS) as a solution to the problem of decision support for the real-time control of the operating theaters, where we detail the behavior of each agent. In addition, we propose a scheduling method based on interactions between services agents and supervisor agents using (CNP), a well-known protocol for coordination, to efficiently allocate resources, and (JADE) in Sect. 4. The implementation of the model is described in Sect. 5. Finally, in Sect. 6 we presented the Conclusion and future works.

2   Related Works

Today, the (MAS) represent a new technology for the design and control of complex systems. They have been applied successfully in the field of health in general and problems related to operating theaters structures in particular, where agents must be able to interact and communicate with one another to resolve differences of opinion and conflict Of interest [3].

In the research literature, there are several agent-based applications reported in the healthcare domain. In particular, one of the earliest examples of work examining the role of multi-agent systems in healthcare is offered by [4] Huang et al., they designed a (MAS) for distributed medical care, facing challenges regarding the distribution of data and control, information uncertainty, and environment dynamism. The coordination mechanism is based on commitments and conventions between different types of agents. The task allocation and coordination is done by managing agents that manage the execution of tasks and by contractor agents that execute the task.

Kim [5] presents a multi-agent system based proactive u-healthcare system which incorporates different functions designed to resolve problems for the sake of rapid and efficient mobile u-healthcare agents. The proposed system allows the system itself to recognize and identify u-healthcare domain problems arising.

Mutingi and Mbohwa [6] propose homecare multiagent system architecture in order to make decision with multiple objectives. The system integrates MAS based on Genetic Algorithm and Web Services that provide decisions in a dynamic multiple-objective environment. The proposed architecture consists of a number of agents that coordinated through efficient communication in homecare dynamic environments.

Decker and Li [7] modeled MAS for hospital patient scheduling with complex medical procedures. They took a function-centered view and modeled nursing wards as autonomous agents. They developed a generalized partial global planning (GPGP) approach as a constraint-based coordination mechanism. It is constructed to avoid resource conflicts and patients are treated as exclusive resources that are handled by a special mechanism.

Nealon and Moreno [8] have discussed various applications of SMA in health care e.g., coordination of organ transplants among Spanish hospitals, patient scheduling, senior citizen care etc.

Riano [9] developed information technology and multi-agent systems to improve the care given to palliative patients.

The works mentioned above are domain specific, e.g. catering to special types of patients. These systems are therefore, not capable of handling problems related to taking care of emergency patients at operation theaters as presented in Sect. 1. More precisely, the efficacy of agents in this application area has still not been explored. So, in the subsequent sections we are introducing and highlighting the concept of using agent technology and the theorem the decision support systems in this regard.

3 Decision Support for the Real-Time Control of the Operating Theaters

3.1 Definition

Operating theaters department and the planning and scheduling of the operating theaters can be regarded as the engine of the healthcare surgery system. The entrance to the surgery system occurs either through the emergency system or the elective surgery [10].

The electives patients who do not need emergency medical treatment and can be served on a pre-agreed time and the emergency patients are served just after arrival. Important decisions of how to allocate the operating theaters capacity between the different surgery specialties. Also considerations have to be taken regarding prioritization, in which emergency cases have the highest priority [11, 12].

3.2 The Agents-Based Approach for Modeling the Operating Theaters

The functioning of the block is determined by a preliminary operating program which specifies the patients to be operated during the day, the various critical resources assigned to each operation and the order in which operations are carried out. However, it is possible that this operating program is not respected because of the different types of hazards that can occur. These uncertainties include uncertainty in the prediction of surgical time, unforeseen complications, the arrival of urgent cases to be performed during the day, before a given time. For this reason, a real-time control tool seemed to be appropriate for the operating theaters. In addition, real-time management complements the proposed decision-making approach, since it allows the decision-making process to be aided when a hazard occurs. Particular attention has been paid to the study of the consideration of urgency in the surgical program. We proposed a real-time approach based on multi agents [13] and a decision support system, the principle of supervision and mathematical modeling to help the piloting of the operating theaters in front of the occurrence of this type of hazards.

The agents used in our approach are cognitive agents; they are diverse by their specific functions. It is a real multidisciplinary co-operative team, participating in the design and implementation of decisions by combining their efforts and constantly adapting to the evolutions of the operating theaters system. Service agents and supervisors are distinguished.

The model of the proposed agents is adapted to the context and the specificities of the operating theaters. In particular, it enables research, selection, negotiation, coordination and cooperation to be carried out in order to realize a real-time control of the operating theaters (Fig. 1).

Fig. 1. The proposed approach for the piloting of the operating theaters (Multi-agent structure)

3.3   Structure of Agents

3.3.1   The Supervisor Agent

The supervising agent (cognitive agent) models the supervision and negotiation system in the operating theaters. Its role is then:

- Manage resources within the operating theaters;
- Receive requests for assignment of urgent operations;
- Receiving and negotiating medical resources available with the surgeon;
- Assist surgeons in decision-making to improve the scheduling of surgical operations;
- The supervisor also has the role and needs of patients.

    A supervisor is composed of several subsystems:

The Analysis and Reaction Module: This module performs a continuous analysis of messages that the supervisor receives from all agents, through its communication interface, and activates the behaviors that correspond to them. It also updates the states of operations in the global agenda following messages sent by surgeons.

Behaviors: The supervising entity has a set of behaviors to accomplish its task:

- The CPC Behavior: is used to find the service agent satisfying the objectives to be achieved for the assignment of urgent surgical operations.
- CPM behavior: aims to find the best medical resources for a surgical operation (in case of unavailability of medical resources).

The Global Agenda: This schedule allows the supervisor to represent and monitor the progress of all operations in the system. This calendar also makes it possible to reconstruct the information of any local diary specific to an agent service.

The Communication Interface: Handles messages in transit between the supervisor agent and all other agents in the system.

The Real-Time Clock: Generates time by the supervisor.

3.3.2 The Service Agents

Each service agent (cognitive agent) has a sufficient level of knowledge to allow decision-making, its role is then:

- Real-time localization of scheduling processes, queues, etc.
- Manage the availability of resources necessary for surgical operations.
- Manage priorities between surgical operations.

The Analysis and Reaction Module: Performs a continuous analysis of the messages received by the supervisor, through its communication interface, and activates the behaviors corresponding to the events received. Thus, the state of surgical operations is updated.

The Behavior of the Service Agent: The behaviors are as follows:

- The Service agent-type1 behavior: is to manage the service agent queue and select the next surgical operation to be performed.
- The agent-service2: agent behavior: searches for resources to replace resources that are not available.

Local Agenda: The agenda being a form of representation of the commitments of every surgeon, obeys the following rules:

- 1st. At each beginning of a surgical operation, the service agent enters his or her agenda as the start of this surgical operation and reports it to the supervisor.
- 2nd. At each end of surgical operation, the service agent enters his or her diary the end of this surgical operation and reports it to the supervisor.

The Expert Interface: Allows the surgeon to view and modify the agent's configuration, to know the current state of the resources, and to monitor the evolution of the operating theaters activity.

The Communication Interface: Allows the management of messages in transit between the service agent and the other entities of the system.

Real Time Clock: Generates the real time factor in the agent service agent.

Each service agent is equipped with two additional modules in the decision subsystem:

The Proposal Generator: Constructs a proposal for a given surgical operation according to initial parameters and surgeon preferences.

The Decision Support Module: Is applied when each agent evaluates the solutions using a multicriteria decision support technique. In our system, Electre III [14] (ELIMINATION AND CHOICE REALITY TRANSLATION) is used to fulfill this function.

Among the ELECTRE methods, ELECTRE III is chosen because it allows us to propose a ranking of resources from the best to the worst, taking into account the effect of each of the criteria (non-compensation between the various criteria) To showcase incomparable resources.

Because of its properties, it is therefore quite appropriate to solve the problem of choosing a resource in an operating theaters management context.

3.3.3 The Coordinating Agent

The coordinating agent ensures that the coordination between the surgeons and the supervisor functions properly, executes the coordination process, processes the messages received and has a mechanism for processing message contents.

A negotiation coordinating agent comprises several types of functional modules such as:

The Module for the Management of Activities in the Operating Theaters: Is the core of this architecture, its role is to break down a complex problem into sub-problems; by its participation it provides valuable assistance to the supervising agent.

The Coordination Module: Allows to manage the stages of the negotiation and synchronizes the different results obtained. This module takes into consideration the overall coordination process.

The Database: Contains information about surgeons, who can perform operations, and all the necessary information and knowledge about the agent himself: his abilities and skills.

The Interface Module: Manages the exchange of information between the coordinating agent and the other agents (Fig. 2).

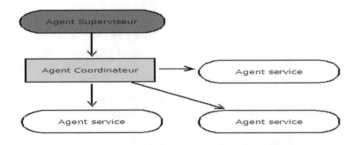

Fig. 2. Negotiation model

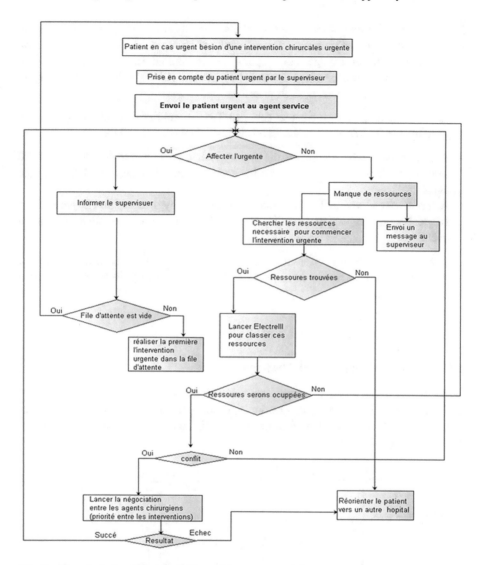

Fig. 3. Flowchart describing the points of integration of the negotiation and the ELECTRE method in the process of solving the problem posed

3.4 Decision-Making Levels

For each operation surgeon who arrives the supervisor must launch a request of execution to the surgeon able to realized it. Following this, it receives an acceptance or refusal response

1. In the first phase, the service agent recognizes the arrival of urgent matters, and triggers local decision- making processes. In case of success (all resources are available) it begins the urgent operation.

2. In the second phase, the service agent fails in the insertion of the urgent caused by a lack of resources or a situation of conflicts over the use of certain common resources. The service agent then opens the negotiation. The protocol is based on the traditional approach of the Contract Net Protocol (CNP) [15]; each service agent expresses their preferences, their priorities.

Decision-making processes use the ELETREIII multicriteria aid method. The surgical agents are in several cases, the most important correspond to:

– A service agent who encounters the problem when inserting the urgent, must make a decision in collaboration with the other service agents, it is called initiating service agent.
– A service agent who suffers the consequences of delay or disruption in performing surgical operations due to a conflict over the common resource or other unforeseen event.

Figure 3 describes the different steps of solving the problem, the unavailability of resources (the resource breaks down or it does not exist or is busy) following the arrival of an urgent order.

4   Agent Negotiation Protocol

4.1   Problem Solving by UML Diagram

At the arrival of the urgent the supervisor starts by assigning the urgent to the services decision-makers, in case of certain resources is not available or broken down, the decision-making service agent launches the Electre III method to prepare its strategy of resolution (Figs. 4 and 5).

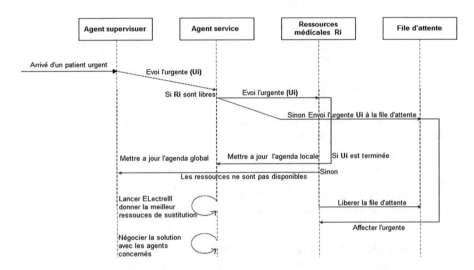

Fig. 4. Sequence diagram

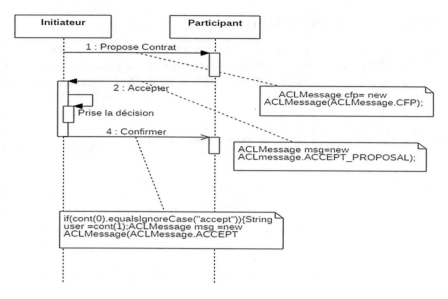

Fig. 5. Sequence diagram of the negotiation based for (CNP)

We describe in this section some scenarios of the negotiation as between the agents

4.2 Resource Allocation Algorithm

The coordinating agent applies the resource allocation algorithm after it receives resource allocation requests from the service agent.

```
Répéter
Réception des demandes d'allocation de ressource Ri par les agents services
RequestList[]=liste des ressources demandés ;
AgentRest[]=liste des agents demandeurs ;
Identificateur de l'état de la ressource Ri ;
      If (Ri.Etat = n'existe pas)
       Envoi le rejet d'initialisation à l'AgentRest[]
      Else
      If(Ri.Etat= occpées)
            Identification de la durée de libération de la ressource
                  D=durée de la libération de la ressource
                  Demande de rappel de demande de ressource après temps =D
      Else
      If Ri.Etat==libre
      //selon la priorité d'AgentRest
      Identification des priorités des opérations d'AgentRest
      AgentRest[]=PrioritéMax(AgentRest)
      If (AgentRest[]= contient un seul demande)
      Allocation faite pour AgentRest[]
```

Algorithm 1. The resource allocation algorithm

4.3 Resource Unavailability Algorithm

During the arrival of emergencies the resources are in constant danger of breakdown or are not available, whereas the realization of surgical operations is imperative. The service agent uses Algorithm 2 to find resources that are similar to resources that do not exist or are not available.

```
L'agent service reçoit la demande de l'urgente
Mise à jour de la base de données de l'agent service
Ressources « R» pour réalisation de l'urgente ne sont pas disponibles
(n'existe pas)
Recherche des ressources identiques à R
Demande de l'aide l'agent coordinateur ;

Traitement de l'aide l'agent coordinateur ;
      If (proposition de ressource)
        Envoi la confirmation de l'affectation de l'urgente  à l'agent
      superviseur
      Else
        Affectation impossible
        Ajout des informations du manque de ressources à la BD de l'agent
      coordinateur
Fin
```

Algorithm 2. Resource unavailability algorithm

4.4 Information Retrieval Algorithm (Request for Help)

The agent uses Algorithm 3 to search for information about resources

```
Réception de demande d'aide de la part de l'agent service
Liste_ressource[]=recherche liste de ressources identiques Libres
If (Liste_ressource[].size==1)
  Envoi proposition Liste_ressource[]d'allocation
  Réception la confirmation
Else
  If (liste_ressource[] !=0)
    R=choisir entre Liste_Ressource[]
    Envoie de R à agent service
  Else
  //Recherche des ressources identiques sous le contrôle des autres agents
  Liste_ressourcesIdentique []
    If (liste_ressourcesIdentiques[] !=0)
      Envoie de ressourcesIdentique [] à chirurgien
    Else
        Envoie pas d'aide ;
Fin
```

Algorithm 3. Information retrieval algorithm (request for help)

5   Experimentation of the Proposed Approach

In this simulation example, we chose, as the scenario, the negotiation between a supervisor and four Agents served who interact with each other in order to reach a compromise on the insertion of urgent surgical operations. These interactions are summarized in the following actions:

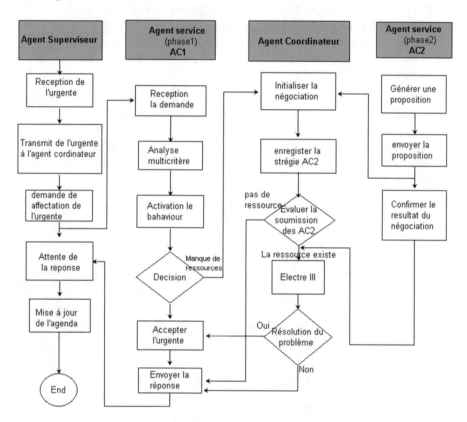

Fig. 6. Organization chart describing the states of the agents during the two decision-making phases

1- The supervising agent asks the service agent 2 (AC2) to insert the urgent operation,
2- If the resources (e.g. the operating theaters or the surgeon…) controlled by the service agent1 are not available or are inoperative, the analysis and reaction module detects this event and determines the associated behavior.
3- If the process fails, Service Agent 2 (AC2) redirects the resource request to the supervisor, triggering the supervisor's type1 behavior.
4- The supervisor sends the request to the other service agents, and processes the answers received in order to solve the problem posed.
5- The result will be communicated to the selected service agent and to the requesting service agent1 (AC2).

6- The service agent4 responds favorably to the supervisor's request (end of the first phase of the decision- making process).

7- Service and service agents find themselves in a situation of conflict.

8- The negotiation is then opened: the service agent 2 (AC2) becomes initiator and the agent serivce4 becomes the participant.

9- The initiating agent provided with its module of generator of propositions, formulates a proposition that it sends to the participating service agent.

10- The agent formulates his contract after consulting his evaluation matrix; Preferences and priorities between operations. Proposal evaluation or counterproposition is facilitated by the results given by ELECTE III in application of the over-ranking algorithm (Fig. 6).

To implement our proposed application, we have chosen JADE (Java Agent Development Framework) as a software framework for developing MAS. JADE presents several advantages [16] such as Interoperability, Portability, Simplicity, Distributed, Programmable interface and FIPA [17] (Foundation for Intelligent Physical Agents) protocols Library Standard. Also there are different reasons for using Jade in our application (Figs. 7, 8, 9, 10 and 11).

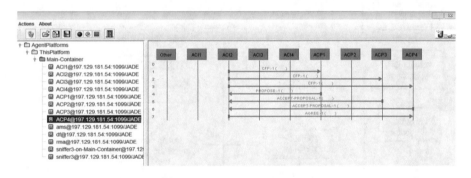

Fig. 7. Scenario negotiation results on the Sniffer

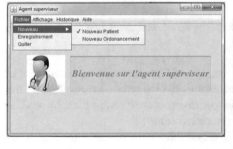

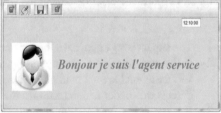

Fig. 8. Supervisor agent   **Fig. 9.** Service agent

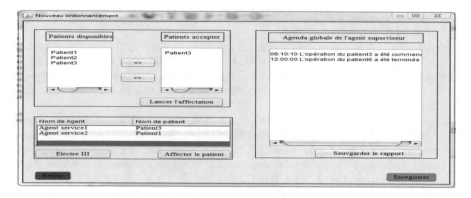

Fig. 10. Description of new ordering request

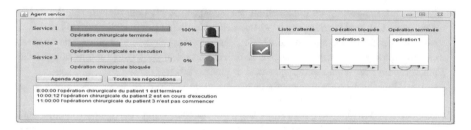

Fig. 11. Negotiation scenario

6   Conclusion

The goals to be attained are an "optimal" and rational use of resources respecting the various constraints of the operating theaters. To do this, we have developed a means of representing a scheduling through a global plan followed by the decision-makers (Service Agents). In practice, the computing time requirement remains an essential element in the construction of scheduling [18–20].

The problems that have been dealt with in this thesis relate to the optimal management of resources by setting up mechanisms to deal with and correct the problems of causing by the occurrence of urgent cases, and lack of resources [21]. This situation can easily lead to management of assignment of urgent surgical operations.

The application developed using the JADE platform enabled us to exploit all the communication management possibilities between the service agents.

References

1. Andrew, P.S.: Decision Support Systems Engineering. Wiley, New York (1991)
2. Phillips-Wren, G., Jain, L.: Recent Advances in Intelligent Decision Technologies. Lecture Notes in Computer Science, vol. 4692, pp. 567–571 (2007)
3. Wooldridge, M.J.: An Introduction to Multi-agent Systems. Wiley, London (2002)

4. Huang, J., Jennings, N.R., Fox, J.: An agent-based approach to health care management. Appl. Artif. Intell. **9**(4), 401–420 (1995)
5. Kim, H.K.: Convergence agent model for developing u-healthcare systems. School of Information Technology, Catholic University of Deagu 712702, Republic of Korea (2013)
6. Mutingi, M., Mbohwa, C.: A home healthcare multi-agent system in a multi-objective environment. In: SAIIE25 Proceedings, Stellenbosch, South Africa SAIIE, 636-1, 9–11 July 2013
7. Dexter, F.: Cost implications of various operating theaters scheduling strategies. Am. Soc. Anesthesiologist's Clin. Update Program **52**(262), 1–6 (2001)
8. Nealon, J., Moreno, A.: Agent-based applications in health care. In: Applications of Software agent technology in the health care domain. Whitestein Series in Software Agent Technologies. Birkhauser Verlag, Basel (2003)
9. Riano, D., Prado, S., Pascual, A., Martin, S., June, A.: Multi-agent system to support palliative care units. In: Proceedings of the 15th IEEE Symposium on Computer-Based Medical Systems (2002)
10. Lafond, N., Landry, S.: La planification des besoins matières pour gérer les stocks du bloc opératoire. Cahier de recherche no. 99-04, HEC, Montréal (1999). ISSN 1485-5496
11. Marcon, E., Kharraja, S., Simmonet, G.: The operating theatre scheduling: an approach centered on the follow-up of the risk of no realization of the planning. Operating theatre planning. In: Proceeding of the Industrial Enginee (2001)
12. Rossi-Turk, D.: Comment garantir la qualité et la sécurité au bloc opératoire par une programmation et logistique innovante. Santé et Systémique **6**, 1–3 (2002)
13. Barbuceanu, M., Fox, M.S.: Cool: a language for describing coordination in multiagent systems. In: Lesser, V., Gasser, L. (eds.) Proceedings of the First International Conference oil Multi-agent Systems (1995)
14. Roy, B.: Electre III, un algorithme de classement fondé sur une représentation floue des préférences en présence de critères multiples. Rapport de recherche (1977)
15. Smith, R.G.: The Contract net protocol: high-level communication and control in a distributed problem solver. IEEE Trans. Comput. **29**, 1104–1113 (1980)
16. JADE: Java Agent Development Framework. http://sharon.cselt.it/projects/jade
17. FIPA: Foundation for Intelligent Physical Agents. http://www.fipa.org
18. Xin, P., Sagan, H.: Digital image clustering algorithm based on multi-agent center optimization. J. Digital Inf. Manage. **14**(1), 8–14 (2016)
19. Lu, L., Zhu, X.: Study of ideological instruction multimedia education thought based on artificial intelligence model. Progress Comput. Appl. **6**(2), 41–46 (2017)
20. Murai, H.: Prototype algorithm for estimating agents and behaviors in plot structures. Int. J. Comput. Linguist. Res. **8**(3), 132–143 (2017)
21. Wang, J.: Research on Japanese digital learning system based on Agent model. J. Intell. Comput. **8**(3), 98–104 (2017)

Applying Data Analytics and Cumulative Accuracy Profile (CAP) Approach in Real-Time Maintenance of Instructional Design Models

Mohamed Housni[1(✉)], Abdelwahed Namir[1], Mohammed Talbi[2], and Nadia Chafiq[2]

[1] Laboratory of Information Technology and Modeling (LTIM), Faculty of Sciences Ben M'Sik, Hassan II University of Casablanca, B.P 7955, Sidi Othmane, Casablanca, Morocco
mohamed.housni.etu@etu.univh2c.ma
[2] Multidisciplinary Laboratory in Sciences and Information, Communication, and Educational Technology (LAPSTICE), Observatory of Research in Didactics and University Pedagogy (ORDIPU), Faculty of Sciences Ben M'Sik, Hassan II University of Casablanca, B.P 7955, Sidi Othmane, Casablanca, Morocco

Abstract. In a constantly changing climate, professors find themselves in a pressing demand to tailor their ways in schooling, according to the needs of the market and students. In conjunction with providing quality learning experiences, whereas the ambition is to attract the millennial generation that constructs a large portion of the student population. However, in an interconnected world with a massive data generated every second, there are opportunities to use it as a tool to gain insights on the seemingly chaotic environment. The aim of this paper is to provide a way to support decision making made by the instructional design engineers, by providing them with an objective method based on the cumulative accuracy profile curve that allows to assess, retrain and rebuild instructional design models used in constructing courses. Moreover, we propose a set of tools for applying this approach in decision making and managing educational resources more efficiently.

Keywords: Learning analytics · Big data · Modeling · Reporting systems
Instructional design

1 Introduction

Implementing an adaptive educational system, that is based on the real-time flow of data is not a simple task to achieve, as it focuses on different parts of the educational system and goes from the ground up to incorporate the best practices from different fields.

The focal point of this paper is to demonstrate the use of a data science approach, that will support first of all teachers in gaining insights about their instructional design method and adapt their teaching style accordingly.

Teaching is a highly complex activity [1]. That requires a high level of skills and knowledge. Before digging inside, there are some assumptions that this work is based on, it will guide you to understand and use the results of this article especially if you are

© Springer International Publishing AG, part of Springer Nature 2019
J. Mizera-Pietraszko et al. (Eds.): RTIS 2017, AISC 756, pp. 17–25, 2019.
https://doi.org/10.1007/978-3-319-91337-7_2

a teacher. First of all, we consider a teacher as facilitator of knowledge and instructional engineer who is acquired to help students amass knowledge, values, and skills [1].

1.1 Instructional Design

Instructional Design is the systematic development of instructional specifications using learning and instructional theory to ensure the quality of instruction. It is the entire process of analysis of learning needs and goals and the development of a delivery system to meet those needs. It includes a development of instructional materials and activities; and tryout and evaluation of all instruction and learner activities [2].

From this definition, we can take away some key elements of an instructional designing process. Fundamentally, it requires a deep understanding of the learning theories, although we won't go over them, it seems necessary to note their importance in modeling proceedings. Furthermore, the definition insists on starting the process by analyzing the needs of the audience and customize the instruction to the environment where the action of "teaching" takes place. As a result, instructional designing is in a way a framework building to surround the actions of teaching and to avoid improvising.

In short, Instructional design models are made for:

- Establishing a plan for courses;
- Making a blueprint for teaching activities and actions;
- Guiding teachers;

It shall be noted, that teachers and instructional engineers are the same in most cases [3]. So, in the end, is up to the teacher to choose which model of instruction may be applied in particular context. Our job in this paper is to provide teachers with a tool based on the Cumulative Accuracy Profile (CAP) approach that uses the data generated during the course period to assess the efficiency of the model chosen.

1.2 Data Matrix

For as much as a minimal instructional design model with 3 fundamental steps: Analysis, development and evaluation, which assigns to each step a X_i attribute ($0 < i \leq d$ with $d \in \mathbb{N}^*$ is the dimension of the model) applied for a course with x_j student ($0 < j \leq n$ with $n \in \mathbb{N}^*$ is the size of the data). Table 1 represent the Data matrix that will be used in building the Cumulative Accuracy Profile curve that will be a visual representation of the instructional design model [4].

$x_{i,j}$ is a punctual data unit taken in t_k time (with $k \in \mathbb{N}^*$), it might be a numerical or nominal type, and sometimes it could be empty because it can not be recorded. Furthermore, all data recorded under X_i must be the same type.

If an attribute has empty data in a t_k moment, it must not be considered as an attribute for the instructional model. The attributes order is meaningless when all of them are independent [4].

Table 1. Data matrix format

$$D = \begin{pmatrix} & X_1 & X_2 & \dots & X_d \\ \hline x_1 & x_{11} & x_{12} & \dots & x_{1d} \\ x_2 & x_{21} & x_{22} & \dots & x_{2d} \\ \vdots & \vdots & \vdots & \ddots & \vdots \\ x_n & x_{n1} & x_{n2} & \dots & x_{nd} \end{pmatrix}$$

1.3 Variables

Variables can be sorted in two distinct types [4]:

1. Categorical: That depicts a quality or a characteristic of a data unit $x_{i,j}$, basically they are answering which category this recorded data belong? and what are its properties? There are two sub-types of categorical variables, Nominal and Ordinal.
   (a) Nominal variables are observations that can not be requested into a sensible arrangement or ordered into a logical sequence, for example: Names, gender, hues,...
   (b) Ordinal variables are observations that can be requested into a sensible arrangement and we can easily order them in a logical sequence as well as we group them, for example: Size, grades,...
2. Numeric: They are basically numbers that depict a quantitative characteristic of a data unit $x_{i,j}$. They are replying, in general when all is said in done, to the question of how much? or how many? There are additionally tow sub-types of numeric variables, discrete and continuous.
   (a) Discrete variables are whole numbers and can not be a fraction of numbers, that portray a data unit, for instance: $\forall x \in \mathbb{N}$ and they can be also negative $\forall x \in \mathbb{Z}$
   (b) Continuous variables unlike discrete variables they can be a portion, in an another sense if we took tow continuous variables there will be always a continuous variable between them, for example: Age, distance, tallness, $\forall x \in \mathbb{R} \dots$

Determine the variables of an instructional design model is a task that require an overview of the methodological suite applied in the instructional design model. One of the most used methodologies is ADDIE [5] an abbreviation of 5 steps (Analysis, Design, Development, Implementation and evaluation) that can go in a linear process or non linear, sometimes with big teams of instructional engineers they can go in parallel. This model is a very descriptive one so it is easy to determine its variables but sometimes we find less descriptive models, like Gradual Release [7], which is based on transferring the responsibility from the teacher to the students in 4 steps: I do (Demonstration), We do (Shared practice), You do (guided practice), You do (Independent practice). This model is noticeably less descriptive and more based on the actions in the course which make the task of extracting the variables very difficult and that imply a high complicated maintenance approach.

1.4   Modeling

As for now, we defined what is an instructional design model, its data matrix and the nature of variables. Now we come to the modeling process, an essential procedure in engineering and science to produce mathematical models. It is defined as an activity, cognitive and mathematical to produce an abstract representation of phenomena or an object in mathematical terms [6]. In modeling we need building blocks and there are two main ones:

- Independent variables or decision variables are the blocks to build the model, model, besides exogenous variable that is constants and error.
- Dependent variable or the study variable is the parameter that varies as the independent variables vary.

In this paper, our goal is to build a linear model as it is one of the easiest to understand by a majority of future users that came from different backgrounds. It can be represented as follow:

$$Y_k = a_1 X_1 + a_2 X_2 + \ldots + a_d X_d + c + \Delta X \tag{1}$$

with $d \in \mathbb{N}^*$

- Y: dependent variable at a t_k moment
- X_i: independent variables
- a_i: Coefficients
- c: Constant
- ΔX: the "error"

As we can notice, the dependent variable Y is the result of a set of decision variables X_i that we defined before as attributes in our data matrix. The attributes as we mentioned before are taken in a t_k time frame (with $k \in \mathbb{N}^*$) and as time goes on, the data records changes. This is called "**model deterioration**", which is a natural phenomena that affects models in general and make them less accurate representation of the reality. There is a lot of possibilities why this happen:

- The expansion of new attributes to the data sets;
- The change of the nature of the existing attributes;
- The change of the instructional model process or the total change of the instructional design model;
- The behavioral changes of the population
- The invention or innovation of new field disrupter
- The change of the needs and demands
- The change of policy and regulations
- The change of the instruments of production
- ...

1.5   The Cumulative Accuracy Profile

The Cumulative Accuracy Profile is defined as the graph of a rating model, it is a curve which plots cumulative population in the X-axis (independent variables) and the corresponding cumulative defaults in the Y-axis (dependent variable), where they are connected by a straight lines (linear interpolation). [8] This is illustrated in Fig. 1 [9].

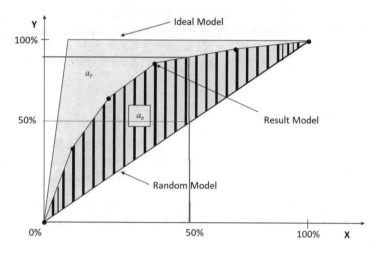

Fig. 1. Cumulative Accuracy Profile

The CAP curve is namely used on assessing model performance with an ideal model, that can predict 100% accuracy the outcomes with minimum cumulative population data, and an arbitrary model which is a linear model with a coefficient of 1. The CAP curve is involving some piece of the range framed by the curve bend.

The territory possessed by the bend of the model is known as the Accuracy Ratio (AR).

$$AR = \frac{a_P}{a_R} \tag{2}$$

An AR with around 51% to 60% of accurate predictions is considered as a model with reasonable prescient power. AR is always less than 100%, if the $AR > 90\%$ the model is over-fitted to the data.

CAP curves are usually between the ideal model and random model, but sometimes it goes under the random model which means that the data is corrupted or the decision variables are not set correctly. With the increase of the time, the Cumulative Accuracy Profile converge toward the random model, this a visual indication as we give the model new sets of data each t_k (with k $\in$ $\mathbb{N}^*$). We can draw a lot of insights from the CAP curve, but as rule of thumb in this paper we will focus on the $X = 50\%$ point to assess the performance of the model each t_k. If the Y_k correspondent is less than the Y_1 the model is deteriorated and it should be rebuild or retrained with a new set of data [10].

2   Method: Building the CAP Curve

A standout amongst the most essential objectives of an instructional design model is answering the needs of the learners. So this will be our comparative variable which is a "YES or NO" response to the inquiry: **Does my instructional model fitted for my population at the t_k time?**

Subsequently, the dependent variable will Y_k will be the translation of the similarity between the instructional model design and the mathematical model. Consequently, the result will be presented in percentage of conformity between the two. We will not use all the attributes in the data set so the first step is fixing the significance level (SL) of an attribute.

2.1   Preparing the Data Set

As specified in the variables section, we have diverse sorts of variables. Numeric ones are easy to use in building the model yet the categorical variables is hard to process by machines. To solve this issue we will create dummy variables by using a straight forward process:

For each X_i attribute with categorical variable we will create new columns that correspond to the number of categories inside the attribute. each column will be populated by "0 or 1":

- 0: If x_j does not belong to the category;
- 1: If x_j does belong to the category, so in each line we will find only 1 in the dummy columns;

After populating the new dummy columns D_i the correspondent X_i is removed from the data sets that will be used in building the model.

Note: If the number of categories inside the attribute is 2 (for example: Attribute categories are PASSED course or FAILED), we will have D_1 and D_2, only one dummy attribute shall be used in building the model because:

$D_2 = 1 - D_1$ so the coefficient of the other will be included in building the model. Furthermore, if there is 3 categories like grade is A, B or C we will use only 2 dummy attributes in the modeling. As result, we always omit one dummy attributes.

Derived variables are also used in preparing the data set, and they are the result of calculation made to normalize the data, like instead using dates, the derived variable will be the number of days since a specified date. Or, instead of using the all the scores of the student x_j we can use the derived variable that represent the median score.

2.2   Attribute Elimination and Model Building

Before eliminating insignificant attributes beside the dummy ones, we should calculate their **P-value** which is the probability of finding the recorded attribute. If $P - Value < SL$ the attribute should not be considered in building the model. the process can be easily done using software like **SPSS** or **Gretl**. that provide scores like $Adj - R^2$ and Akaike criterion to optimize the model. They can also determine the a_i coefficients but we can not use them in the Eq. (1) as we do not have the constant and error.

2.3 Generating the CAP Curve

To present the CAP curve we need two informations:

1. Total population selected percentage (X-axis): it is equal to the ordered x_j by the probability of fitting divided by the total size of the population;
2. Model selection of fitting percentage (Y-axis): it is equal to the percentage of the fitting population selection over the total of fitted population;

After getting these two information we can use any kind of software like excel to present them in a graph form and this is how we get the CAP curve at t_k time.

3 Results

The goal of this paper is not just to determine if a model is fitted or not for student but to help instructional engineers determine which attributes play major role in the model and what affects most the results of fitting.

To do that, suppose the odds of fitting instructional model is the ratio between the number of the students that the instructional design model will be fitting (F) over the not fitting population (N). So the probability of fitting instructional design will be:

$$P_f = \frac{F}{N+F} \leftrightarrow \frac{F}{N} = \frac{P_f}{1-P_f} \tag{3}$$

in the other hand we have:

$$Y = \frac{F}{N} \tag{4}$$

As we noticed multiple times across the paper, managing variables is the key to building robust models, in our case variables are with multiple natures and not always normalized. As consequence, we can not just test evidently the attribute X_i effects by increasing it or decreasing it, as every attribute coefficient is linked to the unit of the attribute. As proposed solution to identify the attributes with most effects of the model we will introduce logarithmic function to the Eq. (3). From (3) and (4) we have:

$$ln(Y) = ln\left(\frac{P_f}{1-P_f}\right) \quad \Rightarrow \quad Y \approx e^c \cdot e^{a_1 X_1} \cdot e^{a_2 X_2} \dots e^{a_d X_d} \tag{5}$$

As result, increasing an independent variable X_i by 1 unit will increase the odds by a multiplicative coefficient of e^{a_i}.

$$e^{a_1 X_1} \Rightarrow e^{a_1(X_1+1)} = e^{a_1} \cdot e^{a_1 X_1} \tag{6}$$

This will help us compare the contribution of each independent variable to increasing the odds of fitting the instructional design to the population.

4   Discussion

Inside the modern classes, the development of information sources and amounts is an open door for teachers and educators, as the growth of data sources is an opportunity to change practices and improve performance. In fact, Applying data analytics and science on learners has emerged in a new field called "learning analytics" which is defined by the measurement, collection, analysis and reporting of data about learners and their contexts, for purposes of understanding and optimizing learning and the environments in which it occurs [11].

Identifying CAP curve as a way to visualize instructional design models is is a decent approach to picture and visualize the performance of models and compare them in an objective matter, but it is limited as we can not conclude which attributes has more effects on the end result. To do that we introduced a formula that standardize attributes and demonstrates those with significant impacts by calculating their exponential coefficient ratio. What's more, is that we can take actions and make the important changes in accordance with the instructional outline display keeping in mind the end goal to boost its execution additional time to maximize its performance.

Correspondingly, we propose this process to apply data analytics and cumulative accuracy profile approach in real-time to assess, maintenance and rebuild the instructional design models overtime (Fig. 2).

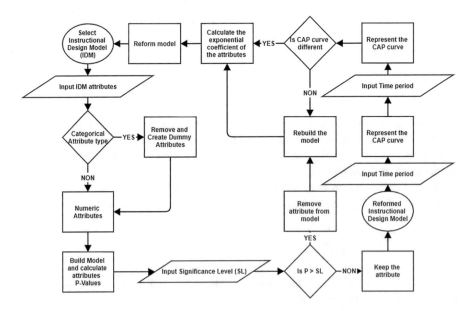

Fig. 2.   Proposed approach to preserve the efficiency of instructional design models

5   Conclusions

In conclusion, we distinguished an approach to gauge the fittingness of instructional design models utilizing cleaned data sets and attributes. Also, making an interpretation of the instructional model into a numerical model to picture it as a CAP curve to recognize and demonstrate the crumbling of models with senectitude.

In the present years, the statistic development in big data created a need for solutions based on exploring data in decision-making. However, leaders are step by step understanding that real-time feedback is the future and they are gradually realizing that data is a wealth to exploit.

References

1. Caena, F.: Literature review Teachers core competences: requirements and development. European Commission, Directorate-General for Education and Culture, 2–28 (2011)
2. Berger, C., Kam, R.: University of Michigan "Training and Instructional Design". Applied Research Laboratory, Penn State University (1996). http://www.umich.edu/ed626/define.html
3. Dee Fink, L., Ambrose, S., Wheeler, D.: Becoming a professional engineering educator: a new role for a new era. J. Eng. Educ. **94**(1), 185–194 (2013)
4. Zaki, M.J., Wagner, Jr., M.: Data Mining and Analysis: Fundamental Concepts and Algorithms Journal of Engineering Education. 1st edn. Cambridge University Press (2014)
5. ADDIE: Online. The Performance Juxtaposition Site (2015). http://www.nwlink.com/~donclark/hrd/sat1.html
6. Fisher, D., Wagner, Jr., M.: Data Mining and Analysis: Fundamental Concepts and Algorithms Journal of Engineering Education. 1st edn. Cambridge University Press (2014)
7. Dee Fink, L., Frey, N.: Better Learning Through Structured Teaching: A Framework for Gradual Release of Responsibility ASCD, 1st edn. (2008)
8. Dabbaghian, V.: Modelling of Complex Social Systems Simon Fraser University, Department of Mathematics (2011). http://www.sfu.ca/~vdabbagh/Chap1-modeling.pdf
9. Engelmann, B., Hayden, E., Tasche, D.: Measuring the Discriminative Power of Rating Systems Discussion paper. Deutsche Bundesbank. Series 2: Banking and Financial Supervision, no. 01 (2003)
10. Mesterton-Gibbons, M.: A Concrete Approach to Mathematical Modelling. Wiley-Interscience Paperback Series, 620 pages (2007)
11. Conference LAK 2011 1st International Conference on Learning Analytics and Knowledge Banff, AB, Canada — February 27 – March 01, 2011. ACM New York, NY, USA (2011). ISBN: 978-1-4503-0944-8

A Meta-model for Real-Time Embedded Systems

Soukaina Moujtahid[✉], Abdessamad Belangour,
and Abdelaziz Marzak

Faculty of Science Ben M'Sik, University Hassan II, Casablanca, Idriss El Harti,
20670 Casablanca, Morocco
soukainamoujtahid@gmail.com, belangour@gmail.com,
marzak@hotmail.com

Abstract. Embedded real-time systems are combinations of hardware and software fully integrated into the systems they control. Due to the continuous technological evolution in the hardware and software and the diversity of the targeted areas of application, these systems have become omnipresent in our professional and personal lives. Thus, various approaches based on model driven engineering (MDE) have been proposed in order to control the inefficiency of the methods of their current design. Each of these approaches has its own meta-model and its corresponding UML profile, specialized or adapted to a particular category of these systems. Indeed, in this paper we will propose a generic meta-model, taking advantage of a large number of these meta-models, which can be adapted to the majority of embedded real-time systems.

Keywords: Real-time embedded system · MDE · Meta-model

1 Introduction

Embedded real-time systems are now omnipresent in various fields. Their intelligence makes them more and more indispensable. They are, however, characterized by complexity arising from their specific characteristics and the high industrial constraints to which they are subjected. To master this complexity, various software abstractions have been implemented. The abstraction offered by model driven engineering (MDE) provides an adequate framework for mastering this complexity. It is a new discipline of software engineering that advocates the massive use of models throughout the software development process. Thus, various approaches based on the MDE have been created by proposing their own process of development of the embedded real-time systems. Our research work lies within of this problem. We aim to propose a new model-based development process for embedded real-time systems. To this end, we have begun by characterizing these systems. We have tried to collect the characteristics of these systems which distinguish them from other computer systems. Then, we studied the different development processes based on models proposed for these systems while carrying out a comparative and detailed analysis of the most well-known meta-models. The aim object of this article is a generic meta-model taking advantage of these different meta-models. Thus, this paper is structured in four parts: The first gives

© Springer International Publishing AG, part of Springer Nature 2019
J. Mizera-Pietraszko et al. (Eds.): RTIS 2017, AISC 756, pp. 26–34, 2019.
https://doi.org/10.1007/978-3-319-91337-7_3

a general view on embedded real-time systems and more precisely on their characteristics. The second part presents the integration of the MDE in the process of development of embedded real-time systems via a panorama of the most known meta-models and on which we were based to create our generic meta-models. This last one which will be the subject of the third section and finally the fourth section concludes with a synthesis of the paper.

2   Characterization of Embedded Real-Time Systems

A real-time system is a computer system to control the behavior of a physical process to which it is connected [1]. To do this, a real-time system consists essentially of a software application controlling the process, a hardware and/or software execution support executing this application, sensors supplying the input data of this application, and actuators executing the orders produced by this application. The application, the execution platforms, the sensors and the actuators are then embedded. They are buried in the process to which they are connected [2]. Embedded systems are combinations of hardware and software [1] completely integrated in the system they control. The complexity of the development of embedded systems stems from the specific characteristics of embedded systems that distinguish them from purely software systems (measurement and control of the physical world, execution on a physical platform limited in resources, autonomy, reliability, reactivity, Etc.), and on the other hand, the high industrial constraints to which these systems are subjected: development and manufacturing costs and delays, multidisciplinary teams, certification and documentation etc. Various methods and languages were proposed in order to master this complexity by emphasizing the modeling of the application and the platform constituting the embedded system [4]. Embedded real-time systems differ from conventional software systems by a set of features. The characterization of these systems will allow us to understand their specificities in order to propose a development approach that is specific to them. However, we can classify these specificities into two categories: characteristics specific to embedded systems and others to real-time systems. Embedded real-time systems are characterized by a memory footprint, energy consumption, weight and volume, autonomy, mobility, communication, security constraints, cost of products in relation to the target sector and other characteristics which will be seen in detail [3, 6]. On the other hand, they have a common characteristic that resides in the existence of temporal constraints to be taken into account. These constraints can take various forms such as deadlines, time intervals, duration of validity, etc. And apply to various objects [1, 2, 4]. The data have a limited lifetime and become obsolete after a certain time, the events appear at special moments and must be taken into account after known delays and the treatments often have moments of beginning, end and fixed execution times. Therefore, these systems work in real-time to manage information and to deduce actions in a controlled time [4]. This means that the accuracy of the results of these systems depends not only on the functional aspects but also on the time in which these results were obtained [3]. Embedded real-time systems are present in several fields [14] such as healthcare, the automotive industry, telecommunications, aeronautics, commerce, household appliances, etc. in addition to their traditional fields such as military or Spatial [1–3, 11]. Thus, for example, an on-board computer is a computer integrated in a vehicle that collects, in real time, information about the condition of the vehicle such as fuel level, oil level, door opening, seat belt buckle, speed control, etc. It is certainly unnecessary if this information is available only after a significant period of time [5]. So the on-board computer is an embedded real-time system.

3   Modeling Embedded Real-Time Systems with the MDE

The development of quality embedded real-time systems at controlled costs represents a very important technological and industrial challenge. In some industrial sectors, such as transport, aeronautics, or telecommunications, the maintenance of these quality/cost objectives implies the use of formal validation and verification tools and techniques [4]. This requires the use of a model-oriented approach. One of the characteristics of this approach is the use of executable models that contain information about the different aspects of the system: functional requirements, static architecture, no-functional requirements, etc. For the use of such models to be relevant, the formalism used must be sufficiently expressive, and its use must be supported, not only by tools for code generation, but also by validation and formal verification tools at the level of the model [10]. The OMG has defined the MDA approach (Architecture Directed by the Models). The key objective of this approach is to develop business models independent of the execution platforms (PIMs) and to transform them into models specific to a given execution platform (PSM) such as Java/J2EE, ASP.Net, etc. This was to preserve information systems against technological changes and increase productivity. The transition from PIM to PSM involves a platform description model (PDM) and model transformation mechanisms [5]. MDE is a discipline that is a continuation of post-object technologies, such as UML, design patterns and the MDA [6]. It generalizes the idea of MDA where the goal is not limited to the PIM-PSM transformation but to the use of the models in all the activities of the software production. Each model conforms to a meta-model expressed by the meta-model MOF (MetaObject Facility) [10]. As part of a model-driven engineering we used the following meta-models to create our own generic meta-model:

SPT (Schedulability, Performance and Time) is a profile adopted by OMG [17] based on version 1.3 of UML [10], taking into account the specificities of real time while retaining the benefits of the object-oriented approach, and provides a framework for temporal modeling [17] of system planning processes that can be used during the making-decision process [12]. SPT's goal was to fill the gaps in AML (Abstract Modeling Level) 1.4 for designers and developers of real-time applications by allowing annotation of model elements by quantitative information about time [17]. This information is then used for performance analyzes based on models, scheduling or verification of compliance with real-time constraints [15]. SPT considers only a metric time that implicitly refers to physical time. SPT introduces the concepts of instant (moment) and duration (duration), as well as those of time-related events and stimuli. SPT also models timing mechanisms and associated services (start, stop, suspend, resume). All this via stereotypes to be applied in UML modeling elements to specify

the time values "RTtime", "RTclock", "RTtimer", "RTtimeout", "RTdelay", "RTin-
tervale" etc. [17]. The general structure of SPT as shown in the figure below consists of
various sub-profiles [17] (Fig. 1).

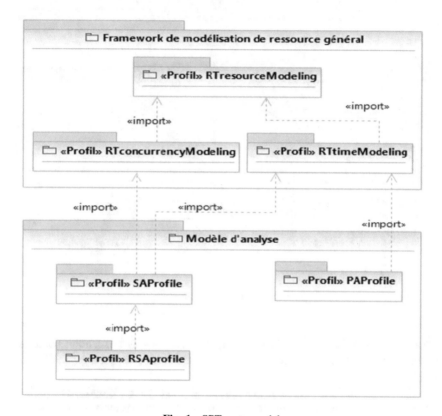

Fig. 1. SPT meta-model

However, SPT has power gaps and flexibility problems. Hence SPT has been
replaced by MARTE [15].

Modeling and Analysis of Embedded Real-Time Systems (MARTE) is a UML
profile [14] which follows the SPT profile [10]. MARTE enriches UML with new
concepts in order to analyze and model software as well as the hardware of embedded
real time systems [6]. The architecture of the profile, as shown in the figure below,
consists of the packages: foundations, design model, analysis model and appendices [9]
(Fig. 2).

The basic concepts in the MARTE meta-model are defined in the CoreElements
sub-package (such as ModelElement, Classifier, Instance, etc.) [9] which is itself
divided into two packages: Foundations and Causality. The first introduces the concepts of Model Elements, Classifer, and Instance. The second deals with behavioral

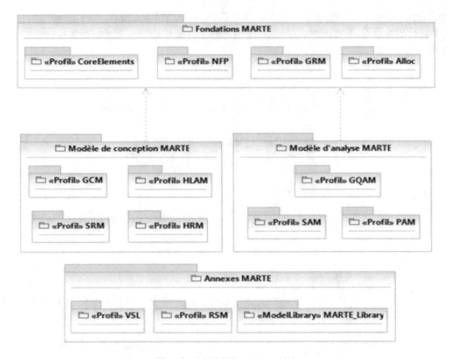

Fig. 2. MARTE meta-model

concepts (commonBehavior) in which events, executions and execution contexts are introduced. Because of these multiple concepts, MARTE presents a great difficulty of use [15].

MASTE (Modeling and Analysis Suite for Real Time Applications) defines a meta-model to describe the temporal behavior of real-time systems to be analyzed by scheduling analysis techniques. MAST also provides a set of open-source tools for performing scheduling analysis or other time analysis to determine whether the system will be able to meet its time requirements. Tools are also provided to assist the designer in assigning scheduling parameters. By having an explicit model of the system and automatic analysis tools, it is also possible to carry out the design of space exploration. A discrete event simulator is also provided to obtain statistical information on the performance of the modeled system [13] (Fig. 3).

For general design and basic concepts, each of these meta-models uses its own concepts based on the UML meta-model. MASTE represents only the time part of the embedded real-time systems, on the other hand MARTE and SPT allows in addition the presentation of the hardware part but with various notions which makes their implementation difficult.

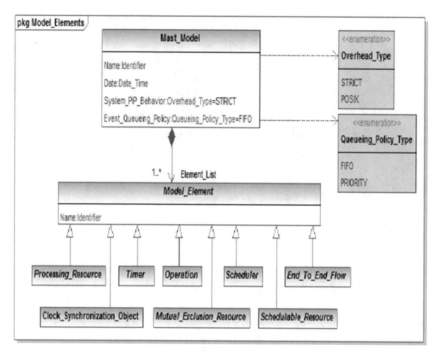

Fig. 3. MASTE meta-model

4 Proposition of a Generic Meta-model for Embedded Real-Time Systems

We have noted in the above that there is not a complete meta-model covering all aspects related to embedded real-time systems and easy to be interpreted. However the coupling between them is possible and can lead to better results. This coupling is possible since most of the profiles are focusing on the process paradigm. What we need is a meta-model, in which we must define rules for the automatic transition from one meta-model to another. This led us to work on a generic meta-model representing an integration of these meta-models: SPT, MARTE and MAST. Below we will give examples of parts of this meta-model. The key elements of our meta-model are resources, services provided by these resources and operations given by them. This resource can be hardware or software. A software resource can be a resource of calculation, communication, backup, time, synchronization etc. [17–19]. The captures below give a general view of the most important parts of our meta-model (Figs. 4, 5 and 6):

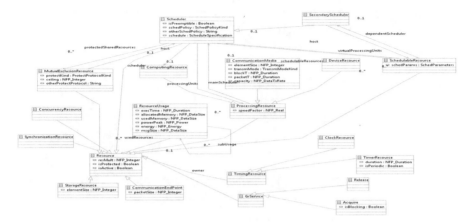

Fig. 4. Resource model

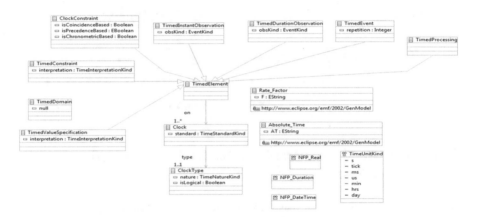

Fig. 5. Time model

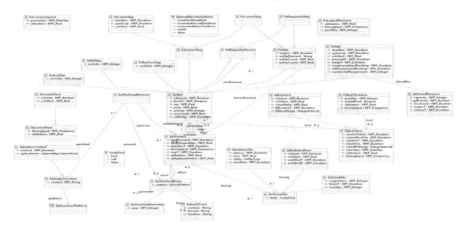

Fig. 6. Analysis operation model

5 Conclusion

We have tried to present in the different sections of this article the contribution of MDE in the modeling process of embedded real-time systems. This is done through its specialized meta-models adapted to its systems [20]. We can find that each of the profiles has these own advantages, although each of them applies to a particular category of embedded real-time systems. It would be more benefit to take advantage of these specificities to have a meta-model suitable for the majority of embedded real-time systems. It is in this context that we presented in the last section our generic meta-model.

References

1. Embedded Systems Design for High-Speed Data Acquisition and Control. Springer International Publishing, Switzerland (2015)
2. Hintenaus, P.: Engineering Embedded Systems. Physics, Programs, Circuits. Springer International Publishing (2015)
3. Dynamic Memory Management for Embedded Systems. Springer International Publishing, Switzerland (2015)
4. Hili, N.: Une méthode pour le développement collaboratif de systèmes embarqués (2014)
5. Model-Driven Engineering of Information Systems. Principles, Techniques and Practice. Apple Academic Press, Inc. (2015)
6. Wiley-ISTE: Model-Driven Engineering for Distributed Real-Time Systems MARTE Modeling, Model Transformations and their Usages (2010)
7. Roques, P.: Modélisation de systèmes complexes avec SysML. Broché (2013)
8. Belloir, N., Bruel, J.-M., Faudou, R.: Modélisation des exigences en UML/SysML (2014)
9. Boukhanoufa, M.-L.: Approche basée sur les modéles pour la conception des systémes dynamiquement reconfigurable: de MARTE vers RecoMARTE (2013)
10. Elsevier Inc.: Modeling and Analysis of Real-Time and Embedded Systems with UML and MARTE. Developing Cyber-Physical Systems. Bran Selic et Sébastien Gérard (2014)
11. Holt, J., Perry, S.: SysML for Systems Engineering. A Model-Based Approach. The Institution of Engineering and Technology (2013)
12. Diaw, S., Lbath, R., Coulette, B.: SPEM4MDE: un métamodèle basé sur SPEM 2 pour la spécification des procédés MDE. MajecSTIC 2009. Avignon, France, 18 novembre 2009
13. Cuesta, C.C., Drake, J.M., Harbour, M.G., Gutiérrez, J.J., Martínez, P.L., Medina, J.L., Palencia, J.C.: Modeling and Analysis Suite for Real Time Applications. Universidad de Cantabria, Spain (2010)
14. Sangiovanni-Vincentelli, A., Zeng, H., Di Natale, M., Marwedel, P.: Embedded Systems Development. From Functional Models to Implementations. Springer Science + Business Media, New York (2014)
15. Aziz, M.W., Mohamad, R., Jawawi, D.N.: Critical evaluation of two UML profiles for Distributed Embedded Real-Time Systems Design. Int. J. Softw. Eng. Appl. 7(3), 137–146 (2013)
16. Zaki, M.Z.B.M., Jawawi, D.N.B.A.: A Review on UML-RT and UML-SPT for Embedded Real-Time Component-Based Development (2013)
17. Barbu, P.-G.: Designing and implementing an embedded bootloader for secure initialization and update of microcontroller applications. J. e-Technology 8(3), 73–84 (2017)

18. Florea, A., Gellert, A.: Developing heuristics for the graph coloring problem applied to register allocation in embedded systems. J. Multimedia Process. Technol. **8**(3), 75–88 (2017)
19. Wang, X., Guo, Y., Wang, Z.: Multi-view discriminative manifold embedding for pattern classification. J. Intell. Comput. **8**(2), 58–63 (2017)
20. Cheng, X., Dang, G.: The research of embedded remote monitoring system based on B/S framework. Int. J. Web Appl. **9**(1), 1–6 (2017)

An Efficient Traffic Monitoring Model Using a Stream Processing Platform Based on Smart Highways Events Generator

Abdelaziz Daaif[✉], Omar Bouattane, Mohamed Youssfi, and Sidi Mohamed Snineh

SSDIA Lab. ENSET Mohammedia, Hassan II University of Casablanca, Casablanca, Morocco
aziz@daaif.net, o.bouattane@gmail.com, med@youssfi.net,
sninehmohamed@gmail.com

Abstract. This paper presents a model of traffic event streams processing. Events are generated by the developed spatiotemporal traffic simulator for real highway networks. The simulator is designed according to a distributed architecture based on mobile agents. It generates a flow of vehicles, assigning them to trips according to a model using geographic data. The highway network is equipped with sensors that generate events when passing vehicles. The event stream is processed in real time by agents to estimate the current traffic state to inform users via traffic message-variable panels. The architecture of the real-time event processing system is based on Kafka Stream Processing. To evaluate the performance of our model, we carried out a simulation of the traffic of a year in 24 h with a constraint of 25 million of Vehicles Kilometer per Day, producing an events density of 9485.2 Events/s. The proposed real-time processing topology shows that the estimation error does not exceed 5% for segments length less than 12 km.

Keywords: Stream processing · Distributed computing · Mobile agents
Highway simulation · Kafka Streams · Traffic events · Traffic monitoring

1 Introduction

Nowadays, highways are increasingly equipped with a set of sensors that detect the passage of vehicles at various strategic points of the infrastructure. Sensors generate immutable events that are collected for real-time or delayed processing as required by applications. The availability of reliable real-time measurements or estimates of traffic conditions is a precondition for successful traffic control on these highways. The main objective of these systems is to improve the safety and convenience of driving, but they are also of great help in alleviating traffic congestion [1].

To improve the efficiency of traffic flows on highway networks, it is essential to develop new methodologies for modeling, simulation, estimation and traffic control. The literature is very rich in terms of approaches related to modeling and traffic control [2–5]. Using a simulator that generates traffic and all associated events is an important step to develop and test traffic management applications before installing physical infrastructures. Such a large platform needs to fuse and harmonize heterogeneous and dynamic data streams. In [6], Authors propose an algorithm based on the fusion of different types of data that come from different sources (inductive loop detectors and toll tickets) and from different calculation algorithms to obtain a fused value more reliable and accurate than any of the individual estimations. In [7], Authors review the big data background by introducing general definition of big data and review on five phases of the value chain of big data such as; the quantity of data (Volume); the rate of data generation and transmission (Velocity); the types of structured, semi-structured and unstructured data (Variety); the important results from the filtered data (Value) and the trust and integrity (Veracity).

In the field of Real-Time Processing, there are more and more real-time stream processing applications, such as traffic monitoring and civil infrastructure monitoring [8]. Recently, several distributed data processing systems have emerged that deviate from the batch oriented approach and tackle data items as they arrive, thus acknowledging the growing importance of Real-Time Data Analytics. Some technologies, such as Apache Flink [9], Apache Storm [10], Yahoo S4 [11], Apache Spark [12] and Apache Kafka [13] have made the big data in real-time stream processing implementation possible. Last year, Apache Software Foundation announced the release of an ultra lightweight library called Kafka Streams that is completely based on Kafka consumers and producers. This library allows the user to build highly scalable, elastic, fault-tolerant and distributed applications by describing the operations to be performed on the streams in a functional way.

In this paper, we propose a model of traffic event streams processing. In this model, events are generated by the developed spatiotemporal traffic simulator for real highway networks. The simulator uses a virtual calendar to speed-up simulations where it generates a flow of vehicles, assigning them to trips (Origin-Destination) according to a model using geographical data. The highways network is equipped with sensors that generate events when passing vehicles. The events stream is processed in real time by central agents to estimate the traffic state and inform users via traffic message panels. The architecture of the real-time event processing system is based on Kafka Stream Processing.

This paper is organized as follows: In the first section we present an overview of smart highway network model. The second one is devoted to describe the architecture of the model components and their interactions. Finally, we show some results obtained using an implementation applied to the Moroccan highway network.

2   Smart Highway Model

A highway network (Fig. 1) consists of several highways that can be interconnected by exchangers. Each highway is viewed as bidirectional graph. Each highway consists of a symmetric set of elements describing it in one direction; it always starts from an entry followed by several intermediate elements and it ends at an exit. The second direction is drawn by inverting the input and the output.

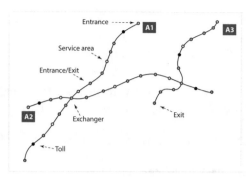

	Type	Km	
🚏	Entrance	0	
🚏	Entrance/Exit	16	
⬜	Exchanger	24	A8
🅿	Service Area	44	
🚏	Entrance/ Exit	64	
TOLL	Toll	92	
🚏	Exit	108	

Fig. 1. Highway components. The table shows an example of a highway description list

In an intelligent highway, all the elements must be instrumented by vehicle traffic sensors (Fig. 2). The sensors delimit segments in which the number of vehicles can be timely determined. Other sensors can be interposed in between these elements at strategic points in order to increase the number of monitored segments.

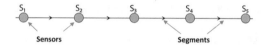

Fig. 2. Segments and sensors

The first oriented graph (IOG) is based on the highway network description list given by the XML file "**highway.xml**". The list elements are represented by the vertices (IOGV) where the edges (IOGE) of the graph represent their succession.

In order to determine all the possible paths (origin-destination) from all entrances of the network to all exits, the IOG must be transformed into a new oriented graph (TOG) describing the entire network in both directions.

From TOG, the Dijkstra Shortest Path Algorithm (DSPA) is performed to determine the list of all possible paths.

As mentioned above, the IOG description is given in an XML file consisting of a collection of vertices and a collection of edges. A vertex is represented by an XML Element "vertex" having the following Attributes:

- name: Sensor identifier (ID)
- type: Element type (Enumeration)
- label: Name of the highway (string)
- locality: The locality name of the sensor position (string)
- long: Longitude (double)
- lat: Latitude (double)
- factor: Attendance factor

The edges are represented by the XML Element "edge" having the following Attributes:

- source: Source node (IDREF)

- target: Destination node (IDREF)
- speed: Segment limit speed (double)
- distance: Distance between the two nodes (double)
- lanes: Number of lanes (int)

The "type" attribute of the "vertex" Element can take one of the following values: {I (Entrance), IO (Entrance/Exit), X (Exchange), R (Service Area), T), S (Sensor), O (Exit)}.

The TOG is obtained by performing an elementary transformation at each vertex of the IOG.

To avoid boundary effects and make all these transformations independent, we have inserted "White" vertices by splitting each edge of the IOG (see Fig. 3).

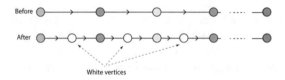

Fig. 3. Nodes isolation before elementary transformations

Depending on the vertex type, an elementary transformation will be provided. The following Table 1 gives a summary of these transformations.

All elementary transformations generate two independent sub-graphs for the two directions except for the exchangers. These sub-graphs are inserted into a temporary intermediate graph (WTOG). Obtaining the final TOG graph is done by linking the WTOG. This operation consists of removing all the "white" vertices and restoring the links between the transformed vertices. Finally, we use DSPA to determine all possible paths in the highway network.

Table 1. Elementary transformations

3   Architecture of the Model

3.1   Platform Overview

The real-time processing unit (RTPU) receives a stream of events from either the simulator or the real-world infrastructure. RTPU deploys multiple instances of Density Agent (DA). Each DA instance is responsible for calculating in real time the traffic density of a certain number of the highway network segments. Each DA has a partial view of the network. It provides the Highway State Agent (HSA) the calculated results to construct the overall state of the network. When a segment density variation exceeds a threshold, HSA alerts the Control Agent (CA) which in turn controls the concerned variable-message panels. Highway users can interact with RTPU through the User Apps Agent (UAA) to inquire about the network state.

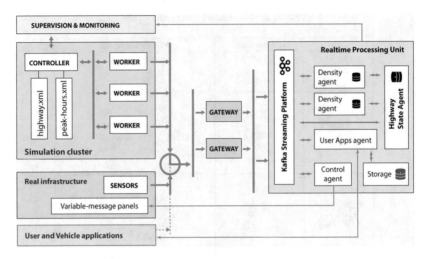

Fig. 4.   Real-time platform architecture

3.2   Worker Module: The Heart of the Simulator

The highway network simulator architecture is based on a macroscopic traffic model. At the heart of this distributed architecture, the Worker module (Fig. 5) runs on multiple nodes and generates traffic that is based on the geographical and temporal data provided in the "**highway.xml**" and "**peak-hours.xml**" files (Fig. 4). At regular time intervals and according to the current date, each scheduler generates a certain number of vehicles, assigns them random paths (Origin-Destination) and then starts them. Vehicles move from segment to another until the end of their path. At the beginning of each segment, the vehicle registers itself in being informed about the next segment density state. When a vehicle arrives at the end of a segment, it generates a time-stamped event and sends it

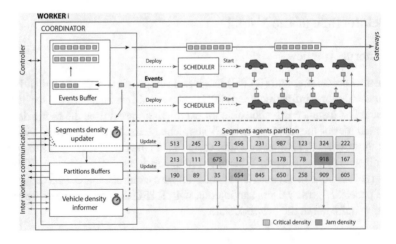

Fig. 5.   Worker architecture - Each worker runs as much scheduler as node processor.

to the coordinator. The latter receives the generated event messages and puts them in the buffer aiming at sending them to the gateways. At the same time, these messages are used to update the segment agent counters.

3.3   Real-Time Processing Topology

Each time a vehicle passes, an event containing at least three basic information is generated; the sensor identifier (previous sensor, current sensor and next sensor), date (Timestamp) and vehicle speed. Here is the general message format:

- Predecessor sensor identifier (null in an entrance)
- Current sensor identifier
- Successor sensor identifier (null in an exit)
- Timestamp
- Speed
- Further information about the vehicle depending on the type of sensor.

We simulated the year-round highways traffic in 24 h with a constraint of 25 million vehicles per day, producing an event density of 9485.2 events/s. Events are ingested in the topic "Events Topic". As shown in Fig. 6, the stream processing topology at the DA level is based on vehicle counting on all segments. The HSA uses a procedure based on the relationship between macroscopic model variables (flow, density, and mean velocity) to compute an estimate of the initial density in all segments.

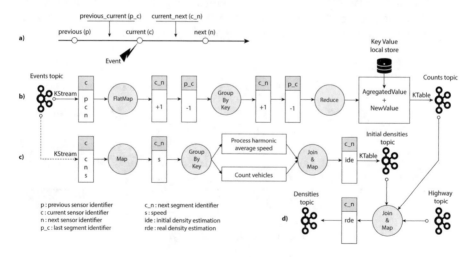

Fig. 6. Stream processing topology.

Figure 6 show the stream processing topology. (a) The generated event contains; the identifiers of the current, previous and next sensor; the timestamp and the vehicle speed. (b) Vehicle counting topology at the level of a segment. (c) Initial estimated density topology. This topology is executed, for a given time before the count begins, to estimate the initial density in all segments. d) Real density processing topology.

4   Applications and Results

The model evaluation was applied to the Moroccan highway network (see Fig. 7).

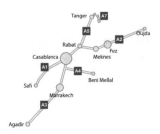

Feature	Value
Number of highways	6
Network total length	1 736 Km
Number of vertices (Nodes)	146
Number of edges (Segments)	286
Service areas	29
Number of entrances	184
Number of exits	184
Number of exchangers	5
Number of paths	2646

Fig. 7. Moroccan highway network features

At the start of the simulation (virtual time: 1 January 2017), with an acceleration factor of 365, the highway network is empty (Fig. 8a). Since the highways network was initially empty, we use the conservation law to determine the exact densities in all segments (Fig. 8b). The events generated by the simulator are logged in the topic "simulation_topic". After 12 h (virtual time: 3 July 2017), we start recording traffic events in the topic "rtpu_topic". The "rtpu_topic" topic records are used in the traffic densities estimation (Fig. 8c). For each segment, the observed error between the actual density and the estimated density is reported every minute (Fig. 8d). For each segment, the observed error is enriched with segment data such as distance and number of lanes (Fig. 8e). Errors are reported in relation to segment distances (Fig. 8f).

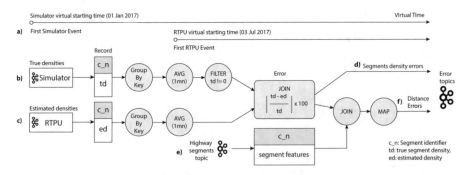

Fig. 8. Evaluation topology.

Figure 9 shows the simulated traffic density during the day of 03 July 2017 in the "Bouznika-Skhirat" section (12 km). During the first ten minutes, the branch "c" (Fig. 6) estimates the initial density in each segment. Figure 10 show that the average error for the "Bouznika-Skhirat" section does not exceed 5%. To obtain globally acceptable results over the whole network, segments whose length is greater than 12 km must be cut by adding intermediate sensors.

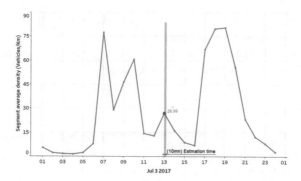

Fig. 9. The density estimation begins on July 3, 2017 at 1:00 pm (Bouznika-Skhirat, 12 km)

Given that the traffic density cannot be negative and must not exceed a maximum threshold for each segment, the topology counting chain "b" of (Fig. 6) uses a corrector to prevent incorrect values. This has the effect of reducing the error after a certain period of operation.

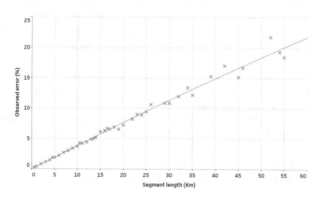

Fig. 10. Errors with respect to segment length

5   Conclusion

In this paper, we proposed a highway traffic monitoring model based on a real-time stream processing platform. The model integrates a macroscopic highway network simulator to facilitate the applications development and testing. The real-time processing module is built around the Kafka streaming platform. Platform agents use the Kafka Streams lightweight library to build stream processing topologies. Although these topologies run outside the Kafka cluster, they can scale horizontally and tolerate failures by using the same group management protocol that Kafka provides for normal consumers. The CA is responsible for controlling the variable-message panels while the UAA provides the drivers and connected vehicles with information about their journeys. We tested the platform using the simulator in the case of the Moroccan motorway network.

We compared the calculated densities with those provided by the simulator for some segments. The results are very encouraging.

In perspective, we plan to extend the model to include events generated by drivers and connected vehicles using these as mobile sensors.

References

1. Diakaki, C., Papageorgiou, M., Papamichail, I., Nikolos, I.K.: Overview and analysis of vehicle automation and communication systems from a motorway traffic management perspective. Transp. Res. A, Policy Pract. **75**, 147–165 (2015)
2. de Fabritiis, C., Ragona, R., Valenti, G.: Traffic estimation and prediction based on real time floating car data," In: Proceedings of IEEE Conference on Intelligent Transportation Systems, Beijing, China, 2008, pp. 197–203 (2008)
3. Ge, J.I., Orosz, G.: Dynamics of connected vehicle systems with delayed acceleration feedback. Transp. Res. C, Emerging Technol. **46**, 46–64 (2014)
4. Kesting, A., Treiber, M., Schonhof, M., Helbing, D.: Adaptive cruise control design for active congestion avoidance. Transp. Res. C, Emerging Technol. **16**(6), 668–683 (2008)
5. Lo, S.-C., Hsu, C.-H.: Cellular automata simulation for mixed manual and automated control traffic. Math. Comput. Modell. **51**(7/8), 1000–1007 (2010)
6. Soriguera Martí, F.: Short-term prediction of highway travel time using multiple data sources. In: Highway Travel Time Estimation With Data Fusion. Springer Tracts on Transportation and Traffic, vol. 11. Springer, Heidelberg (2016)
7. Mohammed, A.F., Humbe, V.T., Chowhan, S.S.: A review of big data environment and its related technologies. In: 2016 International Conference on Information Communication and Embedded Systems (ICICES), Chennai, 2016, pp. 1–5 (2016)
8. Liu, C.H., Zhang, Z., Huang, Y., Leung, K.K.: Distributed and real-time query framework for processing participatory sensing data streams. In: 2015 IEEE 17th International Conference on High Performance Computing and Communications, 2015 IEEE 7th International Symposium on Cyberspace Safety and Security, and 2015 IEEE 12th International Conference on Embedded Software and Systems, New York, NY, 2015, pp. 248–253 (2015)
9. Katsifodimos, A., Schelter, S.: Apache flink: stream analytics at scale. In: 2016 IEEE International Conference on Cloud Engineering Workshop (IC2EW), Berlin, p. 193 (2016). https://doi.org/10.1109/ic2ew.2016.56
10. Batyuk, A., Voityshyn, V.: Apache storm based on topology for real-time processing of streaming data from social networks. In: IEEE First International Conference on Data Stream Mining & Processing (DSMP), Lviv, 2016, pp. 345–349 (2016). https://doi.org/10.1109/dsmp.2016.7583573
11. Xhafa, F., Naranjo, V., Caballé, S.: Processing and analytics of big data streams with Yahoo! S4. In: IEEE 29th International Conference on Advanced Information Networking and Applications, Gwangiu, 2015, pp. 263–270 (2015). https://doi.org/10.1109/aina.2015.194
12. Salloum, S., Dautov, R., Chen, X., et al.: Int. J. Data Sci. Anal. **1**, 145 (2016). https://doi.org/10.1007/s41060-016-0027-9
13. Vohra, D.: Using Apache Kafka. In: Pro Docker. Apress, Berkeley, CA (2016)

Design of Sunflower System Based on Shape Memory Alloy Actuator

Amine Riad[(✉)], Mouna Benzohra, Mohamed Mansouri, and Abdelelah Alhamany

MEET Laboratory, FST of Settat, Hassan 1st University of Settat-Morocco, Settat, Morocco
am.riad@uhp.ac.ma

Abstract. The solar energy provides a renewable energy used to increase the electricity resources which is benefited in many field. The researchers aim to improve the collection of this energy that is using the solar tracker can increase the collection of this energy. There many solar trackers demonstrated its efficiency however, it is not perfect these trackers need electricity. In this work, we propose an intelligent solar system based on the shape memory alloy actuator (SMA) when the collectors acquire sunflowers ability for tracking the sun. These alloys are smart materials that can change forms when they heated or cooled. The thermo-mechanical characteristics of the SMA makes him a good sensor and actuator for tracking application. In this way, Numerical simulations at different temperatures, to show the ability of the solar system in various conditions tracking the sun.

Keywords: Shape memory alloy · SMA actuator · SMA sun tracker
Sunflower system · Intelligent sunflower system

1 Introduction

The Shape Memory Alloys (SMAs) have the ability to return to some previously defined shape when exhibited by heating or cooling over a certain temperature. The shape memory effect happens due to change in the materials crystalline structure. There are two different phases, the Martensite (in low temperature phase) and Austenite (in high temperature phase) [1, 2]. The SMAs used in actuation, and sensing applications, these kind of smart actuators have several advantages for renewable energy such as maintainability, reliability, clean, large deformation easy activation, silent operation of actuator and good recovery compact actuator with high ratio power/mass (Fig. 1).

Specially, the tube SMA are the most adapted in this actuation, due to their forcefulness design and ease fabrication [3]. The SMA actuators are typically of two varieties: one-way and two-way. The one-way actuators designed to actuate only once, and have no opposing force to bring the actuated element back to its original position. However, the Two-way actuators designed to operate in two directions over many cycles that the material remembers two shapes: one at high temperature and the other at low temperature. By using the geometry of tube and the proprieties of the shape memory alloy, in particular the two-way shape memory effect, which is the ability to recover the first shape after a plastic de-formation by sample heat [4]. In general, the sun trackers move the solar systems to keep the best orientation to the sun (Fig. 2).

© Springer International Publishing AG, part of Springer Nature 2019
J. Mizera-Pietraszko et al. (Eds.): RTIS 2017, AISC 756, pp. 45–54, 2019.
https://doi.org/10.1007/978-3-319-91337-7_5

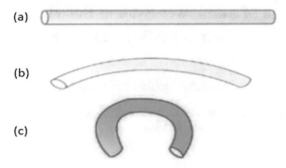

Fig. 1. Shape memory effect (a) austenite phase (b) intermediary phase (c) Martensite phase

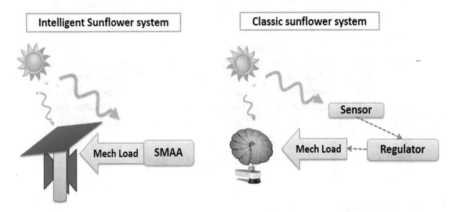

Fig. 2. Comparing the number of working elements with existing sun tracking system versus number of working elements for purposed smart sun tracking system (right) using SMA components that play the role of actuator and sensor simultaneously (figure adapted).

The aim of this paper is to develop a new sun collector system based on shape memory alloy actuator performed the dual functions of sensing and actuating. In this way, the collector can follow the sun like the sunflower and face the sun directly during the day to optimize the production of solar energy. For this purpose, of such applications requires studying the proprieties of material to work with the sun collector playing the solar tracker.

2   Modelling

In this study, the collector has three support made from shape memory alloy actuator (SMAA) in three position. The first one in the east side, the second in south side, and the third in west side. Accordingly, a simple model describes systematically the work of the sun tracker based on shape memory alloy. This smart tracker con-trolled by the change of temperature and strain that used to apply a load can change the direction of the collector.

2.1 SMAA Model

The design of these systems requires behavioral laws to describe the Materials and methods of numerical simulation to solve the structural problem. The aim is to obtain a good description of all the behaviors typical of SMA (superelasticity, memory effect, assisted double memory effect, etc.) and develop thermomechnical model capable to describe the behavior of AMF and adapt to the proposed applications. Whilst This model captures the thermomechanical behaviour of shape memory alloys and especially their proprieties the Super-elasticity and shape memory effect. For that reason, we have relied on the kinetic law to describe the transformation of phase and adjust the initial configurations of the SMA actuator.

The macroscopic deformation can be considered as the sum of the partial deformations associated with the transformation [1]. We can classify them in the following way:

- Elastic deformation ε_e
- The thermal deformation ε_{rad} due to the radiation exposed to the material.
- The transformation deformation ε_{Tr}.

$$\varepsilon_{tot} = \varepsilon_e + \varepsilon_{rad} + \varepsilon_{Tr} \tag{1}$$

The elastic strain:

$$\varepsilon_e = \sigma/E \tag{2}$$

The transformation strain:

$$\varepsilon_{tr} = \left(\varkappa - \frac{1}{2}\right).\gamma_{tw} \tag{3}$$

The thermal strain:

$$\varepsilon_{rad} = \alpha.(\Delta T) \tag{4}$$

Where is E denotes the Young modulus.
Which

$$E = E_a + \varkappa(E_m - E_a) \tag{5}$$

The SMA is characterized by the martensite transformation, which can initiate with applying stress or varying the temperature, that allocate the fraction of martensite into temperature induced martensite ,$\varkappa_T$ include of a self-accommodated of martensitic variants and stress-induced $\varkappa_S$, represents the amount of material treatment in the variant of martensite corresponding to the loading direction [6].
This means:

$$\varkappa = \varkappa_T + \varkappa_\sigma \tag{6}$$

The constitutive equation relates the variables of strain ε, volume fraction of stress induced martensite, and temperature T is:

$$\sigma - \sigma_0 = E(\varepsilon - \varepsilon_0) + \Omega(x - x_0) + \theta\,(T - T_0) \tag{7}$$

The 0 means the initial condition of the SMA [1], θ thermal elastic coefficient Ω is the transformation coefficient as shown in Eq. (8):

$$\Omega = -\varepsilon_L.E \tag{8}$$

ε_L is the maximum transformation strain of shape memory material [6, 7, 9] and is constant at temperatures under austenite finish temperature, Af. The transformation from the martensite to austenite can describe by:

$$if\ T > Ms\ and\ \sigma - Cm(T - Ms) < \sigma < \sigma - Cm(T - Ms) \tag{9}$$

$$x = \frac{1-x_0}{2}\cos\left\{\frac{\pi}{\sigma-\sigma} \times \left[\sigma - \sigma_f^{cr} - Cm.(T - Ms)\right]\right\} + \frac{1+x_0}{2} \tag{10}$$

While the transformation from austenite to martensite describe by:

$$if\ T > As\ and\ C_A\left(T - A_f\right) < \sigma < C_A(T - As) \tag{11}$$

$$x = \frac{x_0}{2}\cos\left\{a_A \times \left[T - As - \frac{\sigma}{C_A}\right]\right\} + 1 \tag{12}$$

$$a_A = \frac{\pi}{Af - As} \tag{13}$$

$$a_A = \frac{\pi}{Ms - Mf} \tag{14}$$

C_m and C_A the material properties that describe the relationship of temperature and the critical stress to induce transformation (Fig. 3).

After the development of this part of the model, the geometrical parameters it is possible to compute the thermal parameters (radiation needed for phase transition) to estimate the power needed for activation. Moreover, from the characteristics of the resulting tube. The beam used helps the tube for can easily slide up and down and can easily control the collector (Fig. 4).

The heating and cooling cycles is an important step in the design, both for the power consumption and for the frequency of the activation cycles. Specially, the heat transfer coefficient relates directly to the temperature reached at constant state and thermal energy furnished during activation [5].

Fig. 3. The shape memory alloy tube (a) In austenite phase, (b) In martensite phase

Fig. 4. The proposed SMA actuator (a) In martensite phase, (b) In austenite phase

The radiation is the transfer of heat by electromagnetic waves. The Stefan-Boltz-mann Law expresses the heat transfer radiation; the thermal energy radiated by a black-body radiator, which for steady state problems is:

$$\frac{P}{A} = \sigma T^2 \left(j/m^2 . s \right) \tag{13}$$

The work according to the radiation heat is a major energy (Fig. 5) that's can have transformed the SMA tube from the martensite phase (compressed) to austenite phase (stretched).

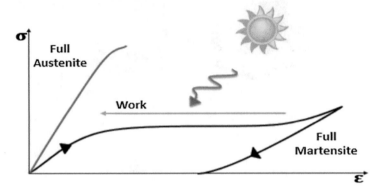

Fig. 5. Work extraction from SMA

The sun tracker works like the sunflowers is controlled by temperature created by sunlight and the shadow of the collector (Fig. 6):

1. The collector starts with the sunrises in the east. Meanwhile, the sunlight heats the SMAA in the east side and compressed unlike the others SMAA in the shadow still stretched (martensite). Hence, the collector faces east.

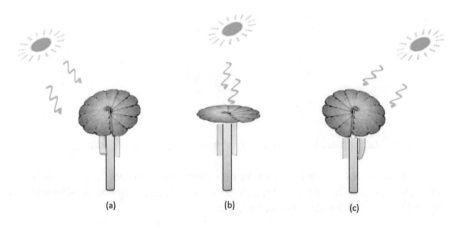

Fig. 6. (a) the collector I the East-side (b) the collector I the South –side (c) the collector I the West –side

2. As the sun moves, the SMAA faces the sun heats by the sunlight that increase the temperature of the SMA that is why transformed to austenite phase (compressed) when the others eclipsed by the collector's shadow.
3. Whenever the sun moves from east to west, the SMAA in the front change gradually the form and extend from one side of the tracker to the other.
4. The collector finished the daily cycle faces west side and can repeat the same process in the next day.

3 Results

In this study, we prepare a simulation by taking a relative value between the minimum temperature and maximum one. We have used the proposed model and used the properties of the material for the simulations (Table 1):

Table 1. Proprieties of the tube actuator used in design calculation

Composition of Ni-Ti alloy				
Moduli	$Da = 67 \times 10^3$ MPa		$Da = 26 \times 10^3$ MPa	
Transformation Temperatures (°C)	$Mf = 5$	$Ms = 20$	$As = 30$	$Af = 40$
Transformation Constants MPa/°C	$C_M = 5$	$C_A = 10$	$\sigma_s^{cr} = 0$	$\sigma_f^{cr} = 242$
Maximum residual strain (%)	$\varepsilon_L = 4$			
The required stroke (S)				
The required stroke(mm)	7			
SSTM driving force (N)	1.18			
The required stroke (S)				
Wire diameter (mm)	0.5			
Average tube diameter(mm)	5			
The number of coils (n)	20			
The maximum allowable shear strain ()	0.06			
Stress-free austenite transformation temperature	40 to 50			
Stress-free Martensite transformation temperature	30 to 20			
SMAA specifications				
Hot test temperature	≥ 45			
Cold test temperature	≤ 25			
Maximum allowable force (N)	1 to 10			
N° of actuation/day	24			
The hour angle tilting rang	$-60°$ to $60°$			
The actuation duration(min)	20			

After the model parameters are inserted in Matlab, the results are obtained as a temperature stress curve as shown in Fig. 7, the model results are in good agreement with literature tests [13, 14].

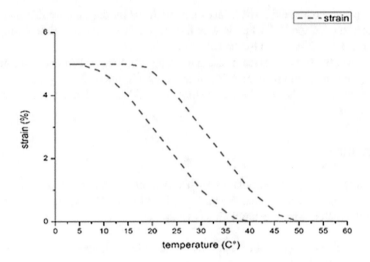

Fig. 7. The Strain-Temperature curve of the SMAA

We also simulate the SMAA in the case of stress Vs strain, in this case, after unloading, the actuator does not recover its initial shape (Fig. 8), but after heating, it can be recovered [15].

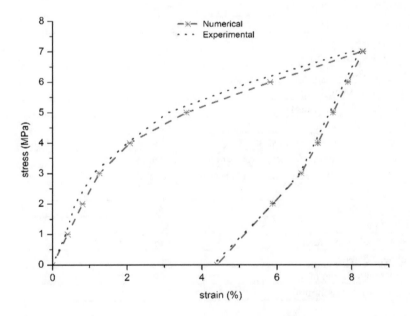

Fig. 8. Stress vs strain curve of the SMAA

It is a sample acquisition structure that requires an analytical framework capable of modelling the interactions between the material phase transition behavior and the thermoelastic mechanics.

4   Discussion

This paper is a modest contribution to the ongoing discussions, that is the results of the simulation, validates the proposed model. In the first phase of the validation process, it has considered that the SMAA wire subjected to a constant load can lift a constant weight that is the collector weight and the temperature change with the position in time of day and seasons. The model output has observed both in terms of variation of temperature and strain in the time.

Figure 7 shows the change of strain with the temperature produced by radiation heat. It is observing that the simulation is able to describe the behavior of these materials and proved in normal day, the SMAA heated by solar radiation achieved 40 °C is, completely austenitic and compressed. However, in the absence of solar radiation the SMAA under 20 °C is completely martensitic, the strain increased and the SMAA stretched hence, the mechanical energy produced lift the weight as shown in Fig. 7 that demonstrate the connection between the stress and strain of the Ni-Ti wire. The stress increase when the strain increase thus the SMAA lift the weight.

In order to theoretically describe the behavior of SMAA processes have to be calculated in space and time hence, we integrate the model finite element software as shown in Fig. 8.

Thus, the model has succeeded in reproducing the shape memory effect. It is possible to simulate the temperature distribution caused by a thermal-induced phase transition for a shape memory effect of SMAA. This figure noticed how the transfer of heat created by the radiation of the sun in the materials and can demonstrate the evolution of the austenite and the martensite volume fraction this indicates that increasing the temperature.

In general, the latent heat released or absorbed by the alloy does allocate the stress–strain response and strains allocation.

5   Conclusion

The study presented in this paper is an ambitious project aims to develop a numerical tool assisting the design of the SMAA used in solar tracker system device. Using both a sensing and an actuating element with direct heating from sunlight. The proposed actuator satisfy certain has many advantage to the studied application, as: simplicity of movement and fabrication, reduce energy consumption, works under different perturbation conditions (wind, rain, important temperature variations); can work in a water environment and can exert enough force to provide the elongation and shortening of the tracker. Have a fatigue life of more than 100 000 cycles according to manufacturer's data sheet [12]. In this way, this study validates numerical and that can validate experimental in the next paper to be similar to the sunflowers process for collecting the energy from the sun.

References

1. Riad, A., Alhamany, A., Benzohra, M.: The shape memory alloy actuator controlled by the sun's radiation. Mater. Res. Express **4**(7), 75701 (2017). http://stacks.iop.org/2053-1591/4/i=7/a=075701
2. Alhamany, A.I., Bensalah, O.M., Fehri, O.F.: Modélisation de l'hétérogénéité dans les alliages à mémoire de forme. Mécanique Industries **6**(6), 575–583 (2005)
3. Alhamany, A., Bensalah, M.O., Fehri, O.F.: Couplage dans les alliages à mémoire de forme. Comptes Rendus - Mec. **332**(11), 941–947 (2004)
4. Majima, S., Kodama, K., Hasegawa, T.: Modeling of shape memory alloy actuator and tracking control system with the model. IEEE Trans. Control Syst. Technol. **9**(1), 54–59 (2001)
5. Gao, F., Deng, H., Zhang, Y.: Hybrid actuator combining shape memory alloy with DC motor for prosthetic fingers. Sens. Actuators, A Phys. **223**, 40–48 (2015)
6. Brinson, L.C.: One-dimensional constitutive behavior of shape memory alloys: thermomechanical derivation with nonconstant material functions and redefined martensite internal variable. J. Intell. Mater. Syst. Struct. **4**(2), 229–242 (1993)
7. Lagoudas, D.: Shape Memory Alloys – Modeling and Engineering Applications. Springer, New York (2008)
8. "SHAPE MEMORY ALLOYS A shape memory alloy (SMA, smart metal, memory metal, memory alloy, muscle wire, smart alloy) is an alloy that 'remembers' its original, cold - forged shape: returning the pre-deformed shape by heating. This material is a."
9. Barcellona, A., Palmeri, D.: Thermo-Mechanical Characterization of Ni-Rich NiTi Shape Memory Alloy, In: Proceedings of the AITEM 2005 Conference (Lecce), September 2005
10. Auricchio, F., Lubliner, J.: A uniaxial model for shape-memory alloys. Int. J. Solids Struct. **34**, 3601–3618 (1997)
11. Rejzner, J., Lexcellent, C., Raniecki, B.: Pseudoelastic behavior of shape memory alloy beams under pure bending: experiments and modelling. Int. J. Mech. Sci. **44**(4), 665–686 (2002)
12. Tabesh, M., Elahinia, M.: Modeling and optimization of shape memory-superelastic antagonistic beam assembly. In: SPIE Smart Structures and Materials + Nondestructive Evaluation and Health Monitoring, p. 76430A. International Society for Optics and Photonics, March 2010
13. Muller, I., Xu, H.: On the pseudo-elastic hysteresis. Acta Metall. Mater. **39**, 263–271 (1991)
14. Poulek, V., Libra, M.: New solar tracker. Sol. Energy Mater. Sol. Cells **51**(2), 113–120 (1998)
15. Laplanche, G., Birk, T., Schneider, S., Frenzel, J., Eggeler, G.: Effect of temperature and texture on the reorientation of martensite variants in NiTi shape memory alloys. Acta Mater. **127**, 143–152 (2017)

Penalized Latent Class Model
for Clustering with Application
to Variable Selection

Abdelghafour Talibi[1(✉)], Boujemâa Achchab[1(✉)], Ahmed Nafidi[1(✉)],
and Ramón Gutiérrez-Sánchez[2(✉)]

[1] Laboratoire d'Analyse et Modélisation des Systèmes et Aide à la Décision,
Ecole Supérieure de Technlogie, Université Hassan 1er, Berrechid, Morocco
a.talibi@uhp.ac.ma, achchab@estb.ac.ma, nafidi@hotmail.com
[2] Departamento de Estadística e Investigación Operativa, Facultad de Ciencias,
Universidad de Granada, Granada, Spain
ramongs@ugr.es

Abstract. Latent class model is becoming a popular clustering algorithm for categorical variables. However when facing modern databases
that are characterized by their large number of available variables and
that only a subset of all existing variables maybe relevant for the clustering, these models, due to the identifiability conditions, make impossible the fit of model with large number of classes. The clustering task
should be therefore made on the basis of the relevant variables, eliminating insignificant variables will also improve at one hand, the clustering
results, on the other hand, the interpretation of the resulting classes
should be mitigated by the meaning of the selected variables, making
essential the selection of relevant variables for clustering. This article
present a penalized approach for latent class model that select the relevant variables, for which we propose a modified EM algorithm and a
modified BIC to select the hyper parameter and the number of classes.

Keywords: Latent class model · Log-linear model
Variable selection · Model-based clustering · Model selection
Penalization · EM algorithm · Newton-Raphson algorithm

1 Introduction

The clustering methods aims to classify the observations of a sample or a population in classes, so that the observations of the same class are similar and different
from the observations of others classes. Unlike the supervised classification where
the number of classes is known in advance, at least for a sample, in the case of
clustering, the classes number is unknown and it remains to be estimated.

There is a very large family of clustering methods. One of these is called partitioning methods that are based on heuristics or geometric procedures defined
by a proximity measure between observations in the same class, or between the

© Springer International Publishing AG, part of Springer Nature 2019
J. Mizera-Pietraszko et al. (Eds.): RTIS 2017, AISC 756, pp. 55–65, 2019.
https://doi.org/10.1007/978-3-319-91337-7_6

observations in different classes, such as hierarchical clustering [1] and K-means [2] for continuous data and K-modes [3] for categorical data, Another family is based on a probabilistic framework called mixture model [4], in this family; the clustering task is treated by a probabilistic approach. This approach, as its name suggests, uses probabilistic modeling. The goal is always the same, to establish an automatic classification of individuals in homogeneous classes, such as Gaussian mixture model [8] for continuous data and latent class model [26] for categorical data. Indeed, since the early 1990s, latent class model had becomes a powerful and a reference when dealing with categorical variables that are highly associated because of some latent variable, which represent the classes in clustering.

In fact, many fields of research used clustering methods, in order to obtain classes of observations that allow understanding and interpreting the phenomenon studied. However, recent databases are often large size described by many variables, and this abundance of descriptive variables may seem an asset to determine a proper classification of data. However, only a subset of these descriptive variables may contain the information necessary for clustering, other variables may be redundant or even non-relevant for the clustering.

In order to consider only the information required for the clustering, the selection of relevant variables must be considered, which will both improve the clustering process and facilitate the interpretation of the clustering results obtained, and also in the case of latent class model, due to the identifiability conditions, removing variables make possible the fit of model with larger number of classes.

Indeed, several recent studies were interested in the variables selection for clustering, which is more difficult than in the supervised classification, due to the fact that it is impossible to construct a criterion based on labels to guide the variables selection. The underlying idea of these works is that only a subset of all existing variables maybe relevant for the clustering, such as [12–16] for K-means, and [17–21] for the Gaussian mixture model for which the data are continuous.

Many of the recent works were interested on variables selection for clustering based on numerical data, contrary to the categorical data, where there are a few proposal methods. In the works of Raftery and Dean (2006) [23], the variable selection problem for latent class model is treated as a comparison problem. They compare two nested subsets which is equivalent to compare two models, given the variables already selected, in one model all variables are informative for the clustering, while in the other, the variables considered for the exclusion are conditionally independent of the clustering. While Silvestre et al. (2015) [24] adopt the same method of [17], in which they propose a solution to the variables selection problem in model-based clustering under the assumption that the irrelevant variables are independent of the relevant variables, by treating it as an estimation problem, which prevents any combinatorial search. Instead of selecting a subset of variables, [24] estimate a set of actual values called features saliencies.

Our method present a penalized log-linear parameterization of latent class model based on the L2-norm penalty, in which the variable selection procedure is included in the clustering process clustering, and for which the Bayesian information criterion is proposed as a model selection criterion.

This article is organized as follow. Section 2 briefly reviews the basics of the finite mixture model and the latent class model. Next, the Sect. 3 presents our penalized latent class model. In the Sect. 4 we present a method for the selection of the tuning parameter and the number of clusters. Finally, our method will be tested on simulated data sets.

2   Latent Class Model and Mixture Model

2.1   Mixture Model

The mixture model presented by Wolfe (1963) [4], Scott and Symons (1971) [5] and Duda et al. (2000) [6], and then studied by Mclachlan and Basford (1988) [7], Mclachlan and Peel (2000) [8], Banfiled and Raftery (1993) [9] or Fraley (1998) [10] and Fraley and Raftery (2002) [11], is a model based clustering, where the clustering task is carried in a probabilistic framework, it's based on the idea that the population is composed of several classes, and that each class is modeled by a probability distribution different from others classes, or by the same distribution but with parameters that are different from the others classes, and the population is modeled as a weighted mixture of these probability distributions.

The general form of the likelihood of a mixture model with K component for a single observation x_i, where x_i (for $i = 1, ...n$) is a vector $(x_{i1}, ...x_{iJ})$ with x_{ij} the variable j value for the observation x_i, is formulated as follows:

$$L(x_i; \theta) = \sum_{k=1}^{K} \pi_k f_k(x_i; \theta_k). \tag{1}$$

where $\pi_1, ..., \pi_K$ are the mixture proportions, with $0 < \pi_k < 1$, for all k and $\sum_{k=1}^{K} \pi_k = 1$, f_k is the probability distribution of the component k and θ_k is the parameters set of f_k.

While the general form of the likelihood of a mixture model with K components for n observations x_i is formulated as follows:

$$L(x; \theta) = \prod_{i=1}^{n} \left[\sum_{k=1}^{K} \pi_k f_k(x_i; \theta_k) \right]. \tag{2}$$

2.2   Latent Class Model

The latent class model [25] which can be formalized by two different and completely equivalent parameterizations; probabilistic and log-linear, was initially introduced by Lazarsfeld and Henry (1968), based on the idea that the dependence between categorical variables is in fact the result of a latent variable, which its modalities represent the classes in clustering.

The traditional latent class model is a model based clustering for multivariate categorical data, for which the classes have a multinomial distributions and the variables are independent given the knowledge of the class label.

The general form of the likelihood of a traditional latent class model with K components for n observations x_i measured on J categorical variables can be formulated as follows:

$$L(x;\theta) = \prod_{i=1}^{n} \left[\sum_{k=1}^{K} \pi_k f_k(x_i;\theta_k) \right]. \tag{3}$$

$$f_k(x_i;\theta_k) = \prod_{j=1}^{J} \prod_{m_j=1}^{d_j} p_{jm_jk}^{1\{x_i=m_j\}}. \tag{4}$$

where d_j is the number of categories that the variable j can takes, $1\{x_i = m_j\}$ is an indicator function that equals 1 if the variable j take the modality m_j as value and 0 otherwise and p_{jm_jk} is the probability that the variable j take the value m_j in the class k,

Closed-form estimates of the parameters of the maximum likelihood does not exist, even thought, a version of the EM algorithm (Dempster, Laird and Rubin 1977) [27,28] based on an iterative approach and on the complete-data log likelihood, expressed as follows;

$$\log L_c(x;\theta) = \sum_{i=1}^{n} \sum_{k=1}^{K} z_{ik} \left[\log \pi_k + \log \prod_{j=1}^{J} \prod_{m_j=1}^{d_j} p_{jm_jk}^{1\{x_i=m_j\}} \right] \tag{5}$$

where $z_{ik} = 1$ if the observation x_i belong to the class k and $z_{ik} = 0$ otherwise, is used to estimate the parameters, which can be used also in combination with Newton-Raphson algorithm (Agresti 1990). The EM approximate the maximum likelihood estimates by iterating between two steps; expectation and maximization, with a stopping criterion, like the maximum number of iterations or the maximum absolute deviation (MAD) of the parameter estimates in two successive iterations.

A not sufficient but a necessary condition for the latent class model to be identified is that there is more known information quantified by the cell frequencies, the sample size and the cells number than the unknown information which is the number of parameters, this condition is expressed as the degree of freedom which should be equal or more than 1.

3   Variable Selection for Latent Class Model

Inspired and with the same motivation as in penalized model-based clustering approach, and by using the relationship between the log-linear model [29,30] we propose a penalized latent class model approach that select the relevant variables and perform clustering by penalizing the log-likelihood function of the log-linear model to minimize.

In fact, the conditional probabilities of the latent class model can be formu-lated by a log-linear model parameters for the complete data which includes as interactions only these between the latent variable and each one of the other variables.

In the case of four indicator variables ($J = 4$), the log-linear model for the expected cell counts $N_{m_1,m_2,m_3,m_4,k}$ of the complete data that include the classes label values can be expressed as:

$$\log(N_{m_1,m_2,m_3,m_4,k}) = \lambda + \lambda_k^{LC} + \lambda_{m_1}^1 + \lambda_{m_2}^2 + \lambda_{m_3}^3 + \lambda_{m_4}^4$$
$$+ \lambda_{m_1,k}^{1,LC} + \lambda_{m_2,k}^{2,LC} + \lambda_{m_3,k}^{3,LC} + \lambda_{m_4,k}^{4,LC} \quad (6)$$

As expressed in (6), the log-linear model includes the first order effect of the indicator variables, the first order effect of the latent variable, and the interaction parameters between the latent variable and each one of the indicator variables. With restrictions,

$$\sum_{k=1}^{K} \lambda_k^{LC} = \sum_{m_1=1}^{d_1} \lambda_{m_1}^1 = \sum_{m_2=1}^{d_2} \lambda_{m_2}^2 = \sum_{m_3=1}^{d_3} \lambda_{m_3}^3 = \sum_{m_4=1}^{d_4} \lambda_{m_4}^4$$
$$= \sum_{m_1=1}^{d_1} \lambda_{m_1,k}^{1,LC} = \sum_{m_2=1}^{d_2} \lambda_{m_2,k}^{2,LC} = \sum_{m_3=1}^{d_3} \lambda_{m_3,k}^{3,LC} = \sum_{m_4=1}^{d_4} \lambda_{m_4,k}^{4,LC}$$
$$= \sum_{k=1}^{K} \lambda_{m_1,k}^{1,LC} = \sum_{k=1}^{K} \lambda_{m_2,k}^{2,LC} = \sum_{k=1}^{K} \lambda_{m_3,k}^{3,LC} = \sum_{k=1}^{K} \lambda_{m_4,k}^{4,LC} = 0$$
$$(7)$$

A general formulation of the log-linear model can be expressed as follows:

$$\log(N) = X\Lambda \quad (8)$$

where N is a vector of expected cell counts, X a design matrix composed by 0s and 1s, depending on the parameters included in the calculation of each one of the expected cell counts, and Λ is the vector of the unknown log-linear parameters.

The relation-ship between the latent class model parameters and the log-linear parameters is formulated in the calculation of the conditional probabilities as follows:

$$p_{jm_jk} = \frac{exp(\lambda_{m_j}^j + \lambda_{m_j,k}^{j,LC})}{\sum_{m_j=1}^{d_j} exp(\lambda_{m_j}^j + \lambda_{m_j,k}^{j,LC})} \quad (9)$$

In latent class model a variable j will be considered as irrelevant if it's dis-tribution is the same across the classes, in the log-linear parameterization, the overall interactions parameters between the latent variable and an irrelevant variable j will be all equal to 0, so then the variable j will have the same dis-tribution across the classes. In order to force the irrelevant variables to have the same distribution across the classes, we propose a penalized function that includes a penalty function on the interactions parameters. The penalized function to be minimized for the estimation of the log linear parameters in the case of four explanatory variables ($J = 4$) have the following form;

$$- \sum_{m_1,m_2,m_3,m_4,k} n_{m_1,m_2,m_3,m_4,k}.log(N_{m_1,m_2,m_3,m_4,k})$$

$$+ \sum_{m_1,m_2,m_3,m_4,k} N_{m_1,m_2,m_3,m_4,k} + P_w(\lambda_{m_j,k}^{j,LC}) \qquad (10)$$

where $n_{m_1,m_2,m_3,m_4,k}$ is the observed cell counts and P_w is a the penalty function on the log-linear interactions parameters which have the following form:

$$P_w(\lambda_{m_j,k}^{j,LC}) = W \sum_{j=1}^{J} w_j \|\lambda_{m_j,k}^{j,LC}\|_2 \qquad (11)$$

where W is an hyper parameter which controls the level of the desired sparsity, w_j is the weight of the variable j estimated by the overall average variance of categories probabilities across the classes and $\|.\|_2$ the l_2-penalty with $\|\lambda_{m_j,k}^{j,LC}\|_2^2 = \sum_{m_j} \sum_k (\lambda_{m_j,k}^{j,LC})^2$. Thus, a small values of the interactions parameters automatically will be regularized to be equal to 0, and if the overall interactions parameters between the latent variable and a variable j are all equal, $\lambda_{m_j,1}^{j,LC} = \lambda_{m_j,2}^{j,LC} = ... = \lambda_{m_j,K}^{j,LC} = 0$ for all $m_j = 1, ..., d_j$, it's distribution will be the same across the clusters and this variable will be considered as irrelevant.

Next we derive an EM algorithm for the above penalized latent class model:

Modified EM algorithm for the Penalized Latent Class Model:

Step 1: initialize the $z_{ik}^{(t=0)}$'s values by the traditional latent class model,
Step 2: Until the reaching the stoping condition, alternate between:

a- Given the $z_{ik}^{(t)}$'s values, initialize the standard log-linear LCM parameters $\lambda^{(t)}$, $\lambda_{m_j}^{j\ (t)}$'s and $\lambda_{m_j,k}^{j,LC(t)}$'s,

b- Using 1 Newton-Raphson algorithm iteration update $\lambda_{m_j,k}^{j,LC(t)}$'s minimizing (10),

c- Given the log-linear parameters $\lambda^{(t)}$, $\lambda_{m_j}^{j\ (t)}$'s and $\lambda_{m_j,k}^{j,LC(t)}$'s calculate the conditional probabilities $p_{jmk}^{(t+1)}$'s

d- Given the categories probabilities $p_{jmk}^{(t+1)}$'s, update the $z_{ik}^{(t+1)}$'s and $\pi_k^{(t+1)}$'s values

3.1   Model Selection

In order to select the tuning parameter W and K the number of classes we use the Bayesian Information Criterion [30]:

$$BIC = -2\log L(\hat{\theta}) + \log(n).p \qquad (12)$$

where p is the number of parameters and n is the sample size.

For the latent class model the BIC criterion can be expressed as:

$$BIC = G^2 - df \cdot \log(n) \tag{13}$$

where G^2 is the likelihood ratio chi-square statistic, df is the number of degrees of freedom and n is the sample size. In our penalized latent class model $df = (I - 1) - M$, where I is the number of cell in the contingency table, and M the number of distinct parameters.

4   Simulations Study

4.1   Binary Data

The simulated data sets consists of generating $n = 1000$ observations of a latent class with 2 classes under the assumption of the local independency between the variables, with 600 observations in class 1 and 400 observations in class 2, measured on 11 binary variables, where 4 variables are relevant for forming the two classes, the other variables are irrelevant for the clustering task.

The relevant variables have different probabilities in the classes while the irrelevant variables have the same probabilities in each class. A summary of the variables probabilities is given in Table 1.

Table 1. Probabilities used to generate the binary data

Variable	Probabilities of modality 1	
	Class 1	Class 2
1	0.3	0.6
2	0.7	0.4
3	0.3	0.7
4	0.8	0.2
5	0.5	0.5
6	0.4	0.4
7	0.8	0.8
8	0.6	0.6
9	0.9	0.9
10	0.5	0.5
11	0.4	0.4
12	0.2	0.2

Our method which set equal to 0 the weight value for the minimum number of variables that are necessary for the identifiability of a latent class with k classes, in this case this number is equal to 3, these variables are choosing to be the variables having the height overall average variance across the classes, select the correct relevant variables. And as summarized in Table 2, our penalized latent class model clustering choose the correct number of classes ($AIC = -6179.469$, $BIC = -26154.03$), and when compared to the traditional latent class model on all the existing variables, which also choose the correct number of classes, our method provide better results in terms of clustering result measured by the classification error, in fact, our method have an classification error equal to the classification error of the standard latent class model on only the 4 relevant variables(Error rate $= 16.2\%$), and our method fit better the data than the traditional latent class model on all the existing variables as measured by the likelihood ratio chi-square statistic $G^2 = 1960.531$ and the Pearson chi-square statistic $Chi^2 = 4033.863$.

Table 2. Latent class model evaluation criteria for the binary data

Method	Chi-square Chi^2	Likelihood ratio G^2	AIC	BIC	Error rate
Tra LCM	4047.383	1965.531	−6174.469	−26149.03	0.168
Tra LCM on Rel var	11.81478	11.76804	−0.23201	−29.67854	0.162
Pen LCM	4033.863	1960.531	−6179.469	−26154.03	0.162

4.2   Non-binary Data

The second simulated data consists of generating $n = 2000$ observations, from a latent class model with 2 classes, with 800 observations in class 1 and 1200 observations in class 2, under the local independency assumption of the traditional latent class model, measured on 9 variables, with 4 relevant variables for the clustering and the others are irrelevant. The relevant variables have different probabilities in the classes while the irrelevant variables have the same probabilities in each class. A summary of the variables probabilities used to generate the data is given in Table 3.

As in the case of the binary data, and as summarized in Table 4, our method select the correct relevant variables and also select the correct number of classes with $BIC = -62538.45$ and $AIC = -13732.19$. In addition, our method, provide a better results than the traditional latent class model when fitted to all the indicator variables in terms of clustering results and model fit, in fact, our method have a classification error equal to 15.5% while the classification error when using the traditional latent class model is equal to 15.55%, and a better fit of the data as measured by the likelihood ratio chi-square statistic $G^2 = 3695.811$ and the Pearson chi-square statistic $Chi^2 = 8127.436$.

Table 3. Probabilities used to generate the non-binary data

Variable	Modality	Probabilities of the modalities	
		Class 1	Class 2
1	1	0.3	0.6
	2	0.6	0.3
	3	0.1	0.1
2	1	0.7	0.7
	2	0.3	0.3
3	1	0.6	0.1
	2	0.3	0.6
	3	0.1	0.3
4	1	0.7	0.2
	2	0.2	0.2
	3	0.1	0.6
5	1	0.5	0.5
	2	0.2	0.2
	3	0.3	0.3
6	1	0.2	0.2
	2	0.6	0.6
	3	0.2	0.2
7	1	0.8	0.8
	2	0.2	0.2
8	1	0.5	0.5
	2	0.3	0.3
	3	0.2	0.2
9	1	0.8	0.8
	2	0.1	0.1
	3	0.1	0.1

Table 4. Latent class model evaluation criteria for the non-binary data

Method	Chi-square Chi^2	Likelihood ratio G^2	AIC	BIC	Error rate
Tra LCM	8178.136	3704.32	-13723.68	-62529.94	0.1555
Tra LCM on Rel var	38.54863	38.15953	-37.84047	-250.6748	0.1545
Pen LCM	8127.436	3695.811	-13732.19	-62538.45	0.155

5   Conclusion

The proposed penalized latent class model algorithm which perform clustering and select the relevant variables, and for which a penalized log linear model was proposed to assist the estimation of variables probabilities in the classes in order to force the irrelevant variables probabilities to be the same across the classes, and for which a modified BIC criterion was proposed to select the number of classes and the tuning parameter, provide good results in terms of both clustering results and variables selection.

References

1. Ward, J.H.: Hierarchical groupings to optimize an objective function. J. Am. Stat. Assoc. **58**, 236–244 (1963)
2. Macqueen, J.: Some methods for classification and analysis of multivariate observations. In: Cam, L.M., Neyman, J. (eds.) Proceedings of the 5th Berkeley Symposium on Mathematical Statistics and Probability, vol. 1, pp. 281–297. University of California Press (1967)
3. Chaturvedi, A., Green, P.E., Caroll, J.D.: K-modes clustering. J. Classif. **18**, 35–55 (2001)
4. Wolfe, J.H.: Object cluster analysis of social areas. Masters thesis. University of California, Berkeley (1963)
5. Scott, A.J., Symons, M.J.: Clustering methods based on likelihood ratio criteria. Biometrics **27**, 387–397 (1971)
6. Duda, R., Hart, P., Stork, D.: Pattern Classification. Wiley, New York (2000)
7. Mclachlan, G., Basford, K.E.: Mixture Models: Inference and Applications to Clustering. Marcel Dekker, New York (1988)
8. Mclachlan, G., Peel, D.: Finite Mixture Models. Wiley Interscience, New York (2000)
9. Banfiled, J., Raftery, A.: Model-based Gaussian and Non-Gaussian clustering. Biometrics **49**, 803–821 (1993)
10. Fraley, C.: Algorithms for model-based Gaussian hierarchical clustering. SIAM J. Sci. Comput. **20**, 270–281 (1998)
11. Fraley, C., Raftery, A.: Model-based clustering, discriminant analysis, and density estimation. J. Am. Stat. Assoc. **97**, 611–631 (2002)
12. Friedman, J.H., Meulman, J.J.: Clustering objects on subsets of attributes (with discussion). J. R. Stat. Soc. Ser. B (Stat. Methodol.) **66**(4), 815–849 (2004)
13. Witten, D.M., Tibshirani, R.: A framework for feature selection in clustering. J. Am. Stat. Assoc. **105**(490), 713–726 (2012)
14. Sun, W., Wang, J., Fang, Y., et al.: Regularized K-means clustering of high-dimensional data and its asymptotic consistency. Electron. J. Stat. **6**, 148–167 (2012)
15. Arias-Castro, E., Xiao, P.: A simple approach to sparse clustering. J. Comput. Stat. Data Anal. **105**, 217–228 (2016)
16. Talibi, A., Achchab, B., Gutiérrez-Sánchez, R.: Variable selection for clustering with regularized K-means algorithm. In: SMC 2017 Tangier (Morocco) (2017)
17. Law, M., Figueiredo, M., Jain, A.: Simultaneous feature selection and clustering using mixture models. IEEE Trans. PAMI **26**(9), 1154–1166 (2004)

18. Raftery, A., Dean, N.: Variable selection for model-based clustering. J. Am. Stat. Assoc. **101**, 168–178 (2006)
19. Maugis, C., Celeux, G., Martin-Magniette, M.L.: Variable selection for clustering with Gaussian mixture models. Biometrics **65**(3), 701–709 (2009)
20. Pan, W., Shen, X.: Penalized model-based clustering with application to variable selection. J. Mach. Learn. Res. **8**, 1145–1164 (2007)
21. Wang, S., Zhu, J.: Variable selection for model-based high dimensional clustering and its application to microarray data. Biometrics **64**, 440–448 (2008)
22. Talibi, A., Achchab, B., Lasri, R.: Variable selection for clustering with Gaussuan mixture models: state of the art. In: VSST 2015 Granada (Spain) (2015)
23. Dean, N., Raftery, A.E.: Latent class analysis variable selection. Ann. Inst. Stat. Assoc. **62**, 11–35 (2010)
24. Silvestre, C., Cardoso, M.G.M.S., Figueiredo, M.: Feature selection for clustering categorical data with an embedded modelling approach. Expert Syst. **32**, 444–453 (2015)
25. Clogg, C.C.: Latent class models for measuring. In: Langeheine, R., Rost, J. (eds.) Latent Trait and Latent Class Models, pp. 173–205. Plenum Press (1988). Chap. 8
26. Clogg, C.C.: Latent class models. In: Arminger, G., Clogg, C.C., Sobel, M.E. (eds.) Handbook of Statistical Modeling for the Social and Behavioral Sciences, pp. 311–360. Plenum Press (1995). Chap. 6
27. Goodman, L.A.: Exploratory latent structure analysis using both identifiable and unidentifiable models. Biometrika **61**, 215–231 (1974)
28. Mclachlan, G.J., Krishnan, T.: The EM Algorithm and Extensions. Wiley, New York (1997)
29. Goodman, L.A.: The analysis of systems of qualitative variables when some of the variables are unobservable. Part I-A modified latent structure approach. Am. J. Sociol. **79**, 1197–1259 (1974a)
30. Haberman, S.J.: Analysis of Qualitative Data. New Developments, vol. 2. Academic Press, New York (1979)

Application Areas of Big Data

Storing RDF Data into Big Data NoSQL Databases

Mouad Banane$^{(\boxtimes)}$, Abdessamad Belangour, and Labriji El Houssine

Faculty of Science Ben M'Sik, University Hassan 2, Casablanca, Morocco
mouadbanane@gmail.com, belangour@gmail.com,
labrigi@yahoo.fr

Abstract. RDF (Resource Description Framework) is a language standardized by the W3C for the exchange of data on the Web. It provides a formal description of Web resources and their metadata. Today a large amount of RDF data is created and is becoming available. The attention of database and semantic Web communities has been shifted to ensure effective and scalable management of RDF data. Several research and solutions have been proposed and realized to efficiently store RDF data. We discuss in this paper the use of NoSQL (Not Only SQL) databases to store massive amounts of RDF data, we also present an overview of state of the art concerning RDF data storage techniques and solutions in different NoSQL database models.

Keywords: RDF · NoSQL · Big Data · Distributed storage

1 Introduction

RDF (Resource Description Framework) is a language standardized by the W3C for the exchange of data on the Web. It provides a formal description of Web resources and their metadata, thus paving the way for the automatic processing of such descriptions. This made it the main language of the semantic Web. An RDF database (called Triplestore) stores RDF documents as a set of triples that associate a subject, a predicate, and an object. The increasing amount of RDF data offers a major performance problem to current RDF data management systems. The research community has carried out research and projects to have efficiency storage and a large-scale query of RDF data sets.

In this paper we discuss how we can store RDF data in different NoSQL database models, the column-oriented, document-oriented, key-value-oriented, and graph-oriented model.

2 RDF and NOSQL

2.1 RDF

Developed by the W3C consortium, the Resource Description Framework (RDF) [1] is a graphical model for the formal description of Web resources and their metadata in

© Springer International Publishing AG, part of Springer Nature 2019
J. Mizera-Pietraszko et al. (Eds.): RTIS 2017, AISC 756, pp. 69–78, 2019.
https://doi.org/10.1007/978-3-319-91337-7_7

order to allow automatic processing of such descriptions. RDF is intended for situations where information needs to be processed by applications, rather than being displayed only to people. Also provides a common framework for expressing this information so that it can be exchanged between applications. The basic idea of RDF is: Every element calls "resource" as a class, property or other. This resource can be identified in the Web by a URI (Uniform Resource Identifier). The RDF is in the form of a triple (subject, predicate, and object). The subject corresponds to a concept, the predicate to a relation and the object to another concept. Let's take the following example:

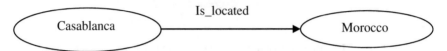

Here we have two knots, "Casablanca" and "Morocco", connected by an arc named "is_located". The subject is the source node so the subject is "Casablanca", the object is the target node i.e. "Morocco" and the predicate is the oriented arc i.e. The property "is_located", also noted that these three elements are resources.

2.2 Principal Approach for Storing RDF Data

To manage RDF data, RDF data must be stored. We can distinguish two different levels of RDF data storage: logical storage and physical storage. In this section we focus mainly on the logical storage of RDF data. RDF data storage approaches that have been reported in the literature are rare. Proposals for the storage of RDF data [2] have been classified into two main categories: native stores and relational stores. Native stores use a custom binary RDF data representation and are built directly on the file system. Relational stores distribute RDF data to appropriate relational databases. From the three basic perspectives (i.e., relational, entity and graph perspectives), proposals for storing RDF data have been classified into three main categories: relational stores, Entity and graph-based stores. In relational stores, an RDF graph is only a particular type of relational data, and thus all RDF triples are stored in a single relational table where the predicates values of the triples are interpreted as column names in a collection of relational schemas. In entity stores, an RDF graph is processed as a collection of entities descriptions. In graph-based stores, the RDF data model can be considered as a graph-based data model. Note that the edge label in an RDF graph is a URI, which may or may not also appear as a node and have a description in the graph. This fact does not make RDF less model based on the model.

Today, to handle this massive amount of RDF data, NoSQL databases are used and a number of RDF data management and data analysis processes can qualify to a Big Data infrastructure [7].

The following figure illustrates a classification of the RDF data stores; we can distinguish two types of RDF data stores from traditional database stores and NoSQL database stores. There is a wealth of approaches and perspectives for storing RDF data sets. Note that this classification is based on database models that are used to store RDF data [8] (Fig. 1).

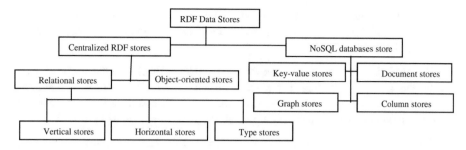

Fig. 1. Classification of RDF data stores

In the past, RDF data is stored in a central database. In addition, the NoSQL databases only appeared as an infrastructure commonly used to manage this Big Data phenomenon. In fact the RDF data stores are divided into two sub-categories. Traditional data stores are divided into two categories of traditional database models that are relational stores and object-oriented stores. For relational RDF data warehouses, we find several different relational schemes, which depend on how to distribute the RDF triples to an appropriate relational schema. We can classify them into three main categories of RDF relational stores, which are type stores, vertical stores, and horizontal stores. NoSQL databases are classified into four categories which are the four NoSQL database models: document-oriented model, key-value oriented, column-oriented and graph-oriented.

2.3 NoSQL

This is the new family of database management systems, NoSQL means "Not Only SQL" which refers to databases that are not based on the traditional relational database architecture. Originally developed to handle this big data explosion, it does not replace relational database management systems, but rather provides an alternative or supplemental functionalities of traditional RDBMSs to provide more interesting solutions and overcome the problems and limitations of these RDBMSs. In order to achieve better performance, NoSQL storage systems have abandoned certain features offered by default by relational database management systems such as transactions or integrity checks.

If relational storage systems have been used for so many years, it is because they also possess many qualities. But there are woods to which NoSQL meets as performance and scalability, the idea started within the giants of the Web like Google, Yahoo…. Through the development of their own database management systems that can run on a distributed architecture and also handle large volumes of data. The first is Google with BigTable then Amazon with its solution Dynamo, Facebook also develope Cassandra. Among the main reasons that led to the creation of these systems, we can cite the following two main points:

- The possibility of using something other than a fixed schema in the form of tables with all the properties fixed in advance;

- The ability to have a system easily distributed across multiple nodes and with which an additional need for storage or scalability simply results in the addition of new servers i.e. Scalability.

The NoSQL databases also meet the CAP theorem or C for Coherence, A availability and P for Partitioning Tolerance. This theorem is more suited to distributed storage systems. Its principle is based on that any distributed system can answer to these constraints of Coherence, High availability and Tolerance to Partitioning:

Consistency: all nodes in the system see exactly the same data at the same time
High Availability: In the event of a failure, the data remains accessible
Partitioning Tolerance: The system can be partitioned

This theorem of the CAP also specifies that only two constraints of these three can be respected simultaneously.

2.4   NoSQL Models

We are now talking about models of NoSQL databases, we can distinguish 4 models that are key-value-oriented, document-oriented, graph-oriented and column-oriented databases

Key/Value: This model can be considered as a distributed hashmap. The data is represented in a very simple way by a key/value pair. The value can be a simple string, an object, etc. This lack of structure or typing has a significant impact at the request level. Indeed, all the intelligence carried previously by the SQL queries must be carried by the application which interrogates the base. To communicate with the database there are the following operations: PUT, GET and DELETE. The best known solutions are the Voldemort project [4] developed by LinkedIn, Redis and Riak.

Column-Oriented: This model, which is similar to a RDBMS with the difference that with a NoSQL-oriented Database, the number of columns is dynamic. It stores data by column and not by row. Indeed, in a relational table, the number of columns is fixed from the creation of the table schema and this number remains the same for all the records in this table. On the other hand, with this model, the number of columns can vary from one record to another, which avoids finding columns with null values. This column orientation makes it easier to add columns to tables (lines do not need to be resized). It also allows column compression, which is effective when the data in the column is similar. As solutions, we mainly find HBase [19] (an open source implementation of the BigTable model published by Google) as well as Cassandra (Apache project which respects the distributed architecture of Dynamo Amazon and the BigTable model of Google).

We choose to talk about HBase as oriented-column NoSQL DB based on a master/slave architecture, it is built on the Hadoop Framework. HBase offers advantages such as providing a scalable way to store data in a column-oriented design. This database is now omnipresent in the Big Data architecture as a means of capturing data in real time according to several factors.

An HBase cluster consists of a master (HMaster) that maintains cluster metadata and manages region servers (RegionServer). Each region contains part of the data.

An HBase value V is located by a Rowkey R, a family of columns F, a column names C and a timestamp T. Indeed we can identify V with (R, F, C, T). In addition HBase requires no space to store null values [10]. As values are identified by their unique set of identifiers (R, F, C, T), it is possible to have a single column containing values, Identified by its line key [11] (Fig. 2).

Row Key	Timestamp	Name Family		Address Family	
		First-name	Last-name	Number	Address
Row1	T1	Jack	Honson		
	T5			13	Casa
	T10			20	Rabat
	T15			8	Tanger
Row2	T20	Lina	Adams		
	T22			88	Taza

Fig. 2. Example of HBase table

Document-Oriented: This model is an evolution of the key-value model. The value, in this case, is a JSON or XML document. Each key is associated with a document whose structure remains free. The advantage is to be able to retrieve, via a single key, a hierarchically structured information set. The same operation in the relational world would involve several joins. For this model, the most popular implementations are CouchDB from Apache and MongoDB.

Graph Oriented: This model of data representation is based on graph theory. It relies on the notions of nodes and arcs, through the relations and properties attached to them.

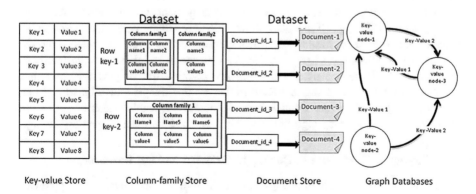

Fig. 3. The different types of NoSQL data models [20]

A graph-oriented database is an object-oriented database. This model facilitates the representation of the real world, which makes it suitable for processing data from social networks. The main solution is Neo4J (Fig. 3).

The column- oriented database and document-oriented database can be seen as an extension of the oriented key/value database.

3   RDF Data Storing Techniques in NOSQL Databases

The NoSQL databases are not specifically intended for RDF data management. But the management of massive RDF data requires and requires the use of NoSQL databases, thereby this Big Data technology offers Scalability and High Performance for Better management of Big RDF Data.

In this section we will see the storage of RDF data in the NoSQL databases models: column-oriented, graph-oriented and finally document-oriented.

3.1   Storage of RDF Data in Column-Oriented NoSQL Databases

To store RDF data in NoSQL-oriented column-oriented databases several jobs have been proposed as [15], which have used HBase for the management of this .HBase data that is a no-column-based NoSQL database management system. Similar to the relational database, HBase uses tables that are named Htable that contains column families. This work is based on the Hexastore [12] schema idea, where RDF triplets are stored in six HBase tables that are possible combinations of RDF triple patterns, and for indexing HBase provides us with an index structure based on The row key. The RDF data modeling is the designation of the row key, in HBase, the row key plays the role of the primary key in SQL databases. A single rowkey will provide quick access to the given dataset. In this case, the rowkey is the subject of the data set, the column name is the predicate, and the object is the value associated with the rowkey and the column for that family. Columns [5].

For a RDF Triplet T = (S, P, 0) of a data set D. the HBase RDF table will be composed of the column family D whose rowkey is S, the column name will be P and The value is O.

In [6] they presented a scalable RDF data management system that uses Accumulo [13] to store RDF data. They also show indexing schemes and query processing techniques that increase billions of triplets across multiple nodes. Through the provision of easy and fast access to data. In [14] they present Jena-HBase, a triplestore based on HBase which can be used with Jena framework. And in [15] they presented Rainbow, an efficient and scalable RDF triplestore that adopts a distributed and hierarchical storage architecture using HBase. To achieve dynamic scalability, they proposed a consistent hashing algorithm, to distribute the RDF data in memory storage.

3.2   Storage of RDF Data in Document-Oriented NoSQL Databases

We know that the RDF triples always contain the three components: object, predicate, subject, so to store this RDF data in document-oriented NoSQL databases it is necessary to represent this data in JSON format. JSON contains only two elements which are the key/value pair. At consistency levels it is obvious that this solution is not consistent, so for the overcome it is necessary to realize a mapping to transform the RDF triplets into JSON. Triple-centered mapping, JSON-LD approach, Subject-centered mapping are three approaches to mapping RDF to JSON.

JSON-LD: This solution that provides by the W3C Consortum for Linked Data as a serialization of data and messaging format for the creation of interoperable web services (Fig. 4).

```
{
  "@context": "http://json-ld.org/contexts/person.jsonld",
  "@id": "http://dbpedia.org/resource/John_Lennon",
  "name": "John Lennon",
  "born": "1940-10-09",
  "spouse": "http://dbpedia.org/resource/Cynthia_Lennon"
}
```

Fig. 4. Simple example of the JSON-LD format [18]

In [3] they used MongoDB [9] as a NoSQL database for storing RDF data, it used the RDF to JSON-LD converter to convert RDF data into JSON-LD format, as well as a translation of SPARQL queries into queries MongoDB.

Avoid combining SI and CGS units, such as current in amperes and magnetic field in oersteds. This often leads to confusion because equations do not balance dimensionally. If you must use mixed units, clearly state the units for each quantity that you use in an equation.

3.3   Storage of RDF Data in Graph-Oriented NoSQL Databases

The graph database allows to address complex problems, the RDF data are mainly graphs, so they can simply map into graph-oriented databases, whose subjects and objects form the nodes and predicates, specify the directed edges And labeled.

We take the example of two graph-oriented database solutions for storing RDF data Dydra and Neo4j, in Dydra which is a cloud-based graph database, the data is stored natively as a property graph So it directly represents the relationships in the underlying RDF data. We present in the following the algorithm used to realize a mapping of the RDF data in Noe4J:

The subjects of the triplets are mapped to the nodes of Neo4j. Each node in Neo4j represents an RDF resource denoted R. $(S, P, O) => (: R \{uri: S\})$

And the predicates are mapped to the properties of the node in Neo4j then if the object of the triple is a literal:

$$(S, P, O) \&\& isLiteral (O) => (: R \{uri: S, P: O\})$$

Second possible case, if the object of the triplet is a resource the predicates of the triplets are mapped to the relations in Neo4j.

$$(S,P,O) \,\&\&\, !isLiteral(O) => (:R \{uri:S\})-[:P]->(:R \{uri:O\})$$

3.4   Storage of RDF Data in Key-Value-Oriented NoSQL Databases

We come now to the storage of RDF data in key-value- oriented NoSQL databases, we can cited CumulusRDF [16] implemented on the open source key-value store Apache Cassandra [17], the authors studied the feasibility of using a key-value distributed database as a data storage component for a Linked Data server it offers unique triple pattern searches, also it provides functionality to serve linked data via HTTP searches. CumulusRDF supports two storage representations for RDF triples patterns: Hierarchical Layout and Flat Layout.

4   Review of NoSQL Systems-Based RDF Stores

This section compares RDF data stores that use NoSQL database systems to store RDF data such as Jena + HBase [14], CumulusRDF [16], Hbase + Hive [5], Rainbow [15], Rya [6], Neo4j [21] and Shard [22].

We compare key features of these RDF stores which uses NoSQL database through execution of Standard RDF benchmarks in distributed and central deployment modes. We measure in this comparison the performance of these approaches to have an adaptive NoSQL system to RDF technology.

The criteria used were: indexing, database model, partitioning method, SPARQL query support and finally execution time on two different deployment modes.

At the index level: Jena + Hbase is the best with indexing of all possible triple patterns followed by Sun and Jin approach [11] and CumulusRDF, and with partial indexing we find Hive + Hbase, SHARD, Rya and Rainbow. Database model for Hive + Hbase, Jena + Hbase, Rainbow, SHARD and Rya is column-oriented since these approaches are based on Hbase, CumulusRDF based on a key-value oriented database, graph-oriented for Noe4J.

The Partitioning method is Sharding for all these bases except Neo4j which uses the cache Sharding.

Cudr and Haque [7] compared Jena + Hbase, Hive + Hbase, Couchbase, 4Store and CumulusRDF, in both deployment modes distributed and on a single machine. The results for a 16-node cluster show that: RDF data blinds using Hbase and Cassandra are not performing at the execution time level when the RDF data quantity is not large and also in the case of execution on a single machine or a small number of nodes.

We can conclude from this work that all these approaches provide efficient and scalable storage of massive RDF data, and each of these approaches has its own advantages. Hive + Hbase, Jena + Hbase and CumulusRDF based on the two famous NoSQL systems Hbase and Cassandra are very efficient when the quantity of the RDF data is very large compared to the other approaches. The advantage of Jena + Hbase is its indexing which takes into account all the triple patterns possible, and for Hive + Hbase uses at the level of querying the Hive language that is similar to SQL and it is very easy to translate a SPARQL query to Hive query.

5 Conclusion

We have presented in this paper RDF data storage into NoSQL databases. NoSQL databases offer very important advantages for efficient storage of RDF data. The diversity of models of these NoSQL data management systems is considered also as an added value and a positive point. In our future work we will deal with on the comparison between the different solutions proposed by the research community for the storage of RDF data in NoSQL.

References

1. RDF. http://www.w3.org/rdf/
2. Sakr, S., Al-Naymat, G.: Relational processing of RDF queries. ACM SIGMOD Rec. **38**(4), 23 (2010). https://doi.org/10.1145/1815948.1815953
3. Aswamenakul, C., Buranarach, M.: A review and design of framework for storing and querying RDF data using NoSQL database, pp. 1–4 (n.d.)
4. Voldemort (n.d.). http://www.project-voldemort.com/voldemort/. Accessed 21 July 2017
5. Haque, A., Perkins, L.: Distributed RDF triple store using HBase and Hive (2012)
6. Punnoose, R., Crainiceanu, A., Rapp, D.: Rya: a scalable RDF triple store for the clouds. In: Proceedings of the 1st International Workshop on Cloud Intelligence, 4 (2012). http://doi.org/10.1145/2347673.2347677
7. Cudr, P., Haque, A., Harth, A., Keppmann, F.L., Miranker, D.P., Sequeda, J.F., Wylot, M.: NoSQL databases for RDF: an empirical evaluation, pp. 310–325 (n.d.)
8. IGI Global & Information Resources Management Association: Big data: concepts, methodologies, tools, and applications (n.d.)
9. Chodorow, K.: MongoDB: the definitive guide (n.d.)
10. Chang, F., Dean, J., Ghemawat, S., Hsieh, W.C., Wallach, D.A., Burrows, M., Chandra, T., Fikes, A., Gruber, R.E.: Bigtable: a distributed storage system for structured data (n.d.)
11. Sun, J., Jin, Q.: Scalable RDF store based on HBase and MapReduce. In: ICACTE 2010 - 2010 3rd International Conference on Advanced Computer Theory and Engineering, Proceedings, vol. 1, pp. 633–636 (2010). http://doi.org/10.1109/ICACTE.2010.5578937
12. Weiss, C., Karras, P., Bernstein, A.: Hexastore. Proc. VLDB Endow. **1**(1), 1008–1019 (2008). https://doi.org/10.14778/1453856.1453965
13. Cordova, A., Rinaldi, B., Wall, M.: Accumulo: application development, table design, and best practices (n.d.)
14. Khadilkar, V., Kantarcioglu, M., Thuraisingham, B., Castagna, P.: /home/vaibhav/Research/ Jena-HBase/Results/LUBM/Query-Time-TDB-Comp-Q10.eps, (ii), 1–4 (n.d.). http://ceur-ws.org/Vol-914/paper_14.pdf
15. Gu, R., Hu, W., Huang, Y.: Rainbow: a distributed and hierarchical RDF triple store with dynamic scalability. In: Proceedings - 2014 IEEE International Conference on Big Data, IEEE Big Data 2014, pp. 561–566 (2015). http://doi.org/10.1109/BigData.2014.7004274

16. Ladwig, G., Harth, A.: CumulusRDF: linked data management on nested key-value stores. In: The 7th International Workshop on Scalable Semantic Web Knowledge Base Systems, pp. 30–42 (2011). http://iswc2011.semanticweb.org/fileadmin/iswc/Papers/Workshops/SSWS/Ladwig-et-all-SSWS2011.pdf

17. Brown, M.: Learning Apache Cassandra: build an efficient, scalable, fault-tolerant, and highly-available data layer into your application using Cassandra (n.d.)

18. JSON-LD - JSON for Linking Data (n.d.). Accessed 21 July 2017. https://json-ld.org/

19. Dimiduk, N., Khurana, A.: HBase in Action. Manning, Shelter Island (2013)

20. Grolinger, K., Higashino, W.A., Tiwari, A., Capretz, M.A.: Data management in cloud environments: NoSQL and NewSQL data stores. J. Cloud Comput. Adv. Syst. Appl. 2(1), 22 (2013). https://doi.org/10.1186/2192-113X-2-22

21. Vukotic, A., Watt, N., Abedrabbo, T., Fox, D., Partner, J.: Neo4j in Action (n.d.). https://www.manning.com/books/neo4j-in-action

22. Rohloff, K., Schantz, R.E.: High-performance, massively scalable distributed systems using the MapReduce software framework: the SHARD triple-store (n.d.)

Mixed-Profiling Recommender Systems for Big Data Environment

Siham Yousfi[1,3(✉)] ⓘ, Maryem Rhanoui[2,3] ⓘ, and Dalila Chiadmi[1] ⓘ

[1] SIP Research Team, Rabat IT Center, EMI, Mohammed V University, Rabat, Morocco
`sihamyousfi@research.emi.ac.ma, chaidmi@emi.ac.ma`
[2] IMS Team, ADMIR Laboratory, Rabat IT Center, ENSIAS,
Mohammed V University, Rabat, Morocco
`mrhanoui@gmail.com`
[3] Meridian Team, LYRICA Laboratory, School of Information Sciences, Rabat, Morocco

Abstract. Recommender systems are intelligent tools that analyze the overall users' interests and tastes and provide them with appropriate recommendations. Such systems face several challenges in a Big Data environment due to the growing size of the recommendation matrix and the huge number of the missing values it contains. Moreover, the inaccuracy of the ratings provided by some users has a negative impact on system's performances. In this paper, we present a Big Data mixed-profiling approach that aims to reduce matrix sparsity by integrating explicit and implicit feedbacks and improve the relevance of recommendations by using a profile for each user that reflects his level of reliability and expertise. Our approach is validated in the context of digital library recommender system using Apache Spark processing engine.

Keywords: Recommender systems · Big Data · Collaborative filtering
User profiling · Apache Spark

1 Introduction

The growth of information technologies has increased the volume and the diversity of available data. Users are therefore confronted to a huge amount of newly created data offering vague and imprecise information that hardly fit their requests. Therefore, the need arises for systems that assist users and guide them towards the items they will probably be interested in. Such tools are named recommender systems. However, the very nature of the data currently being transmitted, which are also known as Big Data, leads to difficulties in developing recommender systems. In addition, these data concern both explicit ratings and implicit traces left by users during their navigation and are influenced by personal feelings, tastes, or experiences. Therefore, one of the main challenges of recommender systems researches is the integration and the combination of collaborative data (as explicit and implicit ratings), user-centric data (as user profiles), and contextual data, into an efficient recommender system [1]. Though, researches have mainly focused on explicit feedbacks, as they reflect accurate user opinions, while neglecting implicit data that require processing steps before being interpreted [2]. Also,

© Springer International Publishing AG, part of Springer Nature 2019
J. Mizera-Pietraszko et al. (Eds.): RTIS 2017, AISC 756, pp. 79–89, 2019.
https://doi.org/10.1007/978-3-319-91337-7_8

the huge number of users and items to be rated creates many issues since the recommendation matrix's is big and sparse and may lack of accuracy.

Therefore, through this paper, we propose a general mixed-profiling approach, for Big Data environment, that combines both implicit and explicit feedbacks. The results are weighted according to the user's profile. Our solution provides more relevant recommendation to the system by integrating both categories of feedbacks and helps handling the increasing volume of data using Spark environment.

The remainder of this paper is organized as follow: Sect. 2 provides the general background of our work, Sect. 3 explains the constraints of recommender systems in a Big Data environment, and Sect. 4 presents an overview of our solution which is validated using a digital library case of study in Sect. 5. Finally, Sect. 6 presents some related works.

2   Background and Context

2.1   Definition and Typology

Recommender systems are a subfamily of artificial intelligence whose main purpose is to help users finding elements of preferences and interests. Indeed, a recommender system compares a user profile with reference characteristics [3], suggests individualized recommendations [5] and seeks to predict the "opinion" that a user would give [4].

There are several approaches [6] that construct a recommender system, among the most popular ones we can find (1) Content-based filtering that aims to suggest to users items that are similar to those he has already viewed or appreciated. (2) Collaborative filtering that aims to measure the similarity between the interests of users in order to make appropriate recommendation. More information about these techniques are available in [6].

2.2   Explicit Data vs Implicit Data

In collaborative filtering, to initiate the recommendation for a specific user, it is essential to build his profile based on the profiles of other users. These profile models describe user preferences based on the two different types of data:

Explicit Data (Active Learning). They are based on interests of users that are explicitly states to the system; According to [7] these interests can be expressed in several ways, including ratings, preferences lists (wishes) and votes.

Implicit Data (Passive Learning). They are based on the traces left implicitly by the user during his navigation. These traces concern the different types of user's behaviors such as purchase history (transactions), navigation habits (historical), time spent consulting a product and mouse movements [8]. These data are important since they allow generating an individual profile for each user without involving him.

Although the explicit data are more accurate, they are more difficult to collect than the implicit data, which may lack of accuracy. Therefore, it is important to combine both methods in order to provide best recommendations.

3   Recommender Systems on Big Data Environment

With the emergence of the web 2.0, more companies are trying to collect the maximum of information about their customers to provide them with suitable products and services. In order to accomplish these objectives, they are using the recommender systems to find similarities between customers by analyzing their products' ratings and behaviors.

However, the growing mass of data and information requires recommender systems to shift to a new generation of technologies capable to handle high volume of rapid data with different levels of reliability. Indeed, as for data volume, the number of customers has grown to millions of users that can provide their opinions, reviews, and comments about several items. Also, the product datasets are continuously becoming larger. Therefore, the matrix (user, product) is becoming bigger and sparser which come up with scalability issues. Moreover, the speed of data generation is also increasing and the traditional recommender systems cannot anymore make fast responses to handle coming data and requests. Finally, at the heart of this growing mass of data and users, it is more difficult to guarantee the accuracy of the input data, which may affect the overall system's reliability.

The technologies that are able to handle data with such characteristics are called "Big Data". Several researchers have shown that Big Data technologies can bring significant added value to recommender systems. In fact, [9] proposes to use HDFS for distributed storage of large-scale data, [9, 10] use respectively Hadoop and spark to improve scalability. In addition, [11] proves that storm can handle velocity issues and improve system performances and [12–14] verifies that users and items profiling can ensure the effectiveness of the recommendations.

4   Mixed-Profiling Approach

The recommender algorithms based on matrix factorization provide more relevant results if the input training matrix is more complete and reliable. Indeed, our approach aims to provide complete and accurate training dataset by combining explicit and explicit users' feedbacks and make them reliable by updating the user's rating according to his profile (expert, newbie, etc.). Figure 1 shows the general process of the proposed approach.

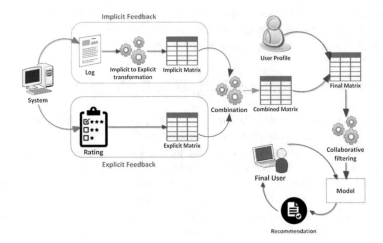

Fig. 1. Mixed Profiling recommender system

4.1 Data Acquisition

The systems collect two different types of feedbacks: (1) the explicit feedbacks represent the ratings that users have assigned to the items on a platform-dependent customizable scale (for example from 1 to 10). These data can be represented as a (user x item) matrix that will be directly exploitable by the recommender system. (2) the implicit feedbacks that characterize the users' behaviors during navigation. They may concern the number of times the users bought, searched for, or displayed the item and so.

4.2 Implicit to Explicit Transformation

The collected implicit data are not directly exploitable since they are not homogeneous (for example the range of values is different for each parameter). The common used method in the literature, to process implicit data is to transform them into explicit values. Therefore, we are combining the implicit parameters in order to find the rating that will express the user's interest in a product.

To convert implicit data to explicit, we need to define, for each implicit value, the corresponding explicit rating as follow:

Calculate Rating of Each Parameter: First we need to calculate the global interval of each parameter (max_parameter_value – min_parameter_value) and choose the scale of the rating (for example 1–10 etc.). Then we divide the data of each parameter into subintervals having or not the same amplitude. The number of intervals is thus conditioned by the scoring scale to be adopted (10 intervals in the case of a scale of 1–10). Finally, we associate to each sub-interval the corresponding value in the predefined scale (1 for the first interval, 2 for the second etc.).

Calculate Global Rating: This step aims to produce a final rating of implicit data based on fusion techniques. For example; we can associate to each implicit parameter a customizable weight that reflects its importance in the recommender system, then use fusion techniques to calculate the global rating.

4.3 Implicit and Explicit Ratings Combination

In Big Data environment, there are a large number of users and items. Therefore, both implicit and explicit matrixes are sparse. In order to reduce this sparsity we propose to combine both matrices using a tuned context-dependent weight that defines the level of equivalence between explicit and implicit data. Therefore, we create a global matrix of feedbacks so as for each user we consider the explicit rating if provided; otherwise, we consider the weighted implicit rating.

4.4 User's Profiling

In order to manage data accuracy, it is important to distinguish between expert users that always have provided reliable reviews and others novice users, non-specialists or who simply have often provided incorrect information. Thus, we will use a profiling matrix built up using one of the users profiling techniques such as [12–14], and multiply it by the (user x item) matrix. The final result will be in the form of a weighted user-item matrix.

5   Case Study: Application to a Digital Library

In order to validate our approach, we apply it to the case of a digital library. In fact, in the era of the information society, libraries ensure intellectual freedom, while making information available to citizens. The digital libraries are logical extensions of physical libraries in the web, providing new levels of access to a wider audience of users. This new generation of libraries needs techniques and tools to better assist their users, this is why they started adopting recommender systems to evaluate and filter the vast amount of available information [15].

The library is open to anyone who needs to use its collections. The categories of users are mainly researchers, teachers, students, professionals and the general public. Students and researchers are the dominant categories. In order to offer the best services to their users, the recommender system of the digital libraries should (1) Allow the user to authenticate and hold his profile (2) Implement a rating mechanism (3) Use the implicit data contained in the logs to customize the navigation by offering resources according to the user's needs and expectations.

Thus, the scenario of recommendation is as follow: A subscriber user can authenticate via his login and password. The system loads his profile and presents recommendations based on its previous needs, and through profile-based similarity analysis, the system makes customized recommendations.

5.1   Implementation of the Recommender System

We have implemented a recommender system in a context of big digital library that aims to provide users with most accurate recommendations by analyzing both users' explicit and implicit ratings and decides for their values considering the user's profiles.

Datasets. We have used a dataset of 278 859 users and 271 380 books containing explicit ratings (249550 records) and implicit (244135 records) information generated from the library log file. The implicit data are structured as follows:

- User ID: The identifier given by the system to a user;
- Last consulted document;
- Number of bounce of the request which refers to the number of search attempts before finding the desired document, this number informs about the interest that this active user has in the found document. The gathered values are from 0 to 8;
- Display of the document in the results: number of times that the document is generated in the results. It informs of the number of consultation of irrelevant documents instead of the found document. The gathered for this parameter vary from 0 to 23;
- Time spent on the document page with values from 10 s to 677 s.

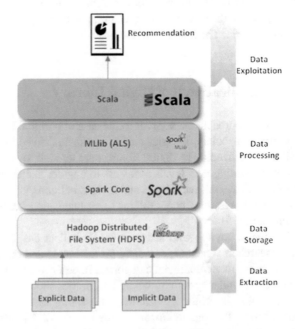

Fig. 2.   Technical Architecture of the proposed recommender system

Technical Architecture. As shown in Fig. 2 our system is built on top of spark engine. The input data that are in from of explicit and implicit feedbacks are stored in HDFS in order to ensure distributed storage of large-scale data. Then, data analysis is performed in spark core engine that guarantees parallel processing. We use the ALS (Alternating

Least Squares) algorithm of the spark's Mllib library for matrix factorization model building that the system will use later for book recommendation.

Implicit to Explicit Data Transformation. We have chosen a scale of 10 to rate books. Table 1 shows the first three ranges we created for each parameter and the corresponding rating. Using the majority voting technique represented by Eq. (1) we merged the rating of these parameters to obtain the final implicit score.

$$H(\mathbf{x}) = \sum_{i=1}^{n} q_i h_i(\mathbf{x})$$

(1)

Where q_i is the weight of to each parameter and $h_i(X)$ is the parameter's value.

Table 1. Ratings of the different implicit parameters

Number of bounce of the request level	Rating	Display of the document in the results	Rating	Time spent on the document page	Rating
[0,1]	1	]0,2]	1	]0,66]	1
]1,2]	2	]2,4]	2	]66,133]	2
]2,3]	3	]4,6]	3	]133,199]	3

Mixed-Profiling Matrix Computation. As shown in Fig. 3 the first step of this process is to merge explicit and implicit data, so as for each user, we select the explicit rating if provided; otherwise, we consider the implicit rating. Therefore, the Combined Matrix $C \in M_{n,m}(\mathbf{R})$ is in fact, the explicit $A \in M_{n,m}(\mathbf{R})$ matrix where the missing values are filled by the implicit matrix $B \in M_{n,m}(\mathbf{R})$ if found. Where n is the number of users and m is the number of books.

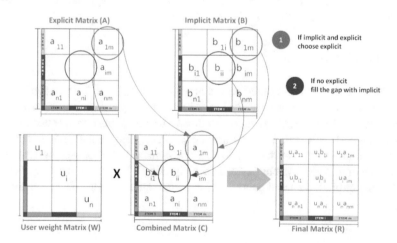

Fig. 3. Mixed profiling process

The second step aims to ensure data reliability by multiplying the C matrix by a User Weight Matrix $W \in M_{nn}(R)$ a diagonal matrix $W = diag(w_{11}, w_{22}, wa_{33}, \ldots w_{nn})$ where $a_{ii} \in]0,2]$ is the weight associated to the user i according to his profile. The values of w_{ii} are chosen as follow: (1) $w = 1$ for experts, (2) $w < 1$ for users that overestimate rating of books (3) and $w > 1$ for users that underestimate rating of books. Many ways have been discussed in the literature concerning how to create users' profiles for more information please refers to [12–14].

5.2   Results

After implementing our solution in the spark architecture described before, we used the metrics Root Mean Squared Error (RMSE), R-squared and precision at 1 to evaluate the performance of our system. The results are shown in Table 2.

Table 2. System performance comparison

	Only explicit	Only implicit	Combined explicit and implicit	Combined and profiling
RMSE	2.64	2.94	1.55	0.97
R-squared	0.472	0.453	0.333	0.806
Precision at 1	0.586	0.447	0.690	0.664

We notice the performances of our system have improved after combining both explicit and implicit matrices and applying user profiling techniques.

The used weights through this work are context-dependent; we believe that choosing optimal weight values will improve the performances of our system. As a perspective, we are planning to calculate this weight automatically.

6   Related Works

A summary of the studied related works is presented in Table 3. In general, the data sources used in literature are usually limited to either explicit or implicit data and are domain specific. Moreover, the papers that discuss user profiling aim to better understand users' needs and recommend the appropriate objects to them. Finally rare are the works that try an implementation in a Big Data environment. Our proposed system aims to improve recommender system performances in Big Data environment by combining implicit and explicit data and using profiling techniques to judge the relevance of the ratings.

Table 3. Summary of related works

Recommender system	Domain specific	Data source		Profiling	Big Data
		Explicit	Implicit		
Guo et al. [16]	Yes	Trade exhibition information User data	-	No	No
Yang and Hwang [17]	Yes	Rating of visited places	-	No	No
Goldberg et al. [18]	No	Document content Annotation Ordinary vote Binary vote	-	No	No
Balabanovi and Shoham [19]	No	Semantic content of web pages Digital vote (0–7)	-	Yes	No
Ruotsalo et al. [20]	Yes	Rating Displayed content	The visitor's position in the museum	Yes	No
Garcia-Crespo et al. [21]	Yes	Rating	Geographical position Social network information	No	No
Konstan et al. [22]	Yes	Rating Film metadata	The number and time of visualizations	No	No
Salter and Antonopoulos [23]	Yes	Rating Metadata of the visualized film	-	No	No
Renda and Straccia [24]	Yes	User ratings Document metadata	Search history	Yes	No
Jiang et al. [25]	Yes	Ranking	Travel history	No	Yes
Rehman et al. [26]	No	Rating	-	No	Yes
Shrote and Deorankar [9]	Yes	Review	-	No	Yes
Deshpande and Shirsath [10]	No	Rating	-	No	Yes

7   Conclusion and Perspectives

This paper presents a mixed-profiling recommendation approach using collaborative filtering in a Big Data environment. The proposed approach is applied to a digital library use case, which demonstrates the increase of the information reliability. As a perspective, we aim to evaluate the system performance on a large-scale datasets on a cluster of high performance machines.

Acknowledgments. The authors gratefully thank Mr. Mohammed Amaadid, Mr. Abdelkrim Tahiri and Mr. Amine Sennouni, graduate students from the School of Information Sciences in Rabat for their cooperation and invaluable contribution to the validation of this work.

References

1. Konstan, J.A., Riedl, J.: Recommender systems: from algorithms to user experience. User Model. User Adap. Inter. **22**(1–2), 101–123 (2012)
2. Jawaheer, G., Weller, P., Kostkova, P.: Modeling user preferences in recommender systems: a classification framework for explicit and implicit user feedback. ACM Trans. Interact. Intell. Syst. **4**(2), 1–26 (2014)
3. Resnick, P., Varian, H.R.: Recommender systems. Commun. ACM **40**(3), 56–58 (1997)
4. Herlocker, J.L., Konstan, J.A., Riedl, J.: Explaining collaborative filtering recommendations. Paper Presented at the Proceedings of the 2000 ACM Conference on Computer Supported Cooperative Work, Philadelphia, Pennsylvania, USA (2000)
5. Burke, R.: Hybrid recommender systems: survey and experiments. User Model. User Adap. Inter. **12**(4), 331–370 (2002)
6. Sharma, R., Singh, R.: Evolution of recommender systems from ancient times to modern era: a survey. Indian J. Sci. Technol. **9**(20) (2016)
7. Elahi, M., Ricci, F., Rubens, N.: A survey of active learning in collaborative filtering recommender systems. Comput. Sci. Rev. **20**, 29–50 (2016)
8. Hu, Y., Koren, Y., Volinsky, C.: Collaborative filtering for implicit feedback datasets. Paper Presented at the Proceedings of the 2008 Eighth IEEE International Conference on Data Mining (2008)
9. Shrote, K.R., Deorankar, A.: Review based service recommendation for Big Data. In: 2nd International Conference on Advances in Electrical, Electronics, Information, Communication and Bio-Informatics (AEEICB), pp. 470–474. IEEE (2016)
10. Deshpande, S.M., Shirsath, R.S.: Ranking of product on Big Data using Apache Spark. IJETT **1**(2) (2017)
11. Huang, Y., Cui, B., Zhang, W., Jiang, J., Xu, Y.: TencentRec: real-time stream recommendation in practice. Paper Presented at the Proceedings of the 2015 ACM SIGMOD International Conference on Management of Data, Melbourne, Victoria, Australia (2015)
12. Blanco, H., Ricci, F.: Acquiring user profiles from implicit feedback in a conversational recommender system. Paper Presented at the Proceedings of the 7th ACM Conference on Recommender Systems, Hong Kong, China (2013)
13. Lacerda, A., Ziviani, N.: Building user profiles to improve user experience in recommender systems. Paper Presented at the Proceedings of the Sixth ACM International Conference on Web Search and Data Mining, Rome, Italy (2013)
14. Lakiotaki, K., Matsatsinis, N.F., Tsoukias, A.: Multicriteria user modeling in recommender systems. IEEE Intell. Syst. **26**(2), 64–76 (2011)
15. Porcel, C., Moreno, J.M., Herrera-Viedma, E.: A multi-disciplinar recommender system to advice research resources in University Digital Libraries. Expert Syst. Appl. **36**(10), 12520–12528 (2009)
16. Guo, X., Zhang, G., Chew, E., Burdon, S.: A hybrid recommendation approach for one-and-only items. Paper Presented at the Proceedings of the 18th Australian Joint Conference on Advances in Artificial Intelligence, Sydney, Australia (2005)
17. Yang, W.-S., Hwang, S.-Y.: iTravel: a recommender system in mobile peer-to-peer environment. J. Syst. Softw. **86**(1), 12–20 (2013)

18. Goldberg, D., Nichols, D., Oki, B.M., Terry, D.: Using collaborative filtering to weave an information tapestry. Commun. ACM **35**(12), 61–70 (1992)
19. Balabanovi, M., Shoham, Y.: Fab: content-based, collaborative recommendation. Commun. ACM **40**(3), 66–72 (1997)
20. Ruotsalo, T., Haav, K., Stoyanov, A., Roche, S., Fani, E., Deliai, R., MaKela, E., Kauppinen, T., HyvoNen, E.: SMARTMUSEUM: a mobile recommender system for the Web of Data. Web Semant. **20**, 50–67 (2013)
21. Garcia-Crespo, A., Chamizo, J., Rivera, I., Mencke, M., Colomo-Palacios, R., Gomez-Berbis, J.M.: SPETA: social pervasive e-Tourism advisor. Telematics Inform. **26**(3), 306–315 (2009)
22. Konstan, J.A., Miller, B.N., Maltz, D., Herlocker, J.L., Gordon, L.R., Riedl, J.: GroupLens: applying collaborative filtering to Usenet news. Commun. ACM **40**(3), 77–87 (1997)
23. Salter, J., Antonopoulos, N.: CinemaScreen recommender agent: combining collaborative and content-based filtering. IEEE Intell. Syst. **21**(1), 35–41 (2006)
24. Renda, M.E., Straccia, U.: A personalized collaborative digital library environment: a model and an application. Inf. Process. Manag. **41**(1), 5–21 (2005)
25. Jiang, S., Qian, X., Mei, T., Fu, Y.: Personalized travel sequence recommendation on multi-source big social media. IEEE Trans. Big Data **2**(1), 43–56 (2016)
26. Rehman, T.U., Khan, M.N.A., Riaz, N.: Analysis of requirement engineering processes, tools/ techniques and methodologies. Int. J. Inf. Technol. Comput. Sci. **5**(3), 40 (2013)

A Review of Scalable Algorithms for Frequent Itemset Mining for Big Data Using Hadoop and Spark

Yassir Rochd$^{(\boxtimes)}$, Imad Hafidi, and Bajil Ouartassi

IPOSI Laboratory, Hassan I University, 26000 Settat, Morocco
y.rochd@gmail.com

Abstract. One of the most principal fields of data mining is finding frequent itemsets. Frequent itemset mine algorithms become resource hungry fast as their search space explodes if we feed them with large datasets. This problem is even more obvious when we try to use them on Big Data. Even if a lot of experiences of trying to apply this kind of techniques to Big Data have been done, some of the implementations proved to efficiently scale to large information's collection. A comparison of a well selected subset of the most extensible and efficient approaches have been presented by this review. By focusing on platforms of Hadoop and Spark, we tend to consider the typical analysis of the data mining's dimension and to value criteria in the Big Data environment as well.

Keywords: Frequent itemset mining · MapReduce · Hadoop · Spark
Data mining

1 Introduction

The importance of Data Mining has radically changed because of the increasing capabilities of recent applications to produce big amounts of information. The interest has risen towards data mining techniques, which focus on extracting effective and usable knowledge from large collections of data in both academic and industrial domains. In this paper, we will be focusing on Frequent Itemset Mining [1], which is a data mining technique that discovers frequently items. Existing itemset mining algorithms revealed to be very efficient on medium-scale datasets but very resourceful in Big Data contexts. Generally, applying data mining techniques to Big Data entails to cope with computational costs that are apt to become bottlenecks when memory-based algorithms are used. Because of this, distributed and parallel approaches that are based on the MapReduce paradigm [2] have been proposed. The main idea of the MapReduce paradigm, which is designed to cope with Big Data, is considering splitting the processing of the data into independent tasks, to the extent that each one working on a chunk of data. Hadoop [3] is known to be the most scalable open-source MapReduce platform. Recently, Apache Spark [4] has grown to become a valid alternative platform that can run on top and exceed the Hadoop resource manager, Yarn. A selected subset of Frequent Itemset Mining algorithms that exploit Hadoop and Spark platforms are being carefully

© Springer International Publishing AG, part of Springer Nature 2019
J. Mizera-Pietraszko et al. (Eds.): RTIS 2017, AISC 756, pp. 90–99, 2019.
https://doi.org/10.1007/978-3-319-91337-7_9

compared in this survey. Before comparing advantages and drawbacks of each approach, the paper presents an overview of the algorithms and the challenges.

2 Frequent Itemset Mining

The attention from the practitioners of all data related industries is now drawn by Data mining techniques for this purpose. The goal of data mining is to verify the results with the detected patterns applied to new subsets after exploring data in research and inter-pretation of unforeseen trends or patterns between variables. Correlation analysis auto-matically became an important foundation for data mining and big data science since data gathered from a variety of data sources are often a series of isolated data. Associ-ation rule mining [5] was proposed to discover some important correlation relationships among the itemset of the data. Furthermore, frequent itemset mining is considered to be an essential step in the process of association rule mining.

Frequent itemset mining is (FIM) an imperative part of data analysis and data mining. FIM's principal goal is to mine information and reveal patterns from massive datasets on the basis of frequent occurrence, i.e. if an event or number of events occur/seem frequently in the data, they are interesting, and this according to a user given minimum frequency threshold. Listless techniques have been invented to mine frequent itemset from databases. These techniques work well in practice on typical datasets [6], unfortu-nately, they aren't applicable for real Big Data. The fact of using frequent itemset mining technique to massive databases is not easy task.

3 Hadoop and MapReduce

Encouraged by benefits of parallel execution in the distributed environment, the Apache Foundation came up with open source platform, Hadoop, for faster and easier analysis and storage of different varieties of data [3]. HDFS and MapReduce programming model are two integral parts of it. Google File System gave birth to HDFS (Hadoop Distributed File system) [7], which mainly deal with storage issues. Contrary to the RDBMs, it follows WORM (write-once read-many) model in order to split large chunk of data to smaller data blocks then join them to the free node available [3]. Stored Input data blocks are kept in more than one node in order to achieve high performance and fault tolerance. MapReduce which is inspired by Google's MapReduce [8, 9] is known to be a linearly adaptable programming model. It contains two main functions a map () function and a reduce () one, both of which work in a synchronous manner in order to operate on one set of key value pairs, and that, to produce the other set of key value pairs. These func-tions are equally valid for any size of data irrespective of the degree of the cluster. MapReduce uses the feature known as data locality to collocate the data with the compute node, so that data access is fast. It follows shared nothing architecture which eliminates the burden from the programmer of thinking about failure. The architecture itself detects failed map or reduce task and assigns it to a healthy node. Figure 1 shows the Hadoop MapReduce architecture.

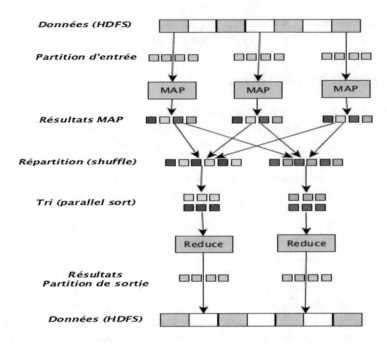

Données (HDFS)

Partition d'entrée

MAP   MAP   MAP

Résultats MAP

Répartition (shuffle)

Tri (parallel sort)

Reduce   Reduce

Résultats
Partition de sortie

Données (HDFS)

Fig. 1. Illustration of the Hadoop MapReduce

4   Spark

Spark is defined as a distributed computing framework developed at the Berkeley AMPLab [10] which offers a certain quantity of features in order to make big data processing fast. The main key feature is its in-memory parallel execution model which memory contains all loaded data. This first key principally benefits the iterative computations. The second feature considers the fact that Spark may offer very scalable DAG-based (directed acyclic graph) data flow in deferring from the nominated two-stage data flow model in MapReduce.

Both of these features may determinately rush the computation for those iterative algorithms just like the Apriori algorithm and other few machine learning algorithms. That way, Spark achieves 1–2 orders of magnitude in a way faster speed than MapReduce. Spark's programming model is working with a new distributed memory abstraction applied on large cluster for in-memory computations named resilient distributed datasets (RDDs).

An RDD which is an immutable collection of data records has the ability to offer a variety of built-in operations in order to change one RDD into another RDD. It's on the worker nodes where Spark caches the contents of the RDDs, and that's what makes data reuse way faster. The fault-tolerance that is based on lineage information may be achieved by RDDs rather than replication. If a node fails, Spark tracks enough information in order to reconstruct RDDs. Figure 2 illustrate the working model of Spark framework.

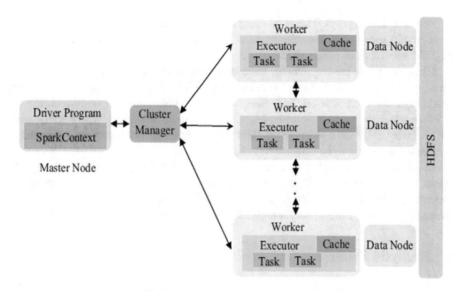

Fig. 2. Working model of Spark framework

5   Algorithms

Five selected Hadoop and Spark implementations based on the most scalable Frequent Itemset Mining algorithms are compared in this paper.

Parallel FP-growth [11] is MapReduce implementations of FP-Growth [5] and it has formed for years the one and only effective and concrete distributed FIM algorithm based on Hadoop. FP-Growth is based on an FP-tree transposition of the transaction dataset and a recursive divide-and-conquer approach. A set of independent FP-trees that are distributed to the cluster nodes are initially built by the parallel version. Then, the complete set of frequent itemset is generated by applying one instance of the FP-growth algorithm on each FP-tree. Knowing that the generated FP-trees are independent, the mining phase can be performed in parallel, i.e., one independent task for every single FP-tree is executed in order to mine a part of the frequent itemset.

BigFIM and Dist-Eclat [12] proposed two methods for frequent itemset mining for Big Data on MapReduce, the first one is DistEclat that we may consider as the distributed version of pure Eclat method that distributes the search space evenly among mappers in order to improve speed, the second one BigFIM works with projected databases that suits memory in order to extract frequent itemset with both Eclat and Apriori based method. BigFIM conquers the issue of Dist-Eclat lifelike mining of sub-trees requires the whole database into main memory and all of dataset should be sent to basically all of the mappers.

BigFIM is one of the most known hybrid approaches that considers Apriori algorithm as a way of generating k-FIs, after that, it applies Eclat algorithm in order to find the frequent item sets. Memory does not suit candidate itemset for more obvious depth is definitly the restriction of using Apriori to produce k-FIs in BigFIM algorithm and speed is really low for BigFIM. The good thing about Dist-Eclat and BigFIM is that it gives respectively speed and extensibility. Dist-Eclat doesn't offer scalability and speed of BigFIM which is less in this case.

HFIM in [13] is known, instead, to be an Apriori distributed implementation, which uses the vertical layout of dataset to solve the problem of scanning the dataset in every single iteration. Moreover, HFIM presents also some enhancements to minimize number of candidate itemset. The proposed algorithm is implemented over Spark framework, which incorporates the concept of resilient distributed datasets and performs in-memory processing to improve the execution time of operation.

In addition, Spark PFP [14] represents a pure transposition of FP-Growth to Apache Spark; it is included in MLLib, the Spark Machine Learning library. The algorithm implementation in Spark is so similar to the Hadoop sibling, i.e., it invokes the mining step on each tree (one independent task for each FP-tree) after building independent FP-trees.

6   Analysis Criteria

In this part we introduce the criteria adopted to evaluate the algorithms. We start with the first set of features which are related to the algorithm implementation. Spark popularity is growing fast while Hadoop is an established platform. We consider that Hence, Spark implementations are more promising and future proof. Then, communication costs and load balancing features are considered. These are two of the most principal and important features in distributed processing but they are usually undervalued. Communication cost is known to be a crucial part of the behavior of a parallelized algorithm. It does often submerge computation costs and it can become a bottleneck for the overall performance. Also, Load balancing, influences performance and limits the parallelization.

Table 1 reports the classification of the five algorithms, based on the criteria described in this section.

Table 1. Algorithm analysis

Name	Framework	Underlying algorithm	Data	Search strategy	Communication cost usage	Load balance usage
PFP	Hadoop	Fp-growth	Dense	Depth first	Yes	Yes
BigFim	Hadoop	Apriori and Eclat	Dense and eparse	Depth and breadth first	Yes	Yes
DisEclat	Hadoop	Eclat	Dense	Depth first	Yes	Yes
Spark PFP	Spark	Fp-growth	Dense	Depth first	Yes	Yes

7   Experimental Evaluation

In this part, we present the results of the experimental comparison. The characters of the algorithm reference applications are compared by considering different data distributions and use cases. The experimental evaluation goes for understanding the connections between the algorithm performance and its parallelization strategies. Algorithms performances are evaluated in terms of efficiency under various conditions: load balancing and communication costs. A typical arrangement of defaults parameter values is characterized for all experiments. Specific experiments with various settings are explicitly mentioned. The default setting of every algorithm was picked by considering the physical characteristics of the Hadoop cluster, to enable each approach to deal with the configuration of the hardware and soft-ware at its best.

We did not determine a default value of minsup, which is a common parameter of all algorithms, because it is very related to the data distribution and the use case, so this parameter value is specifically discussed in each set of experiments.

We considered both synthetic and real datasets. The synthetic ones have been generated by means of the IBM dataset generator [15], commonly used for performance benchmarking in the itemset mining context. The real datasets have been used to simulate real-life use cases. The list of synthetic and real datasets is reported in Table 2. All the experiments were performed on a cluster of 5 nodes running the Cloudera Distribution of Apache Hadoop [16]. Each cluster contains Intel® Core ™ i5 - 3230 M CPU@2.60 GHz processing units and 16.00 GB RAM and SATA 5400-rpm hard disks.

Table 2. Synthetic and real-life dataset

Type of dataset	Name	Num of different items	Avg items per transactions	Size (GB)
Synthetic	Dataset 1	18000	10	0,5
Synthetic	Dataset 2	18000	20	1,2
Real life	Netlogs	160,900,000	15	0,6
Real life	Delicious	57,300,000	4	44

7.1 Load Balancing

Load balancing was analyzed on a 1-hour-long subset of the netlog dataset with a fixed minsup of 1%. We take into consideration the most unbalanced jobs of every algorithm and compare the execution times of the speediest and the slowest tasks. For this raison, we are interested in the normalized execution times, rather than in, the absolute execution time in which the slowest task is assigned a value of 100, and the speediest task is compared to such value as mentioned in Fig. 3.

Spark PFP realizes the best load balancing, with comparable execution times for all tasks over all nodes, whose difference is in the order of 10%. Instead, Parallel FP-growth, demonstrates the worst load balancing issues, with differences as high as 90%. The contrast amongst Spark PFP and Parallel FP-growth can be corresponded to the granularity of the sub problems. Spark PFP enables specifying the number of partitions, i.e., of sub problems, that certainly impacts on the granularity of every sub problem. Thus a good load balancing result is accomplished while setting perfectly this parameter. In a different, Parallel FP-growth naturally sets the number of sub problems and the present heuristic used to set it does not appear to work well on the considered.

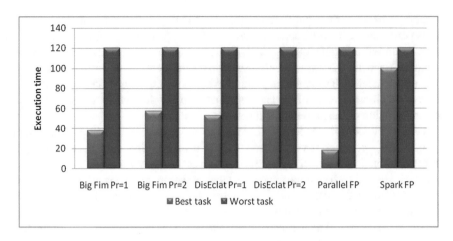

Fig. 3. Execution time of the most unbalanced tasks

BigFIM and DistEclat were included with 2 different first-phase prefix sizes. In accordance to these algorithms, the experiment confirms that a configuration with longer prefixes leads to a more balanced mining tasks than a configuration with short-sized prefixes.

7.2 Communication Costs

The experiments have done on Dataset with a fixed minsup value of 0.1%, which was the most reduced value for which all algorithms completed the extraction. Figure 4 reports, for each algorithm, the average value among transmitted and received traffic, compared to the total execution time. Firstly, the two measures do not appear to be corresponded: higher communication costs are related with low execution times for BigFIM and DistEclat, whereas Spark PFP reports both measures with high values.

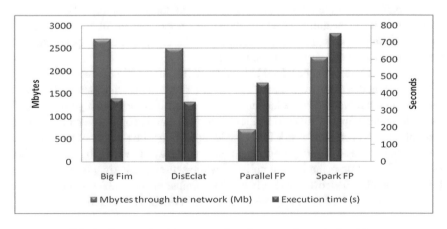

Fig. 4. Communication costs and performance for each algorithm

Parallel FP-growth has a communication cost 4 to 5 times lower than all the others. Parallel FP-growth average communication cost is around 0.5 Gigabytes, which is approximately the dataset size. The difference between DistEclat and BigFIM is not large because with only 2-length prefixes just an extra iteration is done by BigFIM. Although Parallel FP-growth is the most communication cost optimized implementation, the very low amount of data sent through the network is related to the adoption of compression techniques, which lead to higher execution times.

8 Discussion

Seen from an analytical angle, BigFIM and Dist-Eclat [12] are the algorithms according the most attention to communication costs and load balancing. For example, the motivations behind the choice of the prefixes length, generated during the first step of both algorithms are very interesting. In fact, that choice significantly influences both load balancing and communication coast. While the latter would improve with a deeper level of the mining phase before the redistribution of the seeds, the former would benefit of shorter prefixes.

More exactly, BigFIM crashes during its Apriori based phase when low minsup values or dense datasets are taken into account, because to the large number of generated candidate itemset. DistEclat deserves a different consideration: even if it is based on the search space approach, it often runs out of memory, because in its initial job it needs to store the tidlists of all frequent items in main memory and this operation becomes easily unfeasible when large or dense datasets are considered.

Hence, depending on the data distribution and the characteristics of the Hadoop cluster, BigFIM and DistEclat can be tuned to improve communication cost or load balancing that clearly impacts on the overall execution time.

Parallel FP-growth [11] is formed by the generation of independent FPtrees that allow achieving work independence among the nodes. Never the less, the independent FP-trees may have different characteristics (e.g., some are more dense than others) and that's what have impacts significantly on the execution time of the mining tasks that are executed independently on each FP-tree. The tasks are unbalanced and hence the whole mining process is unbalanced when the FP-trees are significantly different. This issue may be potentially solved by splitting complex trees in sub-trees: however, defining a metric to split a tree is not that easy.

The MLLib collection includes the Spark PFP implementation [4]. It is characterized by dynamic and smooth handling of the diverse stages of the algorithm, without any strict division in phases. Its main important advantage over the Hadoop sibling is the low I/O cost, leading potentially to a single read of the dataset, by loading the transactions in an RDD and processing the data in main memory, whereas the Hadoop-based implementation of PFP performs much more I/O operations.

HFIM [13] exploits the Spark architecture and APIs to handle communication costs. It focuses on the problem of scanning the entire dataset in each iteration, which causes the high I/O cost and disk space. HFIM utilizes the concept of vertical layout of dataset [14]. The vertical dataset provides the benefit that there is no need to scan the complete dataset in each iteration and it is smaller in size in comparison with the horizontal data. The vertical dataset carries enough information to generate the possible candidate sets and compute their support count. The vertical data are shared with all the executors of cluster using a special feature of Spark framework called broadcast variable. Therefore, scanning the vertical dataset reduces the cost of I/O and required disk space. To make the algorithm workable in parallel fashion, the horizontal data are stored in form of RDD and distributed among all the executors.

Ultimately, the number of input parameters, that is another principal characteristic of data mining algorithms, is constrained for almost all the considered implementations. With their customizable length of first-phase prefixes, BigFIM and DistEclat could require some experiments to find the proper set of parameters (depending on the dataset distribution and the cluster configuration). Moreover, this parameter allows Big-FIM and DistEclat to handle both communication costs and load balancing. Among Hadoop algorithms, relying on experimental evaluations presented in the papers, we can consider BigFIM as the current baseline in this survey. For future perspective and developments, Spark implementations are considered to be more promising than Hadoop ones, even though the formers currently appear less mature. Spark algorithms have just started to appear in literature, they are expected to find more complete implementations in a really near time.

9   Conclusion

The comparative study presented in this paper mentioned interesting research directions to upgrade distributed itemset mining algorithms for Big Data. As argued in this review, various algorithms have been suggested in literature to discover frequent itemset. Yet, the effective exploitation of every algorithm firmly relies on particular skills and abilities. The analyst is required to select the best The ameliorations in algorithm usability ought to be addressed by designing innovative self-tuning itemset mining frameworks, strategy to effectively treat with the data characteristics, and manually configure it. Therefore, state of the arts algorithms may become inefficient because of the ineffective hand-picked choices of the inappropriate particular implementations, and cumbersome parameter-configuration sessions. The ameliorations in algorithm usability ought to be addressed by designing innovative self-tuning itemset mining frameworks, capable of wisely selecting the most fitting itemset extraction algorithm and automatically configuring it.

References

1. Pang-Ning, T., Steinbach, M., Kumar, V.: Introduction to Data Mining. Addison-Wesley, Boston (2006)
2. Dean, J., Ghemawat, S.: MapReduce: simplified data processing on large clusters. In: OSDI 2004, p. 10 (2004)
3. Apache Hadoop. http://hadoop.apache.org
4. The Apache Spark. https://spark.apache.org/mllib/

5. Han, J., Pei, J., Yin, Y.: Mining frequent patterns without candidate generation. In: SIGMOD 2000, pp. 1–12 (2000)
6. Han, J., Kamber, M., Pei, J.: Data Mining: Concepts and Techniques. Elsevier, New York (2011)
7. Ghemawat, S., Gobioff, H., Leung, S.T.: The Google file system. In: SIGOPS Operating Systems Review, vol. 37, pp. 29–43 (2003)
8. Zhao, W., Ma, H., He, Q.: Parallel K-means clustering based on MapReduce. In: Proceedings of the 1st International Conference on Cloud Computing, pp. 674–679. Springer, Heidelberg (2009)
9. He, Q., Zhuang, F., Li, J., Shi, Z.: Parallel Implementation of classification algorithms based on MapReduce. In: Proceedings of the 5th International Conference on Rough Set and Knowledge Technology (RSKT 2010), pp. 655–662. Springer, Heidelberg (2010)
10. Zaharia, M., Chowdhury, M., Das, T., Dave, A., Ma, J., McCauley, M., Franklin, M., Shenker, S., Stoica, I.: Resilient distributed datasets: a fault-tolerant abstraction for in-memory cluster computing. In: Proceedings of NSDI 2012 of the 9th USENIX Conference on Networked Systems Design and Implementation, p. 2 (2012)
11. Li, H., Wang, Y., Zhang, D., Zhang, M., Chang, E.Y.: PFP: parallel FP-growth for query recommendation. In: RecSys 2008, pp. 107–114 (2008)
12. Moens, S., Aksehirli, E., Goethals, B.: Frequent itemset mining for big data. In: SML: BigData 2013 Workshop on Scalable Machine Learning. IEEE (2013)
13. Sethi, K.K., Ramesh, D.: HFIM: a Spark-based hybrid frequent itemset mining algorithm for big data processing. J. Supercomput. **73**, 3652–3668 (2017)
14. Zaki, M.T., Parthasarathy, S., Ogihara, M., Li, W.: Parallel algorithms for discovery of association rules. Data Min. Knowl. Disc. **1**(4), 343–373 (1997)
15. Agrawal, N., Imielinski, T., Swami, A.: Database mining: a performance perspective. IEEE Trans. Knowl. Data Eng. **5**(6), 914–925 (1993)
16. Cloudera. http://www.cloudera.com. Accessed 16 Oct 2015

An Analytic Hierarchy Process Based Comparative Study of Web Data Extraction Approaches

Said Sadik[✉], Abdessamad Belangour[✉], and Abdelaziz Marzak[✉]

Department of Mathematics and Computer Science, Faculty of Sciences Ben M'Sik,
Casablanca, Morocco
said.sadik.fsb@gmail.com, belangour@gmail.com, marzak@yahoo.com

Abstract. In the air of the Big Data, the extraction of the data from the Web constitutes an unavoidable stage. But unfortunately the data sources on the Web have been designed to be browsed and viewed by human users using a browser. In addition to that with the ease of publishing documents, the number of information increases from where the search for this information turns into a cumbersome and lengthy operation. Yet such information can open the door to many applications (intelligent agents, integration into data mediation systems, etc.) if they are accessible by computer programs. To solve all these problems several approaches are emerging where each approach treats the problematic according to a specific point of view or proposes a method for the efficient extraction of the data. In this paper we propose a comparative study of the most well-known approaches in the bibliography based on the Analytic Hierarchy Process (AHP). This comparative study is made according to a set of criteria whose relevance and reasons for choice will be outlined. The results obtained will constitute the basis for proposing a new method whose objective exceeds data extraction.

Keywords: Data extraction · Analytic Hierarchy Process · World Wide Web
Web pages

1 Introduction

With the flood of information on the World Wide Web, the Internet has become one of the richest information sources in the world and a media of choice to find the answer to a multiple information need. Hence, the extraction of data from the web is a new research question that attracts the interest of many communities. This has given rise to many approaches to addressing the problem of retrieving data from web pages. Some approaches have been designed to solve specific problems. Other approaches, instead, have heavily used techniques and algorithms developed in the field of information mining. In this article, we present a preliminary discussion with a first part describing all the approaches we have chosen, a second part introduces the comparison criteria that we identified, and finally a study based on a method based on matrix calculations to arrive at a matrix allowing to deduce a solution which will be in the form of an approach that satisfies the set of criteria determined.

© Springer International Publishing AG, part of Springer Nature 2019
J. Mizera-Pietraszko et al. (Eds.): RTIS 2017, AISC 756, pp. 100–110, 2019.
https://doi.org/10.1007/978-3-319-91337-7_10

2   Choice of Approaches

In order to carry out our research project, we have initially gathered relevant information to prepare a literature review allowing us to identify the evolution of the concepts, approaches and tools of the extraction data from the web pages in the last few years.

The main objective of this article is the comparative analysis of the different approaches; we have set up search criteria in order to choose the set of approaches that will deal with the problem of extracting data from the web with several methods:

- The approaches must have different extraction methods.
- Approaches should deal with all possible web pages (structured, unstructured, etc.).
- Approaches must belong to different periods.

A bibliographic search and the information we collected led us to target six approaches to extract data from the web, which we find most representative:

- A Conceptual-Modelling Approach [1, 7] (C.M.A): Based on an ontology - an instance of a conceptual model - that describes the data to be extracted, including the lexical appearance, relationships, and keywords of the context. The approach consists in producing a database schema and recognitions for constants and keywords by analyzing the ontology and then invoking methods to recognize and extract data from non-structured documents and structure them according to the basic schema of generated data.
- A Clustering Approach [2, 12] (C.A): The main idea is to represent a Web page in the form of a list of text tokens and to use clustering to group similar clustered tokens of text so that clusters can be used to identify the data. This approach has two main contributions: a new K-neighbors-HAC text token grouping algorithm and a set of similarity measures for efficiently grouping text tokens.
- A Machine Learning Approach [3, 11] (M.L.A): Is based on an encapsulation system that takes as input a set of marked web pages with examples of data to be extracted provided by the user as well as additional pages to mark suggested by the system in order to build wrappers that are very specific. This system then generates extraction rules that describe how to locate the desired information on a web page. Another verification system is implemented to process the extracted data. If a change is detected, the system can automatically repair a wrapper using the same patterns to locate examples on modified pages then reuse the induction wrapper again.
- A Partial Tree Alignment [4] (P.T.A): Consists of two steps, a first step for first identifying data records in the processed page, and a second step for aligning and retrieving the data items from the identified data records. The approach proposes a method for each step, a method based on the visual information for the first step in order to segment the data records, and a partial alignment technique based on tree matching for the second step, Align only the data fields in a data record pair that can be aligned to extract the data more precisely.
- A Gene/Clone with structural prefix-suffix Approach [5] (G.C.A): Evolution of the "Gene\Clone" approach [8] whose objective is to generate the gene which is the smallest repetitive substructure containing all the values of the instances. Then

retrieve the other instances (clones) based on the repetition of this gene in the XML document generated from the web page, implying that it is a search of the nodes concerned instead of a search for relevant information. The addition of the structural suffix and prefix was to avoid the problem that can generate the redundancy of the relevant node that may have several occurrences that do not contain the relevant information. So knowing the node does not solve the problem, but the addition of the structural suffix and prefix notion can solve it by knowing the xpath of the node.

- A Deep Web Approach [6] (D.W.A): Is based on a system that uses two regularities of domain knowledge and interface similarity to assign the tasks proposed by users and selects the most effective set of sites to visit. Then, the model extracts and corrects the data according to the structural characteristics of the acquired pages and the logical rules. Finally, it cleans and commands the raw data set to customize the application layer for later use through its interface.

3   Comparative Study

3.1   Comparison Based on Criteria

After a bibliographic review of the articles and documents describing the approaches, we have defined some basic criteria for evaluating the approaches cited in the first part of the document:

- Extraction of unstructured data: extracting unstructured data, that is, data represented or stored without a predefined format. Typically made up of plain text, but may also contain dates, numbers and facts. This lack of format results in irregularities and ambiguities, which can make it difficult to understand the data, unlike data, stored in spreadsheets or databases for example.
- Extraction of structured data: Extract the structured data framed by specific tags in the page sources and that will allow search engines to interpret that data in a certain way.
- Data storage: The storage of the extraction results must be automatic either in files (XML, Json …) or in a DBMS.
- Automation of the method: The extraction process must be automatic without requiring user interaction.
- User Behavior Analysis: The consideration of the user's behavior during the search, and operation of LOGS file browsing for accurate data extraction.
- User support: Assist the user during the search and retrieval of the data, by proposing suggestions based on the results obtained from the log files of the browser.
- Classification of data: Classify and group extraction results into subgroups, based on criteria to facilitate data analysis by the user.

The following table presents an Approaches-Criteria matrix that takes criteria as rows and approaches as columns. This matrix shows us whether an approach realizes a criterion (+) or not (−) (Table 1).

Table 1. Matrix Criteria-Approaches.

	M.C.A	G.C.A	C.A	M.I.A	P.T.A	D.W.A
Extraction of unstructured data	+	–	–	–	–	–
Extraction of structured data	+	+	+	+	+	+
Data storage	+	+	+	+	+	+
Automatization of the method	+	+	+	+	+	+
User behavior analysis	–	–	–	–	–	–
User support	–	–	–	–	–	–
Classification of data	–	–	–	–	–	–

Based on this table, we note that each approach does not achieve all the given criteria. For example, we can confirm that there is only one approach for extracting unstructured data (wiki page, Google books …). The two criteria concerning the automation of the method and the storage of the data are carried out by all the approaches.

However, no approach performs an analysis of user behavior while browsing web pages, and provides more accurate results based on the data extracted from the navigation caches files. No approach also assists the user during his navigation to suggest additional help.

3.2   AHP (Analytic Hierarchy Process)

Analytic Hierarchy Process (AHP) [9] is a multicriteria decision-making aid based on several criteria in order to make the best decision.

Thomas Saaty, a professor at the Wharton School of Business and a consultant for the US government, founded the AHP in the 1970s [10]. He developed the AHP to optimize the allocation of resources when there are several criteria to consider.

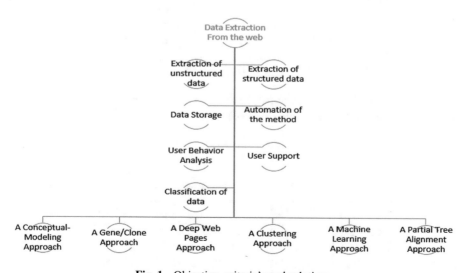

Fig. 1. Objective, criteria's and solutions

A complex decision can be based on dozens of criteria for decisions and potential solutions. AHP facilitates the analysis of solutions by structuring them hierarchically according to your criteria.

In the following, we will compare the approaches based on criteria defined using the AHP method. The process of the analysis is carried out in four steps.

Step 1: We will break down the situation into a hierarchical structure, using our objective, our decision criteria and our alternatives (Fig. 1).

Step 2: In the rest of the analysis, we will do a comparison of the approaches only based on four criteria that are shared between these approaches, and the other criteria are not feasible with the approaches we have, then their coefficient will not have a large importance.

- Extraction of unstructured data.
- Extraction of structured data.
- Data storage.
- Automation of the method.

In this section, we will compare each pair of options against each criterion:

1 = equal importance
2 = moderate importance
3 = high importance
4 = very high importance
5 = extreme importance

We get a first matrix of level 1 comparing the set of criteria that we posed before (Table 2):

Table 2. Matrix Level 1 comparison of criteria.

	E.D.N.S	E.D.S	S.D	A.M
E.D.N.S	1	1/2	4	3
E.D.S	2	1	4	3
S.D	1/4	1/4	1	1/3
A.M	1/3	1/3	3	1
Total	**43/12**	**25/12**	**12**	**22/3**

Then we calculate the relative percentages, to deduce a matrix of level two. For example for the box EDNS/EDNS (Table 3):

$$1/(43/12) = 0,279$$

Table 3. Matrix Level 2 relative percentages.

	E.D.N.S	E.D.S	S.D	A.M	AVG
E.D.N.S	0,27	0,24	0,33	0,40	**0,31**
E.D.S	0,59	0,48	0,33	0,41	**0,44**
S.D	0,07	0,12	0,08	0,04	**0,08**
A.M	0,09	0,16	0,25	0,14	**0,16**

Step 3: In this step, we will create for each decision criterion a matrix allowing comparing the approaches based on each criterion, then we will compute from the decision matrix another decision matrix also but in relative percentage (Tables 4, 5, 6, 7, 8, 9, 10 and 11).

Table 4. Decision matrix, Extraction of unstructured data.

Extraction of unstructured data						
	C.M.A	G.C.A	C.A	M.L.A	P.T.A	D.W.A
C.M.A	1	4	4	4	4	4
G.C.A	1/4	1	1/3	1/2	1/2	2
C.A	1/4	3	1	3	3	3
M.L.A	1/4	2	1/3	1	1/2	3
P.T.A	1/4	2	1/3	2	1	3
D.W.A	1/4	1/2	1/3	1/3	1/3	1
Total	9/4	25/2	19/3	65/6	28/3	16

Table 5. Decision matrix in relative %, Extraction of unstructured data.

Extraction of unstructured data							
	C.M	C.A	C.G	M.L	P.T	D.W	AVG
C.M.A	0,44	0,63	0,63	0,37	0,43	0,25	**0,41**
G.C.A	0,11	0,05	0,05	0,05	0,05	0,12	**0,08**
C.A	0,11	0,16	0,16	0,28	0,32	0,19	**0,22**
M.L.A	0,11	0,05	0,05	0,09	0,05	0,19	**0,11**
P.T.A	0,11	0,05	0,05	0,18	0,11	0,19	**0,13**
D.W.A	0,11	0,05	0,05	0,03	0,03	0,06	**0,05**

Table 6. Decision matrix, Extraction of structured data.

Extraction of structured data						
	C.M.A	G.C.A	C.A	M.L.A	P.T.A	D.W.A
C.M.A	1	2	2	2	2	3
G.C.A	1/2	1	1/2	1	1	3
C.A	1/2	2	1	2	2	3
M.L.A	1/2	1	1/2	1	1	3
P.T.A	1/2	1	1/2	1	1	3
D.W.A	1/3	1/3	1/3	1/3	1/3	1
Total	10/3	22/3	29/6	22/3	22/3	16

Table 7. Decision matrix in relative %, Extraction of structured data.

Extraction structured data							
	C.M	C.A	C.G	M.L	P.T	D.W	AVG
C.M.A	0,30	0,27	0,41	0,27	0,27	0,19	**0,29**
G.C.A	0,15	0,14	0,10	0,14	0,14	0,19	**0,14**
C.A	0,15	0,27	0,21	0,27	0,27	0,19	**0,23**
M.L.A	0,15	0,14	0,10	0,14	0,14	0,19	**0,14**
P.T.A	0,15	0,14	0,10	0,14	0,14	0,19	**0,14**
D.W.A	0,10	0,04	0,07	0,04	0,04	0,06	**0,06**

Table 8. Decision matrix, Data storage.

Data storage						
	C.M.A	G.C.A	C.A	M.L.A	P.T.A	D.W.A
C.M.A	1	3	2	2	2	3
G.C.A	1/3	1	1/3	1/2	1/3	2
C.A	1/2	3	1	2	1/3	2
M.L.A	1/2	2	1/2	1	1/2	3
P.T.A	1/2	3	3	2	1	2
D.W.A	1/3	1/2	1/2	1/3	1/2	1
Total	19/6	25/2	22/3	47/6	14/3	13

Table 9. Decision matrix in relative %, Data storage.

Storage of data							
	C.M	C.A	C.G	M.L	P.T	D.W	AVG
C.M.A	0,31	0,24	0,27	0,25	0,43	0,23	**0,29**
G.C.A	0,10	0,08	0,04	0,06	0,07	0,15	**0,09**
C.A	0,16	0,24	0,14	0,25	0,07	0,15	**0,17**
M.L.A	0,16	0,16	0,07	0,13	0,11	0,23	**0.14**
P.T.A	0,16	0,24	0,41	0,25	0,21	0,15	**0,24**
D.W.A	0,10	0,04	0,07	0,04	0,11	0,08	**0,07**

Table 10. Decision matrix, Automation of the method.

Automation of the method						
	C.M.A	G.C.A	C.A	M.L.A	P.T.A	D.W.A
C.M.A	1	3	2	2	2	3
G.C.A	1/3	1	1/2	1/3	1/2	2
C.A	1/2	2	1	1/2	1/2	2
M.L.A	1/2	3	2	1	2	3
P.T.A	1/2	2	2	1/2	1	3
D.W.A	1/3	1/2	1/2	1/3	1/3	1
Total	19/6	23/2	8	14/3	19/3	14

Table 11. Decision matrix in relative %, Automation of the method.

Automation of the method							
	C.M	C.A	C.G	M.L	P.T	D.W	AVG
C.M.A	0,31	0,26	0,25	0,43	0,31	0,21	**0,30**
G.C.A	0,10	0,09	0,06	0,07	0,08	0,14	**0,09**
C.A	0,16	0,17	0,12	0,11	0,08	0,14	**0,13**
M.L.A	0,16	0,26	0,25	0,21	0,31	0,21	**0,23**
P.T.A	0,16	0,17	0,25	0,11	0,16	0,21	**0,18**
D.W.A	0,10	0,04	0,06	0,07	0,05	0,07	**0,07**

Step 4: We will now create a matrix of solutions by inserting the solutions into the first column and the decision criteria in the first row. We will use the averages calculated in step 3 to complete our solution matrix (Table 12):

Table 12. Final solution matrix.

Solution matrix				
	Extraction of unstructured data	Extraction of structured data	Data storage	Automation of the method
C.M.A	0,41	0,29	0,29	0,30
G.C.A	0,09	0,14	0,09	0,09
C.A	0,22	0,23	0,17	0,13
M.L.A	0,11	0,14	0.14	0,23
P.T.A	0,13	0,14	0,24	0,18
D.W.A	0,05	0,06	0,07	0,07

Step 5: In this step we will multiply, final solution matrix with the matrix of the relative percentages calculated in step 2 to determine the best solution and that meets our need.

$$\begin{pmatrix} 0,41 & 0,29 & 0,29 & 0,30 \\ 0,09 & 0,14 & 0,09 & 0,09 \\ 0,22 & 0,23 & 0,17 & 0,13 \\ 0,11 & 0,14 & 0,14 & 0,23 \\ 0,13 & 0,14 & 0,24 & 0,18 \\ 0,05 & 0,06 & 0,07 & 0,07 \end{pmatrix} \begin{pmatrix} 0,31 \\ 0,44 \\ 0,08 \\ 0,16 \end{pmatrix} = \begin{pmatrix} 0,33 \\ 0,11 \\ 0,20 \\ 0,14 \\ 0,15 \\ 0,06 \end{pmatrix}$$

The results (Table 13):

Table 13. Results by approaches.

A Conceptual-Modeling Approach to Extracting Data from the Web	0,33
Generate Tools through a Data Extraction Web	0,11
Data Extraction from Deep Web Pages	0,20
Data Extraction from Semi-structured Web Pages by Clustering	0,14
Accurately and Reliably Extracting Data from the Web: A Machine Learning Approach	0,15
Web Data Extraction Based on Partial Tree Alignment	0,60

From the table above, the optimal solution and the approach that satisfies the first four criteria is the approach based on the conceptual model for extracting data from the web.

In this study, we used the AHP method to determine the approach that gives results that are more relevant to the criteria chosen, we have made several calculations and passed through stages to finally arrive at a matrix of solutions that gives us a ranking of the approaches cited before based on the calculations (Fig. 2).

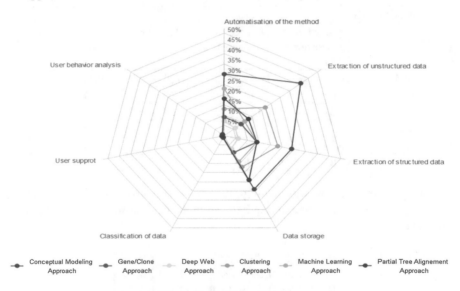

Fig. 2. Spider draw of results

The study allowed us to arrive at mixed results, we can confirm that the approach closest to our objective and which gives results more relevant compared to others is that using the conceptual model of the data, with a score of 0.326, this approach proposes an entire system based on an ontology that is generated according to the user's search. However, unfortunately, this proposed approach does not take into consideration other criteria such as behavior analysis and user support during navigation. Moreover, this approach does not deal with the deep web case and does not propose a complete classification of data after their extraction from the web.

Our study confirms that despite the large number of approaches proposed to extract data from the web, each approach addresses the problem on one side without taking into consideration the other sides. So there is not an approach that treats all the possibilities (type of data, type of web pages, analysis of the user behavior etc.), for this purpose we will propose in our next work, our new method which will treat any The possibilities concerning the subject of data extraction from the web. Our approach will be an entire system implemented to ensure a good extraction of data regardless of the type of data or the type of web pages.

4   Conclusion

The aim of this paper was to present the approaches to extracting data from the web pages, with a detailed presentation of each approach, and finally a comparison of all the approaches based on well-defined criteria precise.

References

1. Embley, D.W., Campbell, D.M., Jiang, Y.S., Liddle, S.W., Ng, Y.-K., Quass, D.W., Smith, R.D.: A conceptual-modeling approach to extracting data from the web. In: International Conference on Conceptual Modeling, 16–19 November 1998, pp. 78–91. Springer, Heidelberg (Singapore) (1998)
2. Vuong, L.P.B., Gao, X., Zhang, M.: Data extraction from semi-structured web pages by clustering. In: Proceedings of the 2006 IEEE/WIC/ACM International Conference on Web Intelligence, 18–22 December 2006, pp. 374–377. IEEE Computer Society, New Zealand (2006)
3. Knoblock, C.A., Lerman, K., Minton, S., Muslea, I.: Accurately and reliably extracting data from the web: a machine learning approach. In: Intelligent Exploration of the Web, pp. 275–287. Springer, Heidelberg (2003)
4. Zhai, Y., Liu, B.: Web data extraction based on partial tree alignment. In: Proceedings of the 14th International Conference on World Wide Web, 10–14 May 2005, pp. 76–85. ACM, New York (2005)
5. Aziz, C., Habib, B., Tragha, A., et al.: Generate tools through a data extraction web. Int. J. Sci. Adv. Technol. **3**, 60–69 (2012)
6. Yang, J., Shi, G., Zheng, Y., Wang, Q.: Data extraction from deep web pages. In: International Conference on Computational Intelligence and Security (CIS 2007), 15–19 December 2007, pp. 237–241 (2007)
7. Adelberg, B.: NoDoSE—a tool fors semi-automatically extracting structured and semistructured data from text documents. In: ACMSIGMOD International Conference on Management of Data, 01–04 June 1998, pp. 283–294 (1998)
8. EL Habib, B., Aziz, D.S., Asmaa, E.O.: A new solution for data extraction: gene/clone Method. Int. J. Comput. Sci. Netw. Secur. 60–69 (2006)
9. Saaty, T.L.: How to make a decision: the analytic hierarchy process. Eur. J. Oper. Res. **48**, 9–26 (1990)
10. Saaty, T.L.: Analytic hierarchy process. In: Encyclopedia of Operations Research and Management Science, 23 January 2016, pp. 52–64 (2016)

11. Muslea, I., Minton, S., Knoblock, C.A.: Hierarchical wrapper induction for semistructured information sources. Autonom. Agents Multi-Agent Syst. **4**, 93–114 (2001)
12. Bar-Yossef, Z., Rajagopalan, S.: Template detection via data mining and its applications. In: Proceedings of the 11th International Conference on World Wide Web, 07–11 May 2002, pp. 580–591 (2002)

A New Vision for Multilingual Architecture

Kawtar J'Nini[(✉)], Faouzia Benabbou[(✉)], and Nawal Sael[(✉)]

Laboratoire Traitement de l'Information et Modélisation,
Faculté des Sciences Ben M'SIK, Casablanca, Morocco
{Kawtar.jnini,faouzia.benabbou,nawal.sael}@univh2m.ma

Abstract. No one can deny the fact that orientation is a big help for students to determine their future. This leads us to the obligation to make it automatic and accessible, both in terms of immediate accessibility throughout the world and in the offering of several languages that correspond to the language used by the majority of students or persons in need of guidance. Our project aims to solve these problems by offering a multilingual e-orientation platform accessible from the web. In this paper, we are going to present a multilingual orientation system architecture based on a domain ontology that takes in the entry a request expressed in any language, and the result is a number of records extracted from a multilingual source and presented in the language of the user.

Keywords: CLIR · Multilingual · Query translation · Document translation
Multilingual architecture

1 Introduction

Student's orientation is a crucial step in constructing one's future; unfortunately, it has not been given much importance. Orientation should not only be for graduating students, it should be available for everybody as there is no specific age or class to start building a career. Currently, in some schools, an expert of orientation helps the students to choose the right orientation based on their grades and on their career preferences. These students tend to have the best grades later and a higher rate of succeeding. The necessity of orientation comes from the fact that many choices are available and choosing the right course and the right school turn out to be the hardest task. If not well orientated, the student can be confused, this leads to an inability to make the right choice. Thus, we are confronted with the necessity of finding a good way to assure a specific orientation, that will help students to be aware of all the courses and all the possibilities offered for them, this kind of orientation has to be available everywhere and for anyone; this will give everybody equal chances of succeeding their lives. The use of information systems became more frequent in Morocco, and internet is available everywhere. This availability makes internet the best way to offer orientation everywhere. This can be possible if we have a large data base that will take in information about schools, the courses proposed, diplomas, the outlets of each course and also testimonials from people who have followed a specific course. Based on these points, came the idea to set up a platform for e-orientation providing all the information for guidance. Only having an online

© Springer International Publishing AG, part of Springer Nature 2019
J. Mizera-Pietraszko et al. (Eds.): RTIS 2017, AISC 756, pp. 111–119, 2019.
https://doi.org/10.1007/978-3-319-91337-7_11

platform requires making it available for everybody no matter what language they use. This implies the necessity of using a multilingual system that will allow the translation of the database information to let it match with the user's chosen language. This paper proposes an e-orientation system that has the added advantage of providing uniformity so all users will see the same interface. We will expose in start by giving a brief introduction to the orientation and the idea behind using an e-orientation system, then we will talk about multilingual data systems, after that we are going to present the architecture that we propose with a focus on the approach used to treat the multilingual data issue.

2 Background

2.1 E-orientation

Student's orientation is constructed while they are still studying with regular conversation with other students, theirs parents or teachers, and the guidance counselors. And also, while surfing on websites that are relevant enough.

Among the platforms that can be found on the internet, the Diplomeo [12] platform is a free service dedicated to the orientation in higher education. Using referral forms to define their path and their project, it allows visitors to find the higher education that best matches their course and their project. They will then be contacted by those responsible for these formations to receive more information and choose to study in the best conditions. Another example is the French FRATELI [13] platform. It provides the ability to view different formations using two approaches: according to various existing curricula or depending on the profile of visitors, regardless of type of diploma.

In Morocco, nowadays orientation is available through the web on websites dedicated for this purpose, the most known site as much for its content as its publications is Etudiant.ma. This portal is an orientation platform and exchanges for young people in training. In addition to provide information on the curriculum and training in Morocco and abroad, it offers more interactive applications, videos, chat, orientation tests, IQ tests, FAQs, surveys and blogs.

There is also, the platform 9rayti.Com which includes a comprehensive database of Moroccan schools (private and public) and aspires to facilitate access to documentation, allow the exchange between members, and provide advice and guidance.

These various platforms help providing information - from a database - or work on stereotypes of professional performances. These websites based on games of questions and answers nevertheless suffer from some problems. The biggest one is that they require serious accompaniment and thoughtful approach to be used.

In addition, from all mentioned platforms; only the site 9rayti.ma offers some pages written in Arabic. The rest of the platforms are exclusively in French. A student who does not speak this language won't be able to benefit from these platforms. Worse, a misunderstanding could lead to a bad decision.

2.2   Multilingual Data Systems

It refers to computer program that allows user interaction with the computer in one or more languages, where the language can be selected dynamically, either at the time of invocation of the program or subsequently during its execution.

Multilingual Systems are not only feasible at the level of an application program but can also present a truly localized environment for a user desiring to interact with computers in regional languages.

The multilingual information access is critical for the acquisition, dissemination, exchange, and understanding of knowledge in the global information society. The accelerated growth in the size, content and reach of Internet, the diversity of user demographics and the skew in the availability of information across languages, all point to the increasingly critical need for Multilingual Information Access (MIA) [2].

The MIA technologies such as Cross-language Information Retrieval, Multilingual Information Extraction, and Machine Translation bridge the gap between available information and the user needs transparently across languages.

When working on multilingual data, we are always confronted to work on having multilingual linguistic information for generating and interpreting the multilingual realization of semantic definitions in web ontologies and linked data, and multilingual mapping between web ontologies that establish cross lingual connections as well as between linked datasets that use different languages in their lexical representation [3].

2.3   CLIR: Cross Lingual Information Retrieval

Most internet users can use more than one language; this gives them the possibility of querying in any possible language. So, they expect to have the query's response retrieved with a multilingual system. Cross language text retrieval can efficiently help them doing so. If the user is only able to read one language, then cross language information retrieval will take the request in its original language, then, based on the approach used to treat this request, the retrieval will be done on all the available documents and we will get an answer corresponding to the user's needs in the user's language.

Cross language information retrieval gives the possibility to assess the information regardless of the language in which it is authored. It is based on the techniques used in the monolingual information retrieval, with modifications to make it work with multilingual information [2, 4].

If we look at a monolingual information retrieval system, we have a request expressed in a mono langue, and a collection of documents that represent the documents base. These documents are processed into derived representations. After their indexation, we get a set of terms, phrases and concepts that are recorded in a format that permits rapid access. The same thing goes with the request, and then the resulted representations are compared to determine the best match to pass it to the user.

Same thing goes with the multilingual information access, but with the addition of extra steps, the first step will be the pre-translation. This step consists on defining and identifying, and processing the language in the source text.

After that, comes the translation, here is where we find three possible approaches, Query translation, Document translation, and Dual translation. We are going to present all three of them in the next section.

Finally, it's the post translation phase; it consists on shaping the output of the translation into a coherent product. The post translation can also help in expanding the requests to make it more meaningful while extracting data.

2.4   Multilingual Information Access

In the article "Toward multilingual system in E-orientation domain", we present three approaches that maybe used to retrieve an answer to a user's request; these approaches are based on the treatment of the query and the document each time differently (Query translation, document translation, translate all in one) [4].

The first approach, "query translation" being based on the requests translation is considered as the less expensive approach. Only the request is translated, so the translation part is done before the retrieval which reduces the cost and makes the task more economical. Only this approach has also disadvantages. The request is indeed translated to a language understandable by the system where we have the documents base, but the retrieval in made each time from the documents in the user's language. So in the last phase of the post translation, we may not have a complete answer extracted from all the documents available.

The second approach, "translate documents" consists on translating the documents in the documents base each time to make them match with the request. This approach is expensive, as it consists on keeping the request in the user's language, and translating a whole base. This base, is most of the time a big one that uses many resources for the translation. And then after the retrieval, the data will be translated again to the users preferred language. But in the end of the process it gets a full answer to the user's request, even though it demands a lot of resources and time.

The third approach, "translate all in one language" where both the request and the documents are translated the translation is done in a unique language, and then the data access is processed as monolingual data, and the resulted documents are translated again to the users language. The approach's results are very accurate as the data access is done on all documents and the retrieved data is translated to one language depending on the user's preference. But this approach is still the most difficult one as the translation process is done on both the request, and also on all the documents available in the documents base [6].

The difference in the speed to access information, the time and the efforts consumed in the translation process in each one of those approaches, make the difference notable. The first approach "query translation" technique, being the less resources consuming is the approach we chose to translate multilingual information [4].

3 Related Works

Despite being an important step to help students make choices regarding their professional life choices, orientation has not been given the right amount of attention. A huge number of universities omit the fact that tracking a student's individual educational trajectory is the key to raise the rate of success.

Marie-Jo Wilson and al present a study on online orientation. The experience was done on students from the University of Auckland, the principle is that they implemented a face to face orientation program and an online orientation course for undergraduate, then they studied means to improve the transition experience between both courses and how to meet the needs of students. The result was that the online orientation exceeded expectations in being on time, on budget and receiving very positive feedback from students and staff [8].

Zeera and Streltsovb in their article Technological Platform for Realization of Students' Individual Educational Trajectories in a Vocational School discuss what an orientation program can do and what it needs. So, the key elements of a new technological education platform are: a massive open online course (MOOC); learning management systems (LMS); the ecosystem of support for a new education; a new education infrastructure (Konanchuk 2013) [10]. Then explained how two important functions will provide the realization of a new technological platform: firstly, the alignment of individual educational trajectories, their control and ensuring the continuity of the educational experience; secondly, the evaluation of education results according to a single coordinate system understandable for the representatives of various interest groups - teachers, employers, students themselves (Potocnik 2005) [11]. External condition for the realization of individual educational trajectories is a technological education platform that unites separate Ed Tech-projects into a single system. This system allows to move from the logic of selecting an educational program or institution to the logic of the "competences formation chain" management, and to make a significant step towards the change of the "institutional core" of education: to move from the institution management to individual trajectory management throughout the whole life (Konanchuk 2013; Harre 1983) [9].

4 Proposition

By our project, we aim to build a system that will allow any user by simple click to obtain all necessary information in relation with orientation in the language desired and depending on his profile.

To complete this task, many requirements have to be respected. First, the choice of the sources we are going to get the information data from, second, it's having a strong automated extraction system that allows the update constantly. Then, choosing a translation system that can be adapted to our needs.

BDR, PDF, txt, web pages, forum, and comments… are multiple resources to obtain multilingual information.

The extraction will be done by using proper algorithms for data extractions (the data extraction will be explained in another part of the project).

While working on giving an answer to a user's request, the main focus should not only be on the request but also on the language in which the answer can be retrieved and on the user's profile.

These elements will be used to extend the request and give it a more meaningful form.

5   Architecture Proposed

See Fig. 1.

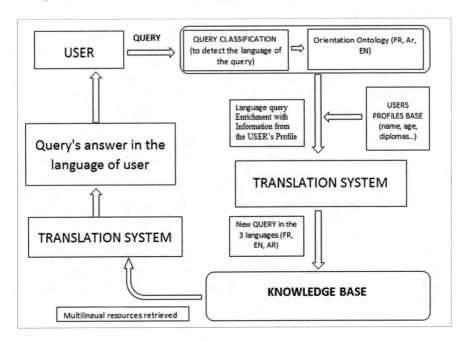

Fig. 1. Multilingual system

5.1   Classification and Pretreatment

The system is composed of, orientation ontology, a user's profile base, a knowledge database and a translation system.

– **Orientation Ontology**: It's the first step in the query's path toward obtaining an answer, the system expects to receive a query expressed in a specific language; our system will take in charge at the beginning tree languages which are: French, English and Arabic.

Before it's redirected to the orientation ontology, the query will be classified to detect the language used then through a pretreatment phase will be revised to correct any possible errors and go up to the concepts used in the ontology.

As a start, we will use a predefined ontology that will be developed progressively. The ontology will serve in offering the query compliment.

Once the query reaches the ontology base, the system will do the linking with the corresponding ontology based on the query keywords. This will modify the query to make it more meaningful while extracting data.

5.2   Query Enrichment with Profile User

The user of the system will be asked to create a profile as a first step. To do so, it's necessary to fill in some information to complete the inscription, ex: name, age, interests, level of study, subjects studied, and of course the preferred language for communication. This information is stored in a profile base.

At the reception of the request, the matching is done with the profile, and these elements are added to the resulting query to enrich it and make it more specific for each user.

5.3   Translation System

The approach that we are using to treat the multilingual data is to translate the query and then do the extraction of data, so we need a translation system to make sure that once we start supplying the knowledge data base the translation will be possible.

The query is translated, this gives us as an outcome, three requests in the three languages used in our system R1 in French, R2 in English and R3 in Arabic.

Now we have 3 enriched requests in three different languages and we can proceed with retrieving data to answer these queries.

Knowledge database: The data retrieval will be made from our knowledge database, it's filled with data extracted from different sources (i.e.: web, documents, data bases...) and in relation with our subject which is the orientation. We should note that the data extraction will be done in another project and we are going to use the results to build our database.

As the three requests reach the knowledge database, they are identical with the only difference of being in three languages. From the knowledge database, we will retrieve all documents corresponding to these requests. By the end of the retrieval, we will have documents in three languages, the system will move then back to the information stored in the users profile, and check the preferred language. After that, comes the translation step, documents that are in the user's preferred language will be put aside to be sent directly as a response, without wasting time on their translation, and the other documents will be translated to the right language.

The user at the end of the process receives a list of links to the documents that were retrieved. The results are in one language, while the data retrieval was in Multilanguage.

For the translation system; we propose to use machine translation approach which is known as the process that enables the translation from a language to another.

A machine translation (MT) system first analyses the source language input and creates an internal representation. This representation is manipulated and transferred to a form suitable for the target language. Then at last the output is generated in the target language [7].

The process goes by:

– Decoding the signification of the source text
– Recode the signification in the target language.

The ontology concepts translated to match the user's language will be sent to the user and also sent to the ontology base to be stocked for another usage. This will help reduce the processing time of new requests and enhance the performances of the system.

6   Conclusion

With the fast growing role of the orientation in the assistance of the students to choose the right path to take in their studies, we are confronted with the necessity to set up a process that can produce the necessary assistance to ensure correct orientation. This led us to think about creating an e-orientation platform available, easy to access and multi-lingual.

To ensure the availability part, we opted for an online system; the users can create their profiles by giving the essential information about their educational background, hobbies and personal information. They can then express by using a request, their needs. The request will then be redirected inside our system, where it will be reformulated and enriched with information from the user's profile and then treated inside our knowledge database to extract the available documents corresponding to the request; multiple possible propositions will be extracted and transmitted to the user. And to give more accessibility to the solution, we will work on making it accessible in different languages, based on the language preference on the user's profile and using translation systems and multilingual data source to extract knowledge.

7   Perspective

Our project is about the treatment of users request in a e-orientation system, in order to help them get the right information. We are focused on the multilingual part, so the users subscribed in the system do not have to only express their requests in a single language, and we can have a large community of users. The documents that we are going to access to retrieve data from, are also in many different languages. To retrieve data in such environment, the use of a multilingual information retrieval data is necessary. We opted for the translation of the request and then the translation of the resulted data after extraction.

Our next step will be the construction and alimentation of the knowledge database to enable it to receive the reformulated multilingual query, and then choose the right extraction technique to extract data and translate it using the translation system.

After that we will have to construct a multilingual ontology, and put our system to use by some students (small faculty population) in order to test and validate results obtained.

References

1. Wachira, R.S.: A Parallel Corpus Based Translation Using Sentence Similarity, P58/73542/2009
2. Bian, G.-W., Chen, H.-H.: Cross-language information access to multilingual collections on the internet. J. Am. Soc. Inf. Sci. **51**, 281–296 (2000)
3. Gracia, J., Montiel-Ponsoda, E., Cimiano, P., Gómez-Pérez, A., Buitelaar, P., McCrae, J.: Challenges for the multilingual web of data. Web Semant. Sci. Serv. Agents World Wide Web **11**, 63–71 (2012)
4. Zhou, D., Truran, M., Brailsford, T., Wade, V., Ashman, H.: Translation techniques in cross-language information retrieval. ACM Comput. Surv. **45**(1), Article 1 (2012)
5. Fujii, A., Ishikawa, T.: Japanese/English cross-language information retrieval: exploration of query translation and transliteration. Comput. Humanit. **35**(4), 389–420 (2001)
6. McCarley, J.S.: Should we translate the documents or the queries in cross-language information retrieval (1999). dl.acm.org
7. Okpor, M.D.: Machine Translation Approaches: Issues and Challenges. IJCSI Int. J. Comput. Sci. Issues **11**(5), Article no 2 (2014)
8. Wilson, M.-J., Minhas-Taneja, V.: Online Orientation – a blended learning approach to student transition. J. Aust. N. Z. Stud. Serv. Assoc. **47**, 61–68 (2016)
9. Zeera, E.F., Streltsovb, A.V.: Technological platform for realization of students' individual educational trajectories in a vocational school. IEJME Math. Educ. **11**(7), 2639–2650 (2016)
10. Konanchuk, D.S.: Methodological aspects of the projecting the individual educational routes. Psychol. Sci. Educ. **4**, 15–21 (2013)
11. Potocnik, J.: European Technology Platforms: Making the Move to Implementation (2005). Accessed 29 July 2016
12. http://www.diplomeo.com/, April 2015
13. http://frateli.org/frateli_orientation/, April 2015

Evolutionary Optimization Algorithms and Artificial Intelligence

Using Continuous Hopfield Neural Network for Choice Architecture of Probabilistic Self-Organizing Map

Nour-Eddine Joudar$^{(\boxtimes)}$, En-naimani Zakariae, and Mohamed Ettaouil

Department of Mathematics, Faculty of Sciences and Technology,
University of Sidi Mohamed Ben Abdellah, 30000 Fes, Morocco
`noureddine.joudar@usmba.ac.ma`, `z.ennaimani@gmail.com`,
`mohamedettaouil@yahoo.fr`

Abstract. The probabilistic self-organizing map (PRSOM) is an improved version of the Kohonen classical model (SOM) that appeared in the late 1990's. Mathematically, PRSOM gives an estimation of the density probability function of a set of samples. And this estimation depends on parameters given by the architecture of the model. Therefore, the main problem of this model, that we try to approach in this paper, is the architecture choice (the number of neurons). In summary, in the present paper, we describe a recent approach of PRSOM trying to find a solution to the problem below. For that, we propose an architecture optimization model that is a mixed integer non-linear optimization model under linear and quadratic constraints. Resolution of suggested model is carried out by continuous Hopfield neural network (CHN). The performance of the technique is supported by the use of the proposed model in data analysis, notably, classification of iris dataset.

Keywords: Probabilistic self-organizing map
Continuous Hopfield neural network · Optimization problem
Architecture choice · Classification

1 Introduction

In real research domains, it's rare to find observations that could be modeled by only one classical probabilistic law. The majority of observed data comes from sources containing many sub-populations, each sub-population is modeled in a separate way, knowing that the population is a mixture of these sub-populations. In this case, we model by using a finite mixture of probabilistic laws [1].

In the present paper, we chose to work on the Probabilistic Self Organizing Map, which is the probabilistic version of the Self Organizing Map (Kohonen network). The SOM is a deterministic NN that has been introduced by Kohonen in 1982 [2]. The central property of this model is to form a nonlinear projection of a high-dimensional data on low-dimensional (usually 2D) grid i.e. to perform a quantification of the input space. In 1998, Anouar et al. [3] presented a probabilistic variant of the SOM called Probabilistic SOM. This latter gives a maximum approximation (likelihood) of the density distribution and a clustering of the data space. In the PRSOM, density

© Springer International Publishing AG, part of Springer Nature 2019
J. Mizera-Pietraszko et al. (Eds.): RTIS 2017, AISC 756, pp. 123–133, 2019.
https://doi.org/10.1007/978-3-319-91337-7_12

distribution is a Gaussian mixture [3] depending on parameters (weights and covariance matrix) estimated by a learning algorithm based on likelihood maximum. And this estimation depends on the parameters given by the model's architecture [4].

The main purpose of this work is, thus, to model the problem of choosing the neural Architecture of the PRSOM, using a mixed-integer nonlinear problem with linear and quadratic constraints. When the model is set, the important coming step is its resolution, which could be done by two different approaches: exact approach or heuristic approach. For the reason of model complexity, we opt for the heuristic approach and use the continuous Hopfield network [5, 6], that proved its effectiveness in solving the optimization problems.

Recently, artificial neural networks (ANNs) have been widely developed to solve optimization problems. One of the most popular networks which have proved its effectiveness is the Hopfield Neural Network (HNN). The first official appearance of this network for solving Linear Programming problems (LP) was on 1985 by Hopfield and Tank [7]. In that work, authors have investigated the computational power of the HNN to solve the Travelling Salesman Problem (TSP) [8]. Although many types of research have clarified some drawbacks of this approach, several works have been made to correct these problems [9]. Thereafter, different extensions have been investigated to solve mathematical problems such as linear programming problems, nonlinear programming (NLP) problems, and mixed-integer linear programming (MILP) problems [9]. As enunciated above, in the present paper we will try to approach the problem in four Sections. Section 2 presents all the scientific background needed for the problem resolution, especially the definition of the following: Probabilistic Self Organizing Map and continuous Hopfield neural network. In Sect. 3, we present the proposed optimization model of PRSOM architecture and its resolution via CHN. Finally, in Sect. 4 we expose experimental results of the model, applied to the clustering problem.

2   Background and Related Works

2.1   Probabilistic Self-Organizing Map

In the real research domains, we rarely find observations that could be modeled by only one classical probability distribution. The majority of observed data $\{x_1, \ldots, x_n\} = D \subset \mathbb{R}^d$ comes from sources that contain many sub-populations modeled each in a different way. Seeing that the entire population is a mixture of these sub-populations, we consequently model it by using a finite mixture of probability distributions [1], such as the probabilistic self-organizing map. In order to define the PRSOM model, Anouar et al. propose to associate each neuron k of the map, to a Gaussian density function f_k [3]. Furthermore, the neighborhood notion allows introducing a mixture of Gaussians. These two notions enable to introduce a density function defined as a mixture of probabilistic mixtures where each neuron of the map is assigned to a component of the mixture. Probabilistic self-organizing map model is used to approach distribution density of the observations of a set D which elements are supposed to be i.i.d. using a mixture of normal probability densities:

$$\forall \theta = (W, \Sigma), L(\theta) = p(x_1, \ldots, x_n, W, \Sigma) = \prod_{i=1}^{n} \sum_{j=1}^{K} p(n_j^{(2)}) p(x_i / n_j^{(2)}) \qquad (1)$$

With:

- $p(n_j^{(2)})$: the initial probability of each neuron in the map $C^{(2)}$.
- $p(x / n_j^{(2)}) = \sum_{k=1}^{K} p(n_k^{(1)} / n_j^{(2)}) p(x / n_k^{(1)})$: a mixture of probabilities for the conditional generation of the observation x_i.
- $p(n_k^{(1)} / n_j^{(2)}) = \dfrac{K_T(\delta(n_k^{(1)}, n_j^{(2)}))}{\sum\limits_{k=1}^{K} K_T(\delta(n_k^{(1)}, n_j^{(2)}))}$: the probability that generates the neighborhood

function between the neurons of the map (the influence of the map $C^{(1)}$ neurons on the map $C^{(2)}$ neurons).

In sum, this model is characterized by the two sets $W = \{w_1, \ldots, w_K\}$ and $\Sigma = \{\Sigma_1, \ldots, \Sigma_K\}$ that have to be estimated during the network's learning phase. We use the maximization estimator of the likelihood function given by the Eq. (1), which is equivalent to estimate the optimal parameters (W^*, Σ^*) that maximize this likelihood. The learning is performed by the algorithm cited in [3], via the two following phases: the assignation phase and the maximization phase.

2.2 Continuous Hopfield Neural Network

In the beginning of the 1980s, Hopfield published two scientific papers, which have been drawn the researchers attention. In [7], Hopfield and Tank presented the energy function approach in order to solve several optimization problems. Their results encouraged a number of researchers to apply this network to different problems such as image restoration [10]. The Continuous Hopfield Networks (CHN) consist of N interconnected neurons with a smooth sigmoid activation function (usually a hyperbolic tangent).

The differential equation which governs the dynamics of the CHN is:

$$\frac{du}{dt} = -\frac{u}{\tau} + Tv + I^t \qquad (2)$$

where u, v and I are, respectively, the vectors of neuron states, the outputs and the biases. The output function $v_i = g(u_i)$ is a hyperbolic tangent, which is bounded below by -1 and above by 1. The matrix T represents the weights of the synaptic connection from the neuron i to neuron j.

Based on (2), Hopfield has introduced the energy function E on $[0, 1]^n$ which is defined by

$$E(v) = -\frac{1}{2}v^tTv - (i)^tv + \frac{1}{\tau}\sum_{i=1}^{n}\int_0^{v_i} g^{-1}(v)dv$$

The philosophy of this approach is that the objective function, which characterizes the combinatorial problem, is associated with the energy function of the network when $\tau \to +\infty$. Thus, the new H-T energy function associated to combinatorial problems becomes:

$$E(v) = -\frac{1}{2}v^tTv - (I)^tv \qquad (3)$$

A point v^e is called an equilibrium point of the system (1) if for an input vector v^0, v^e satisfies $v(t) = v^e \quad \forall t \ge t_e$, for some $t_e \ge 0$.

A point $v^e \in H$, will be an equilibrium point for the CHN if and only if, the three following conditions are satisfied:

$$E_i(v^e) = \frac{\partial E(V^e)}{\partial v_i} \ge 0 \ \forall i = 1,..,n \ such \ that \ v_i^e = 0 \qquad (4)$$

$$E_i(v^e) = \frac{\partial E(V^e)}{\partial v_i} \le 0 \ \forall i = 1,..,n \ such \ that \ v_i^e = 1 \qquad (5)$$

$$E_i(v^e) = \frac{\partial E(V^e)}{\partial v_i} = 0 \ \forall i = 1,..,n \ such \ that \ v_i^e \in \,]0,1[\qquad (6)$$

It should be noted that if the energy function (or Layapunov function) exists, the equilibrium point exists as well. Hopfield [8] proved that the symmetry of matrix of the weight is a sufficient condition for the existence of Lyapunov function.

3   Architecture Choice of PRSOM: Modelling and Resolution

Despite it is power to model a better estimate of the density distribution for a data mixture of a space, PRSOM has disadvantages, with the most important being the dependence of this density on the choice of components number of used mixtures. The random choice of many components can lead to bad solutions. Talking about the number of neurons constituting the map is equivalent to talking about the architecture choice problem. In fact, this problem has not only a significant impact on the convergence, but it also affects the quality of the obtained results. To overcome this problem, we propose in this paper a new mathematical model of PRSOM that controls the size of the map. In this section, we will describe the construction steps of our model. The first one consists of integrating the special term which controls the size of the map. The second step gives in a first time the constraints which ensure the allocation of each data to only one neuron (component), and in a second time it gives another constraint that ensures the subsistence of space observations representative components.

3.1   Modeling

We propose a new model of neural architecture optimization problem of probabilistic self-organizing map as an optimization problem represented by a mixed-integer non-linear problem with linear and nonlinear constraints.

To formulate this model we need to define some parameters as follows:

Parameters

- n: number of observation data;
- d: Vector dimension of observation data;
- N: Optimal number of neurons (components) in the topological map of PRSOM;
- N_{max}: Maximal number of neurons in the topological map of PRSOM.

Variables

- BD: Matrix of Training base elements;
- W: Matrix of referent vectors;
- Σ: Matrix of covariance;
- V: Matrix of binary variables;

with $v_{i,j}$, is the assignment variable that define the relationship between data and neuron, $v_{1,j}$ the control variable which allows controlling the size of PRSOM map.

Objective Function

Basing on the work of Bishop for the Gaussian mixture model [11], we will define the objective function of the probabilistic self-organizing map mathematical model as:

$$\text{Max } p(W, \Sigma, V) = \prod_{i=1}^{n} \prod_{j=1}^{N_{max}} \left(\pi_j * \left(\sum_{k=1}^{N_{max}} K^T(\delta(j,k)) f_k(x_i, w_k, \sigma_k) \right) \right)^{v_{i+1,j} v_{1,j}} \quad (7)$$

For reasons of convenience, the log level helps to reduce the volume of digits representing a series. Moreover, the linear logarithm is a multiplicative relationship i.e. we transform a multiplicative series to an additive one. The log function is strictly increasing. It is then better to maximize $\log(p)$ than p.

As well the function becomes:

$$Max \ \ln(p(W, \Sigma, V)) = \sum_{i=1}^{n} \sum_{j=1}^{N_{max}} v_{i+1,j} v_{1,j} \log[(\pi_j) \sum_{k=1}^{N_{max}} K^T(\delta(j,k)) f_k(x_i, w_k, \sigma_k)] \quad (8)$$

The research for a maximum can always be transformed to the research of a minimum, the objective function is thus as following:

$$Min \ E(W, \Sigma, V) = -\sum_{i=1}^{n} \sum_{j=1}^{N_{max}} v_{i+1,j} v_{1,j} \log[(\pi_j) \sum_{k=1}^{N_{max}} K^T(\delta(j,k)) f_k(x_i, w_k, \sigma_k)] \quad (9)$$

Constraints

Each data element must be allocated to one neuron (component). In consequence we obtain the following n constraints:

$$\sum_{j=1}^{N_{max}} v_{i,j} = 1 \quad \forall i \in \{2, \ldots, n+1\} \tag{10}$$

Besides assignment constraints, we add another one called transmission constraint. If the neuron j is not used $v_{1,j} = 0$, i.e., $\sum_{i=2}^{n+1} v_{i,j} = 0$; else $v_{1,j} = 1$, i.e. $\sum_{i=2}^{n+1} v_{i,j} \geq 1$, then the constraint is:

$$\sum_{j=1}^{N_{max}} (1 - v_{1,j}) \sum_{i=2}^{n+1} v_{i,j} = 0 \tag{11}$$

Proposed Model

A general formulation for the mixed integer nonlinear problem (MINLP) is given by:

$$P = \begin{cases} Min\ E(W, \Sigma, V) = -\sum_{i=1}^{n}\sum_{j=1}^{N_{max}} v_{i+1,j} v_{1,j} C_{i,j} \\ \qquad\qquad Subject\ to: \\ \sum_{j=1}^{N_{max}} v_{i,j} = 1 \quad \forall i \in \{2, \ldots, n+1\} \\ \sum_{j=1}^{N_{max}} (1 - v_{1,j}) \sum_{i=2}^{n+1} v_{i,j} = 0 \\ w_j \in \mathbb{R}^d \quad \forall j \in \{1, \ldots, N_{max}\} \\ \sigma_j \in \mathbb{R}^+ \quad \forall j \in \{1, \ldots, N_{max}\} \\ V \in \{0,1\}^{(n+1) \times N_{max}} \end{cases} \tag{12}$$

where $C_{i,j} = \log[(\pi_j) \sum_{k=1}^{N_{max}} K^T(\delta(j,k)) f_k(x_i, w_k, \sigma_k))]$

3.2 Using CHN to Solve the Model (P)

The model (P) is a mixed-variables problem with a complex quadratic objective function, therefore, resolution by exact methods is very difficult. In this context, we use the continuous Hopfield neural network to solve (P) basing on two steps:

- Assignment phase: we fix the weight vectors and we solve the obtained problem: the polynomial assignment problem of integer variables.
- Minimization phase: we fix the assignment vectors and we solve the obtained problem: the nonlinear optimization problem with continuous variables.

3.2.1   Resolution of Assignment Phase

Since the resolution is performed sequentially using the two phases: expectation and minimization, we fix the obtained W and σ at the iteration $t - 1$ and we solve the following optimization problem (P1) with decision variables using the CHN:

$$P1 = \begin{cases} \underset{V \in \{0,1\}^{(n+1)N_{max}}}{Min} E(V) = \sum_{i=1}^{n} \sum_{j=1}^{N_{max}} v_{1j} v_{i+1,j} C_{ij} \\ e_{i+1}(V) = \sum_{j=1}^{N_{max}} v_{i+1,j} = 1 \quad i = 1,..,n+1 \\ H(V) = \sum_{j=1}^{N_{max}} \left((1 - v_{1j}) \sum_{i=1}^{n} v_{i+1,j} \right) = 0 \end{cases} \tag{12}$$

First, we handle the constraints by defining the following energy function:

$$E(V) = \alpha \sum_{i=1}^{n} \sum_{j=1}^{N_{max}} v_{1j} v_{i+1,j} C_{ij} + \rho H(V) + \frac{\beta}{2} \sum_{i=1}^{n} (e_i(V) - 1)^2 \\ + \gamma_0 \sum_{j=1}^{N_{max}} v_{1j} (1 - v_{1j}) + \gamma_1 \sum_{i=1}^{n} \sum_{j=1}^{N_{max}} v_{i+1,j} (1 - v_{i+1,j}) \tag{13}$$

where the scaling parameters $\alpha, \beta, \rho, \gamma_0$ and γ_1 are positive parameters which will be specified next. The network proposed for solving (P) consists of $(n+1)N_{max}$ neurons mutually interconnected. It is characterized by:

- The set of the network states:

$$V = \{v_{i,j} / i = 1,..,n+1 \text{ and } j = 1,..,N_{max}\} \tag{14}$$

- The weight matrix:

$$T = \{T_{ij,kl} / i,k = 1,..,n+1 \text{ and } j,l = 1,..,N_{max}\} \tag{15}$$

where T_{ij} denotes the weight that connects the ith and j^{th} neurons.

Using the identification between the Eq. 3 and Lyapunov functions 13, we obtain the following terms:

$$S = \begin{cases} T_{1b,1d} = 2\gamma_0 \delta_{bd} \\ T_{(a+1)b,(c+1)d} = -\beta + 2\gamma_1 \delta_{ac} \delta_{bd} \\ T_{1b,(c+1)d} = -2(\alpha C_{c,b} - \rho) \delta_{bd} \\ I_{1j} = -\gamma_0 \\ I_{(a+1)b} = -\rho + \beta - \gamma_1 \end{cases} \tag{16}$$

Where $a, c = 1,..,n+1$; $b,d = 1,..,N_{max}$ and δ_{ij} is equal to 1 if (i = j) and 0 (otherwise).

3.2.2   Parameter Settings

A feasible solution is guaranteed by the CHN from a stability no linear analysis of the Hamming hypercube corners set: $H_C = \{0, 1\}^{(n+1)N_{max}}$. The parameter-setting procedure is based on the partial derivatives of the generalized energy function:

- For $b = 1, \ldots, N_{max}$

$$E_{1b}(V) = \frac{\partial E(V)}{\partial v_{1b}} = \alpha \sum_{i=1}^{n} v_{i+1,b} C_{ib} + \gamma_0 (1 - 2v_{1b}) - \rho \sum_{i=1}^{n} v_{i+1,b} \qquad (17)$$

- For $a = 1, \ldots, n + 1$

$$E_{a+1,b}(V) = \frac{\partial E(V)}{\partial v_{a+1,b}} = \alpha v_{1,b} C_{ab} + \rho(1 - v_{1b}) + \beta(e_i(V) - 1) + \gamma_1(1 - 2v_{a+1,b}) \qquad (18)$$

In order to force the minimization and penalizing the constraints we impose the following conditions:

$$\alpha, \gamma_0, \gamma_1, \varepsilon \geq 0 \qquad (19)$$

The feasible solution set is defined by

$$F = \{V \in \{0, 1\}^{(n+1)N_{max}} \setminus v_{1,j} = 1 \ and \ \sum_{i=1}^{n} v_{i+1,j} \geq 1 \ and \ \sum_{j=1}^{N_{max}} v_{i+1,j} = 1\} \qquad (20)$$

Replacing the feasible state in the Eqs. 4 and 5, we get

$$v_{1b} = 0 \ \Rightarrow \ -\rho + \gamma_0 \geq \varepsilon \qquad (21)$$

$$v_{a+1,b} = 0 \ \Rightarrow \ \rho - \beta + \gamma_1 \geq \varepsilon \qquad (22)$$

In the case of $v_{1,b} = 1$, there exist $k \in 1, .., n$ such that $\sum_{i=1}^{n} v_{i+1,b} = k$, therefore, we obtain the following condition:

$$v_{1b} = 1 \ \Rightarrow \ \alpha k C_{ib} - k\rho - \gamma_0 \leq -\varepsilon$$

By maximizing C_{ib}

$$v_{1b} = 1 \ \Rightarrow \ \alpha k \ \max C_{ib} - k\rho - \gamma_0 \leq -\varepsilon \qquad (23)$$

The last condition is given as

$$v_{a+1,b} = 1 \implies \alpha C_{ab} + \beta(k-1) - \gamma_1 \leq -\varepsilon \tag{24}$$

Thus, we summarize all the constraints in the following system:

$$S = \begin{cases} \alpha, \gamma_0, \gamma_1, \varepsilon \geq 0 \\ -\rho + \gamma_0 \geq \varepsilon \\ \rho - \beta + \gamma_1 \geq \varepsilon \\ \alpha k \max C_{ib} - k\rho - \gamma_0 \leq -\varepsilon \\ \alpha C_{ab} + \beta(k-1) - \gamma_1 \leq -\varepsilon \end{cases} \tag{25}$$

Given the size of the description space and the size of the Kohonen topological map, we can determine the parameters by resolving the later System S. We need to fix α and ε, and compute the other parameters.

3.2.3   Resolution of Minimization Phase

At this level, we fix the obtained V from the previous phase and we solve the following optimization problem:

$$P1 = \begin{cases} Min \ E(w,\sigma) = \sum_{i=1}^{n} \sum_{j=1}^{N_{max}} v_{1j} v_{i+1,j} C_{ij} \\ w_j \in IR^d \\ \sigma_j \in IR^{+} \qquad j = 1, .., N_{max} \end{cases} \tag{26}$$

The presented objective function is a convex quadratic, so, the solution of (P2) is given by the critical point that verifies:

$$\begin{cases} \frac{\partial E(w,\sigma)}{\partial w_j} = 0 \\ \frac{\partial E(w,\sigma)}{\partial \sigma_j} = 0 \end{cases} \qquad \forall j = 1, .., N_{max} \tag{27}$$

During the minimization phase, V is chosen as the solution founded in the assignment phase.

4   Computer Simulations

In the present section we aim to increase the ability of our method to reduce PRSOM architecture and give satisfactory results in terms of classification. Method performance is then measured via experiences realized using Iris Dataset, well known in the classification literature. This data comes from California-Irvine University machine learning repository [12].

The Table 1 shows total number of instances in the database, with the useful number of data for learning and test phases, attributes number and classes' number. Database is divided into two parts: 50% for learning and 50% for tests.

Table 1. Iris dataset caracteristiques

Data set	Size	Nr.Tr.D	Nr.Ts.D	At.Nr.	Cl.Nr.
Iris	150	75	75	4	3

The Table 2 presents in the mean of remaining neurons associated with different size of maps. So, the necessary number of neurons to clusters data Iris converges approximately to 7 neurons. The numerical results are presented in Table 2. For example, a map contains 30 neurons, the mean of remaining neurons is approximately 7 neurons.

Table 2. Optimal topological maps sizes of Dataset

Data set/N_{max}	10	15	20	25	30	40	50
Iris	8	7	7	9	7	7	8

Table 3 presents the obtained clustering results of training and testing data. This table shows that our method gives the good results, because all the training data were correctly classified except two. In addition, this table shows the numerical results obtained from the data classification of the testing data. We see that, the important results are obtained because we have, only, one misclassified data among 75 testing data.

Table 3. Numerical results of the training and testing data classification via CHN-PRSOM

Data set	Nr.Tr.D.	C.C.Tr	MC.Tr	Overall
Training set	75	73	2	97.3
Testing set	75	74	1	98.6

5 Conclusion

New model of PRSOM architecture is proposed. It consists of a main objective function that describes the PRSOM and constraints that support the suggested method. Obtained model is a nonlinear mathematical program with mixed variables. Resolution phase was performed by continuous Hopfield neural network which is adapted technically to the problem according to specific parameters extracted from stability conditions. Model was efficiently accomplished to Iris data classification. Indeed, obtained results are satisfactory in term of the optimal architecture and data classification.

References

1. McLachlan, G.J., Peel, D.: Finite Mixture Models. Wiley, New York (2000). https://doi.org/10.1002/0471721182
2. Kohonen, T.: Self-Organizing Maps, 3rd edn. Springer, Heidelberg (2001). http://www.springer.com/in/book/9783540679219
3. Anouar, F., Badran, F., Thiria, S.: Probabilistic self-organizing map and radial basis function networks. Neurocomputing **20**, 83–96 (1998)
4. En-Naimani, Z., Lazaar, M., Ettaouil, M.: Architecture optimization model for the probabilistic self-organizing maps and speech compression. Int. J. Comput. Intell. Appl. **15**, 1650007 (2016)
5. Hopfield, J.J.: Neural networks and physical systems with emergent collective computational abilities. Proc. Natl. Acad. Sci. **79**, 25542558 (1982)
6. Hopfield, J.J.: Neurons with graded response have collective computational proper-ties like those of two-state neurons. Proc. Natl. Acad. Sci. **81**, 30883092 (1984)
7. Hopfield, J.J., Tank, D.W.: Neural computation of decisions in optimization problems. Biol. Cybern. **52**, 141152 (1985)
8. Talavan, P.M., Yez, J.: Parameter setting of the Hopfield network applied to TSP. Neural Netw. **15**(3), 363–373 (2002)
9. Wen, U.P., Lan, K.M., Shih, H.S.: A review of Hopfield neural networks for solving mathematical programming problems. Eur. J. Oper. Res. **198**(3), 675–687 (2009)
10. Joudar, N., El Moutouakil, K., Ettaouil, M.: An original continuous Hopfield network for optimal images restoration. WSEAS Trans. Comput. **14**, 668–679 (2015)
11. Bishop, C.M.: Pattern Recognition and Machine Learning. Springer, New York (2006)
12. UCI Machine Learning Repository. http://archive.ics.uci.edu/ml/datasets.html

The Hybrid Framework for Multi-objective Evolutionary Optimization Based on Harmony Search Algorithm

Iyad Abu Doush[1]($\boxtimes$) (iD), Mohammad Qasem Bataineh[2],
and Mohammed El-Abd[3]

[1] Computer Science and Information Systems,
American University of Kuwait, Salmiya, Kuwait
`idoush@auk.edu.kw`
[2] Computer Science Department, Yarmouk University, Irbid, Jordan
`m.bataineh88@yahoo.com`
[3] Electrical and Computer Engineering, American University of Kuwait,
Salmiya, Kuwait
`melabd@auk.edu.kw`

Abstract. In evolutionary multi-objective optimization, an evolutionary algorithm is invoked to solve an optimization problem involving concurrent optimization of multiple objective functions. Many techniques have been proposed in the literature to solve multi-objective optimization problems including NSGA-II, MOEA/D and MOPSO algorithms. Harmony Search (HS), which is a relatively new heuristic algorithm, has been successfully used in solving multi-objective problems when combined with non-dominated sorting (NSHS) or the breakdown of the multi-objectives into scalar sub-problems (MOHS/D). In this paper, the performance of NSHS and MOHS/D is enhanced by using a previously proposed hybrid framework. In this framework, the diversity of the population is measured every a predetermined number of iterations. Based on the measured diversity, either local search or a diversity enhancement mechanism is invoked. The efficiency of the hybrid framework when adopting HS is investigated using the ZDT, DTLZ and CEC2009 benchmarks. Experimental results confirm the improved performance of the hybrid framework when incorporating HS as the main algorithm.

Keywords: Multi-objective optimization · Harmony Search
Multi-objective optimization evolutionary algorithms

1 Introduction

In Multi-objective Optimization Problems (MOPs), the aim is to find an optimal solution to a problem having multiple objectives [4]. Almost all engineering

---

I. A. Doush—Dr. Iyad Abu Doush, Department Computer Science and Information Systems, American University of Kuwait, Salmiya, Kuwait.

© Springer International Publishing AG, part of Springer Nature 2019
J. Mizera-Pietraszko et al. (Eds.): RTIS 2017, AISC 756, pp. 134–142, 2019.
https://doi.org/10.1007/978-3-319-91337-7_13

problems are multi-objective. Moreover, these objectives are often conflicting (i.e., maximize performance function, minimize cost function, maximize reliability function, etc.). Hence, an optimal solution of one objective function will not provide the best solution for other objective(s). Therefore, different solutions offer trade-offs between different objectives and a set of solutions is maintained [10,12].

Population-based meta-heuristics were proven to be very effective in handling MOPs. This class covers a wide range of algorithms including EAs, swarm intelligence, and foraging algorithms. These algorithms rely on updating a population of solutions in each iteration. Hence, such algorithms have the ability to maintain a set of non-dominated solutions. These algorithms are generally characterized by having less sensitivity to the shape of the pareto front, more robustness, and are easily parallelized [1]. Well-known algorithms applied to MOPs include NSGA-II [5], MOEA/D [18], and MOPSO [15].

The work in [17] proposed a hybrid evolutionary multi-objective optimization framework having a modular structure. Sample implementations of this framework were implemented using NSGA-II and MOEA/D (referred to from now on as Hybrid NSGA-II and Hybrid MOEA/D). In this work, the same hybrid framework is implemented using the Harmony Search (HS) algorithm and performances are compared against Hybrid NSGA-II and Hybrid MOEA/D.

HS [11] is a meta-heuristic algorithm simulating a group of musicians searching for the right harmony. A single problem variable is a *pitch* of a different musical instrument and a complete solution is a *harmony vector*. If a specific pitch results in an improved harmony, it will be stored in the *Harmony Memory* (HM). Initially, HM is filled with randomly initialized harmonies. In every iteration, the generated harmony replaces the worst harmony in HM if it is better than it [7,8].

In HS, new harmonies are generated by developing a sequence of new pitches. A pitch is produced by playing a pitch from memory, picking then perturbing a pitch from memory, or playing a random pitch. One operation is applied based on two parameters: the *Harmony Memory Considering Rate* (HMCR) and the *Pitch Adjusting Rate* (PAR). HS and its variants were successfully applied for single objective continuous optimization as in [3,9]. Previous work tackled the problem of MOPs using HS [2,14,16].

The rest of paper is divided as follows: the implemented hybrid framework is introduced in Sect. 2. Experimental results are presented in Sect. 3. Finally, the paper is concluded in Sect. 4.

2   Methodology

In this section, the previously proposed NSHS and MOHS/D [2] are briefly explained. This is followed by a brief explanation of the hybrid framework.

NSHS and MOHS/D: In [2], Abu Doush and Bataineh followed similar frameworks in [5,13,17]. For NSHS, HM is randomly initialized with HMS solutions. Non-dominated sorting is used to sort the population based on Pareto optimality, where the best solution is ranked number 1. HS operators are used to create additional HMS solutions. The two populations are then combined to form a large population of size $2 \times$ HMS and the sorting procedure is run to rank the two populations. An elitism process is adopted to select the best HMS solutions for the next generation. If the number of rank 1 solutions is less than HMS, we select all individuals to be inserted in the new HM. Next we select individuals from the set of ranked 2 solutions, and so on. This procedure is continued until no more solutions can be accommodated. In the last set having more solutions than needed, remaining solutions are selected using a crowding comparator. NSHS steps are repeated until the stopping criterion is reached.

For MOHS/D, parameters to be tuned are: N (the number of sub-problems of MOP), N weight vectors uniformly generated, and the neighborhood size T of each weight vector. An empty External Population (EP) is created to be used as the set of non-dominated solutions. MOHS/D then finds, for each solution, the indexes of the closest T weight vectors formulating the B sets. An initial population HM is randomly generated. For all solutions in HM, F(x) is computed, which consists of a set of f objective functions with size m. In the last step of the initialization process, MOHS/D assigns a target point in the objective space Z using a problem-specific method in order to guide the solutions towards the Pareto front. This Z vector is subsequently modified in every iteration.

For each weight vector, HS operators are used to create a new Harmony (y) by randomly selecting two indexes from the B set associated with this weight vector. MOHS/D then applies a problem-specific repair on y to produce y'. The objective values of y' is then compared to the target vector Z. If f(y'_i) is better than z_i for each $i = (1, \ldots, m)$, then $z_i = \mathrm{F}(y'_i)$. After that the subproblem updates its neighborhood weight vector and if the Tchebycheff value of y' is better than its neighbor, the neighbor becomes y'. Finally, MOHS/D removes from EP all the vectors that are dominated by F(y') and add F(y') to EP if no vectors in EP dominate F(y'). This process is repeated until the stopping criterion is satisfied.

Hybrid Framework: In this framework, a clustering process is run on the population every a predetermined number of iterations. The process clusters the population into k clusters and then $Q_{current}$ is computed as follows:

$$Q_{current} = \sum_{i=1}^{k} \frac{1}{s_i} \sum_{j=1}^{s_i} D\left(c_i^j, \sigma_i\right) \tag{1}$$

where k is the number of clusters, σ_i is the centroid of cluster i, and C_i^j is a point j in cluster i. Furthermore, $D(C_i^j, \sigma_i)$ is the Euclidean distance of point j in cluster i to its centroid and s_i is the number of individuals in the cluster i.

The equation is used to evaluate the diversity for a specific generation (in our case each 5 generations). The calculated value of $Q_{current}$ is used to be our new lower bound Q_{bound} of the next generation. Starting from the next generation, the value of Q_{bound} value will be compared with $Q_{current}$ until we reach to the next 5th generation. In case $Q_{current}$ is found to be lower than Q_{bound} then the population has a bad diversity otherwise, it has a good diversity. In case the diversity is good, local search (LS) is triggered, the local search that used in this step is already implemented in Jmetal framework. In case the diversity is bad, the diversity enhancement module is triggered. In this work, this hybrid framework is implemented using NSHS and MOHS/D (referred to from now on as Hybrid NSHS and Hybrid MOHS/D) and compared against Hybrid NSGA-II and Hybrid MOEA/D.

3 Experimental Procedure

To evaluate the Hybrid NSHS and Hybrid MOHS/D algorithms, experiments are conducted using the ZDT [20], DLTZ [6], and CEC09 [19] benchmarks. The performance is compared against Hybrid NSGA-II and Hybrid MOEA/D.

Comparison are based on minimizing the **Inverted Generational Distance (IGD)** measure [17]. The smaller values of IGD, the closer the generated solutions are to the true PF. IGD is calculated as follows:

$$IGD(P, P^*) = \frac{\sum_{v \in P^*} d(V, P)}{|P^*|}$$

where P is the set of obtained approximated solutions in objective space, P^* is the true PF in objective space, $d(V, P)$ is the minimum Euclidean distance between v and points in P, and $|P^*|$ is the number of points in P^*.

Maximum number of function evaluations (FE) is different for different test instances. As in [17] FE is set to 20,000 for ZDT family and DTLZ1 and DTLZ7, 5,000 for DTLZ2, DTLZ4, and DTLZ5, 15,000 for DTLZ6, and 50,000 for DTLZ3 and UF. Results reported based on 15 independent runs. For HS, *HMS* (i.e., the population size) is 100 for bi-objective problems and 200 for three objectives, $MCR = 0.95$, $PAR = 0.4$, and $bw = 0.01$. Finally, Sequential Quadratic Programming SQP is used as the LS with n of $P_{local} = 3$ for Hybrid NSHS and n of $P_{local} = 20$ for Hybrid MOHS/D.

3.1 Results and Discussion

Hybrid NSHS and Hybrid NSGA-II: Table 1 shows the minimum, median, and maximum of the IGD-metric values for both Hybrid NSHS and Hybrid NSGA-II algorithms. The bold font represents the algorithm with the best result. Figure 1 presents a random sample of PF of some test problems for both algorithms. In a previous study [17], Hybrid NSGA-II was tested on ZDT4, DTLZ1, DTLZ2, DTLZ3, DTLZ4, DTLZ5 and DTLZ6 only, the experimental results showed that the Hybrid NSGA-II outperformed NSGA-II in these test problems.

From Table 1, Hybrid NSHS outperforms Hybrid NSGA-II in 14 out of 22 problems. Hybrid NSHS outperformed Hybrid NSGA-II in solving DTLZ1 and DTLZ3, this emphasizes that Harmony operators have the advantage in solving multiple local fronts problems. However, Hybrid NSGA-II algorithm showed lower IGD values in ZDT4 as Hybrid NSHS is stuck in local minima.

Table 1. The minimum, median, and maximum of the IGD-metric values for Hybrid NSHS and Hybrid NSGA-II.

Test Problem	Type	Hybrid NSHS	Hybrid NSGA-II
ZDT1	Best	1.42e-03	6.98e-04
	Median	**4.13e-03**	7.33e-03
	Worst	9.19e-03	8.18e-03
ZDT2	Best	8.83e-04	5.39e-04
	Median	**1.15e-03**	2.62e-03
	Worst	1.26e-03	7.09e-03
ZDT3	Best	5.88e-04	5.30e-04
	Median	**5.11e-04**	2.65e-03
	Worst	1.03e-03	3.06e-03
ZDT4	Best	2.55e-02	4.89e-04
	Median	4.42e-02	**1.90e-03**
	Worst	1.32e-01	1.06e-02
ZDT6	Best	6.04e-04	1.02e-03
	Median	**2.65e-03**	9.44e-03
	Worst	8.27e-03	2.50e-02
DTLZ1	Best	8.82e-04	1.14e-02
	Median	**6.24e-03**	2.37e-02
	Worst	1.24e-02	4.47e-02
DTLZ2	Best	5.93e-04	6.29e-04
	Median	8.31e-04	**5.04e-04**
	Worst	9.39e-04	8.08e-04
DTLZ3	Best	1.38e-02	1.37e-01
	Median	**6.74e-02**	2.87e-01
	Worst	1.31e-01	5.82e-01
DTLZ4	Best	8.49e-04	8.53e-04
	Median	1.05e-03	**9.58e-04**
	Worst	1.34e-03	2.32e-03
DTLZ5	Best	3.25e-05	3.19e-05
	Median	2.46e-04	**4.79e-05**
	Worst	6.25e-04	6.14e-05
DTLZ6	Best	4.65e-04	2.11e-02
	Median	**8.34e-04**	1.23e-02
	Worst	5.63e-03	2.40e-02
DTLZ7	Best	2.86e-03	1.68e-03
	Median	3.42e-03	**1.87e-03**
	Worst	3.72e-03	2.00e-03
UF1	Best	3.05e-03	3.98e-03
	Median	**3.93e-03**	7.07e-03
	Worst	5.25e-03	2.57e-02
UF2	Best	4.33e-03	6.54e-03
	Median	**5.58e-03**	9.56e-03
	Worst	7.75e-03	1.18e-02
UF3	Best	7.82e-03	4.60e-03
	Median	8.69e-03	**8.33e-03**
	Worst	1.01e-02	1.18e-02
UF4	Best	3.02e-03	3.01e-03
	Median	**3.22e-03**	7.44e-03
	Worst	3.62e-03	1.75e-02
UF5	Best	4.33e-02	1.12e-01
	Median	**1.49e-01**	8.15e-01
	Worst	9.05e-01	1.08e+00
UF6	Best	6.13e-03	6.20e-03
	Median	**8.80e-03**	9.10e-03
	Worst	1.31e-02	1.34e-02
UF7	Best	2.15e-03	3.88e-03
	Median	**3.48e-03**	1.34e-02
	Worst	1.70e-02	1.89e-02
UF8	Best	2.76e-03	2.16e-03
	Median	2.91e-03	**2.81e-03**
	Worst	3.19e-03	2.92e-03
UF9	Best	3.59e-03	1.27e-03
	Median	4.31e-03	**2.43e-03**
	Worst	4.84e-03	3.65e-03
UF10	Best	2.85e-03	2.87e-03
	Median	**3.46e-03**	3.56e-03
	Worst	3.83e-03	5.10e-03

Table 2. The minimum, median, and maximum of the IGD-metric values for Hybrid MOHS/D and Hybrid MOEA/D.

Test Problem	Type	Hybrid MOHS/D	Hybrid MOEA/D
ZDT1	Best	1.49e-03	2.67e-04
	Median	1.97e-03	**3.28e-04**
	Worst	2.70e-03	4.43e-04
ZDT2	Best	3.30e-04	3.37e-02
	Median	**2.85e-04**	2.96e-03
	Worst	4.82e-04	3.58e-03
ZDT3	Best	9.41e-04	3.00e-04
	Median	1.27e-03	**3.51e-04**
	Worst	1.49e-03	4.28e-04
ZDT4	Best	8.70e-03	4.92e-02
	Median	**2.45e-02**	1.39e-01
	Worst	5.89e-02	3.57e-01
ZDT6	Best	1.56e-04	1.40e-04
	Median	1.73e-04	**1.42e-04**
	Worst	7.87e-03	6.08e-04
DTLZ1	Best	3.82e-03	4.56e-03
	Median	**1.82e-02**	3.00e-02
	Worst	4.73e-02	9.20e-02
DTLZ2	Best	3.36e-03	3.36e-03
	Median	4.37e-03	**3.75e-03**
	Worst	4.66e-03	4.63e-03
DTLZ3	Best	1.03e-01	1.63e-01
	Median	**1.39e-01**	4.71e-01
	Worst	1.88e-01	1.17e+00
DTLZ4	Best	8.41e-03	8.89e-03
	Median	**1.00e-02**	1.08e-02
	Worst	1.09e-02	1.27e-02
DTLZ5	Best	1.36e-03	1.42e-03
	Median	**1.41e-03**	1.43e-03
	Worst	1.46e-03	1.48e-03
DTLZ6	Best	7.96e-03	6.84e-03
	Median	2.08e-02	**1.03e-02**
	Worst	2.86e-02	3.15e-02
DTLZ7	Best	2.04e-02	2.79e-01
	Median	**8.98e-02**	1.80e+00
	Worst	3.09e-01	5.90e+00
UF1	Best	1.65e-03	1.22e-03
	Median	2.77e-03	**2.68e-03**
	Worst	3.80e-03	7.30e-03
UF2	Best	1.40e-03	8.00e-04
	Median	1.79e-03	**1.30e-03**
	Worst	2.74e-03	6.99e-03
UF3	Best	5.54e-03	4.73e-03
	Median	**5.53e-03**	8.38e-03
	Worst	8.59e-03	1.14e-02
UF4	Best	1.91e-03	1.79e-03
	Median	**2.01e-03**	2.09e-03
	Worst	3.37e-03	2.55e-03
UF5	Best	3.86e-02	8.25e-02
	Median	**4.89e-02**	1.35e-01
	Worst	7.00e-02	1.74e-01
UF6	Best	6.92e-03	3.46e-03
	Median	**1.03e-02**	1.14e-02
	Worst	1.25e-02	2.54e-02
UF7	Best	1.48e-03	9.28e-04
	Median	1.22e-02	**2.33e-03**
	Worst	1.56e-02	2.17e-02
UF8	Best	5.35e-03	8.52e-03
	Median	**5.35e-03**	9.57e-03
	Worst	8.91e-03	9.93e-03
UF9	Best	6.36e-03	5.23e-03
	Median	**8.99e-03**	9.22e-03
	Worst	6.69e-02	6.78e-02
UF10	Best	6.57e-03	8.10e-03
	Median	**8.33e-03**	1.04e-02
	Worst	9.57e-03	1.32e-02

Hybrid NSGA-II is better in solving DTLZ2, DTLZ4 and DTLZ5. The reason for the higher median IGD values for the Hybrid NSHS algorithm could be the excessive diversity caused by inappropriate pitch adjustment rate. In contrast, Hybrid NSHS algorithm produces better results in DTLZ6. Figure 1 presents a visual evidence on approximating the two algorithms for PF of sample functions. Hybrid NSHS approximates the DTLZ6 and UF5 PF slightly better than Hybrid NSGA-II while Hybrid NSGA-II approximate DTLZ1 PF slightly better.

In Fig. 1, we can see that Hybrid NSHS plots ZDT6 and UF5 better than the Hybrid NSGA-II. Since Harmony operators are better suited to deal with non-convex and discontinuous problems. Figure 1 shows that Hybrid NSGA-II was more successful in plotting PF of ZDT1. However Hybrid NSGA-II and Hybrid NSHS have relatively the same approximation for ZDT1 PF.

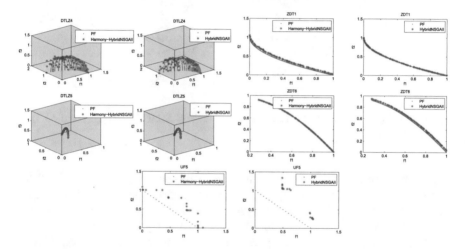

Fig. 1. Pareto fronts for lowest IGD values of Hybrid NSHS and Hybrid NSGA-II algorithms

Hybrid MOHS/D and Hybrid MOEA/D: Table 2 shows the minimum, median, and maximum of the IGD-metric values for both Hybrid MOHS/D and Hybrid MOEA/D. The bold font represents the algorithm with the best result. Figure 2 presents a random sample of PF of some test problems for both algorithms. In the same study [17], Hybrid MOEA/D was tested on ZDT4, DTLZ1, DTLZ2, DTLZ3, DTLZ4, DTLZ5 and DTLZ6 only. The experimental results showed that the Hybrid MOEA/D outperformed MOEA/D in these test problems except DTLZ3 and DTLZ6.

Hybrid MOHS/D outperformed Hybrid MOEA/D in 14 out of 22 problems. Hybrid MOHS/D outperformed Hybrid MOEAD in solving DTLZ1 and DTLZ3, again due to harmony operators being better suited for solving multiple local fronts problems. In ZDT4, Hybrid MOHS/D algorithm showed lower median IGD value as the Hybrid MOEA/D is stuck in local minima.

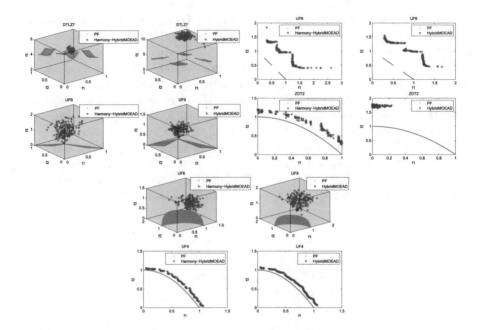

Fig. 2. Pareto fronts for lowest IGD values of Hybrid MOHS/D and hybrid MOEAD algorithms

Figure 2 presents a visual evidence on approximating the two algorithms for PF of sample functions. Hybrid MOHS/D approximate the PF of DTLZ7 and UF8 significantly better than Hybrid MOEAD. The figure also shows that Hybrid MOHS/D plots ZDT2 and UF6 better than Hybrid MOEAD.

The Friedman test is the non-parametric alternative to the one-way ANOVA with repeated measures. It is used to test for differences between groups when the dependent variable being measured is ordinal.

Table 3 shows mean ranks for the proposed algorithm in terms of quality indicators, the lowest mean rank is the best (bold font). From Table 3 we can see that Harmony Hybrid MOEAD algorithm outperformed all proposed algorithms.

Table 3. Friedman mean ranks for the proposed algorithm in terms of quality indicators.

Indicator	Harmony NSGA-II	Harmony MOEAD	Harmony Hybrid NSGA-II	Harmony Hybrid MOEAD
IGD	2.64	3.00	2.36	**2.00**

4   Conclusions

In this paper, we present an enhanced version of the multi-objective harmony search algorithm NSHS and MOHS/D using a previously proposed hybrid framework. The two new algorithms are called Hybrid NSHS and Hybrid MOHS/D. The applied framework investigate the solutions diversity on each predetermined number of iterations. The two proposed algorithms are compared with the Hybrid NSGA-II and Hybrid MOEA/D using well-known datasets from the literature. The experimental results show that both Hybrid NSHS and Hybrid MOHS/D outperforms Hybrid NSGA-II and Hybrid MOEA/D respectively in terms of IGD. The proposed algorithms provide a better vehicle for solving problems with multiple local fronts. However, for a few benchmark functions, the proposed algorithms show shortcoming because of the excessive diversity caused by the high exploration due to the pitch adjustment rate. As a future work, we will study the effect of the different operators on the performance of the proposed algorithms.

References

1. Abraham, A., Jain, L.: Evolutionary multiobjective optimization. Springer (2005)
2. Abu Doush, I., Bataineh, M.Q.: Hybedrized NSGA-II and MOEA/D with Harmony Search Algorithm to Solve Multi-objective Optimization Problems, pp. 606–614. Springer (2015)
3. Al-Betar, M.A., Doush, I.A., Khader, A.T., Awadallah, M.A.: Novel selection schemes for harmony search. Appl. Math. Comput. **218**(10), 6095–6117 (2012)
4. Deb, K.: Multi-objective Optimization Using Evolutionary Algorithms. Wiley, Chichester (2001)
5. Deb, K., Pratap, A., Agarwal, S., Meyarivan, T.: A fast and elitist multiobjective genetic algorithm: Nsga-ii. IEEE Trans. Evol. Comput. **6**(2), 182–197 (2002)
6. Deb, K., Thiele, L., Laumanns, M., Zitzler, E.: Scalable multi-objective optimization test problems. In: Proceedings of the Congress on Evolutionary Computation (CEC-2002), (Honolulu, USA), pp. 825–830 (2002)
7. Doush, I.A.: Harmony Search with Multi-Parent Crossover for Solving IEEE-CEC2011 Competition Problems, pp. 108–114. Springer, Heidelberg (2012)
8. Doush, I.A., Al-Betar, M.A., Khader, A.T., Awadallah, M.A., Mohammed, A.B.: Analysis of takeover time and convergence rate for harmony search with novel selection methods. Int. J. Math. Model. Numer. Optim. **4**(4), 305–322 (2013)
9. El-Abd, M.: An improved global-best harmony search algorithm. Appl. Math. Comput. **222**, 94–106 (2013)
10. Esfe, M.H., Hajmohammad, H., Toghraie, D., Rostamian, H., Mahian, O., Wongwises, S.: Multi-objective optimization of nanofluid flow in double tube heat exchangers for applications in energy systems. Energy **137**, 160–171 (2017)
11. Geem, Z.W., Kim, J.H., Loganathan, G.: A new heuristic optimization algorithm: harmony search. Simulation **76**(2), 60–68 (2001)
12. Gutjahr, W.J., Pichler, A.: Stochastic multi-objective optimization: a survey on non-scalarizing methods. Ann. Oper. Res. **236**(2), 475–499 (2016)
13. Li, H., Zhang, Q.: Multiobjective optimization problems with complicated Pareto sets, MOEA/D and NSGA-II. IEEE Trans. Evol. Comput. **13**(2), 284–302 (2009)

14. Pavelski, L.M., Almeida, C.P., Gonalves, R.A.: Harmony search for multi-objective optimization. In: 2012 Brazilian Symposium on Neural Networks (SBRN), pp. 220–225. IEEE (2012)
15. Reyes Sierra, M., Coello Coello, C.A.: Multi-objective particle swarm optimizers: a survey of the state-of-the-art. Int. J. Comput. Intell. Res. 2(3), 287–308 (2006)
16. Ricart, J., Hüttemann, G., Lima, J., Barán, B.: Multiobjective harmony search algorithm proposals. Electron. Notes Theor. Comput. Sci. 281, 51–67 (2011)
17. Sindhya, K., Miettinen, K., Deb, K.: A hybrid framework for evolutionary multi-objective optimization. IEEE Trans. Evol. Comput. 17(4), 495–511 (2013)
18. Zhang, Q., Li, H.: Moea/d: A multiobjective evolutionary algorithm based on decomposition. IEEE Trans. Evol. Comput. 11(6), 712–731 (2007)
19. Zhang, Q., Zhou, A., Zhao, S., Suganthan, P.N., Liu, W., Tiwari, S.: Multiobjective optimization test instances for the CEC 2009 special session and competition. University of Essex, Colchester, UK and Nanyang Technological University, Singapore, Special Session on Performance Assessment of Multi-Objective Optimization Algorithms, Technical report (2008)
20. Zitzler, E., Deb, K., Thiele, L.: Comparison of multiobjective evolutionary algorithms: empirical results. Evol. Comput. 8(2), 173–195 (2000)

New Prior Model for Bayesian Neural Networks Learning and Application to Classification of Tissues in Mammographic Images

Hassan Ramchoun[✉] and Mohamed Ettaouil

Department of Mathematics, Faculty of Sciences and Technology,
University of Sidi Mohamed Ben Abdellah, 30000 Fes, Morocco
ramchounhassan@gmail.com, mohamedettaouil@yahoo.fr

Abstract. The Multilayer Perceptron (MLP) is the most useful artificial neural network to estimate the functional structure in the non- linear systems, but the determination of its architecture, weights and hyperparameters is a fundamental problem due to their direct impact on the network generalization ability and convergence. The Bayesian approach provides a naturel way to adjust the weights decay parameters automatically that's give the best generalization. This paper presents an improvement of a prior model to construct a new objective function for learning neural network in Bayesian perspectives with the Hybrid Monte Carlo (HMC) algorithm. The proposed model is applied to classification of Normal, Benign and Malignant Tissues in Mammographic images. Compared to the other regularization model the numerical results illustrate the advantages of our approach.

Keywords: Multilayer perceptron · Bayesian learning · Prior
Hybrid Monte Carlo · Classification · Mammographic images

1 Introduction

One of the most common causes for women mortality all over the world is considered to be the breast cancer, early detection is the key for improving breast cancer prognosis. To reduce the high number of unnecessary breast biopsies, several computer aided diagnosis systems have been proposed. The use of artificial neural networks has shown great potential in this field. So, it is very important to develop a Computer aided Diagnosis (CAD) system, which can distinguish normal–abnormal as well as benign–malignant mammogram. The main objective of CAD system is to increase diagnosis accuracy and enhancing the mammogram interpretation. Thus, CAD system can reduce the variability in judgments among radiologists by providing an accurate diagnosis of digital mammograms.

H. Ramchoun—Ph.D Student, Modeling and Scientific Computing Laboratory, USMBA, Fez, Morocco.

The most effective method for early detection and screening of breast cancers is mammography [1]. Micro-calcifications and masses are two important early signs of the diseases [2]. In the last two decades, researchers proposed various types of classifier in order to create an effective and optimal CAD for mammograms. Artificial neural network always have a good performance in pattern recognition rather than the other without implementing ANN [3]. Neural networks (NNs) have frequently been proposed for Classification of Normal, Benign and Malignant Tissues in Mammographic Images, because of their capabilities for nonlinear modeling of large multivariate datasets. The family of NN models known as multilayer perceptron (MLPs) are probably the most frequently used, since they have been shown to be universal approximator of functions (Haykin, 1999).

Recently, Bayesian approaches have shown considerable potential in such problems. Bayesian learning of Artificial Neural Networks (ANN) is equivalent to find the values of all weights and hyperparametrs of his distribution with this way we treat the regularization problem automatically [4] and we provide a naturel way to find the functional structure in the non linear system, In the Bayesian approach prior information from the problem is combined to the evidence from the data, giving the posterior probability of solutions.

Predictions are made by integrating over this posterior distribution. In case of insufficient data the prior dominates the solution, and the effect of the prior diminishes with increased evidence from the data. With MLPs the main difficulty is in controlling the complexity of the model. Another problem of standard MLP models is the lack of tools for analyzing the results (confidence intervals, etc.). Bayesian methods have become a viable alternative to the older error minimization based (ML or MAP) approaches (MacKay, 1992; Bishop, 1995; Neal, 1996) [4–7].

The next section present and discuss related works on Bayesian neural network and mammography. Section 3 describes the Hybrid Monte Carlo (HMC) algorithm for artificial neural networks. In Sect. 4, we describe the Gray Level Co-occurrence Matrix for feature extraction. And before concluding, a proposed Bayesian neural network trained by HMC is used to predict the classification for each tissue sample.

2   Related Works

A number of approaches in the literature have taken account the use of artificial neural network and Bayesian approach for classification or feature selection in mammographic images. This section describes only those works that are more or less similar to our work.

The classification step is very important for the performance of the computer aided diagnosis (CAD). There are many classifiers techniques which are performed well in mass classification such as artificial neural network and linear discriminants analysis. The medical image system uses other classifiers like binary decision tree, support vector machine and Bayesian network.

Yu and Huang [8] proposed a wavelet filter method to detect cancer region by calculating the mean pixel value. In order to extract the features of mammograms, Markov random field (MRF), fractal models, and statistical textural feature were used. Then, for the classification process, all extracted features were inputted into Back-propagation neural network with three layers.

The combination of Gray-level Co-occurrence Matrix (GLCM) and Artificial Neural Network model had been done by Wong et al. [9], GLCM is used to produce twelve texture feature of mammograms. The four significant output of GLCM was used as the input for ANN model to classify either mass or non-mass region. Martins and all [10] proposed the GLCM and Bayesian neural network trained by MCMC and Gibbs sampling method.

Halkiotis, Botsis, and Rangoussi (2007) [11] used a MLP type neural network on the MIAS database. They obtained a good classification rate with an average of 0.27 false positives per image for microcalcifications.

In 2010, Islam et al. [12] Introduced an effective method for benign – malignant classification of digital mammograms. The authors used seven texture features as the input for Multi-layer Perceptron classifier with 7 units in input layer, 5 units in hidden layer and 1 unit in the output layer.

3 Bayesian Learning by Hybrid Monte Carlo for MLP

Bayesian methods use probability to quantify uncertainty in inferences and the result of Bayesian learning is a probability distribution expressing our beliefs regarding how likely the different predictions are. Bayesian paradigm offers consistent way to do inference using models with even very large number of parameters. See, e.g., Gelman et al. (1995) [13] for good introduction to Bayesian methods.

3.1 Bayesian Learning for MLP

In this paper, we focus exclusively on feedforward networks with a single hidden layer with m nodes and k outputs, besides do not allow direct connections from the inputs and outputs. The particular form of the neural model is

$$u_k(x, w) = w_{k0} + \sum_{j=1}^{m} w_{kj} \phi_j (w_{j0} + \sum_{i=1}^{p} w_{ij} x_i) \tag{1}$$

where x is the input vector and ϕ represents the activation function and the set of all weights (parameters), represented by the vector w, including input-hidden weights, biases and hidden-output weights. If the neural model is used for classification problem, the output $f_k(x, w)$ is the final value used for classification process. For many class problem, the probability that a class target, y, has value j is

$$f_j(x, w) = p(y = j/x, w) = \frac{\exp(u_j(x, w))}{\sum_k \exp(u_k(x, w))} \tag{2}$$

The Bayesian learning consists of considering the networks weights as random variables with prior distribution $p(w/\alpha)$ expresses our initial beliefs about parameter values, before any data has been observed. After observing new data $D = \left((x_1, y_1), \dots, (x_n, y_n) \right)$ prior distribution is updated using Bayes' rule to obtain the posterior distribution $p(w, \alpha/D)$ which is then written in the following form

$$p(w, \alpha/D) = \frac{p(D/w)p(w/\alpha)}{p(D)} \tag{3}$$

Assuming that $(x_1, y_1), \dots, (x_n, y_n)$ are independents and identically distributed, we have the following likelihood function for the training data

$$p(D/w) = \prod_{i=1}^{n} \prod_{k=1}^{c} f_k(x_i, w)^{y_i^k} \tag{4}$$

To find the weight vector w^*, corresponding to the maximum of the posterior distribution, is equivalent to minimize the error-function E(w), which is given by

$$E(w) = -\ln(p(D/w)) - \ln(p(w/\alpha)) \tag{5}$$

Where $p(w/\alpha)$ is the prior distributions on networks weights before receiving any data and correspond to regularizer or penalty terms. This uncertainty about the network parameters plays an important role to avoid over-fitting and minimize the generalization error. Different prior models are proposed in the literature such as Gaussians, Cauchy, Laplace,... In this paper we propose to use the Laplace prior with some improvement see Sect. 5.

We thus obtain from Eq. (3)

$$p(w, \alpha/D) = \frac{1}{Z} \exp(E(w)) \tag{6}$$

Where Z is a normalization constant. Then predictions on new data can be made by

$$\hat{y}_k^{n+1} = p\left(y_{n+1}/x_{n+1}, D\right) = \iint p\left(y_{n+1}/x_{n+1}, w, \alpha\right) p(w, \alpha/D) dw d\alpha \tag{7}$$

In general, this integral is analytically intractable. To solve it numerically, the most used method is the Markov Chain Monte Carlo (MCMC) approach, which consists in generating a Markov chain that satisfies the ergodicity condition and converges to the posterior distribution. The integral in Eq. (7) can then be approximated by

$$\hat{y}_k^{n+1} \approx \frac{1}{M} \sum_{t=1}^{M} f_k(x_{n+1}, w^{(t)}) \tag{8}$$

The posterior distribution is very complex with many mode, The integral in (7) can be also approximated using Gaussian approximation introduced by Mackay 1992 [4], or with numerical integration involved by Neal 1993 [14] such as the Hybrid Monte Carlo algorithm described in the following section.

3.2   Hybrid Monte Carlo Algorithm

Monte Carlo simulation methods are useful computational algorithm, it can compute the integration relying on random sampling from the integral field. Metropolis algorithm is a basic MCMC algorithm, it has been used with great success in many applications fields.

In Bayesian neural network learning, Metropolis algorithm was used to compute the integration over the high dimensional weight vector. Although Metropolis algorithm can work better than the simple Monte Carlo method, it cannot good deal with the random walk problem which always occurs in sampling progress. The gradient information can help us to deal with random walk problem, the gradients directs us to choose search directions which favor regions with high posterior probability. In order to use gradient information in random sampling, Duane [15] and Neal [7] presented Hybrid Monte Carlo algorithm which is a combinatory algorithm. In HMC algorithm, the candidate states of the Markov chain were considered as the variables of the Hamiltonian dynamical system, hence, the candidate state is sampled following the Hamiltonian equations.

A Hamiltonian system is a system of 2n ordinary differential equations called Hamiltonian equations of the form

$$\dot{q}_i = \frac{\partial}{\partial p_i} H(t, q, p) \quad \dot{p}_i = \frac{\partial}{\partial q_i} H(t, q, p) \quad i = 1, \dots, n \tag{9}$$

Where $H(t, q, p)$ called the Hamiltonian function, is a smoothreal-valued function defined for an open set in $\mathbb{R}^1 \times \mathbb{R}^n \times \mathbb{R}^n$.

The vectors $q = (q_1, \dots, q_n)$ and $p = (p_1, \dots, p_n)$ are traditionally called the position and momentum vectors, respectively, and t is called time. In the special case when H is independent of t, the differential equations (9) are autonomous, and the Hamiltonian system is called conservative, and the set of 2n dimension vectors (q, p) is called the phase space. The Hamiltonian H is usually considered as the total energy of a physical system, it is the sum of its potential energy and kinetic energy, then

$$H(q, p) = E(q) + K(p) \tag{10}$$

The quadratic kinetic energy function $K(p) = \frac{1}{2} p^T M^{-1} p$ corresponds to the negative log-density of a zero-mean multivariate Gaussian distribution with the covariance M. Here, M is known as the mass matrix.

In practical application, it is difficult to get the analytical solution of the Hamiltonian equations (9), usually we only get the approximate numerical solution. Leapfrog algorithm was popular used to solve Hamilton equations, it can update discrete–time approximation $\hat{q}$ and $\hat{p}$ to the position and momentum variables using following equations:

$$\begin{cases} p_* = p_{l-1} \frac{\varepsilon}{2} \frac{\partial E}{\partial q}(q_{l-1}) \\ q_l = q_{l-1} + \varepsilon M^{-1} p_* \\ p_l = p_* - \frac{\varepsilon}{2} \frac{\partial E}{\partial q}(q_l) \end{cases} \tag{11}$$

Suppose the configuration at the nth iteration of HMC is (q_n, p_n), then the next state (q_{n+1}, p_{n+1}) is obtained as follows:

- Generate a new momentum vector p from the Gaussian distribution $\pi(p) \propto \exp(-K(p))$
- Run the leapfrog algorithm for L steps to reach a new configuration in the phase space (q^*, p^*)
- Let $(q_{n+1}, p_{n+1}) = (q^*, p^*)$ with probability $u < \min(1, \exp(H(q, p) - H(q^*, p^*)))$ and let $(q_{n+1}, p_{n+1}) = (q_n, p_n)$ otherwise, were u is generated from the uniform distribution in [0, 1].

In learning process, the Bayesian neural network is considered as a special Hamiltonian dynamical system, and the weights vector as the system position variable, during the learning process, the Hamilton function is defined as

$$H(w, p) = E(w) + K(p) \tag{12}$$

4   Feature Extraction by Gray-Level Co-occurrence Matrix (GLCM)

The co-occurrence matrix or Spatial Gray Level Dependence Method is a texture analysis technique that has been frequently used in 2D image segmentation and identification [16–18]. It is used to extract the texture in an image by doing the transition of gray level between two pixels. Specific applications to medical images can be found in [17, 19].

The GLCM gives a joint distribution of gray level pairs of neighboring pixels within an image [18]. The co-occurrence matrix of the region of interest (ROI), Fig. 1, is useful in classification of types of breast tissues by extracting descriptors from the matrix. For the computation of GLCM, first a spatial relationship is established between two pixels, one is the reference pixel, and the other is a neighbor pixel. This process forms the GLCM containing different combination of pixel gray values in an image. GLCM are constructed by observing pairs of image cells distance d from each other and incrementing the matrix position corresponding to the grey level of both cells. This allows us to derive four matrices for each given distance and four different directions.

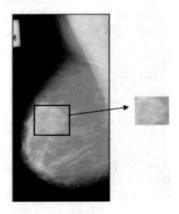

Fig. 1.   Illustration of ROI example

Based on above co-occurrence matrix, many different textures are produced. Therefore, in this paper we used just 5 of the 13 measures proposed by Haralick et al. [18] to perform pattern recognition based on co-occurrence matrix P. The measures used in this work are: Contrast, Homogeneity, Inverse Difference Moment, Entropy and Energy.

For the development and evaluation of the proposed system, we used the mini MIAS [20] database. This database contains left and right breast images for a total of 161 patients with ages ranging from 50 to 65. The spatial resolution of the image is 50 μm × 50 μm. All images are digitized at a resolution of 1024 × 1024 pixels and at 8-bit grey scale level. All images also include the locations of any abnormalities that may be present. The existing data in the collection consists of the location of the abnormality (like the center of a circle surrounding the tumor), its radius, breast position (left or right), type of breast tissues (fatty, fatty-glandular and dense) and tumor type if it exists (benign or malignant).

The (ROI) were manually extracted from each image based on the information provided by the MIAS database. From this database, we selected 181 normal tissues samples from each group and 119 abnormal (68 benign and 51 malignant mammograms). To each mammogram, a ROI was manually selected, containing the lesion, in the case of the benign and malignant mammograms. For the normal mammograms, the ROI was randomly selected. Only the pectoral muscle was not considered as a possible ROI, although tissue and fatty tissue were. If the tissues had different sizes, it was rescaled each ROI. Therefore, they were resized to 50 × 50 pixels see [10].

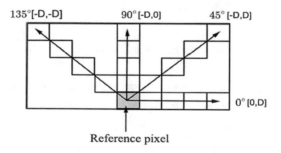

Fig. 2. Directions used for co-occurrence matrix

To compute the co-occurrence matrix, we used four directions 0, 45, 90 and 135 such as in Fig. 2, For each direction we compute the measures for the 1-pixel distance. We used the five measures defined in the previous Section. This method had the number of gray levels equal to 8, 16, 32, 64, 128 and 256. Using these parameters, the co-occurrence matrix generated 120 measures for each ROI (4 directions × 1 distance × 5 measures × 6 number of quantization levels) [10].

To make feasible the computation we need to select from all the obtained measures which were the minimum set that has the power to discriminate benign from malignant tissues. To do this, we used the method described in [10], Thus, the numbers of variables decreased to 8.

5 Bayesian Neural Network for Image Classification

In this section, we report results on using Bayesian MLPs for classifying the data obtained in the previous section, for this we generated a MLP with 8 inputs, one hidden layer with a fixed number m of hidden neurons and three output neurons. Several values for m = 4, 5..., 16 were used in the experiments, and we report just the best one, that was obtained with m = 10. The nonlinear activation function used for the hidden units was the logistic sigmoid, which produces an output between 0 and 1. For the output units, we used the SoftMax function.

Assuming a Laplacian prior distribution on the parameters which employs an L_1 norm penalty on the parameters see (Vidaurre, Bielza, and Larrañaga, [21]) for a review, This is equivalent to the LASSO regularization (Tibshirani) [22], The LASSO's implicit prior naturally tends to encourage large values or zero values more than ridge regression's implicit Gaussian prior [23], within this context we can guaranties more generalization.

The prior over network weights and bias takes the following form:

$$p(w/\alpha) = \left(\frac{\alpha}{2}\right)^W \exp(-\alpha \sum_{i=1}^{W} |w_i|) \tag{13}$$

A convenient form for the hyper-prior $p(\alpha)$ is vague Gamma distribution with mean $\mu = 0.1$ and shape parameter $a = 0.01$

$$p(\alpha) \sim Gamma(\mu, a) \propto \alpha^{\frac{a}{2}-1} \exp(-\alpha \frac{a}{2\mu}) \tag{14}$$

The penalty term $E_w = \sum_{i=1}^{W} |w_i|$ involves absolute values and is not differentiable at the origin to overcame this drawback, we propose to modify the usual L_1 regularization term by smoothing it at the origin using this function

$$g(x) = \begin{cases} |x| & if\ |x| \geq c \\ \frac{-1}{8c^3}x^4 + \frac{3}{4c}x^2 + \frac{3}{8} & if\ |x| < c \end{cases} \tag{15}$$

Where c is a small positive constant, $(c = 0.005)$ then we have

$$g'(x) = \begin{cases} -1 & if\ x \leq -c \\ \frac{-1}{2c^3}x^3 + \frac{3}{2c}x & if\ -c < x < c \\ 1 & if\ x > -c \end{cases} \tag{16}$$

Now we can express the function energy used to implement the HMC as follow
$H(w, p) = -\ln(p(D/w)) - \ln p(w/\alpha) + K(p)$

$$= -\ln\left(\prod_{i=1}^{n} \prod_{k=1}^{c} f_k(x_i, w)^{y_i^k}\right) - \left(\frac{\alpha}{2}\right)^W \exp\left(-\alpha \sum_{i=1}^{W} g(w_i)\right) + \frac{1}{2}p^T M^{-1}p \tag{17}$$

In our experiments, we have used Flexible Bayesian Modeling (FBM) software [7], which implements our proposed approach for the Bayesian learning to neural network described in (Neal, 1996) [7]. The configuration of the machine used are Intel (R) Core (TM) i3-2348M CPU @ 2.3 GHz (CPUs) and 4 Go of RAM.

The experiment was configured with 250 training samples and 50 samples for tests. For the training, we obtained an accuracy of 100.0%. Table 1 shows the results obtained with the proposed method for the tests. Based on the Table 1, we can see that the method obtained a mean success rate of 88% on discriminating malignant from benign and normal tissues.

Table 1. Detailed accuracy analysis in the breast tissue characterization

Tissue	Benign	Malignant	Normal
Total	12	13	25
Correct	10	12	22
Correct (%)	83,33	92,30	88

The problem of classification of mammographic images exists in the literature, but with different proposed models and different training methodology. We describe some of them and we compare it with our proposed method:

- **Leonardo de O. Martins *et al.*** propose to classify mammographic image using Bayesian neural network learned by MCMC and Gibbs sampling method [10]
- **OMLP:** Our previous work where we deal with optimizing weights number and hidden layers in the network [24]
- **BP:** This method allows training a fully connected MLP with Back-propagation algorithm [24]
- **EF:** The Evidence Framework for learning neural network proposed by Mackay [4]
- **HMC:** The method proposed by Neal [7], It is training neural network with Gaussian prior by HMC algorithm, see Table 2.

In most of the simulations, the best classification rate was obtained by the proposed method. We notice that the EF method is expansive in term of execution time because it based on the calculus of hessian matrix and his eigenvalues. The Bayesian neural network used in L.O. Martins and all does not take advantages of the information provided by the gradient contrary to the proposed approach for learning NN such as HMC.

Table 2. Results comparison

Method	Accuracy (%)	CPU time (s)
Proposed method	**88**	37,36
EF	87,23	50,12
HMC	87,40	38,45
OMLP	85,97	35,23
L. O. Martins *et al.*	86,84	–

The use of Laplace prior with function approximate to absolute value gives better pruning and removes the oscillation of the gradient which allows to reduce the time and to gain in term of generalization. For the MIAS database we can conclude that our proposed method for classification is the good one in term of accuracy and execution time.

6   Conclusion

In this paper, we investigate a usual HMC method for learning Bayesian neural network, Here, employing a function approximate to absolute value which appears in Laplace prior allows to solve the problem of non-differentiability thus to improve the speed of convergence which proves that Laplace prior lowers prediction errors below those of the Gaussian. The classification rate obtained for the classification of mammographic images is encouraging, we wish to test our approach on other problems of regression and classification.

References

1. Bovis, K., Singh, S., Fieldsend, J., Pinder, C.: Identification of masses in digital mammograms with MLP and RBF nets. In: Proceedings of the IEEE-INNS-ENNS International Joint Conference on Neural Networks, pp. 342–347 (2000)
2. Cheng, H.D., Cai, X.P., Chen, X.W., Hu, L.M., Lou, X.L.: Computer-aided detection and classification of microcalcifications in mammograms: a survey. Pattern Recogn. **36**, 2967–2991 (2003)
3. Basu, J.K., Bhattacharyya, D., Kim, T.-H.: Use of artificial neural network in pattern recognition. Int. J. Softw. Eng. Appl. **4**(2), 23–34 (2010)
4. MacKay, D.J.C.: The evidence framework applied to classification networks. Neural Comput. **4**(5), 698–714 (1992)
5. Bishop, C.M.: Neural Networks for Pattern Recognition. Oxford University Press, New York (1995)
6. Neal, R.M.: Bayesian training of back-propagation networks by the hybrid Monte Carlo method, Technical report CRG-TG-92-1, Department of Computer Science, University of Toronto (1992)
7. Neal, R.M.: Bayesian Learning for Neural Networks. Lecture Notes in Statistics, vol. 118. Springer, New York (1996)
8. Yu, S.-N., Huang, Y.-K.: Detection of microcalcifications in digital mammograms using combined model based and statistical textural features. Expert Syst. Appl. **37**(7), 5461–5469 (2010)
9. Wong, M.T., He, X., Nguyen, H., Yeh, W.-C.: Mass classification in digitized mammograms using texture features and artificial neural network. In: The 19th International Conference on Neural Information Processing (ICONIP), vol. 5, Doha, Qatar, pp. 151–158 (2012)
10. de Martins, L.O., dos Santos, A.M., Silva, A.C., Paiva, A.C.: Classification of normal, benign and malignant tissues using co-occurrence matrix and bayesian neural network in mammographic images. IEEE (2006)
11. Halkiotis, S., Botsis, T., Rangoussi, M.: Automatic detection of clustered microcalcifications in digital mammograms using mathematical morphology and neural networks. Sig. Process. **87**, 1559–1568 (2007)

12. Islam, M.J., Ahmadi, M., Sid-Ahmed, M.A.: An efficient automatic mass classification method in digitized mammograms using artificial neural network. Int. J. Artif. Intell. Appl. (IJAIA) **1**(3), 1–13 (2010)
13. Gelman, A., Carlin, J.B., Stern, H.S., Rubin, D.R.: Bayesian Data Analysis. Texts in Statistical Science. Chapman & Hall, London (1995)
14. Neal, R.M.: Probabilistic inference using Markov Chain Monte Carlo Methods, Technical report CRG-TR-93-1, Dept. of Computer Science, University of Toronto (1993)
15. Duane, S., Kennedy, A.D., Pendlton, B.J., Roweth, D.: Hybrid Monte Carlo. Phys. Lett. B **195**, 216–222 (1987)
16. Jain, A.K.: Fundamentals of Digital Image Processing. Prentice Hall, Englewood Cliffs (1989)
17. McNitt-Gray, M.F., Hart, E.M., Wyckoff, N., Sayre, J.W., Goldin, J.G., Aberle, D.R.: The effects of co-occurrence matrix based texture parameters on the classification of solitary pulmonary nodules imaged on computed tomography. Comput. Med. Imaging Graph. **23**, 339–348 (1999)
18. Haralick, R., Shanmugam, K., Dinstein, I.: Textural features for image classification. IEEE Trans. Syst. Man Cybern. SMC **3**(6), 610–621 (1973)
19. McNitt-Gray, M.F., Hart, E.M., Wyckoff, N., Sayre, J.W., Goldin, J.G., Aberle, D.R.: A pattern classification approach to characterizing solitary pulmonary nodules imaged on high resolution CT: preliminary results. Med. Phys. **26**(6), 880–888 (1999)
20. Suckling, H.J., Parker, J., Dance, D., Astley, S., Hutt, I., Boggis, C., et al.: The mammographic images analysis society digital mammogram database. Exerpta Med. **375–378**, 1994 (1069)
21. Vidaurre, D., Bielza, C., Larrañaga, P.: A survey of L1 regression. Int. Stat. Rev. **81**, 361–387 (2013)
22. Tibshirani, R.: Regression shrinkage and selection via the lasso. J. Roy. Stat. Soc. Ser. B. Stat. Methodol. **73**, 267–288 (1996)
23. Williams, Peter M.: Bayesian regularization and pruning using a Laplace prior. Neural Comput. **7**, 117–143 (1995)
24. Ramchoun, H., Janati Idrissi, M.A., Ghanou, Y., Ettaouil, M.: New modeling of multilayer perceptron architecture optimization with regularization: an application to pattern classification. IAENG Int. J. Comput. Sci. **44**(3), 261–269 (2017)

Multilayer Perceptron: NSGA II for a New Multi-objective Learning Method for Training and Model Complexity

Kaoutar Senhaji[✉], Hassan Ramchoun, and Mohamed Ettaouil

Modeling and Scientific Computing Laboratory, Faculty of Sciences and Technology,
University Sidi Mohammed Ben Abdellah, Fes, Morocco
kaoutar.senhaji@usmba.ac.ma, ramchounhassan@gmail.com,
mohamedettaouil@yahoo.fr

Abstract. The multi-layer perceptron has proved its efficiencies in several fields as pattern and voice recognition. Unfortunately, the classical training for MLP suffers from a poor generalization. In this respect, we have proposed a new multi-objective training model with constraints, satisfies two objectives. The first one is the learning objective: minimizing the perceptron error and the second is the complexity objective: optimizing number of weights and neurons. The proposed model will provide a balance between the multi-layer perceptron learning and the complexity to get a good generalization. Our model has been solved using an evolutionary approach called the Non-Dominated Sorting Genetic Algorithm (NSGA II). This approach has led to a good representation of the Pareto set for the MLP network, from which an improved generalization performance model is selected.

Keywords: Multi-objective training · Multilayer perceptron
Supervised learning · Non-linear optimization
Non-dominated Sorting Genetic Algorithm II (NSGA II) · Pareto front

1 Introduction

The multi-layer Perceptron is an efficient neural network capable to approximate any continuous function or classify any data, as long as he have enough neurons number and the adequate weight, What is knowing as optimization and learning of the MLP's. The learning of a neural network is mainly based on the search for the adequate weights value allowing to have a better result without really worrying about the best network topology, thus intervenes the optimization of the neural network. Several works treat each problem individually on the other using classical learning methods [1], optimization methods [2, 3], unfortunately learning method that use only training data error do not necessarily yield to good generalization models for noisy data since they does not

---

K. Senhaji—Ph'D student, laboratory of modeling and scientific computing, USMBA, Fez, Morocco.

© Springer International Publishing AG, part of Springer Nature 2019
J. Mizera-Pietraszko et al. (Eds.): RTIS 2017, AISC 756, pp. 154–167, 2019.
https://doi.org/10.1007/978-3-319-91337-7_15

control flexibility during the training process. These two subjects allows having a good generalisation performance, which can be obtained with techniques such as validation [4], pruning [5] or constructive algorithms [6]. It has been shown in Reference [7] that generalization is strongly related to the norm w of the weight vectors. In this fact, a few theory basing on a multi-objective optimization such as model MOBJ [8, 9] and LASSO (last absolute shrinkage and selection operator) [10] appears but proposes to solve the problem via mono-objective optimization. Our approach consists in modelling and solving the generalization problem of the MLP by a purely multi-objective methodology including two objectives: to minimize the general error of the perceptron and the sum of the absolute values of the weights under constraints. In this case, the solution of the problem is a set of undominated solutions by the Pareto concept [11], what is the set name from, "PARETO FRONT". We propose to solve the model by the NSGA II (Non-dominated Sorting Genetic Algorithm II) [12] one of the most efficient multi-objective genetic algorithms. This approach will make it possible to imply the generalization of the MLP's and reduce the topology.

The remaining part of this article is organized as follows. In Sect. 2, we introduced some related works on multi-objective and regularization neural networks. In Sect. 3 we describe the training of multilayer perceptron. In Sect. 4, we present the proposed model. Section 5 is about the multi-objective resolver, NSGA II, for MLP and before concluding, experimental results are given in Sect. 6.

2   Related Work

Generalization is the artificial neural network (ANN) ability to properly answer to unknown patterns. In this fact, for a good generalization solution, it is necessary to fit the ANN to problem complexity.

Recently many researchers have introduced the multi-objective evolutionary strategy; Inspired from neural network regularization, the training error and the sum of the absolute weights were minimized using an Epsilon-constraint-based multi-objective optimization method.

Many approaches in the literature take into account the generalization concept of MLP. This section describes works that are more or less similar to our proposed approach. The first proposed model in this context was by Liu and Kadirkamanathan [13]. Takahashi et al. [8] presents a new learning scheme for improving generalization of multilayer perceptron's within multi-objective optimization approach to balance between the error of the training data and the norm of network weight vectors to avoid over-fitting. Takahashi, Saldanha [9] introduces a new scheme for training MLPs, which employs a relaxation method for multi-objective optimization. The algorithm works by obtaining a reduced set of solutions, from which the one with the best generalization is selected. Costa et al. [14] gives a new sliding mode control algorithm that is able to guide the trajectory of a multi-layer perceptron within the plane formed by the two objective functions: training set error and norm of the weight vectors. Costa et al. [15] proposed an approach that explicitly considers the two objectives of minimizing the squared error and the norm of the weight vectors. The learning task is carried on by minimizing both objectives simultaneously, using vector optimization methods. This leads to a set of solutions that is called the Pareto optimal set, from which the best network for modeling the data is selected. Costa et al. [16] Improve generalization of MLPs with sliding mode control and the Levenberg–Marquardt algorithm. In addition, our previous works [17, 18].

3   Multilayer Perceptron and Training

A Multilayer Perceptron is a variant of the original Perceptron model proposed by Rosenblatt in the 1950 [19]. It has one or more hidden layers between its input and output layers, the neurons are organized in layers, the connections are always directed from lower layers to upper layers, the neurons in the same layer are not interconnected.

The choice of layers number and neurons in each layers and connections called architecture problem, the neurons number in the input layer equal to the number of measurement for the pattern problem and the neurons number in the output layer equal to the number of class.

The Learning for the MLP is the process to adapt the connections weights in order to obtain a minimal difference between the network output and the desired output, for this raison in the literature some algorithms are used such as Ant colony [20], but the most used called Back-propagation witch based on descent gradient techniques [21].

Assuming that we used an input layer with n_0 neurons $X = (x_0, x_1, \ldots, x_{n_0})$ and a sigmoid activation function $f(x)$ where:

$$f(x) = \frac{1}{1 + e^{-x}} \tag{1}$$

To obtain the network output we need to compute the output of each unit in each layer.

Now consider a set of hidden layers $(h_1, h_2, \ldots, h_N)$, assuming that n_i are the neurons number in each hidden layer h_i.

For the output of first hidden layer

$$h_1^j = f\left(\sum_{k=1}^{n_0} w_{k,j}^0 x_k\right) \qquad j = 1, \ldots, n_1 \tag{2}$$

The outputs h_i^j of neurons in the hidden layers are calculated as follows:

$$h_i^j = f\left(\sum_{k=1}^{n_{i-1}} w_{k,j}^{i-1} h_{i-1}^k\right) \qquad i = 2, \ldots, N \ and \ j = 1, \ldots, n_i \tag{3}$$

Where $w_{k,j}^i$ is the weight between the neuron k in the hidden layer i and the neuron j in the hidden layer $i + 1$, n_i is the number of the neurons in the ith hidden layer, The output of the ith layers can be formulated by:

$$h_i = (h_i^1, h_i^2, \ldots, h_i^{n_i}) \tag{4}$$

The network outputs are computed by

$$y_i = f\left(\sum_{k=1}^{n_N} w_{k,j}^N h_N^k\right) \tag{5}$$

$$Y = \left(y_1, \ldots, y_j, \ldots, y_{N+1}\right) = F(W, X) \tag{6}$$

Where $w_{k,j}^N$ is the weight between the neuron k of the Nth hidden layer and the neuron j of the output layer, n_N is the number of the neurons in the Nth hidden layer, Y is the vector of output layer, F is the transfer function and W is the weights matrix, it's defined as follows:

$$W = \left[W^0, \ldots, W^j, \ldots, W^N\right]$$

$$W^i = \left(w_{j,k}^i\right) \begin{array}{l} 0 \leq i \leq N \\ 1 \leq j \leq n_{i+1} \\ 1 \leq k \leq n_i \end{array} \text{ where } w_{j,k}^i \in \mathbb{R} \tag{7}$$

Where X is the input of neural network and f is the activation function and W^i is the matrix of weights between the ith hidden layer and the $(i+1)^{th}$ hidden layer for $i = 1, \ldots, N-1$, W^0 is the matrix of weights between the input layer and the first hidden layer, and W^N is the matrix of weights between the Nth hidden layer and the output layer.

4 Proposed Model

Choosing a suitable topology for the MLP neural network is a difficult problem as reasons that an unsuitable topology increases the training time or even causes non-convergence and that it usually decreases the MLP generalization capability [22], another important criterion in the MLP neural network efficiency. Therefore, we propose to find the tradeoffs between the network size and generalization performance by a multi-objective optimization model [11] under constrains.

The proposed model aims to satisfy two objective, the learning and complexity one. The constraints aims to eliminate any unfunctional neuron in the MLP architecture, what produces a large weight elimination. In this case, the nodes pruning does a good job. While weights pruning can give better results at cost of more complex ANN structure and higher computational time. Combining the both approaches will end up with great pruned neural network.

The network studied in this paper consists of a single hidden layer, in this fact we use the following formula:

4.1 Notation

N: The number of neurons in the hidden layer.
n_0: The number of neurons in input layer.

n: The number of neurons in output layer.

X: Input data of the neural network.

Y: Calculated output of the neural network.

d: Desired output.

f: Activation function.

h: The hidden layer.

w: Network weights.

C_{ij}^k: Binary variable as:

$$C_{ij}^k = \begin{cases} 1 & \text{if the connection between the neuron } i \text{ and } j \text{ is kaped} \\ 0 & \text{if not} \end{cases}$$

k: is the connection index between two successive layers.

U_i: Binary variable as:

$$U_i = \begin{cases} 1 & \text{if the neuron } i \text{ is kaped in the hidden layer} \\ 0 & \text{if not} \end{cases}$$

4.2 Output of the Hidden Layer

As we have a single hidden layer in the neural network, he is directly connected to the input layer, so the suppression of the hidden layer neurons and the weights connecting the input layer to the hidden layer is represented in output of each neuron by:

$$h = \begin{pmatrix} h_1 \\ \vdots \\ h_i \\ \vdots \\ h_N \end{pmatrix} = \begin{pmatrix} U_1\left(1 - \prod_{j=1}^{n_0}\left(1 - C_{j1}^0\right)\right)f\left(\sum_{j=1}^{n_0} C_{j1}^0 w_{j1}^0 x_j\right) \\ \vdots \\ U_i\left(1 - \prod_{j=1}^{n_0}\left(1 - C_{ji}^0\right)\right)f\left(\sum_{j=1}^{n_0} C_{ji}^0 w_{ji}^0 x_j\right) \\ \vdots \\ U_N\left(1 - \prod_{j=1}^{n_0}\left(1 - C_{jN}^0\right)\right)f\left(\sum_{j=1}^{n_0} C_{jN}^0 w_{jN}^0 x_j\right) \end{pmatrix} \qquad (8)$$

4.3 Output of the Neural Network

The output network is calculated using the weights connecting the hidden layer to the output layer or at this point, we remove some connection, so the network output is calculated by:

$$Y = \begin{pmatrix} Y_1 \\ \vdots \\ Y_i \\ \vdots \\ Y_n \end{pmatrix} = \begin{pmatrix} \left(1 - \prod_{j=1}^{N}\left(1 - C_{j1}^1\right)\right)f\left(\sum_{j=1}^{N} C_{j1}^1 w_{j1}^1 h_j\right) \\ \vdots \\ \left(1 - \prod_{j=1}^{N}\left(1 - C_{ji}^1\right)\right)f\left(\sum_{j=1}^{N} C_{ji}^1 w_{ji}^1 h_j\right) \\ \vdots \\ \left(1 - \prod_{j=1}^{N}\left(1 - C_{jn}^1\right)\right)f\left(\sum_{j=1}^{N} C_{jn}^1 w_{jn}^1 h_j\right) \end{pmatrix} \qquad (9)$$

4.4   Objectives Functions

The model is a multi-objective model in order to satisfy two objectives:

The first one is the global error of the multi-layer perceptron training defined as the error between calculated output and desired output, presented in the following form:

$$\left\| F(U, C, X, w)^2 \right\| \tag{10}$$

The second objective aims to reduce the network weights number in order to control the weights variations during the training for a better generalization.

$$\sum_{i=1}^{n_0} \sum_{j=1}^{N} \left| C_{ij}^0 w_{ij}^0 \right| + \sum_{i=1}^{N} \sum_{j=1}^{n} \left| C_{ij}^1 w_{ij}^1 \right| \tag{11}$$

4.5   Constraints

The first constraint guarantees the existence of hidden layer by the existence of; at least, one neuron in it, the constraints is expressed by:

$$\sum_{i=1}^{N} U_i \geq 1 \tag{12}$$

Since, we remove connections and neurons of the network. We must ashore the communication between them. A neuron, which has no connections, should be removed from the network architecture. The same, as we delete a neuron, we must remove its connections too. The constraint, then, is expressed by:

- The communication between the input layer and hidden layer:

$$U_j \cdot \prod_{i=1}^{n_0} \left(1 - C_{ij}^0 \right) = 0 \qquad \forall j = 1 \dots N \tag{13}$$

- The communication between the hidden layer and the output layer:

$$U_i \cdot \prod_{j=1}^{n} \left(1 - C_{ij}^1 \right) = 0 \qquad \forall i = 1 \dots N \tag{14}$$

4.6   Obtained Model

The multi-layer perceptron optimization problem can be formulated as:

$$
\begin{cases}
\min \|F(U,C,X,w)\|^2 \\
\min \sum_{i=1}^{n_o} \sum_{j=1}^{N} \left| C_{ij}^0 w_{ij}^0 \right| + \sum_{i=1}^{N} \sum_{j=1}^{n} \left| C_{ij}^1 w_{ij}^1 \right| \\
\text{Subject to} \\
\sum_{i=1}^{N} U_i \geq 1 \\
U_j \cdot \prod_{i=1}^{n_o} \left(1 - C_{ij}^0 \right) = 0 \quad \forall j = 1 \dots N \\
U_i \cdot \prod_{j=1}^{n} \left(1 - C_{ij}^1 \right) = 0 \quad \forall i = 1 \dots N \\
w = \left(w_{ij}^k \right)_{\substack{k=\{1,2\} \\ 1 \leq i \leq N \\ 1 \leq j \leq n}} \quad \text{where } w_{ij}^k \in \mathbb{R} \\
C = \left(C_{ij}^k \right)_{\substack{k=\{1,2\} \\ 1 \leq i \leq N \\ 1 \leq j \leq n}} \quad \text{where } C_{ij}^k \in \{0,1\} \\
U = \left(U_i \right)_{1 \leq i \leq N} \quad \text{where } U_i \in \{0,1\}
\end{cases}
\tag{15}
$$

Many methods was proposed to solve a multiobjective optimization model [11], in this paper we proposed to solve the multiobjective learning method for training and model complexity by an evolutionary method called non dominated sorting genetic algorithm II (NSGA II). The next section discuses in details the mentioned method.

5   NSGA II for MLP Multi-objective Training Model

5.1   Non Dominated Sorting Genetic Algorithm II (NSGA II)

The genetic algorithm (GA) is a metaheuristic approach based in evolutionary population, proposed by Holland [23]. During the ten last years the genetic algorithm was developed and adapted to resolve multi-objective optimization problem, so many variation appeared [24, 25]. One of the variation of GA multi-objective is NSGA [26]. It is a very effective algorithm but has been generally critiqued for its computational complexity, lack of elitism and for choosing the optimal parameter value for sharing parameter σshare. A modified version, NSGAII [12] was developed to avoid the NSGA shortcoming. The NSGA II was classified as one of the most efficient multi-objective evolutionary algorithm [27, 28], which has a better sorting algorithm, incorporates elitism and no sharing parameter needs to be chosen a priori. NSGA-II is discussed in detail in this section.

5.1.1   Algorithm
The general algorithm steps are presented as follows:

Step 1:   Create a random parent population P_t of size Z.

Step 2:   Sort the random parent population based on non-domination concept.

Step 3:   For each non-dominated solution assign a rank equal to its non-domination level, 1 is the best level, the second one is the next best level and so on.

Step 4:   Create on offspring population Q_t using selection and production operators as following:

Selection.   The comparison step applied to choose a parents solutions is defined by the rank value, if n solutions have the same rank value, the crowded comparison operator based on the crowding distance is applied, the solution with the best crowding distance Is kept;

Crossover.   Choose the parents solutions for crossover;

Mutation.   Choose the parents solutions for mutation;

Step 5:   Create the mating pool R_t by combining the parent population P_t and the offspring population Q_t.

Step 6:   Sort the combined population R_t according to the fast non-dominated sorting procedure to identify all non-dominated fronts $(F_{r1}, F_{r2}, \cdots, F_{rm})$.

Step 7:   Generate the new parent population P_{t+1} of size Z by choosing non-dominated solutions, starting from the first ranked non- dominated front F_{r1}. When the population size z is exceed, reject some of the lower ranked non-dominated solution.

Step 8:   Repeat from step 3 until the stopping criteria is reached

The NSGA II considers all non-dominated solutions of the combined populations as the next generation member. If the population size isn't reached, the next front is caped, until the population is completed.

To select the candidates of the next generation through the crowding distance criteria, the mean advantage of this criterion is maintaining the solutions diversity in the population. This procedure prevent premature convergence. The crowding distance is detailed as follow:

5.1.2   Crowding Distance
The crowding distance serves as an estimate of the perimeter of the cuboid formed by using the neighbors as the vertices; an estimate of the density of solutions surrounding a particular solution in the population.

The algorithm used to calculate the crowding distance of each point in the set F_r is given by:

Step 1:   For each solution in the set, F_r assign 0 to the crowding distance corresponding;

Step 2: for each objective function $f_m, m = 1, 2, \ldots, M$, sort the set in worse order of f_m;
Step 3: assign a large distance to the boundary solutions $d^r(1) = d^r(l) = \infty$ (1 is the first solution and l is the last one in the front F_r. On the other hand, for all other solutions $i = 2, \ldots, l - 1$ assign:

$$d^r(i) = d^r(i) + \frac{\left(f_m^r(i+1) - f_m^r(i-1)\right)}{f_m^{max} - f_m^{min}} \tag{16}$$

$f_m^r(i+1)$ is the mth objective function value of the $(i+1)$ solution in the set F_r;
$f_m^r(i-1)$ is the mth objective function value of the $(i-1)$ solution in the set F_r;
f_m^{max} is the maximum value of the mth objective;
f_m^{min} is the minimum value of the mth objective;

5.2 NSGA II Adaptation to MLP

To resolve the multi-objective training model for the multi-layer perceptron we use the NSGA II. As result, we well get a solutions set, each of which satisfies the objectives at an acceptable level without being dominated by any other solution. In follow we will adequate the NSGA II to the proposed model.

Solution Coding. The coding step is a very important one; each individual in the population is a presentation of a possible solution of the problem, which can be represented as an integer, continuous, binary or mixed chromosome. For our problem a single individual, contain tree chromosomes presenting the tree problem variables: w_{ij}^k, C_{ij}^k and U_i; so the NSGA II individual is a structure in Fig. 1.

$w_{11}^0\ w_{12}^0\ \cdots\cdots\ w_{1N-1}^0\ w_{1N}^0$	$w_{11}^1\ w_{12}^1\ \cdots\cdots\ w_{1n-1}^1\ w_{1n}^1$
$w_{21}^0\ w_{22}^0\ \cdots\cdots\ w_{2N-1}^0 w_{2N}^0$	$w_{21}^1\ w_{22}^1\ \cdots\cdots\ w_{2n-1}^1 w_{2n}^1$
$\vdots\ \ddots\ \vdots$	$\vdots\ \ddots\ \vdots$
$w_{n_01}^0\ w_{n_02}^0\ \cdots\cdots\ w_{n_0N-1}^0\ w_{n_0N}^0$	$w_{N1}^1\ w_{N2}^1\ \cdots\cdots\ w_{Nn-1}^1\ w_{Nn}^1$
$C_{11}^0\ C_{12}^0\ \cdots\cdots\ C_{1N-1}^0\ C_{1N}^0$	$C_{11}^1\ C_{12}^1\ \cdots\cdots\ C_{1n-1}^1\ C_{1n}^1$
$C_{21}^0\ C_{22}^0\ \cdots\cdots\ C_{2N-1}^0 C_{2N}^0$	$C_{21}^1\ C_{22}^1\ \cdots\cdots\ C_{2n-1}^1 C_{2n}^1$
$\vdots\ \ddots\ \vdots$	$\vdots\ \ddots\ \vdots$
$C_{n_01}^0\ C_{n_02}^0\ \cdots\cdots\ C_{n_0N-1}^0\ C_{n_0N}^0$	$C_{N1}^1\ C_{N2}^1\ \cdots\cdots\ C_{Nn-1}^1\ C_{Nn}^1$
$U_1, U_2, \cdots\cdots, U_i, \cdots\cdots, U_{N-1}, U_N$	

Fig. 1 The NSGA II individual for the multilayer perceptron multi-objective learning model

Initial Population. The genetic algorithm population is the pool of the initial solution with what the algorithm will start. In most cases, it is chosen randomly, it is the same for our population. The vector U_i and the matrix C_{ij}^k are a binary variable so we generate it 0 or 1. However, for w_{ij}^k is a continuous variable generate between $[-0, 5, 0, 5]$.

Fitness Assignment. To evaluate the solutions, the algorithm classify each solution in the adequate front using the ranking philosophy. The algorithm assign a rank = 1 if the individual isn't dominate by any solution in the population, in this fact the individual belongs to the first front, the optimal front and removed from the unclassified solution. The process is repeated with an increment, each time, of the rank until all individual is classified.

Parent Selection. Is the process of selecting parents [29] (mate and recombine) to create offspring for the next generation. This step drive individuals to better and fitter solutions. It exist many different method for the parent selection, for our case, we select the parent using the Tournament Selection [30]. The method consists to select randomly a set of k individuals, then, ranked according to their relative fitness (rank and crowding distance) and the fittest individual is selected for reproduction. The whole process is repeated n times until the selection proportion is reached.

Operator Production. We applied the crossover operation in two points chosen randomly. On the other hand, the mutation point is randomly chosen.

Correction. Every children generated is not necessarily a possible solution of the problem, Therefore a verification-correction is required.

Replacement. We used a steady-state population replacement strategy. With this strategy, each new child is placed in the population as soon as it is generated and corrected. The best solutions are chosen to be caped in the population.

6 Implementations and Numerical Results

6.1 Description of Used Data Set

In this section, a number of experiments were conducted with standard benchmark data sets of the University of California Irvine (UCI) machine learning repository [31] to test the performance of our methodology. Five classification problems are used: Fisher's iris data set, Seed data set, Breast cancer Wisconsin, Wine data set and Thyroid data.

The Table 1 chow the summary of the used data sets along with the number of examples, number of attributes and class.

Table 1. Characteristics of used data set

Database	Examples	Attributes	Class
Iris	150	4	3
Wine	178	13	3
Cancer	699	9	2
Seed	210	7	3
Thyroid	7200	21	3

6.2 Numerical Results

To illustrate the advantages of the proposed approach, we tested it on databases described above. Since we adapt the NSGA II algorithm to solve the obtained model. We have use a MLP with a single hidden layer containing 15 neurons. The adaptation of the NSGA II algorithm requires, also, a setting of the algorithm parameter, the following table (Table 2) shows the proposed setting:

Table 2. NSGA II parameter setting

Parameter	Setting
Population size	100
Selection rate	80%
Crossover rate	70%
Mutation rate	30%
Stopping criteria	100*1000 iteration

At first, we obtain the right shape of the Pareto front, but it isn't enough, for our case, the users need only a single solution. It is, from the algorithm point of view, all the solution in the Pareto front are equally important. For this, many approaches was proposed [31]. We have chosen the must adequate solution between the obtained optimal solutions by an approach basing on the crowding distance. The solution obtained is illustrated in Table 3.

Table 3. The obtained solution

Method	Data	MLP Error	$\sum |C_i w_i|$	$\sum |w_i|$	n_w	n_{neuron}	Class (%)
BP+LASSO	Iris	0,8	–	30,25	105	15	2,66
	Wine	0,38	–	51,29	240	15	3,37
	Cancer	0,91	–	42,38	165	15	2,28
	Seed	0,56	–	49,21	150	15	5,38
	Thyroid	0,74	–	69,33	360	15	3,53
P. Method	Iris	0.85	16,43	–	30	6	1,33
	Wine	0,39	34,56	–	80	8	2
	Cancer	0.94	10,12	–	6	8	2
	Seed	0,64	16,58	–	36	7	4,71
	Thyroid	0,76	22,19	–	208	10	2,87

The class presented in Table 3 is calculate by the following formula:

$$Class(\%) = \frac{Mal\,classified \times 100}{Card(Testingset)} \tag{18}$$

The P. Method in Table 3 represent the proposed model and BP+LASSO represent the Back propagation method with a regularized error. Well, in order to prove the efficiency of our approach, we compare it to BP+LASSO. The back propagation is applied to minimize the MLP error with a second term, aiming to regulate the current error, the error is presented as:

$$\left\| F(U, C, X, w)^2 \right\| + \beta \sum |w_i| \tag{19}$$

The β is a parameter fixed by testing several time with different value (0,05; 0,1; 0,15 ...), the best results was find at $\beta = 0, 1$. Which to presented results in Table 3.

From the Table 3 we can see that our approach obtain a better results in the regulation term, compared to the BP+LASSO, as we have add a decision variable. The number of the connections and neurons are also considerably reduces compared to the BP+LASSO Method. The Table 3, also chows a better calculate class rate for the proposed model. In this fact, we can say that the proposed model is able to provide a good generalization and reduce the MLP topology.

7   Conclusion

In this paper, a new multiobjective learning model is presented to improve the MLPs generalization by minimizing the training error and controlling the weights variation of the network during the learning process in order to prune the network efficiently. Furthermore, we have proposed two new constraints to ensure the communication between connection and neurons. The model was solved by adapting the NSGA II algorithm to the current model. The results show that the proposed method allow to reduce unnecessary nodes and connections, compared to algorithm that minimize only the regularized error, such as standard backpropagation with Lasso regularizer, our approach was able to reach higher generalization solutions for different Data problems, well as a reduced topology.

References

1. Rumelhart, D.E., Hinton, G.E., Williams, R.J.: Learning representations by back-propagation error. Nature **323**, 533–536 (1986)
2. Ramchoun, H., Janati Idrissi, M.A., Ghanou, Y., Ettaouil, M.: New modeling of multilayer perceptron architecture optimization with regularization: an application to pattern classification. IAENG Int. J. Comput. Sci. **44**(3), 261–269 (2017)
3. Abdelatif, E.S., Fidae, H., Mohamed, E.: Optimization of the organized KOHONEN map by a new model of preprocessing phase and application in clustering. J. Emerg. Technol. Web Intell. **6**(1), 80–85 (2014)
4. Arlot, S., Celisse, A.: A survey of cross-validation procedures for model selection. Stat. Surv. **4**, 40–79 (2010)
5. Reed, R.: Pruning algorithms-a survey. IEEE Trans. Neural Netw. **4**(5), 740–747 (1993)
6. Kwok, T.Y.: Constructive algorithms for structure learning in feedforward neural networks (Doctoral dissertation). (1996)
7. Bartlett, P.L.: For valid generalization the size of the weights is more important than the size of the network. In: Advances in Neural Information Processing Systems, pp. 134–140 (1997)

8. De Albuquerque Teixeira, R., Braga, A.P., Takahashi, R.H., Saldanha, R.R.: Improving generalization of MLPs with multi-objective optimization. Neurocomputing **35**(1), 189–194 (2000)
9. De Albuquerque Teixeira, R., Braga, A.P., Takahashi, R.H., Saldanha, R.R.: Recent advances in the MOBJ algorithm for training artificial neural networks. Int. J. Neural Syst. **11**(03), 265–270 (2001)
10. Costa, M.A., Braga, A.P.: Optimization of neural networks with multi-objective lasso algorithm. In: International Joint Conference on Neural Networks, IJCNN 2006, pp. 3312–3318. IEEE, July 2006
11. Chankong, V., Haimes, Y.Y.: Multiobjective Decision-Making: Theory and Methodology. Courier Dover Publications, Mineola (2008)
12. Deb, K., Agrawal, S., Pratap, A., Meyarivan, T.: A fast elitist non-dominated sorting genetic algorithm for multi-objective optimization: NSGA-II. In: International Conference on Parallel Problem Solving From Nature, pp. 849–858. Springer, Heidelberg, September 2000
13. Liu, G.P., Kadirkamanathan, V.: Learning with multi-objective criteria. In: 4th International Conference on Artificial Neural Networks, pp. 53–58 (1995)
14. Costa, M.A., Braga, A.P., Menezes, B.R., Teixeira, R.A., Parma, G.G.: Training neural networks with a multi-objective sliding mode control algorithm. Neurocomputing **51**, 467–473 (2003)
15. Braga, A., Takahashi, R., Costa, M., Teixeira, R.: Multi-objective algorithms for neural networks learning. In: Multi-objective Machine Learning, pp. 151–171. Springer (2006)
16. Costa, M.A., de Pádua Braga, A., de Menezes, B.R.: Improving generalization of MLPs with sliding mode control and the Levenberg–Marquardt algorithm. Neurocomputing **70**(7), 1342–1347 (2007)
17. Janati Idrissi, M A., Ramchoun, H., Ghanou, Y., Ettaouil, M.: Genetic algorithm for neural network architecture optimization. In: IEEE Proceeding of the 3rd International Conference of Logistics Operations Management 2016. IEEE, Morocco, 23–25 May 2016
18. Senhaji, K., Ettaouil, M.: Multi-criteria optimization of neural networks using multi-objective genetic algorithm. In: 2017 Intelligent Systems and Computer Vision (ISCV), 4-pp. IEEE, April 2017
19. Rosenblatt, F.: The Perceptron, A Theory of Statistical Separability in Cognitive Systems, Cornell Aeronautical Laboratory. Tr. No. VG-1196-6-1, January 1958
20. Ghanou, Y., Bencheikh, G.: Architecture optimization and training for the multilayer perceptron using ant system. Architecture **28**, 10 (2016)
21. Salomon, D.: Data Compression: The Complete Reference. Springer Science & Business Media, London (2004)
22. Sietsma, J., Dow, R.J.: Creating artificial neural networks that generalize. Neural Netw. **4**(1), 67–79 (1991)
23. Holland, J.H.: Adaptation in Natural and Artificial Systems. An Introductory Analysis with Application to Biology, Control, and Artificial Intelligence. University of Michigan Press, Ann Arbors (1975)
24. Coello, C.A.C., Lamont, G.B., Van Veldhuizen, D.A.: Evolutionary Algorithms for Solving Multi-objective Problems, vol. 5. Springer, New York (2007)
25. Jaimes, A.L., Coello, C.A.C.: An introduction to multi-objective evolutionary algorithms and some of their potential uses in biology. In: Applications of Computational Intelligence in Biology, pp. 79–102. Springer, Heidelberg (2008)
26. Srinivas, N., Deb, K.: Muiltiobjective optimization using nondominated sorting in genetic algorithms. Evol. Comput. **2**(3), 221–248 (1994)

27. Zitzler, E., Deb, K., Thiele, L.: Comparison of multiobjective evolutionary algorithms: Empirical results. Evol. Comput. **8**(2), 173–195 (2000)
28. Zitzler, E., Thiele, L.: Multiobjective optimization using evolutionary algorithms-a comparative case study. In: International Conference on Parallel Problem Solving from Nature, pp. 292–301. Springer, Heidelberg, September 1998
29. Jebari, K., Madiafi, M.: Selection methods for genetic algorithms. Int. J. Emerg. Sci. **3**(4), 333–344 (2013)
30. Hingee, K., Hutter, M.: Equivalence of probabilistic tournament and polynomial ranking selection. In: IEEE World Congress on Computational Intelligence. IEEE Congress on Evolutionary Computation, CEC 2008, pp. 564–571. IEEE, June 2008
31. Hwang, C.L., Masud, A.S.M.: Multiple Objective Decision Making-Methods and Applications: A State-of-the-Art Survey, vol. 164. Springer Science & Business Media, Heidelberg (2012)

A New Quasi-Cyclic Majority Logic Codes Constructed from Disjoint Difference Sets by Genetic Algorithm

Karim Rkizat[1], Said Nouh[2], Mohammed Lahmer[3], and Mostafa Belkasmi[1(✉)]

[1] ENSIAS, Mohammed V University in Rabat, Rabat, Morocco
{karim.rkizat,m.belkasmi}@um5s.net.ma
[2] Faculty of Sciences Ben M'sik, University Hassan II of Casablanca,
Casablanca, Morocco
said.nouh@univh2m.ma
[3] High School of Technologie, Moulay Ismail University, Meknes, Morocco
mohammed.lahmer@gmail.com

Abstract. In this paper, we deal with the construction of disjoint difference set as combinatorial optimisation problem, which allows us to construct Quasi-cyclic majority logic decoding (QC MLD) codes. We will propose an algorithm based on genetic algorithm to construct many of these designs. Our algorithm was able to construct many new Quasi-cyclic MLD codes based on the constructed Disjoint Difference Set.

Keywords: Quasi-cyclic codes · OSMLD · Genetic algorithm
Combinatorial design

1 Introduction

Self-Orthogonal Quasi-cyclic codes as presented by Townsend and Weldon [1] are one-step majority logic decodable (OSMLD) codes which enables fast and parallel decoding [2]. These codes are powerful because they can be decoded by the majority logic decoder elaborated by Reed [3] for decoding Reed-Muller codes [4], this decoder is very simple and has low complexity, and also the encoding method can be done by the simple shift register. But the difficulty is to construct this code family that could be decoded with this decoder.

The problem of designing this family of codes is equivalent to the problem of designing Disjoint Difference Sets (DDS) [1]. The construction of combinatorial designs like DDS has been and remains a very active research area in discrete mathematics and statistics. Many algorithms have been given to construct certain types of these designs, for example: balanced incomplete block designs (BIBDs) [5], partially balanced incomplete block designs with m associate classes(PBIBD) [6], Steiner triples systems [7]. However, there is no efficient algorithm for this purpose. Though designs can be constructed for a series of particular parameters, the general solution is unknown so far. These designs are known to exist for a limited number of parameter situations.

© Springer International Publishing AG, part of Springer Nature 2019
J. Mizera-Pietraszko et al. (Eds.): RTIS 2017, AISC 756, pp. 168–177, 2019.
https://doi.org/10.1007/978-3-319-91337-7_16

Several approaches have been proposed to construct DDS. Of these, the least explored and most interesting to us is the combinatorial optimization approach. In order to construct designs some optimization techniques have been applied: A simulated annealing algorithm has been used in the construction of block designs [8]. A Tabu search algorithm was used by Morales [9] to find difference families. In this paper the construction of DDS designs is formulated as a combinatorial optimization problem. For this a heuristic procedure based on the genetic algorithm technique is described, to construct various instances of DDS. The optimal solution is reached at 100% of the runs for a small design parameters. However for larger ones, our implementation must turn many time to produce an optimal solution. Genetic algorithm could find many new DDS, and others known that were obtained before by other methods. Section 2 gives the definition of DDS, and some of their properties. Section 3 provides the reader with a concise description of not only the Quasi-Cyclic OSMLD codes but also the majority logic decoding algorithm. In Sect. 4, a brief review of the basic principles of Genetic algorithm is given, then an implementation of Genetic algorithm for the optimization problem. Computational results are reported in Sect. 5, and the latter section consists of conclusions.

2   Cyclic Difference System

2.1   Cyclic Disjoint Difference Sets

A difference set of order S and modulo $m \geq S(S - 1) + 1$, as presented in [5], is defined as a collection of S integer specified from the set $\{0, 1, ..., m - 1\}$ such that no two of the $S(S - 1)$ ordered differences modulo m are identical. If $m = S(S - 1) + 1$, then for any non-zero integer $p < m$ there is exactly one pair of elements in the difference set such that their difference is congruent to p modulo m. Such a set is called a perfect difference set.

The order of the set of differences associated with a difference set of order S is $S(S - 1)$. If two sets of differences have no element in common they are said to be disjoint. Two difference sets are said to be disjoint if their sets of differences are disjoint.

2.2   Cyclic Disjoint Difference Families

Difference family [5] is a generalisation of difference sets. Let $B = \{b_1, ... , b_k\}$ be a subset of an additive group G. The list of differences from B is the multiset $\triangle B = \{b_i - b_j \mid i, j = 1, . . . , k; i \neq j\}$.

Let G be a group of order v. A collection $\{B_1, ..., B_t\}$ of k-subsets of G form a (v, k, λ) difference family (or difference system) if every nonidentity element of G occurs λ times in $\partial B_1 \cup ... \cup \partial B_t$. The sets B_t are base blocks. A difference family having at least one short block is partial.

Example 1

1. $B_1 = \{0, 1, 4\}$, $B_2 = \{0, 2, 8\}$, $B_3 = \{0, 5, 10\}$ form a partial $(15, 3, 1)$ difference family over Z_{15}.
2. $B_1 = \{0, 1, 3, 24\}$, $B_2 = \{0, 4, 9, 15\}$, $B_3 = \{0, 7, 17, 25\}$ form a $(37, 4, 1)$ difference family over Z_{37}.

3   Quasi-Cyclic OSMLD Codes

3.1   OSMLD Codes

Consider an (n, k) linear code C with parity-check matrix H. The row space of H is an (n, n-k) code, denoted by $C^{\perp}$, which is the dual code of C or the null space of C. For any vector v in C and any vector w in $C^{\perp}$, the inner product of v and w is zero [2]. Now let consider that a codeword vector in C is transmitted over a binary symmetric channel. Taking into consideration that $e(e_1, e_2,..., e_n)$ and $r(r_1, r_2,..., r_n)$ are the error vector and the received vector, respectively. Then r = v + e. The construction of the below linear sum of the received vector for any vector w in the dual code $C^{\perp}$:

$$A = \sum_{p=1}^{n} r_p w_p \tag{1}$$

Which is called a parity-check sum. Using the fact that $\langle w, v \rangle = 0$, the following relationship between the parity-check sum A and error digits in e is obtained:

$$A = \sum_{p=1}^{n} e_p w_p \tag{2}$$

Suppose that there exists J vectors in the dual code $C^{\perp}$, which have the following properties:

1. The jth component of each vector w_i is a 1.
2. For $i \neq j$ there is at most one vector whose i^{th} component is a 1.

These J vectors are said to be orthogonal on the jth digit position. They are called orthogonal vectors. Now, let us form J parity-check sums from these J orthogonal vectors, For each i in 1,.., J $A_i = \sum_{p \neq 1} e_p + e_j$ the error digit e_j is checked by all the check sums above. Because of the second property of the orthogonal vectors, any error digit other than e_j is checked by at most one check sum. These J check sums are said to be orthogonal on the error digit e_j. If all the error digits in the sum A_i are zero for $i \neq j$, the value of A_i is equal to e_j. Based on this fact, the parity-check sums orthogonal on e_i can be used to estimate e_i, or to decode the received digit r_i.

Their are many class of codes which are OSMLD codes, like Difference set cyclic (DSC)codes Reed-Muller codes and Geometry codes [10], and Quasi-cyclic codes constructed from Disjoint Difference Set which we present in the next section.

3.2   Quasi-Cyclic Codes

A code is said to be quasi-cyclic if every cyclic shift of a codeword by p positions results in another codeword [10]. Therefore, a QC codes are a generalization of cyclic codes with p = 1. A QC code (mn_0, mk_0) with a minimum distance d based on difference set can be specified with k_0 disjoints difference sets $\{D_1, D_2, ..., D_{k_0}\}$ such that $D_i(d_{i0}, d_{i1}, d_{i2}, ..., d_{i(S-1)})$ of order S, chosen from the set $\{0, 1, 2, ..., mk_0\}$ [1]. The parity check matrix H in the systematic form of such code is completely defined as follows:

$$H = [P_1 P_2 ... P_{k_0} I_{n-k}] \tag{3}$$

The circulant matrix P_i is deducted from the difference set D_i; the elements of D_i can specify the position in the matrix header P_i with one, while d_{ij} represents one in the position j, the others rows are obtained by a cyclic shift of the header. Where I represents the identity matrix.

The majority logic decoding algorithm for QC codes is the same as cyclic codes. However, there is a little bit difference between them. Hence, in cyclic codes each error digit e_i can be decoded by cyclically permuting the received word r, but in QC codes in systematic form, shift is done cyclically by one position of each (n-k) bits simultaneously.

Example 2: Let consider the QC code $C(6, 3, 3)$. This code is of the rate 1/2 and based on the Singer difference set DS$\{0,1\}$ of order 2.

The parity check matrix H in systematic form is [P I_3]

$$\begin{pmatrix} 1 & 1 & 0 & 1 & 0 & 0 \\ 0 & 1 & 1 & 0 & 1 & 0 \\ 1 & 0 & 1 & 0 & 0 & 1 \end{pmatrix}$$

The parity-check sum orthogonal on e_3 is obtained from the parity check matrix H:
$A_1 = e_2 + e_3 + e_5$
$A_2 = e_1 + e_3 + e_6$

4   Genetic Algorithm

4.1   Genetic Algorithm Technique

Genetic algorithms [11] are evolutionary meta-heuristic search algorithms inspired by the process of natural selection. They yield generally to high-quality solutions to optimization and search problems by their powerful operators such as selection, mutation and crossover. The first operator allows selecting the best individuals to insert in the intermediate generation. The second one allows creating some children by crossing some parents with a crossover probability pc. The third one allows muting some genes according to the mutation probability pm. In [12–14], some evolutionary algorithms are used to find best error correcting codes. In [15,16], genetic algorithms have permits to successfully decode some linear code.

4.2   The Proposed Algorithm

In the sequel of this paper, the number M represents the length of the Total Difference Set and N represents the length of the Sub Difference Set. In this paper, genetic algorithms will be used to find a set difference S of M/N subsets with the following characteristics:

1. For each subset X and each elements X[i] and X[j], there are no elements u and v: (X[u]-X[v]) mod m=(X[i]-X[j]) mod m.
2. Each subset X and Y form S are disjoint: X∩Y=∅.
3. The set of all integers less than the parameter m should be recovered by S. Otherwise, for each integer d in $\{0, 1, 2, 3, ..., m - 1\}$, there exists a subset X of S and two elements X[i] and X[j] from X: (X[i]-X[j]) mod m = d. For representing the proposed genetic algorithm GADiffSet, we give the following three preliminary functions. These functions allow measuring the quality of an individual according to the three criteria given

For representing the proposed genetic algorithm GADiffSet, we give the following three preliminary functions. These functions allow measuring the quality of an individual according to the three criteria given above.

The proposed genetic algorithm GADiffSet works as follows:

Inputs

– The parameter m:
– M: The number of elements of the Total Difference Set
– N: The number of elements of the Sub Difference Set
– The crossover rate Pc
– The mutation rate Pm

1. Generate the initial population, of Ni individuals; each individual is a word of length M and each gene contains a random value between 0 and m − 1.
2. For Ng from 1 to Ngmax:
   (a) Compute the quality of each individual by the fitness function.
   (b) Sort the population by increasing order of the fitness.
       If there is an individual X of fitness 0 then return X and skip this genetic algorithm.
   (c) Copy the best Ne individuals (of small fitness) in the intermediate population.
   (d) For i = Ne to Ni:
       i. Select and cross a couple of parents (p1,p2) to generate ch1 and ch2 according to the crossover probability pc.
       ii. Mute ch1 and ch2 according to the mutation probability pm.
       iii. f1 ← fitness(ch1); f2 ← fitness(ch2).
       iv. if (f1 < f2) then insert ch1 in the intermediate population else insert ch2. end if; End for;
   (e) Ng ← Ng + 1.

Algorithm 1. GADiffSet

1. 1: **function** PARTIALFITNESS($indiv, T$) ▷ Where indiv - array of N integer, T - array
    of S integer
    2:     $pos \leftarrow 1$
    3:     $S \leftarrow N * (N - 1)$
    4:     **for** $i = 1$ to N **do**
    5:         **for** $j = i + 1$ to N **do**
    6:             $T[pos] \leftarrow (indiv[i] - indiv[j]) mod m$
    7:             **if** $(T[pos] < 0)$ **then**
    8:                 $T[pos] \leftarrow T[pos] + m$
    9:             **end if**
    10:            $pos \leftarrow pos + 1$
    11:            $T[pos] \leftarrow (indiv[j] - indiv[i]) mod m$
    12:            **if** $(T[pos] < 0)$ **then**
    13:                $T[pos] \leftarrow T[pos] + m$
    14:            **end if**
    15:            $pos \leftarrow pos + 1$
    16:        **end for**
    17:    **end for**
    18:    $f \leftarrow 0$
    19:    **for** $i = 1$ to S **do**
    20:        **for** $j = i + 1$ to S **do**
    21:            **if** $(T[i] = T[j])$ **then**
    22:                $f \leftarrow f + 1$
    23:            **end if**
    24:        **end for**
    25:    **end for**
           return f
    26: **end function**

27: **procedure** DECOMPOS($indiv, Dec$)   ▷ Where indiv - array of M integer, Dec -
    array of M/N rows and N columns of integers
28:    $pos \leftarrow 1$
29:    **for** $i = 1$ to M/N **do**
30:        **for** $j = 1$ to N **do**
31:            $Dec[i][j] \leftarrow indiv[pos]$
32:            $pos \leftarrow pos + 1$
33:        **end for**
34:    **end for**
35: **end procedure**

36: **function** FITNESS(*indiv*)   ▷ Where indiv - array of M integer
37:    T : *array of M/N rows and $N * (N - 1)$ columns of integers*
38:    Dec : *array of M/N rows and N columns of integers*
39:    TL : *array of $M * (N - 1)$ integers*
40:    $pos \leftarrow 1$
41:    Decompos(indiv,Dec)
42:    $f \leftarrow 0$
43:    **for** $i = 1$ to M/N **do**
44:       $f \leftarrow f + m * PartialFitness(i^{th}rowofDec, i^{th}rowofT)$
45:    **end for**
46:    **for** $i = 1$ to $M * (N - 1)$ **do**
47:       **for** $j = 1$ to $M * (N - 1)$ **do**
48:          **if** $(TL[i] = TL[j])$ **then**
49:             $f \leftarrow f + 1$
50:          **end if**
51:       **end for**
52:    **end for**
53:    **for** $i = 1$ to M **do**
54:       **for** $j = i + 1$ to M **do**
55:          **if** $(indiv[i] = indiv[j])$ **then**
56:             $f \leftarrow f + 1$
57:          **end if**
58:       **end for**
59:    **end for**
60:    **for** $i = 1$ to $M * (N - 1)$ **do**
61:       **if** $TL \not\ni i$ **then**
62:          $f \leftarrow f + 1$
63:       **end if**
64:    **end for**
65: **end function**

Outputs: The first individual in the last population

The results given in this paper are obtained by using the parameters shown in the Table 1. We note that even if the number of generations and that of individual are chosen to have big values, this choice doesn't affect the rapidity of the proposed genetic algorithm GADiffSet because in the step 2.2 it stops once a difference set of the three characteristics is found. So, GADiffSet stops generally after a small number of iterations.

Table 1. Default parameters of the proposed genetic algorithm

GA-GPC parameter	Parameter value
Crossover probability p_c	0.75
Mutation probability p_m	0.05
Population size N_i	10000
Number of generations	10000
Elite number N_e	2

5   Numerical Results

The genetic algorithm described above were implemented in C language. Our heuristic algorithms were used to construct DDS with the objective of constructing QC OSMLD codes. The Table 2 represent some of the constructing DDS and correspondent codes. For each DDS constructed we give the parameters the modulo m the L and r the number of DDS, also we give the length n and the dimension k of the associated QC-OSMLD code. The QC-OSMLD codes constructed by GADiffSet algorithm are new codes or are equivalent to other codes constructed by mathematical methods [1, 17].

Table 2. Some DDS constructed by genetic algorithm and associated QC MLD codes

m	L	r	DDS	(n, k)	Origin
26	4	2	$(0, 5, 15, 19)(3, 4, 6, 12)$	$(78, 52)$	T.W
41	5	2	$(4\ 25\ 33\ 36\ 38)(7\ 8\ 22\ 26\ 32)$	$(123, 82)$	Chen
7	2	3	$(0\ 3)\ (5\ 6)(2\ 4)$	$(28, 21)$	T.W
19	3	3	$(0\ 3\ 12)(5\ 7\ 13)(4\ 8\ 9)$	$(76, 57)$	GADiffSet
37	4	3	$(21\ 6\ 2\ 12)(30\ 16\ 19\ 14)(34\ 17\ 10\ 9)$	$(148, 111)$	Chen
9	2	4	$(7\ 8)(2\ 4)(3\ 6)(1\ 5)$	$(45, 36)$	T.W
25	3	4	$(10\ 13\ 20)(11\ 12\ 23)(0\ 4\ 9)(1\ 18\ 24)$	$(125, 100)$	T.W
49	4	4	$(10, 17, 21, 46)(6, 18, 24, 34)(8, 11, 13, 30)(10, 36, 44, 45)$	$(245, 196)$	Chen
11	2	5	$(0\ 6)(7\ 9)(2\ 10)(1\ 8)(3\ 4)$	$(66, 55)$	T.W
31	3	5	$(1\ 4\ 13)(3\ 19\ 29)(14\ 15\ 28)(9\ 11\ 17)(16\ 23\ 27)$	$(186, 155)$	T.W
61	4	5	$(2\ 25\ 29\ 43)\ (5\ 8\ 40\ 50)\ (3\ 16\ 31\ 56)\ (41\ 52\ 53\ 58)$ $(10\ 12\ 19\ 49)$	$(366, 305)$	GADiffSet
13	2	6	$(4\ 11)(10\ 12)(0\ 9)(5\ 8)(1\ 6)(2\ 3)$	$(91, 78)$	GADiffSet
37	3	6	$(1, 19, 25)(16, 21, 24)(23, 34, 35)(5, 26, 33)(4, 27, 31)$ $(3, 18, 20)$	$(259, 222)$	GADiffSet
15	2	7	$(7, 10)(0, 11)(3, 4)(2, 8)(1, 6)(5, 13)(12, 14)$	$(120, 105)$	GADiffSet
43	3	7	$(10, 24, 42)(5, 21, 28)(3, 15, 16)(12, 29, 34)(25, 33, 35)$ $(4, 19, 23)(11, 17, 20)$	$(344, 301)$	GADiffSet
85	4	7	$(1, 62, 71, 76)(10, 22, 26, 68)(0, 40, 47, 60)(21, 24, 50, 56)$ $(42, 64, 75, 83)(19, 36, 37, 67)(18, 20, 41, 69)$	$(680, 595)$	GADiffSet

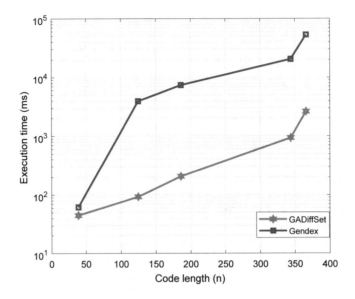

Fig. 1. GADiffSet vs Gendex

In Fig. 1 we give the time complexity of our algorithm GADiffSet with the evolution of the code length n, and we compared it with the interchange algorithm proposed by Nguyen [18] to produce incomplete block designs (IBD), this algorithm is about equally effective as the simulated annealing algorithm of Venables and Eccleston [19]. We have used Gendex Toolkit [20] implementing the Nguen's algorithm. The IBD produced are equivalent to DDS constructed by GADiffSet. The Fig. 1 show that in small code length the time complexity are close, but for larger code length the GADiffeSet is less complex then Gendex.

6   Conclusions

In this paper, we used a recent result on difference system to formulate the construction of CDDS s as a optimization problem where we require optimal solutions rather than close approximations to them. We have shown that it is possible to find an optimal solution to this problem using a procedure based on genetic algorithm. From the computational experiments we have seen that there are a design parameters which are easy, difficult and hard to construct, and others where genetic algorithm produces no optimal solutions. So, GA procedures construct some new CDDS which are not available in the catalogues of experimental designs. Moreover GA was able to construct other designs whose existence already was known. GA was able to construct "number" CDDS, many of them with high probability of successful runs. The constructed CDDS allow us to construct many new QC-MLD codes with optimal parameters, and others equivalents to known codes.

References

1. Townsend, L., Weldon, E.: Self-orthogonal quasi-cyclic codes. IEEE Inf. Theo. **IT-13**(2), 183-195 (1967)
2. Lin, S., Costello, D.J.: Error Control Coding: Fundamentals and Applications. Prentice-Hall, Englewood Cliffs (1983)
3. Reed, I.S.: A class of multiples error correcting codes and the decoding scheme. IRE. Trans. **IT-4**, 38-49 (1954)
4. Muller, D.E.: Application of boolean algebra to switching circuit design and error detection. IRE. Trans **EC-3**, 6–12 (1954)
5. Colbourn, C.J., Dinitz, J.H.: Handbook of Combinatorial Designs, 2nd edn. Chapman and Hall/CRC (2007). ISBN-13 978-1584885061
6. Stinson, D.R.: Combinatorial Designs, Construction and Analysis. Springer (2004)
7. Rosa, A.: Poznámka o cyklických Steinerových systémoch trojíc. Mat. Fyz. Časopis **16**, 285–290 (1966)
8. Bofill, P., Guimera, R., Torras, C.: Comparison of simulated annealing and mean field annealing as applied to the generation of block designs. Neural Netw. **16**, 1421–1428 (2003)
9. Morales, L.B.: Constructing difference families through an optimization approach: six new BIBDs. J. Comb. Des. **8**, 221–309 (2000)
10. Peterson, W.W., Weldon, E.J.: Error-Correcting Codes, 2nd edn. MIT Press, Cambridge (1972)
11. McCall, J.: Genetic algorithms for modelling and optimization. J. Comput. Appl. Math. **184**(1), 205–222 (2005)
12. Azouaoui, A., Askali, M., Belkasmi, M.: A genetic algorithm to search of good double-circulant codes. In: IEEE International Conference on Multimedia Computing and Systems (ICMCS 2011) Proceeding, pp. 829- 833, 07–09 April, 2011, Ouarzazate, Morocco (2011)
13. Askali, M., Nouh, S., Azouaoui, A., Belkasmi, M.: Discovery of good double and triple circulant codes using multiple impulse method. Adv. Comput. Res. **5**(1), 141–148 (2015). ISSN: 0975-3273, E-ISSN: 0975-9085
14. Aylaj, B., Belkasmi, M., Nouh, S.: A method to search good quasi-cyclic codes using new simulated annealing. In: International Workshop on Codes, Cryptography and Communication Systems, El Jadida, Morocco (2014)
15. Nouh, S., El khatabi, A., Belkasmi, M.: Majority voting procedure allowing soft decision decoding of linear block codes on binary channels. Int. J. Commun. Netw. Syst. Sci. **5**(9) (2012)
16. Nouh, S., Chana, I., Belkasmi, M.: Decoding of block codes by using genetic algorithms and permutations set. Int. J. Commun. Netw. Inf. Secur. **5**(3) (2013)
17. Zhi, C., Pingzhy, F., Fan, J.: On optimal self-orthogonal quasi-cyclic codes. In: IEEE International Conference on Communications, ICC 1990, Including Supercomm Technical Sessions. SUPERCOMM/ICC 1990. Conference Record, vol. 3, 16–19 April 1990, pp. 1256–1260 (1990)
18. Nguyen, N.-K.: An algorithm for constructing optimal resolvable block designs. Commun. Statist. B **22**, 911–923 (1993)
19. Venables, W.N., Ecclston, J.A.: Randomized search strategies for finding near-optimal block and row-column designs (1992)
20. Gendex. http://designcomputing.net/gendex/

Prediction of Coordinate Measuring Machines Geometric Errors by Measuring a Ball Step-Gauge

Loubna Laaouina[1]([✉]) [ID], Abdelhak Nafi[1] [ID], and Ahmed Mouchtachi[2] [ID]

[1] Laboratory of Advanced Mechanics Research and Industrial Applications, ENSAM, BP 4024, Meknes Ismailia, Morocco
loubna.laaouina@gmail.com, abdelhaknafi@hotmail.fr
[2] Laboratory of Renewable Energy and Sustainable Development, ENSAM, Casablanca, Morocco
ahmedmouchtachi@yahoo.fr

Abstract. The coordinate measuring machine (CMM) is among the powerful tools used to measure the conformity of the mechanical parts. The performance of this machine is partly affected by geometric errors associated to its articulations. To provide remedial actions to the CMM in case of non-conformity, we propose a method to identify geometric errors characterizing the machine using the inverse method, when measuring a ball step-gauge in seven locations as defined in the ASME 89.4.10360.2-2008 standard and nineteen positions proposed by G. Zhang.

Keywords: Coordinate Measuring Machine · Geometric errors · Measurement Metrology · Volumetric errors · Ball step-gauge

1 Introduction

Coordinate Measuring Machines (CMMs) are widely used in manufacturing industry, for quality control and inspection of complex shaped parts quickly and with great precision. However, the measurement results, on a CMM, are affected by twenty-one geometric errors related the machine errors, including eighteen kinematic errors and three out-of-squareness. Therefore, it is essential to carry out regular calibration tests and compensate these geometric errors. Several techniques have been developed either by using laser or artifacts such as gauge block, tetrahedron, step gauge, ball plate, hole plate and ball bar [1–10].

Zhang et al. [5] determined the twenty-one geometric errors of a CMM by measuring the displacement errors along twenty-two lines in the machine volume, using the laser interferometer.

The ASME 89.4.10360.2-2008 standard [4] suggested the use of five lengths to provide an indication of the CMM's performance.

Nafi et al. [7] proposed a method to identify scale errors and out of squareness errors, based on the measurement of a step gauge in seven positions proposed by ASME 89.4.10360.2-2008 standard [4].

© Springer International Publishing AG, part of Springer Nature 2019
J. Mizera-Pietraszko et al. (Eds.): RTIS 2017, AISC 756, pp. 178–186, 2019.
https://doi.org/10.1007/978-3-319-91337-7_17

This study presents a model for the identification of geometric errors characterizing the CMM. The proposed approach is based on the measurement of a ball step gauge in seven locations described by the ASME 89.4.10360.2-2008 standard [6] and in nineteen locations proposed by Zhang [7].

2 Modeling of Geometric Errors

2.1 Position of the Stylus Tip Relative to the WFYXZT Machine

In this study, we use a CMM with a topology WFYXZT (Fig. 1). The machine has a serial structure. The coordinates of the stylus tip T in the reference foundation {F} are nominally defined by the moving axes coordinates X, Y and Z, by the length Ls (distance from the pivot of the articulated system to the centre of the stylus tip T) and by the orientations angles of the articulated system (angles A and B).

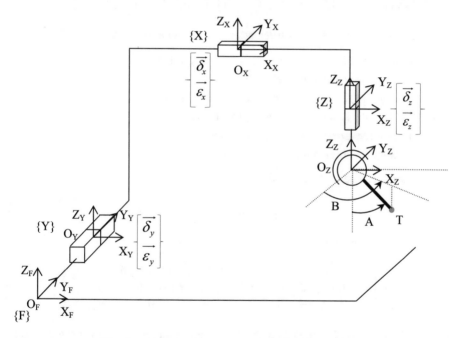

Fig. 1. The stylus tip position in the coordinate system

In reality, the position of the Stylus tip does not correspond exactly to its nominal position T, because the articulations that control the movement of the CMM are affected by geometric errors, There is an error torsor composed of two elements of reduction δ (linear error) and ε (angular error) at each articulation, Fig. 1 shows the linear and angular errors associated to each axis X, Y and Z.

For example $\vec{\delta_x}$ and $\vec{\varepsilon_x}$ are respectively the linear and the angular errors associated to the X axis.

2.2   Error Position of the Stylus Tip

Geometric errors, associated to articulations, propagate in the structure of CMM by the Abbe's effect and Bryan offsets.

The position of the stylus tip is affected by these errors. The error of stylus's tip position is defined by a volumetric error, such as a vector who characterizes the difference between the nominal and the real position of the stylus tip. The volumetric error throughout the machine measurement volume at each point M is [2]:

$$\overrightarrow{e_M} = \overrightarrow{\delta_Y} + \overrightarrow{\varepsilon_Y} \wedge \overrightarrow{O_Y T} + \overrightarrow{\delta_X} + \overrightarrow{\varepsilon_X} \wedge \overrightarrow{O_X T} + \overrightarrow{\delta_Z} + \overrightarrow{\varepsilon_X} \wedge \overrightarrow{O_Z T} \tag{1}$$

After decomposition of vectors $\overrightarrow{O_Y T}$, $\overrightarrow{O_X T}$ and $\overrightarrow{O_Z T}$ in frame {F}

$$\overrightarrow{e_M} = \begin{pmatrix} \delta_x(Y) \\ \delta_y(Y) \\ \delta_z(Y) \end{pmatrix} + \begin{pmatrix} \varepsilon_x(Y) \\ \varepsilon_y(Y) \\ \varepsilon_z(Y) \end{pmatrix} \wedge \begin{pmatrix} X + t_x \\ t_y \\ Z + t_z \end{pmatrix} + \begin{pmatrix} \delta_x(X) \\ \delta_y(X) \\ \delta_z(X) \end{pmatrix} + \begin{pmatrix} \varepsilon_x(X) \\ \varepsilon_y(X) \\ \varepsilon_z(X) \end{pmatrix} \wedge \begin{pmatrix} t_x \\ t_y \\ Z + t_z \end{pmatrix} + \begin{pmatrix} \delta_x(Z) \\ \delta_y(Z) \\ \delta_z(Z) \end{pmatrix} + \begin{pmatrix} \varepsilon_x(Z) \\ \varepsilon_y(Z) \\ \varepsilon_z(Z) \end{pmatrix} \wedge \begin{pmatrix} t_x \\ t_y \\ t_z \end{pmatrix} \tag{2}$$

Six errors associated to each articulation, for example for the Y articulation:

$\delta_x(Y)$: Straightness error of the articulation Y in x direction;

$\delta_y(Y)$: Scale error of the articulation Y;

$\delta_z(Y)$: Straightness error of the articulation Y in x direction;

$\varepsilon_x(Y)$: Pitch of the articulation Y around x direction;

$\varepsilon_y(Y)$: Roll of the articulation Y around y direction;

$\varepsilon_z(Y)$: Yaw of the articulation Y around z direction.

Equation 2 can be written as follows

$$\overrightarrow{e_M} = J\delta P \tag{3}$$

Where J is the Jacobian matrix.

$$J = \begin{bmatrix} 1 & 0 & 0 & 0 & z_t & -t_y & 1 & 0 & 0 & 0 & z_t & -t_y & 1 & 0 & 0 & 0 & t_z & -t_y \\ 0 & 1 & 0 & -z_t & 0 & t_x & 0 & 1 & 0 & -z_t & 0 & t_x & 0 & 1 & 0 & -t_z & 0 & t_x \\ 0 & 0 & 1 & y_t & -t_x & 0 & 0 & 0 & 1 & t_y & -t_x & 0 & 0 & 0 & 1 & t_y & -t_x & 0 \end{bmatrix}$$

δP Column matrix (18 × 1) composed of all linear and angular errors associated to the machine axes

$$\delta P = \begin{bmatrix} \delta_x(X) & \delta_y(X) & \delta_z(X) & \varepsilon_x(X) & \varepsilon_y(X) & \varepsilon_z(X) & \delta_x(Y) & \delta_y(Y) & \delta_z(Y) & \varepsilon_x(Y) & \varepsilon_y(Y) & \varepsilon_z(Y) & \delta_x(Z) & \delta_y(Z) & \delta_z(Z) & \varepsilon_x(Z) & \varepsilon_y(Z) & \varepsilon_z(Z) \end{bmatrix}^T$$

3   Ball Step-Gauge

Figure 2 illustrate a ball step-gauge measurement. It has the form of an H beam, on which the balls are mounted rigidly, each ball has a diameter of 22.22 mm with a sphericity less than 0.54 μm. there are seven balls, the nominal distance between the adjacent balls is 83 mm [11]. The ball step-gauge has an advantage compared to a step gauge and gauge-block; in the measurements, it's not necessary to assess the probe errors, most of measurement errors come from geometric errors related to machine axes.

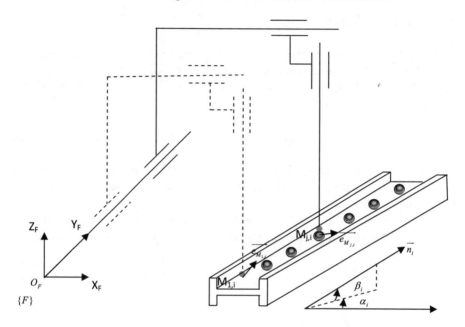

Fig. 2. Ball step-gauge placed in one direction $\vec{n}_i$ in the reference {F}

4   Measurement Error in the Ball Step Gauge

The ith position of the ball step-gauge in the coordinate system of CMM is defined by angles α_i and β_i and the coordinates of the identified center $M_{1,i}$ of the first measured ball as shown in Fig. 2. Each ball will be probed on nine points to identify its center.

The measurement error is the difference between the distance measured by the machine and the conventionally true distance. This error can be estimated using the volumetric error model. The error between the experimentally measured centers of the balls $M_{1,i}$ and $M_{j,i}$ is the projection along $\vec{n}_i$ of the difference between the volumetric error vectors calculated at the centers $M_{j,i}$ and $M_{1,i}$.

$$E_{1j,i} = (\vec{e}_{Mj,i} - \vec{e}_{M1,i}) \cdot \vec{n}_i \tag{4}$$

Substituting Eq. (3) into Eq. (4), the distance error is given by

$$E_{1j,i} = \left[(J_{j,i}\delta P_{j,i}) - (J_{1,i}\delta P_{j,i}) \right]\vec{n}_i \tag{5}$$

Where $J_{j,i}$ is the jacobian matrix calculated at $M_{j,i}$ point.
$J_{1,i}$ is the jacobian matrix calculated at $M_{1,i}$ point.

j identifies the measured ball of the ball step gauge and i indicates the artefact position.

5 Simulated Machine

In this section, WFYXZT virtual machines are developed in Matlab. Each machine is characterized by geometric errors associated to its axes. Each geometric error is modeled by a polynomial of degree 3 and represented by 4 coefficients, for example the error $[\delta_x(X_{j,i})]$ is written as follows

$$[\delta_x(X_{j,i})] = \left[\delta_{x,X,0} + \delta_{x,X,1}.X_{j,i} + \delta_{x,X,2} \cdot (X_{j,i})^2 + \delta_{x,X,3} \cdot (X_{j,i})^3\right] \tag{6}$$

For a full model, the measured errors on the ball step gauge depend on 72 coefficients (72 = 18*4), 18 is the number of geometric errors, 4 is the number of polynomial coefficients.

By applying the Vandermonde extension of the vector of geometrics errors. The measurement error between $M_{1,i}$ center and $M_{j,i}$ center of ball step gauge in $\vec{n}_i$ direction is:

$$E_{1j,i} = \left[\left(J_{j,i}^T.\vec{n}_i\right)^T . MVDM_j - \left(J_{1,i}^T.\vec{n}_i\right)^T . MVDM_1\right] . \delta P \tag{7}$$

We pose $H_{1j,i} = \left[\left(J_{j,i}^T.\vec{n}_i\right)^T . MVDM_j - \left(J_{1,i}^T.\vec{n}_i\right)^T . MVDM_1\right]$
So, we have

$$E_{1j,i} = H_{1j,i} . \delta P \tag{8}$$

$H_{1j,i}$: A line identification matrix, corresponding to the jth ball center's measurement of the ith position of the ball step-gauge.
δP: Vector composed of the 72 polynomials coefficients modeling the geometric errors.

The ball step-gauge is measured in seven positions suggested by ASME B89.4.10360.2-2008 standard [6] and in nineteen positions suggested by Zhang et al. [7]. The evolution of the rank of identification matrix is studied, as a function of the number of accumulated positions, (Fig. 3). It shows that the maximum rank of the identification matrix is 45, it becomes stable after the accumulation of 20 positions. Therefore, the identification matrix is deficient by (72–45) and the optimal number of positions is 20.

The program of the virtual machine takes as input the 72 geometric errors randomly generated, the positions of the ball step-gauge, the machine volume and the true distances between ball's centers. It gives as output the measured errors in different positions of the ball step gauge for different measured distances. The results of a simulation example are presented in Fig. 4.

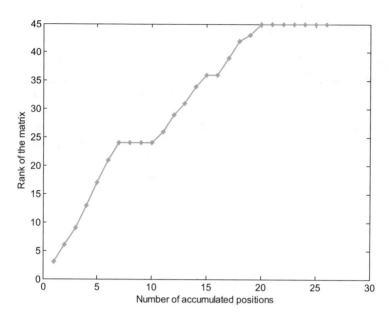

Fig. 3. Evolution of the rank of matrix formed by adding positions

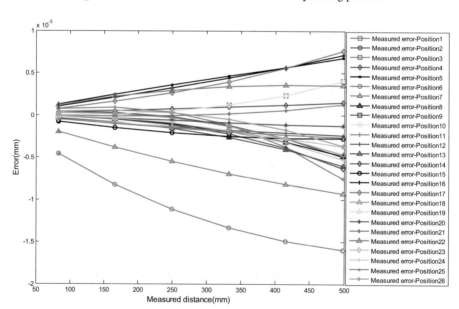

Fig. 4. Measured errors on ball step gauge in 26 positions

6   The Inverse Method

The identification matrix composed by measuring the ball step gauge in the 20 positions (the optimal number of positions) is singular and it has a high condition number tend to infinity. In order to improve its condition number, the SVD decomposition is applied. Figure 5 presents the flowchart used to reduce the identification matrix. The SVD decomposition method reduce the condition number of the matrix, which allow to determine the 45 geometrics errors characterizing the CMM. Thus, volumetric errors can also be predicted with only 45 geometrics errors. Figure 6 shows a comparison between the measured errors using the virtual machine and the predicted errors using the 45 geometrics errors found by the inverse method in position 24.

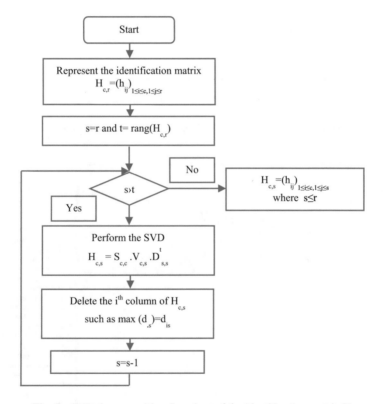

Fig. 5. SVD decomposition flowchart of the identification matrix H

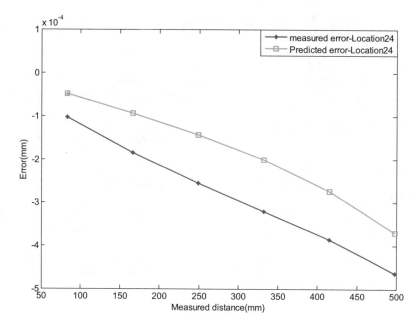

Fig. 6. Measured and Predicted errors using estimated coefficients of ball step gauge for position 24

7 Conclusion

The proposed model is based on measuring a ball step-gauge in seven positions described by the ASME 89.4.10360.2-2008 standard [6] and in nineteen positions proposed by Zhang [7]. Virtual machines are created to simulate the measurement and calculate the measurement errors which are used in the inverse method. The SVD decomposition is applied to improve the condition number of the identification matrix. The developed method allows identifying geometrics errors characterizing the CMM. It allows also predicting volumetric errors at each point in the machine volume with less geometric errors found by the inverse method.

References

1. Cauchick-Miguel, P., Kinga, T., Davis, J.: CMM verification: a survey. Measurement **17**(1), 1–16 (1996)
2. Sartori, S., Zhang, G.X.: Geometric error measurement and compensation of machines. CIRP Ann. Manuf. Technol. **44**(2), 599–609 (1995)
3. Lim, C.K., Burdekin, M.: Rapid volumetric calibration of coordinate measuring machines using a hole bar artefact. Proc. Inst. Mech. Eng. Part B J. Eng. Manuf. **216**(8), 1083–1093 (2002)
4. ASME89.4.10360.2-2008: Acceptance test and reverification test for CMMs – Part 2: CMMs used for measuring linear dimensions (2008)

5. Zhang, G., Ouyang, R., Lu, B., Hocken, R., Veale, R., Donmez, A.: A displacement method for machine geometry calibration. CIRP Ann. Manuf. Technol. **37**(1), 515–518 (1988)
6. Curran, E., Phelan, P.: Quick check error verification of coordinate measuring machines. J. Mater. Process. Technol. **155–156**, 1207–1213 (2004)
7. Nafi, A., Mayer, J.R.R.: Identification of scale and squareness errors on a CMM using a step gauge measured based on the ASME 89.4.10360.2-2008 standard. Trans. North Am. Manuf. Res. Inst. SME **38**, 325–332 (2010)
8. Schwenke, H., Franke, M., Hannaford, J., Kunzmann, H.: Error mapping of CMMs and machine tools by a single tracking interferometer. CIRP Ann. Manuf. Technol. **54**(1), 475–478 (2005)
9. Schwenke, H., Knapp, W., Haitjema, H., Weckenmann, A., Schmitt, R., Delbressine, F.: Geometric error measurement and compensation of machines—an update. CIRP Ann. Manuf. Technol. **57**(2), 660–675 (2008)
10. Kruth, J.-P., Zhou, L., Van den Bergh, C., Vanherck, P.: A method for squareness error verification on a coordinate measuring machine. Int. J. Adv. Manuf. Technol. **21**, 874–878 (2003)
11. Osawa, S., Takatsuji, T., Noguchi, H., Kurosawa, T.: Development of a ball step-gauge and an interferometric stepper used for ball-plate calibration. Precis. Eng. **26**(2), 214–221 (2002)

Cloud Computing and Internet of Things

Hierarchical Load Balancing Strategy
in Cloud Environment

Amal Zaouch[✉] and Faouzia Benabbou

Information Technology and Modeling Laboratory, Science Faculty Ben M'sik,
Casablanca, Morocco
zaouch_amal@yahoo.fr, faouzia.benabbou@univh2m.ma

Abstract. Cloud balancing provides an environment with the ability to distribute application requests across any number of application deployments located in different data centers. This paper, proposes a load balancing strategy for distributed use of a cloud data center and apply it on two levels control: PMs (Physicals Machines) and Clusters. Load balancing is done by two managers; they ensure exchange information and decide afterwards the level concerned with load balancing. Measurement of load is based on load information including CPU utilization and memory utilization. VMs (Virtual Machines) are allowed to be migrated between different federations to distribute loads while the communication costs are also incurred. Therefore, the objectives of this model are twofold: reducing the load of the overloaded hosts and decreasing the communication costs among different federations. Proposed method not only gives good Cloud balancing but also ensures reducing response time and communication cost and by result enhancing throughput of the whole system.

Keywords: Cloud computing · Load balancing · Hierarchical · Communication Overhead

1 Introduction

Cloud computing is the delivery of computing as a service rather than a product, whereby shared resources, software, and information are provided to computers and other devices as a utility over the Internet. A Cloud datacenter can be considered a hierarchical system in structure, which is composed of many clusters and each cluster contains one or more physical machines (PMs), in every PM runs some virtual machines (VMs). Virtualization and the live migration of VMs between PMs are key enablers of efficient resource allocation in data centers. Live migration of a VM from one PM to another makes it possible to react to the changing resource requirements of the VMs [1]. When the resource demand of the VMs increases, they can be migrated to other PMs with a lower load, thus avoiding SLA violations. For these reasons, VM migration is a key ingredient of load balancing problem [2]. On the other hand, VM migrations take time, create overhead, and can have adverse impact on SLA fulfillment [1]. A VM migration may increase the load of both the source and the PM receiver, puts additional burden on the network, and makes the migrated VM less responsive during migration [12]. Therefore,

© Springer International Publishing AG, part of Springer Nature 2019
J. Mizera-Pietraszko et al. (Eds.): RTIS 2017, AISC 756, pp. 189–195, 2019.
https://doi.org/10.1007/978-3-319-91337-7_18

it is important to keep the number of live VM migrations at a reasonable level. Understanding the exact impact of live migration is a difficult problem on its own. Hence, VM allocation must find the optimal balance between Quality of Services and cost of communication.

Load balancing is the mechanism of distributing the load among various nodes of any system [6]. Major goal of load balancing is optimal utilization of available resources. With virtualization, Cloud datacenters should have ability to balance the load at each level in a hierarchical manner. Such agility becomes a key in modern cloud computing infrastructures that aim to efficiently distribute the load among resources.

The remainder of the paper is organized as follows: Sect. 1 an introduction, Sect. 2 describe load balancing problem, Sect. 3 a detailed description of proposed strategy. Finally, implementation of proposed mechanism is outlined in Sect. 4 as a future work.

2 Load Balancing Problem

The load is an abstract concept describing the busyness of the system, and the distribution balancing of load of all resources of parallel system is called load balancing. The load balancing contributes to assure the high efficiency of task assignment algorithm, and it can adjust the load assignment at all times to keep all resources in the system in balanced state. There are two kinds of techniques for the load balancing of all resources in the system: load assignment and load migration. The load assignment is to properly assign the user tasks to all resources to make the system load on all resources roughly equal. The load migration is to migrate the tasks from heavy-loaded resources to light-loaded, so that the system load gets balanced and the overall performance is improved. The load balancing algorithm proposed there in is used in load migration.

Based on spatial distribution of nodes, we can find three types of algorithms that specify which node is responsible for balancing load in cloud computing environment; In centralized load balancing technique all the allocation and scheduling decision are made by a single node, in distributed load balancing technique, no single node is responsible for making resource provisioning or task scheduling decision; Every node in the network maintains local knowledge base to ensure efficient distribution of tasks in static environment and re-distribution in dynamic environment, and finally the type of algorithm, adopted here, the hierarchical load balancing involves different levels of the cloud in load balancing decision. Such load balancing techniques mostly operate in master slave mode. These can be modeled using tree data structure wherein every node in the tree is balanced under the supervision of its parent node. Master or manager can use light weight agent process to get statistics of slave nodes or child nodes. Based upon the information gathered by the parent node provisioning or scheduling decision is made [3]. One of the challenges of load balancing algorithms is overhead which determines the amount of overhead involved while implementing a load balancing system. It is composed of overhead due to VM migration cost or communication cost [15]. A well designed load balancing algorithm should reduce overhead. The main objective of load balancing methods is to speed up the execution of applications on resources whose workload varies at run time in unpredictable way [16]. Hence, it is significant to define metrics to measure the resource workload. Several load indices have been proposed in the literature, like CPU queue length, average CPU queue length, CPU utilization, etc. The success of a load balancing algorithm depends from stability of the number of messages (small overhead), support environment, low cost update of the workload, and short mean response time which is a significant measurement for a user [17]. It is also essential to measure the communication cost induced by a load balancing operation.

There are quite many research conducted in the area of load balancing solution. Most of them focus on load balancing for one level. One of the challenging scheduling problems in Cloud datacenters is to take into consideration workload in different level composed Cloud Architecture and the integrated characteristics of clusters and hosting physical machines composed it. Unlike traditional load-balancing scheduling algorithms which consider only Virtual Machines with one factor such as CPU load, our proposed model treats CPU, and memory integrated for both physical machines (PM) and clusters. In [5] Tian et al. introduced a dynamic and integrated resource scheduling algorithm (DAIRS) for balancing VMs in Cloud. This algorithm treats CPU, memory and network bandwidth as integrated resource with weights. They also developed a new metric, average imbalance level of all the hosts, to evaluate the performance under multiple resource scheduling. DAIRS is one of the earliest algorithms that explored the multiple types of resources and treated them as integrated value. The main drawback of DAIRS is that it ignores the communication cost of migrations. Malhotra et al. [6] proposed a framework based on intelligent agents at two levels in cloud computing model. The first one is at the data center level and the other at the global level. Hao et al. [7] proposed a load balancing scheme that works at three levels: Datacenter, host and processing elements. Cloud balancing is controlled by the physical resource entity including datacenters, hosts and PEs using the minimum value of standard deviation. The standard deviation load may be more or less than before scheduling. Wang et al. [8] proposed a load balancing in a three level of Cloud Computing architecture. The third level is the service node that used to execute subtask. The second level is the service manager that used to divide the task into some logical independent subtasks. Load Balance Min-Min (LBMM) scheduling algorithm takes the characteristic of the Min-Min scheduling algorithm as foundation. But the biggest weakness of Min-Min scheduling algorithm is it only considers the completion time of each task at the node but does not consider the work loading of each node. So, LBMM will improve the load unbalance of the Min-Min and reduce the execution time of each node effectively. The multi-level hierarchical network topology can decrease the data store cost. Another Cluster-Based load balancing algorithm in Cloud Computing is proposed in [9] and is designed for multiple masters and multiple slave architecture (MMMS), where network is divided into clusters using a clustering algorithm such that every node (master) in the network belongs to exactly one cluster. New nodes entering the network are assigned an existing cluster or a new cluster in case the cluster is completely occupied based on the clustering strategy used. Every cluster has at least one Inter Cluster Communication (ICC) node. The scalability of the architecture is therefore dependent on the clustering technique. The clustering is done in the initial stages while the network is being initialized. The types of nodes and their functions are as follows. A slave is the computing element of the network. The processing of the tasks is done in these elements. Every slave is connected to exactly one master directly. Proposed load balancing algorithm is divided into two parts depending on the type of nodes and the communication between them. These parts are referred to as: (1) Load distribution among masters; and (2) Load distribution from master to slave. A master is a computing machine that decides the load balancing policy of the tasks among the slaves and selects a slave node for the execution of the tasks. In this paper, we propose a model for cloud computing environment, where a hierarchical load balancing strategy is developed. Characteristics of the proposed strategy can be summarized as follows:

(i)   It is a balancing strategy at physical machines' level at first.
(ii) It proceed to a load balancing at the cluster's level to reduce communication costs if the system became in a saturation state.

A dynamic model is proposed with the global manager at higher level and the local manager at next. It's a technique that can be used to improve the performance of cloud computing by balancing the workload across all the nodes in the cloud with maximum resource utilization, in turn reducing overhead and cost of communication.

3   System Design

3.1   Architecture

As shown in Fig. 1 cloud architecture is a set of data centers and each data center is a finite set of G clusters Ck, interconnected, where each cluster contains one or more physical machine (PM) interconnected by switches and in every PM runs some virtual machines (VM). Physicals machines in clusters are heterogeneous so the load in one cluster may be very high while the other clusters may have nothing running on it. A higher throughput can be achieved if load balancing is added between the clusters as well as balancing loads between the physical machines within a cluster.

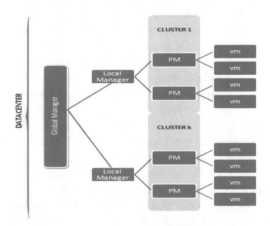

Fig. 1.   Architecture of proposed strategy

Referring to the structure of proposed model in Fig. 1, we propose a hierarchical load balancing strategy based on two managers: local and global.

- Local managers collect the load information from the physicals machines PM and balance the load in that local area. They balance the load for the physicals machines.
- Global manager collects periodically the load information from every cluster. The global manager balances the load for clusters and communicates periodically with local managers to collect information load of each cluster.

This approach is designed to improve the overall resources utilization of the cluster and to reduce the Inter-process Communication overhead in one cluster. Therefore, the objectives of this algorithm are twofold: reducing the load of the overloaded machines by migrating load of clusters, and decreasing the communication costs by reducing number of migrations among different federations if the system is in saturated status.

3.2 Proposed Strategy

The proposed work allows us to develop a hierarchical strategy at two levels and are designed: intra-cluster and inter-Clusters: Intra cluster load balancing: in this level load balancing is launched only when some VM's managers fail to balance locally the overload of their VMs. Knowing the global state of each PM, the PMs manager can evenly distribute the global overload between its physicals machines.

Inter-cluster load balancing: in this second level it performs a global load balancing among all clusters of a cloud datacenter. It is executed only if the other level are failed to achieve a complete load balance. It is significant to remark that at this level, the algorithm always succeeds load balancing all these clusters. It is thus useless to test if some clusters are still imbalanced.

The main advantage of this strategy is to prioritize local load balancing first (within a cluster, then within a datacenter). The goal of this neighborhood strategy is to decrease the amount of messages between physicals machines. As consequence of this goal, the overhead induced by our strategy is reduced.

3.2.1 Load Estimation and Information Exchange Policy

Ideally the load information should reflect the current CPU utilization, memory utilization and network traffic of a node. Traditionally, the load of a node at given time was described simply by CPU utilization. In the proposed work, CPU utilization and memory utilization are used to measure the load.

Suppose that one cluster k has n physicals machines as $\{pm_{k,1} \ldots pm_{k,n}\}$. Suppose that the load of PM is l_i and $\bar{l}$ is the average load of each PM. When we calculate standard deviation of load, we pay attention to physicals machines which is belongs to the cluster.

The standard deviation of load in one cluster is given by:

$$\sigma = \sqrt{\frac{1}{n} \sum_{i=1}^{n} (li - \bar{l})^2} \qquad (1)$$

If the standard deviation value is small, it means that the difference of each load is small. The small standard deviation tells that the load of the entire system is balanced. The lower value the standard deviation has, the more load balanced the system has.

Definition of State's Cluster
Physicals machines composed one cluster can become overloaded or saturated even if live migration of Vms is proceeded to avoid bottlenecks or overloading some resources; in this case cluster is considered in saturated state so another cluster's level is considered to perform resource utilization and distribute the workload evenly across all the machines composed the data center. Two thresholds values T1 and T2 are considered to decide the process transfer policy and according to the formula (1); cluster's state and are as follow:

Balanced state:

```
If  σ  ≤ T1: Cluster is balanced
Else
        If  σ ≥ T1 and  σ  ≤ T2
        Cluster is overloaded.
          Else
        Cluster is saturated.
        End
    End
```

The information exchange policy chosen for this research is a periodic policy with a time interval that will be set in the simulation experiments.

3.2.2 Process Transfer Policy
In the proposed strategy, this determination includes two steps. First the physicals machines are classified according to their load. The decision is made whether to start a process on another machine. Second, clusters are classified to work like receiver or sources in case of saturation state of one of them.

Load Classification of Physical Machine
According to the formula (1) and when a cluster is overloaded, we can consider the triggering of a load balancing operation. To determine whether a Physical Machines in a cluster is in a suitable condition to participate in a process transfer like a server or receiver of tasks according to its availability status. We classify physical Machines to overloaded (sources), normal (neutral) and the under loaded one (receivers).

Load Classification of Cluster
When a cluster is in saturated state, Vms migration will be moved to another level, so clusters are classified into overloaded, normal and under loaded to participate respectively like sources neutral or receivers.

4   Conclusion and Future Work

This paper proposed a hierarchical strategy load balancing for Cloud datacenters. The main objective is to achieve a significant benefit in mean response time and a minimum communication cost between clusters. Since this work is just a conceptual model, more work is needed to implement the algorithm and resolve problems.

References

1. Amar, M., Anurag, K., Rakesh, K., Rupesh, K., Prashant, Y.: SLA driven load balancing for web applications in cloud computing environment. Inf. Knowl. Manage. **1**(1), 5–13 (2011)
2. Mann, Z.Á.: Allocation of virtual machines in cloud data centers—a survey of problem models and optimization algorithms. ACM Comput. Surv. (CSUR), **48**(1), 11 (2015)
3. Jung, G., Hiltunen, M.A., Joshi, K.R., Schlichting, R.D., Mistral, C.P.: Dynamically managing power, performance, and adaptation cost in cloud infrastructures. In: IEEE 30th International Conference on Distributed Computing Systems (ICDCS), pp. 62–73 (2010)
4. Katyal, M., Mishra, A.: A comparative study of load balancing algorithms in cloud computing environment. arXiv preprint arXiv:1403.6918 (2014)
5. Tian, W., Zhao, Y., Zhong, Y., Xu, M., Jing, C.: A dynamic and integrated load-balancing scheduling algorithm for cloud datacenters. In: 2011 IEEE International Conference on Cloud Computing and Intelligence Systems (CCIS), pp. 311–315. IEEE (2011)
6. Malhotra, M., Singh, A.: Adaptive framework for load balancing to improve the performance of cloud environment. In: 2015 IEEE International Conference on Computational Intelligence & Communication Technology (CICT), pp. 224–228. IEEE (2015)
7. Hao, Y., Liu, G., Lu, J.: Three levels load balancing on cloudsim. Int. J. Grid Distrib. Comput. **7**(3), 71–88 (2014)
8. Wang, S.C., Yan, K.Q., Liao, W.P., Wang, S.S.: Towards a load balancing in a three-level cloud computing network In: 2010 3rd IEEE International Conference on Computer Science and Information Technology (ICCSIT), vol. 1, pp. 108–113. IEEE (2010)
9. Dhurandher, S.K., Obaidat, M.S., Woungang, I., Agarwal, P., Gupta, A., Gupta, P.: A cluster-based load balancing algorithm in cloud computing. In: 2014 IEEE International Conference on Communications (ICC), pp. 2921–2925. IEEE (2014)
10. Culler, D.E., Karp, R.M., Patterson, D.A., Sahay, A., Santos, E.E., Schauser, K.E., Subramonian, R., von Eicken, T.: LogP: a practical model of parallel computation (1993)
11. Culler, D.E., Karp, R.M., Patterson, D.A., Sahay, A., Santos, E.E., Schauser, K.E., Subramonian, R., von Eicken, T.: LogP: a practical model of parallel computation. Commun. ACM **39**(11), 78–85 (1996)
12. Xu, G., Pang, J., Fu, X.: A load balancing model based on cloud partitioning for the public cloud. Tsinghua Sci. Technol. **18**(1), 34–39 (2013)
13. Smith, J., Nair, R.: The architecture of virtual machines. Computer **38**(50), 32–38 (2005)
14. Chaoqun, X., Yi, Z., Wei, Z.: A load balancing algorithm with key resource relevance for virtual cluster. Int. J. Grid Distrib. Comput. **6**(5), 1–16 (2013)
15. Xu, M., Tian, W., Buyya, R.: A survey on load balancing algorithms for VM placement in cloud computing. arXiv preprint arXiv:1607.06269 (2016)
16. Eager, D.L., Lazowska, E.D., Zahorjan, J.: Adaptive load sharing inhomogeneous distributed systems. IEEE Trans. Softw. Eng. **12**(5), 662–675 (1986)
17. Badidi, E.: Architecture and services for load balancing in object distributed systems. Ph.D. thesis, Faculty of High Studies, University of Montreal, Mai (2000)

Impact of Hybrid Virtualization Using VM and Container on Live Migration and Cloud Performance

Oussama Smimite[(✉)] and Karim Afdel

Laboratory Computing Systems and Vision - LabSiV, Ibn Zohr University, Agadir, Morocco
oussama.smimite@edu.uiz.ac.ma, k.afdel@uiz.ac.ma

Abstract. Hypervisor-based virtualization and containerization offer abstraction capabilities that make applications independent of the cloud's hardware infrastructure for migration. Live migration is a widely used technology for load balancing, fault tolerance and energy saving in Cloud Datacenters. In this paper, we will be converging the two types of virtualization (Container and VMs together) into a hybrid architecture that promotes their complementarities without altering or degrading the quality of service (QoS) of the Cloud infrastructure. For this, we studied the live migration of a monolithic application and precisely the video streaming installed in a container nested in a virtual machine. We examined the contribution of this new type of virtualization in terms of processor utilization, disk usage, total migration time and downtime. The results presented show the advantages of Hybrid virtualization especially in terms of total migration time. Indeed, the absence of a real virtualization layer in the containers leads to a rapid interaction with the operating system processes, which affects the total migration time.

Keywords: Cloud · Virtualization · Live migration · VM container
Hybrid virtualization · Performance migration · Availability

1 Introduction

Cloud computing has multiple advantages for Software as a Service: portability, upgradability, scalability, high availability, cost efficiency, and resource sharing [1]. For these reasons, operators and software providers are switching to cloud computing for their infrastructure to gain more flexibility and optimize their costs.

For the cloud data centers, virtualization technologies as Xen [2], Hyper V [3] and VMware [4] are widely used for their ease of use, cost efficiency, flexibility of resource management, scalability and the simplicity of enabling the high availability.

More precisely, virtualization offers the possibility of a fine-tuning of resources allocation by associating processors, RAM, disk space and network bandwidth to a specific Virtual Machine [5, 6]. This approach has allowed the development of solutions such Software as a Service (SaaS) and Platform as a Service (PaaS), where services providers can quickly make available virtual machines with the required resources to their customers almost in no time, and not burden them with infrastructure management.

© Springer International Publishing AG, part of Springer Nature 2019
J. Mizera-Pietraszko et al. (Eds.): RTIS 2017, AISC 756, pp. 196–208, 2019.
https://doi.org/10.1007/978-3-319-91337-7_19

Sometimes, for servers' maintenance period, load balancing across nodes, fault tolerance and energy management on the datacenter, administrators are required to perform migration from one node to another. Accordingly, migration has become of the most important aspects of virtualization and even a powerful selling argument for many cloud services providers.

Operating Systems (OS) instances migration across different physical hosts is a useful feature for Cloud datacenters administrators. By performing migration while OS continue to operate, service providers can guarantee high performances with minimal application downtime. It can actually be so small so that a virtual machine user does not even notice it. However, as small as the downtime might be, it can cause some serious issues for time-sensitive applications such as online gaming, video streaming and critical web servers. Therefore, this degradation is simply not acceptable for these service and cause harm to the quality of service and service level agreement offered by the provider. Thus, it is vital to understand the migration procedure and techniques in order to integrate them efficiently and reduce their impact on the quality of service of the cloud applications [1].

This paper is organized as follow: Sect. 2 presents the direct migration mechanisms of VM and containers. Hybrid migration is illustrated in Sect. 3 where we describe the selected architecture to achieve complementarity between hypervisor-based VM and containers and how it positively affects the performances and management of the cloud. Section 4 presents the experimental part of our research to assess the performances of direct migration using hybrid virtualization. Conclusions and perspectives are described in Sect. 5.

2   Live Migration

2.1   Migration of Virtual Machine (VM)

Migration allows a clear isolation between Hardware and Software and optimize portability, fault tolerance, load balancing and low level system maintenance [7] (Fig. 1).

Fig. 1. Impact of live migration on user's request

VM migration can be classified into two categories [8]:

- Cold migration: when an administrator is planning to upgrade a piece of hardware or perform a maintenance task on a cluster, he/she must inform the applications owners (the customers) in advance about the maintenance schedule and its duration. Then, the hardware is turned off and back on only when the system and data are successfully restored. This approach is bother-some as it is not compliant with High Availability as the downtime is generally important.
- Live migration: it based on the process of transferring a virtual machine or a container while running from a physical node to another with a minimal interruption time. The migration of the physical host can be achieved by moving the state of processor, network and storage toward a new one and even the content of the RAM.

The process of transporting a virtual machine from a physical server A to a new one, B, consists of five steps [8, 9]:

Step 1: Reservation. Making sure that the destination server (B) has enough resources to host the VM to migrate and allocate them.

Step 2: Iterative pre-copy. Transferring data from memory pages of server (A) toward server (B). This action is repeated then only for the update pages (Dirtied) during the previous transfer.

Step 3: Stop-and-Copy. Halt the Virtual Machine on the source server (A) then copy the remaining data to destination server (B).

Step 4: Engaging. Send a notification message from the server (B) to (A) to inform about the end of migration. The VM instance is deleted from the source sever (A).

Step 5: Activation. Enabling the VM that has been successfully migrated to server (B).

2.2   Containers Migration

Containers technologies [10–13] is generally very interesting considering that it simplifies portability of cloud applications: it allows the execution of an application across multiple platform and platform without worrying about low level dependencies or OS compatibility. This explains their attractiveness for multiple businesses as they offer unprecedented flexibility for application deployment (Fig. 2).

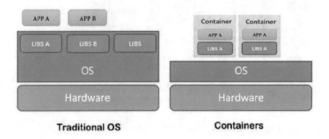

Fig. 2. Difference between deployment on traditional OS and containers

Unlike hypervisor-based virtualization, container-based virtualization is performed at a low-level system and does not aim to emulate a full hardware environment. It relies on the Linux Kernel capabilities to isolate applications. With this level of virtualization, multiple isolated Linux systems (containerized) can run on a control host while sharing a single instance of the OS kernel. Each container has its own processes and own network. The isolation is achieved thank to the name spaces. In fact, the processes of a container have unique identifier within their name space and cannot interact with those of another name space directly. A container can be seen as a set of processes (binaries, associated libraries, config files) that has its own lifecycle and dependencies and separated from other containers. Containerization is done today using tools like LXC (LinuX Containers) [11], Docker [12] or OpenVZ (Fig. 3).

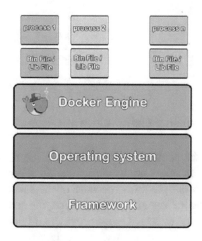

Fig. 3. Containerization is a set of processes sharing a namespace

Container migration is relatively a new technology on the market. Generally speaking, containerization is getting a lot of attractiveness from many companies thanks to its many benefits: almost no downtime maintenance, low effort for configuration and deployment, flexibility and the simplicity to achieve high availability. Unlike virtualization, containerization allows virtual instances (containers) to share a single host operating system, including binary files, libraries, or drivers. Thanks to that, it allows a server to potentially host many more containers than virtual machines. The difference in hosting can be considerable, so a server can accommodate 10 to 100 times more container instances than virtual machine instances. Because an application container does not depend on the OS, it is significantly smaller, easier to migrate or download, faster to backup or restore. Moreover, by isolating software packages from each other in containers, it [11–13] ensures a higher safety of sensitive applications.

Container migration is relatively complex, because containers are rather "processes" belonging to a namespace. Currently, container migration is based on the CRIU tool (Checkpoint/Restore In Userspace). CRIU performs this by freezing all process states on the Host machine and then restores them in the destination machine according to the sequence diagram of Fig. 4.

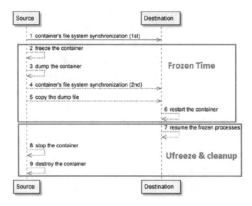

Fig. 4. The sequence diagram showing the live migration of the container

3   Hybrid Virtualization Using Container and VM

Because containers on the same physical host share the operating system kernel, a container security violation might compromise and affect other applications that share the same physical host. More precisely, because of the software nature of isolation in containers, unlike VM where it is of a hardware type, the security aspect still a bit of a challenge. Because of these aspects, service providers do not offer the same migration capabilities, as it is the case for virtual machines (Fig. 5).

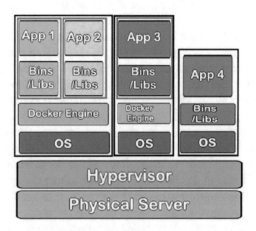

Fig. 5. Nesting of containers integrating applications in VMs

To remedy these problems, we will start by simply nesting a container in a virtual machine. This allows administrators to manage each container separately using a simple model: one container per VM. This allows using the existing virtualization management software and keeping the existing process. In addition, because each container relies on a VM as an additional layer of abstraction, administrators can avoid the security issues of multiple containers sharing the same OS kernel and maintain administrative consistency.

For Containerization, we focus mainly on Docker. This Hybrid architecture [14–19] raises questions about the efficiency of combining containers and VMs, and how to optimize the capacity usage of the hardware infrastructure. In other words, if virtual machines and Docker containers can coexist, and if so, can the Docker container services interact with VM services properly.

In fact, the experience described below shows that VM machines and containers can coexist and that container services can interact with the services of the VM virtual machine without any particular difficulty.

4   Experimental Setup

4.1   Hardware and Software Used

The experimental setup is as follow: for the hardware and software requirements, it consists of a client node (VLC media player) and two virtual servers (A and B) installed on VMware Workstation. We used a Laptop powered by a 2.4 GHz, i5 Intel processor, using a 8 GB of RAM. The operating system used is Ubuntu 14.04. We installed XenServer 7.0 [8] on servers A and B. Figure 6 shows the specifications used.

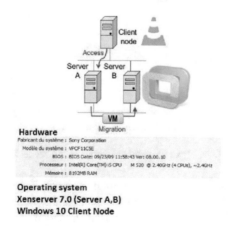

Fig. 6. Experimental configuration and hardware and software requirements

4.2 Experimentation

To study the impact of hybrid virtualization on live migration, we carried out the following scenarios:

Scenario 1: baseline. To have the ability to compare the impact of live migration of virtual machine VMs, we started with a machine that contains just the Ubuntu operating system in order to have a reference time. After the two servers were started, we started the VM in Server A and then migrated it from VM to Server B and recorded the processor and memory status during the operation.

Scenario 2: introducing VM virtualization. In this scenario, we have installed and configured Streaming Video Server [20, 21] in a VM machine. The Nginx server and the RTMP modules allow you to broadcast different types of video files. We started by installing the necessary software and dependencies and then configuring the server.

After we begin streaming the video, we start the video using VLC Player and we started the migration of the VM machine from server A to server B and recorded the state of the processor and memory during this operation. Figure 7a shows the specifications of Server A. In scenario 2 where we installed and configured a nginx [9] streaming server in a VM machine using Ubuntu 14.04 [22].

Fig. 7a. Specifications of Server A. In scenario 2

Scenario 3: Introducing Hybrid virtualization. In Scenario 3, we integrated the Docker container in which we installed the Nginx video streaming server in the VM machine.

After the running Docker Container, we start the video using VLC Player and we started the migration of the VM machine from server A to server B and recorded the state of the processor and memory during this operation. Figure 7b shows the specifications of Server A.

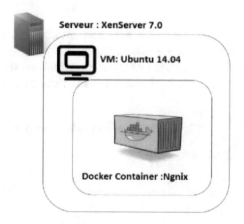

Fig. 7b. Specifications of Server A. In scenario 3

4.3 Results

Scenario 1. The migration of the VM machine between server A and B lasted 350 s. The virtual machine does not contain a streaming video server or a docker container. The following results show the status of the two servers A and B during the migration.

Server A status monitoring:

The CPU activity of the server A in all migration was normal except increments in the pre-copy iterative phase when the transfer of pages updated (dirtied) is repeated. The copy cycle stops when a number of Maximum iterations has been achieved.

The memory consumption remains maximum during the migration, and this decreases at the time of transfer from VM to the server B, where the memory used by the VM is released. Network traffic was low during all the migration. The migration time of scenario 1 is considered the reference migration time that will be used for comparison with the migration times of scenarios 2 and 3 (Fig. 8a and 8b).

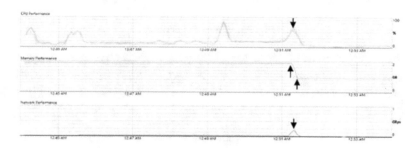

Fig. 8a. Status of Server A

Server B status monitoring:

The CPU activity of Server B was average and almost stable throughout the migration and started to rise when the VM booted to Server B.

Memory activity was low during most migration time and started to rise when the VM was started on server B. Network traffic was low during all migration.

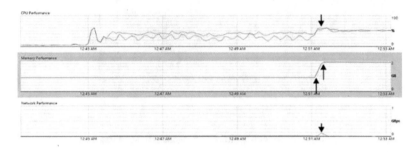

Fig. 8b.   Status of Server B

Scenario 2. The migration between server A and B took 420 s. The Nginx video streaming server is installed and configured directly on the VM machine of server A. The following results shows the state of the two servers A and B during the migration (Fig. 9a and 9b).

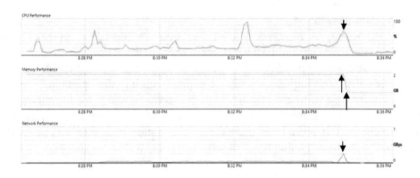

Fig. 9a.   Status of Server A

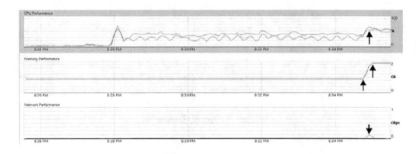

Fig. 9b.   Status of Server B

Server A. The CPU activity of the server A in all migration was normal except increment in the pre-copy iterative phase when the transfer of pages updated (dirtied) is repeated, The copy cycle stops when a number of Maximum iterations has been achieved.

The memory consumption remains maximum during the migration, and this decreases at the time of transfer from VM to the server B, where the memory used by the VM is released. Network traffic was low during all the migration.

Server B. The CPU activity of Server B was average and almost stable throughout the migration and started to rise when the VM booted to Server B. Memory activity was low during most migration time and started to rise when the VM was started on server B. Network traffic was low during all migration.

Scenario 3. Migration of the VM between server A and B lasted 390 s. The Nginx video streaming server is installed and configured in a Docker container nested in a VM machine. The following results show the status of the two servers A and B during the migration (Fig. 10a and 10b).

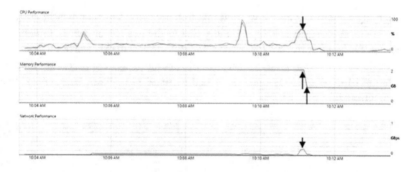

Fig. 10a. Status of Server A

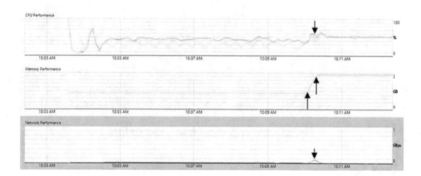

Fig. 10b. Status of Server B

In order to better compare the two scenarios we rewrite both scenarios 2 and 3 but increase the number of video streaming clients to study the impact of client overload on the server state and also the migration time (Fig. 11).

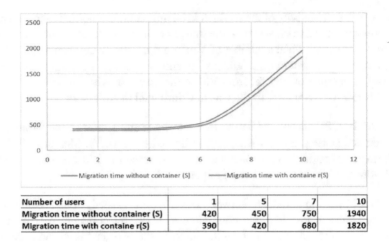

Number of users	1	5	7	10
Migration time without container (S)	420	450	750	1940
Migration time with containe r(S)	390	420	680	1820

Fig. 11. User overload impact on migration time

Comparing scenarios 2 and 3, we find that the migration time is lower in the case of scenario 3 than in scenario 2. This is mainly due to the nesting of the containers in the VM. We can explain this by the lack of a virtualization layer in the case of the containers, which translates into a rapid interaction with the operating system processes that affect the migration time. We also noticed a marked increase in network activity and use of CPU as mentioned in Fig. 12.

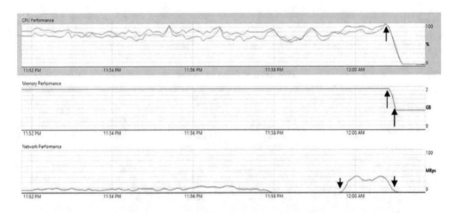

Fig. 12. Server A status with a 10 user load

5   Conclusion

In this work, we conducted a study on the value of Hybrid virtualization and its impact on direct migration. For this, we studied the direct migration of a monolithic application and precisely the video streaming installed in a container nested in a virtual machine, as such, application is sensitive to timing and performance. The results presented showed the advantages of Hybrid virtualization especially in terms of total migration time. Indeed, the absence of a real virtualization layer in the case of containers results in a rapid interaction with the operating system processes, which affects positively the total migration time. As a perspective, we will study the contribution of the Hybrid architecture on the migration and the administration of layers of N-tiers applications (J2EE application) where each third party will be hosted in a container.

References

1. Vaquero, L.M., Rodero-Merino, L., Caceres, J., Lindner, M.: A break in the clouds: towards a cloud definition. Comput. Commun. Rev. **39**(1), 50–55 (2009). SAP Res., Belfast, United Kingdom, Telefonica Investig. y Desarrollo, Madrid, Spain
2. Site web de Xen Hypervisor. http://www.xen.org/
3. HyperV. https://msdn.microsoft.com/fr/library/hh846766(v=ws.11).aspx
4. VMware. http://www.vmware.com
5. Vaughan-Nichols, S.J.: New approach to virtualization is a lightweight. Computer **39**(11), 12–14 (2006)
6. Bachu, R.: A framework to migrate and replicate VMware virtual machines to amazon elastic compute cloud: performance comparison between on premise and the migrated virtual machine. Master thesis, Blekinge Institute of Technology (BTH) (2015)
7. Liu, H., He, B.: VMbuddies: coordinating live migration of multi-tier applications in cloud environments. IEEE Trans. Parallel Distrib. Syst. **26**(4), 1045–1205 (2015)
8. Kikuchi, S., Matsumoto, Y.: Impact of live migration on multi-tier application performance in clouds. In: 2012 IEEE Fifth International Conference on Cloud Computing. University of California at Berkley, USA (2009)
9. Clark, C., Fraser, K., Hand, S., Hansen, J.G., Jul, E., Limpach, C., Pratt, I., Warfield, A.: Live migration of virtual machines. In: Proceedings of the 2nd Conference on Symposium on Networked Systems Design & Implementation, vol. 2, pp. 273–286. USENIX Association (2005)
10. Kotikalapudi, S.V.N.: Comparing live migration between linux containers and kernel virtual machine: investigation study in terms of parameters. Master of Science in Computer Science, February 2017
11. Joy, A.M.: Performance comparison between linux containers and virtual machines. In: International Conference on Advances in Computer Engineering and Applications (ICACEA), pp. 342–346 (2015)
12. Li, W., Kanso, A.: Comparing containers versus virtual machines for achieving high availability. In: IEEE International Conference on Cloud Engineering (IC2E), 9–13 March 2015 (2015). https://doi.org/10.1109/ic2e.2015.79
13. Felter, W., Ferreira, A., Rajamony, R., Rubio, J.: An updated performance comparison of virtual machines and linux containers. In: IEEE International Symposium on Performance Analysis of Systems and Software (ISPASS), pp. 171–172 (2015). https://doi.org/10.1109/ispass.2015.7095802
14. Sahni, S., Varma, V.: A hybrid approach to live migration of virtual machines. In: IEEE International Conference on Cloud Computing in Emerging Markets (CCEM), vol. 6 (2012)
15. Biancheri, C., Dagenais, M.R.: A hybrid approach to live migration of virtual machines. J. Cloud Comput. **5**, 19 (2016). ISSN 2192-113X
16. Bigelow, S.J.: Quelle différence entre conteneurisation et virtualisation? http://www.lemagit.fr/conseil/Quelle-est-la-difference-entre-la-conteneurisation-et-la-virtualisation

17. von Eicken, T.: Docker vs. VMs? Combining Both for Cloud Portability Nirvana, 02 September 2014. http://www.rightscale.com/blog/cloud-management-best-practices/docker-vs-vms-combining-both-cloud-portability-nirvana
18. Coleman, M.: Container and VMS Together, 8 April 2016. https://blog.docker.com/2016/04/containers-and-vms-together/
19. Hines, M.R., Deshpande, U., Gopalan, K.: Post-copy live migration of virtual machines. ACM SIGOPS Oper. Syst. Rev. **43**(3), 14–26 (2009)
20. Rattanaopas, K., Tandayya, P.: Performance analysis of a multi-tier video on demand service on virtualization platforms. In: International Computer Science and Engineering Conference (ICSEC), 23–26 November 2015 (2015). https://doi.org/10.1109/icsec.2015.7401437
21. Site Web du serveur vidéo Nginx. https://nginx.org/en/
22. Site web Système d'exploitation LinuxUbuntu 14.04. https://ubuntu-fr.org/

IoT Interoperability Architectures: Comparative Study

Rachida Ait Abdelouahid[✉], Loubna Chhiba, Abdelaziz Marzak,
Abdelaziz Mamouni, and Nawal Sael

Laboratory of Technology of Information and Modeling, Faculty of Sciences Ben M'sik,
Hassan II University of Casablanca, Casablanca, Morocco
rachida.aitbks@gmail.com, chhibaloubna@gmail.com,
marzak@hotmail.com, mamouni.abdelaziz@gmail.com,
saelnawal@gmail.com

Abstract. Internet of things or IoT for short is a global infrastructure of the information society which provides advanced services by interconnecting physical and virtual objects. This interconnection will be existed or evolved interoperable information and communication technologies. Currently, different existing platforms provide several such technologies with a special complexity. This generates a high cost at interoperability level. In order to mitigate this drawback, in this paper, we present and describe in detail a comparative study of different existing IoT interoperability architectures according to protocols and technologies used as well as their advantages and limitations.

Keywords: Internet of Things · Distributed architecture · Interoperability
Virtual objects

1 Introduction

Currently, most of IoT platforms allows, via systems of electronic identification normalized (standardized) and wireless, to identify and to communicate numerically the virtual world with physical objects to be able to measure and exchange data between them. This communication bases itself on a distributed architecture by communicating with messages through the network. IoT is thus allowed, using systems of electronic identification normalized and unified, and wireless mobile devices to identify without ambiguity of the digital entities and the physical objects directly and so to be able to recover, store, transfer and deal without discontinuity between the physical and virtual worlds, the data being connected with it. Currently, different existing platforms provide several technologies with a special complexity. This generates a high cost at interoperability level with the other IoT platforms. The internet of things, in which the objects of the everyday life can be equipped with capacities of identification, detection and treatment, offer important advantages both in terms of efficiency and new services. To benefit from the full potential of the IoT, objects do not just have to be simply connected to Internet, they must also be found, accessible, managed and connected potentially to other objects. To allow this interaction, one degree of interoperability is necessary which goes beyond the simple interoperability protocol such as supplied by the Internet. To mitigate this issue, this paper aims to present a clear view of the connected objects and their interoperability architectures as well as adapted technologies and fields of application. This paper is organized as follows. The second section consists of a balance sheet regarding the already existing IoT architectures of interoperability. The third section is dedicated to present a comparative study of the IoT interoperability architectures. The fourth section is dedicated to present a synthetic study. Finally, the last section presents a conclusion of the recapitulative of the study realized and future perspectives.

2   Related Works

Classically, the interoperability is the connection of the people, the data and the diversified systems [1]. Interoperability is of great importance and relevance in large systems and should be seen as a requirement. To be interoperable means to be able to exchange streams of various kinds and to share the elements realized by these flows with confidence in order to carry out an action that is independent of the environment with which these flows exchange [1]. In literature, many works were conducted in this area; most of them focus on proposing new architectures and new approaches to support IoT interoperability. Desai et al. [2], have proposed a semantic Web-based IoT architecture to ensure interoperability between systems, using established communication and data standards. This architecture allows the discovery of physical sensors and the interpretation of messages between objects, using the SGS (Semantic Gateway as a Service), which allows the translation between messaging protocols such as XMPP (Extensible Messaging and Presence Protocol), CoAP (Constrained Application Protocol) and MQTT(Message Queue Telemetry Transport) via an architecture multiprotocol proxy. As well as the use of W3C's Semantic Sensor Network (SSN) ontology for semantic annotations of sensor data provides semantic interoperability between messages and supports semantic reasoning to obtain knowledge that can be used at higher levels from low-level sensor data. However, this solution is not secure in term of confidentiality of data exchanged and limited in term of functionality and cost to the technologies implemented. Blackstock et al. [3], have proposed a new architecture of conception and super imposing which meets the requirements of the IoT. This approach is mainly based on protocols distributed by hash table to supply the required flexibility. For that, Blackstock et al. have partitioned networks in two main parts: points of accesses and intelligent objects, which are based on two protocols viz; system chord and virtual cord protocol. The proposed approach treats effectively cases of mobility. However, this solution is not able to optimize resources consumption and their cost. Ghofrane [4], has proposed an approach based on distributed hash table systems and the VCP protocol (Virtual Cord Protocol). Since intelligent objects are placed on a massive scale in all parts of the world, they have divided the network into two main parts: the first part consists of access points and the second part consists of intelligent objects. Access points have significant memory sizes and processing capabilities. They are static and have no energy constraints because they have access to power supply. This allows them to have extensive coverage. Each access point is responsible for intelligent objects in its coverage. It stores important information related to these intelligent objects. This architecture provides a higher degree of interoperability as well as some flexibility and mobility taking advantage of technologies such as REST and JSON architecture. However, the cost of the technologies used is very high. Aloi et al. [5], have proposed and deployed complete mobile gateway software architecture to support IoT interoperability through a Smartphone-centric application. It is based on a multi-standard, multi-interface and multi-technology communication framework capable of integrating different communication standards and radio interfaces into an entire platform. The presented solution allows the continuous collection and transmission of data from wireless IoT devices and sensors transmitted on different interfaces and communication standards. In addition, it can send control messages or data streams, such as streaming video, to neighboring IoT devices in an opportunistic manner. The results obtained using a real test bench on various common smart phones show that the proposed software architecture is capable of acting as a data collector, disseminating and managing by allowing new ubiquitous and transient services without make excessive use of hardware resources in terms of CPU (Central Process Unit) and memory. However, this architecture is not efficiency in term of energy consumption level. Antonio et al. [6], have proposed a relevant approach which present a distributed software-controlled sector aware spectrum sensing architecture, in order to store and analyze the spectrum usage information. This approach has been integrated to a future Internet architecture called Nova Genesis. This later makes it possible to develop more efficient platforms which optimize the consumption of resources, with a low cost of detection and low power consumption, as well as it is based on a heterogynous technologies like cognitive radios network, bands RF, Restful and is mainly based on different protocols like HTTP, TCP/IP/Ethernet, SCC/SSS, SeroMQ which makes it possible to provide platforms with better confidentiality and a simple access to data exchanged. However, the proposed solution is not secure at the data confidentiality level.

3   Comparative Study

In this section, we present a comparative study of the IoT interoperability architectures studied in the previous section in tabular format. To compare these different architectures, we have based on the following criteria:

- Protocols: Many M2M (Machine to Machine) devices for IoTs use different protocols for communication (ex, COAP, XMPP, RESTFUL, MQTT, RDF, XML, WIFI, ZWANE, NFC, ZIGBEE and 4G);
- Technologies: Most IoT interoperability architectures are focused on certain technologies (ex, cognitive radios, RF bands, REST, hub and JSON);
- Application domain: Each IoT interoperability architectures is designed to be used in a field of application such as transport, intelligent vile, health, home automation, and more;
- Confidentiality: Each IoT platform is protected against unauthorized access such as access control and access audit;
- Efficiency: This feature is related to the ability of an IoT solution to provide desirable performance at the level of usage and minimization of resources (CPU, memory,

disk, etc.) consumption, as well as how to run, speed archiving. The effectiveness of time and resource behaviour is distinct. IoT devices suffer from limited battery life and dominate power consumption. Hence the need for energy efficiency that can be increased by judiciously adjusting transmission power or optimizing the use of technologies. Therefore, there is a need for solutions that limit the energy consumption of these IoT devices;

- Mobility: This feature is simply used in IoTs applications; it refers to the ease of connecting users to IoT application information in a quick and efficient manner.
- Flexibility: (Adaptability): Ease of adding/modifying/removing functionality: Modularity, generality, extensibility, auto descriptively;
- Distributed model: A distributed system is a model in which components located on networked computers communicate and coordinate their actions by passing messages [7]. The components interact with each other in order to achieve a common goal. Three significant characteristics of distributed systems are: concurrency of components, lack of a global clock, and independent failure of components [8]. Examples of distributed systems vary from SOA-based systems to massively multiplayer online games to peer-to-peer applications;
- Cost: It mains the cost of energy consumed and resources used in different IoT platforms and devices;
- Functionality: The ability of a proposed architecture to deliver functions that respond to explicit and implicit needs in a given context, for the IoTs applications, functional specifications means the ability to provide the appropriate functions to meet the specific tasks and the needs of the user with the degree of accuracy expected, for IoTs platforms it is the ability to interact with one or more specified systems; finally, the cost.

Table 1 below summarizes the various IoT interoperability architectures and Table 2 summarizes the advantages and limitations of these architectures. We have defined seven criteria for comparing these different architectures, viz; the criteria of confidentiality, efficiency, mobility, flexibility, distributed model, cost, and functionality.

Table 1. Comparative study of IoT interoperability architectures

References	Features						
	Confidentiality	Efficiency	Mobility	Flexibility	Distributed Model	Cost	Functionality
Ghofrane [4]	x	x	x	x	x		
Aloi et al. [5]	x	x	x	x	x	x	
Desai et al. [2]		x	x	x	x		
Mike et al. [3]	x		x	x	x		
Antonio et al. [6]		x	x	x	x	x	x

Table 2. Advantages and limitations of IoT interoperability architectures

References	Features				
	Protocols	Technologies	Application domains	Advantages	Limitations
Ghofrane [4]	- Système chord - VCP - DHT117/ MOBYDHT	- Smart Bluetooth	Smart Cities	- Distributed architecture - Flexible - mobile	- High cost - Do not optimize the resources
Aloi et al. [5]	- WIFI - ZWANE - NFC - ZIGBEE - 4G	- Smart phone	Traffic Management	- Reduce use of material resources - acts as a flexible interface and transparent-mobile	- Gourmand at level of energy
Desai et al. [2]	- CoAP - XMPP - MQTT - RDF - XML	- SGS - SemSOS - SSN - Ontology - REST - JSON	Smart Home Health Care	- Ensure connectivity between silos - provides semantic interoperability between messages supports semantic reasoning	- Luck of security at the data confidentiality level - Does not optimize resources -functionality limited
Mike et al. [3]	- HTTP	- HUB - JSON - REST	Smart Streets	- Mobile - Flexible - Efficiency	- Gourmand at the level of energy
Antonio et al. [6]	- HTTP - TCP/IP - SeroMQ	- RRC - Bandes RF - RESTful	Transport network	- allow detecting cost - Low power consumption	- Do not secure at the data confidentiality level

4   Discussion

As shown in Tables 1 and 2 above, there is no IoT interoperability architecture that strongly meets all predefined criteria. Moreover, despite the diversity of architectures and their usefulness, these solutions are often very limited, they deal with problems in isolation, most proposals do not meet the requirements and specificities requested and do not take into account all areas of application, see also that many works do not meet the new requirements which have emerged in the field of software engineering, in particular, interoperability, mobility and confidentiality. Most of these architectures allow to answer certain interoperability problems but according to a single application domain. More specifically, the architecture proposed by Desai et al. [2] makes it possible to develop more efficient platforms which optimizes the consumption of resources, with efficient execution and storage. As well as this architecture is based on the technologies of the web service like CoAP, XMPP, MQTT, JSON, RD and XML, which makes it possible to provide platforms with better confidentiality and a simple access to data exchanged. The architecture proposed by Blackstock et al. [3] makes it possible to develop platforms protected against unauthorized access with the ease of changing the execution environment (operating system, hardware, etc.), this is due to the use of JSON and REST web-service technologies. Also, this architecture is deployed in a distributed environment. The architecture proposed by Ghofrane [4], in the domain of smart cities, is mainly based on tree core technologies viz; Hub, Json and Restful, and makes it possible to develop mobile and flexible platforms that are accessible at any time using HTTP protocol. The architecture developed by Aloi et al. [5], makes it possible to develop mobile platforms that are accessible at any time using mobile technologies such as Smart phones. The field of communication, they allow anybody, wherever it is, to remain in constant contact, thus making it possible to take advantage of all the protocols that are available in this technology like Wifi, Zwane, NFC, Zigbee and 4G. This allows a better optimization of the cost of developing platforms based on this architecture. The architecture proposed by Antonio et al. [6] allows the developed platforms to deliver the functions responding to explicit and implicit needs in a given context. It is based on Restful web-service technology and cognitive radios networks that enable low-cost detection. Therefore, from these two tables, we find after the analysis that despite the diversity of these architectures and their usefulness, they have shown some crucial limitations and weaknesses namely:

- Most of these architectures do not meet the requirements of mobility, functionality, efficiency and cost optimization;
- The vast majority of these architectures are limited to a specific application domain;
- These architectural proposals generally limited to the level of energy consumption;
- Most of them do not offer the ability to deliver platforms that interact with one or more specified systems, which also offer relevant, accurate, secure and compliant functionality;
- None of these architectures presents a generic model taking into account all protocols and technologies and their use cases.

5   Conclusion and Future Work

In this paper, we have presented and described in detail a comparative study of different existing IoT interoperability architectures according to protocols and technologies used as well as their advantages and limitations. This study have identified a set of original seven criteria viz; confidentiality, efficiency, cost, flexibility, mobility, model of architecture, and functionality. Basing on these criteria, this study shows after the analysis that despite the diversity of these architectures and their usefulness, they have shown some crucial limitations and weaknesses namely: Most of these architectures do not meet the requirements of mobility, functionality, efficiency, and cost optimization; the vast majority of them are limited to a specific application domain; most of them do not offer the ability to deliver platforms that interact with one or more specified systems, which also offer relevant, accurate, secure and compliant functionality; finally, none of these architectures presents a generic model taking into account all protocols and technologies and their use cases. In order to mitigate the limitations of existing architectures, we planning in our future work to propose a new architecture allowing to improve and to support IoT interoperability.

References

1. Chapurlat, V., Daclin, N.: System interoperability: definition and proposition of interface model in MBSE Context. In: the Proceedings of the 14th IFAC Symposium on Information Control Problems in Manufacturing, IFAC, Bucharest, Romania, 23–25 May (2012)
2. Desai, P., Sheth, A., Anantharam, P.: Semantic gateway as service architecture for IoT interoperability. In: The IEEE International Conference on Mobile Services, New York, NY, pp. 313–319 (2015) https://doi.org/10.1109/MobServ.2015.51
3. Blackstock, M., Lea, R.: IoT interoperability: a hub-based approach. In: The International Conference on the Internet of Things (IOT), Cambridge, MA, pp. 79–84 (2014). https://doi.org/10.1109/IOT.2014.7030119
4. Ghofrane, F.: A distributed and flexible architecture for Internet of Things. In: The International Conference on Advanced Wireless, Information, and Communication Technologies, AWICT, pp. 130–137 (2015)
5. Aloi, G., Caliciuri, G., Fortino G., Gravina, R., Pace, P., Russo, W., Savaglio, C.: A mobile multi-technology gateway to enable IoT interoperability. In: The First International Conference on Internet-of-Things Design and Implementation (IoTDI), Berlin, pp. 259–264. IEEE (2016). https://doi.org/10.1109/IoTDI.2015.29
6. Antonio, L., Albertia, M., Daniel, M., Bontempo, M., Lucio, H.O., Rodrigo, R.R., Arismar Jr., C.S.: Cognitive radio in the context of Internet of Things using a novel future internet architecture called Nova Genesis. Comput. Electr. Eng. **57**, 147–161 (2017)
7. George, C., Jean, D., Tim, K., Gordon, B.: Distributed Systems: Concepts and Design, 5th edn. Addison-Wesley, Boston (2011). ISBN 0-132-14301-1
8. Kashif, D., Amir, T.K., Harun, B., Frank, E., Kurt, G.: A resource oriented integration architecture for the Internet of Things: a business process perspective. In: Pervasive and Mobile Computing (2015)

Collaborative and Communicative Logistics Flows Management Using the Internet of Things

Loubna Terrada$^{(\boxtimes)}$, Jamila Bakkoury, Mohamed El Khaili, and Azeddine Khiat

Laboratory Signals, Distributed Systems and Artificial Intelligence, ENSET Mohammedia,
Hassan II University, Casablanca, Morocco
loubna.terrada@gmail.com, jamila.bakkoury@gmail.com,
elkhailimed@gmail.com, azeddine.khiat@gmail.com

Abstract. Companies primarily use tools based on information systems such as ERP, WMS, APS, TMS or other similar existing systems to keep their logistics flows under control. The Internet of Things arrival (IoT) was a revolution in the field of Information and Communication Technologies (ICT), with the aim of extracting, transferring, storing, processing and sharing the necessary information at every logistics activity. In addition, for a better collaboration and interoperability improvement in the Supply Chain, it is important to automatically communicate and share each operation related to the logistics flows to the actors involved. In this paper we give a general overview of the use of IoT in the supply Chain Management to ensure the convenience of its activities and that it is thus collaborative and communicative.

Keywords: Internet of things · Supply chain management · Communication
Collaboration · Interoperability

1 Introduction

This paper is the subject of a review on the exploitation of the IoT in the logistics flows management. The set of the information logistics systems have become imperative for supply chain management, it is also subjected to several challenges and struggles requiring improvements to existing operating practices. The heterogeneity of the platforms and technologies used by the various links in the supply chain has made it difficult to identify, track and control the flows in the chain in real time. Therefore, The Internet of things advent brings a new approach to collect, transfer, store and share information throughout the SC for better collaboration and communication between supply chain stockholders in order to optimize the overall costs and to increase the revenue through enhancing services [1]. In this study, we present the set of definitions related to the concepts of the Supply chain aggregated to the IoT. The last section will deal with the exploitation of connected objects in the management of logistics flows within the framework of a value system.

© Springer International Publishing AG, part of Springer Nature 2019
J. Mizera-Pietraszko et al. (Eds.): RTIS 2017, AISC 756, pp. 216–224, 2019.
https://doi.org/10.1007/978-3-319-91337-7_21

2   Overview of the Supply Chain Management

The Supply Chain Management concept has appeared since 1980s [2], in order to abolish the old concepts - which must be put into questioned– that limited logistics in few functions (i.e. warehousing and Transport). Since, the SCM has become a crucial pillar of innovation in the management of material, financial and informational flows from supplier level to production, distribution until the final customer as shown in the Fig. 1.

Fig. 1. An example of Supply Chain structure [3]

The SCM's principle is to maintain cooperative relationships between stockholders by developing structured logistical links to achieve overall performance up to the end customer. This supply chain is essentially set up to efficiently and efficiently produce and make products available to end consumers by creating values throughout the whole process, based on the performance of each stakeholder, but each entity directs the supply chain to its own account in order to achieve its own goals and promote its interests - this problem and generally spread among SMEs. The second common problem in the logistical process concerns uncertainty in forecasting and planning, as each stage in the Supply chain requires a high level of stock to avoid stock-outs. And to get rid of inventory changes that is constantly going to generate over-stock, thus, this phenomenon is called "Bullwhip effect" [4].

SCM assumes the integration and collaboration of the set logistics activities, whose purposes is to plan, control and manage material or non-material flows. Thus, companies basically get hold of some tools, such as: Enterprise Resource Planning (ERP), Advanced Planning System (APS), Warehouse Management System (WMS), and Transport Management System (TMS). Most of them are struggling and defeated by the current SC challenges (e.g. coordination and overall supply chain governance, collaboration...), these tools must deal with risk management and decision-making at the local and global level for a decentralized supply chain; which require the interoperability of logistics networks with the constraints of standards heterogeneity. [5]. Hence the necessity of using new ICTs (developed or under-development) linked to IoT. Researchers estimate that the IoT will reach billion units by 2020, and will consider all supply chain partners and linked operations, from production line and warehousing to retail and delivery. Industrial enterprises tend to invest in the IoT to set up and optimize their workflows, to reduce their factory costs and improve supply efficiency [1]. In the next section we will describe in detail the concept and functioning of the IoT.

3   Internet of Things (IoT)

The Internet of Things (IoT), also called the Internet of Everything or the Industrial Internet, is a new technology paradigm ideated as a global network of machines and devices able to interact with each other [1]. Recently, the world has experienced an impressive development of the multimedia world. This is due to the technical and technological progress and major innovations that have revolutionized the world of telecommunication, IT cloud (i.e. Cloud Computing), social media, Internet of Things…

The latter represents the extension of the internet of things and places in the physical world. While the Internet usually is not extended beyond the electronic world, the Internet of Things represents the exchange of information and data from devices present in the real world to the Internet. Indeed, the Internet of Things is regarded as the third evolution of the Internet, known as Web 3.0 [6]. The objects constituting the "Internet of objects" called "connected", "communicating" or "smart".

Currently connected or intelligent objects are used everywhere, they invaded the world and impacted our personal and professional lives. They generate billions of information that must be processed and analyzed then stored to make them usable. According to Cisco [7] 50 to 80 billion connected devices will be in circulation worldwide in 2020. In fact, a connected object is an object whose primary purpose is not to be computing devices or web access interface but that the addition of an internet connection has added additional value in terms of features, information and interaction with its environment [8]. Today, connected objects begin to take part in our daily lives and are translated into several and different objects in multiple fields of application [9].

3.1   The Various Forms of Internet of Things

The Internet of Things has a universal character to design objects connected to various uses. There are 3 types of IoT [10]:

- Connected objects through Chips and Smart labels
- M2M [11]
- Computers, Smartphone, Tabs

Since 2010, these three families have experienced a strong evolution with a number of connected objects in the world not exceeding 5 billion according to IDATE statistics in 2013 [11]. Five years after, the world of connected objects has evolved greatly to reach the 40 billion intelligent devices in circulation around the world. According to the predictions, this number could be between 50 and 80 billion in the world by 2020 [12].

3.2   Architecture of IoT

The Internet of Things architecture can be illustrated by different models with various supports IoT technologies [13]. It serves to illustrate how they are interconnected in different scenarios. The Fig. 2 illustrates the role of the various processes of the architecture of IoT:

- Sensors to transform a physical quantity analog to a digital signal.
- Connect allows interfacing a specialized object network to a standard IP network (LAN) or consumer devices.
- Store calls made to aggregate raw data produced in real time, Meta tagged, arriving in unpredictable ways.
- Present indicates the ability to return the information in a comprehensible way by humans, while providing a means to do it and/or interact.

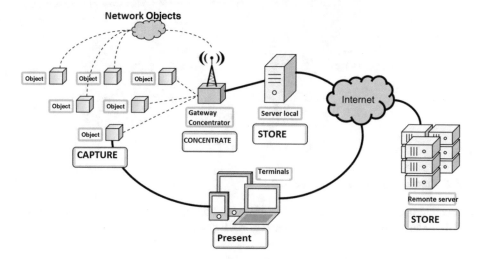

Fig. 2. Architecture modelling of IoT

In industrial firms, five IoT technologies are basically used for the deployment of successful IoT-based products and services [1], namely:

- Radio frequency identification (RFID);
- Wireless sensor networks (WSN);
- Middleware;
- Cloud computing;
- IoT application software.

3.3 Functionning of IoT

The Internet of Things (IoT) concept allows objects to the relationship between what happens online and the physical world. Christened in 1999 by researcher Kevin Ashton center at MIT (Massachusetts Institute of Technology), his team launched the promotion of open connectivity of all objects using RFID (Radio Frequency IDentification). The first solutions in this area seemed simple calling itself Machine To Machine (M2M), they are based on the fact that a single type of hardware/sensor connects via a service gateway (Internet) to a single application. Then these solutions have become more complex while using several sensor/display/actuator using a multi serving gateway (Internet TCP/IP) to a single application. Since the work more complicated because there are other standards taken into consideration as the geographic dispersion and increased device data consumers [10]. The functioning of the Internet of Things via cloud is shown in the illustration below (Fig. 3):

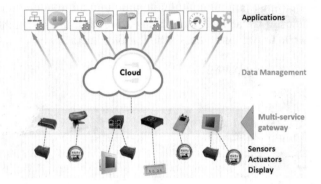

Fig. 3. IoT functioning via Cloud

3.4   The Application Areas of IoT

The Internet of Things can be used in plenty and different domains, such as: Healthcare [14], Industry, sport, Smart cities, Smart grids, logistics…

3.5   Benefits and Drawbacks of IoT

The Internet of Things provides many benefits for people and companies. However, this new innovation is still limited by sensitive safety issues. So what are these different advantages and disadvantages? (Table 1).

Table 1.   Benefits and drawbacks of IoT

Benefits	Drawbacks
• Lead time processing • Accessibility & mobility: « anytime, anywhere, any device» • Tracking ability • Modern life style	• Data security • Personal privacy • Data massivity • Environment issues

4   Using the Internet of Things for Logistics Performance Improvement

4.1   A New Concept of Logistic Management

An enhanced real-time visibility into the product physical flows is a crucial step to ensure a proper Business Management, from this necessity derives the idea to design new applications able to ensure and enhance the tracking function and to avoid stock-out and over-stock generated by increasing swings in inventory i.e. using RFID tags or GSM tags [15]. Nowadays, firms are trying to make their processes intelligent by improving the inventory function in order to avoid stopping the activities of the warehouse when the quantities in stocks can be obtained directly by census of the number of products present in a perimeter. This technology is already widely deployed in several industries (e.g. Decathlon and Amazon). The Fig. 4 shows an example of the overall architecture of the solution where RFID technology is used to ensure the tracking in the SC of a wine wholesaler. This section presents the main components of the platform and related technologies. Indeed, we will focus on the main functions such as RFIDentification, tracking and labelling, communication, transmission and data sharing [16].

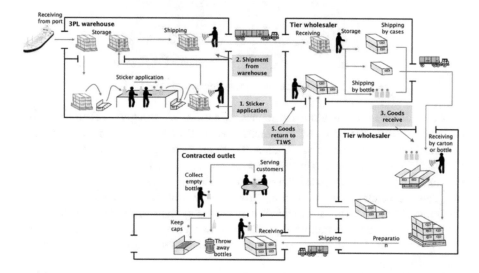

Fig. 4. Using RFID in a SC of wine

This model is an IoT based wine bottles-routing inventory and SCM information sharing system that involves: RFID tags, position, handled readers and other similar kind of devices. The installed database has the central position in the divided system, it can insure the communication between suppliers and distributors whereby loading and inventory workers. As we can see the RFID helps the company to optimize pick-up and delivery routes between delivery workshops and warehouses. It also optimizes the resources of the handling by determining the most appropriate equipments, and aim to reverse logistics as well.

4.2   Identification and Tracking Using RFID

In the literature, researchers focus on the emphasis of the RFID in the Supply chain field to increase the availability of stock in the warehouse and to optimize the overall costs that represent an important indicator for performance SC improvement. RFID technologies are emerging in areas as diverse as logistics. The incorporation of such devices into industrial products could lead to a world in which objects communicate and interact with each other and with humans, i.e. "the Internet of things" (Fig. 5).

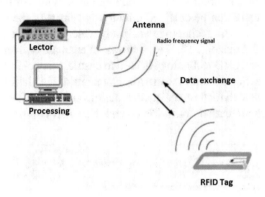

Fig. 5. RFID functioning [5]

According to our literature researches, Mitton et al. categorize RFID tags and describe the different types of this mutant technology coming along with burning research topics with sample application. There are two types of RFID tags; Passive RFID: it is a tag that retro-modulates the wave coming from the interrogator to transmit information. It does not include RF transmitters. The passive tag generally uses the wave (magnetic or electromagnetic) coming from the interrogator to supply the on-board electronic circuit. And the active RFID tags known by their autonomy and embark an RF transmitter. The communication with the interrogator in this case is peer-to-peer type. This type of tag usually has a power source and is often coupled with a temperature or humidity sensors.

Thus, the main issue of the industrial manufacturing is the high cost of the RFID tags that limits the use of this technology. While companies strive to label each product to ensure better traceability and easy inventory, and make the use of barcodes an outdated application, which opens up new avenues for research to explore[17].

The RFID is basically used to provide solutions to problems related to Bullwhip effect mentioned in the second section, thus the benefits of using such a technology in the SCM can be summed up as following [18]:

– Inventory accuracy
– Diminishing error rate
– Customer relationship management
– Security
– Productivity gains
– Tracking enhancement
– Real time visibility of the overall SC components

According to surveys, the RFID solution has helped enterprises to increase the availability of stock in their stores, improve the efficiency of their retail and logistics platforms. For instance, Decathlon saw an 11% increase in sales from July 2014 to July 2015, and the company attributes part of that growth to the RFID deployment [19]. In fact, Decathlon began a global rollout of an RFID solution for tracking throughout its supply chain. Today, RFID is improving efficiencies in all Decathlon facilities with 1,030 stores and 43 warehouses. The company has tagged 1.4 billion items.

A communicative supply chain is based on a set of performance indicators (KPI) aggregated to the deployment of RFID in warehousing and management inventory, such as [18] (Table 2):

Table 2. Performance indicators linked to RFID solution in the SCM

KPI in warehousing and management inventory	KPI in production	KPI in transport
- Costs	- Cost	- Route optimization
- Inventory level	- Lead time	- Quality
- Inventory turns	- Quality	- Operating costs
- Delivery time	- Productivity	- Delivery time
- Good returns	- Service level	
- Stock-out condition		
- Service level		
- Resource optimization		

It is noteworthy that major issue is related to the warehousing and management inventory. Consequently, the majority of researchers in this field has focused on this weak point, thus highlighting the leading causes of its failures and the effects of using IoT to get over the inventory inaccuracy. However, this IoT technology is still facing struggles, inasmuch as a good RFID functioning, IT infrastructure must be able to deal with the huge amount of data generated by millions of transactions and transformations [19]. The amount of information is not only huge; it is also accessible "instantly". In addition, several labels can be read simultaneously.

5 Conclusion

Logistics today is an integrative philosophy of flows management technologies. Indeed, Supply Chain Management is an important part of the scope of new technologies and concepts related to IoT. However, there are still a lot of challenges and struggles to adopting firms in terms of privacy, security [18], data explosion, integration and sharing on Cloud platforms [14].

Sensors and devices used in the IoT generate massive amount of data that need to be processed and stored as well [18]. Consequently, researchers and practitioners aim today to improve the performance of networks in terms of throughput and energy consumption to convey data from sensors as well as the improvement of algorithms in the processing of large volumes and heterogeneous data.

This research is contributing to the improvement of logistics performance by using the new technologies aggregated into the Internet of Things which present today a new trend in the architecture of the information systems whose standards and protocols of communication is subjected to the voluminous mass and diversity of data sources.

References

1. Lee, I., Lee, K.: The Internet of Things (IoT): Applications, Investments, and Challenges for Enterprises. Kelley School of Business, Elsivier (2015)
2. Giordano-Spring, S., Travaille, D.: Chaîne logistique intégrée et intégration des informations comptables. Université Montpellier, Logistique & Management (2010)
3. Helo, P., Szekely, B.: Logistics information systems: an analysis of software solutions for supply chain co-ordination, Industrial Management & Data Systems (2005)
4. Ducq, Y., Berrah, Y.: Supply Chain Performance Measurement: management models, performance indicators and interoperability. University of Bordeaux, IFAC (2009)
5. Nait-Sidi-Moh, A., Durand, D., Fortin, J.: Internet des objets et interopérabilité des flux logistiques: état de l'art et perspectives. Université de Picardie Jule Verne (2015)
6. LeMag Web4. 10 Juillet (2015). https://lemag.agenceweb4.ch/site/fr/lemag-web4?tag=1
7. EVANS, Dave. Cisco. L'internet des objets, Avril 2011. http://www.cisco.com/web/CA/solutions/executive/assets/pdf/internet-of-things-fr.pdf
8. COULON, Alain. L'internet des objets un gisement à exploiter. Hiver (2010). http://www.adeli.org/document/23-l78p26pdf
9. Vineela, A., Rani, L.S.: Internet of Things – overview. Int. J. Res. Sci. Technol. 2(4), 8–12 (2015)
10. International Telecommunication. The Internet of Things. ITU (2005). https://www.itu.int/net/wsis/tunis/newsroom/stats/The-Internet-of-Things-2005.pdf
11. BUSINESS. Objets Connectés Et Machines Communicantes (M2M). 27 Novembre 2015
12. Nemri, M.: France Stratégie, Janvier 2015 N°22 Source: IDATE (2013). http://www.strategie.gouv.fr/sites/strategie.gouv.fr/files/atoms/files/notes_danalyse_22.pdf
13. Plouin, G., Colomer, N.: Colomer Modèles d'architectures de l'Internet des Objets, 13 September 2011. http://blog.octo.com/modeles-architectures-internet-des-objets
14. Jean.caelen. ObjetsConnectes (2011). http://www-clips.imag.fr/geod/User/jean.caelen/Publis_fichiers/ObjetsConnectes.pptx
15. Gnimpieba, Z.D.R., Nait-Sidi-Moh, A., Durand, D., Fortin, J.: Using Internet of Things technologies for a collaborative supply chain: Application to tracking of pallets and containers. Université de Picardie Jule Verne, Elsivier B.V. (2015)
16. Comment l'Internet des objets va révolutionner la logistique et le transport, Supply Chain Magazine - Mars 2015
17. Mitton, N., Simplot-Ryl, D.: De l'Internet des objets à l'Internet du monde réel. Elsevier. Inria, Lille (2011)
18. Brahim-Djelloul, S.: Impact de l'utilisation de la technologie RFID sur la performance d'une Supply Chain int´egrant le transport, Ecole centrale des arts, Paris (2014)
19. Violino, B.: Decathlon scores a big win with RFID. RFID J. (2016)

Performance Analysis of Internet of Things Application Layer Protocol

Manel Houimli[(✉)], Laid Kahloul, and Siham Benaoune

LINFI Laboratory, Computer Science Department, Biskra University, Biskra, Algeria
maneling@gmail.com, kahloul2006@yahoo.fr, s.benaoune@gmail.com

Abstract. This paper deals with the modeling and performance analysis of the widely used protocol in the Inernet of Things (IoT). Developing a successful IoT application or protocol is still not an easy task due to multiple challenges such as mobility, reliability, scalability, management, availability, interoperability, and security. Therefore, the implementation of protocols for IoT must comply the standards and satisfy the good properties. The formal methods can meet these conditions. The MQTT protocol is currently considered one of the most serious candidates for transporting data within IoT architectures. The contribution of this original paper is outlined in the formal modeling and performance evaluation of MQTT 3.1.1 using timed and probabilistic automata and statistical model checker, respectively, provided in UPPAAL SMC tool-set.

Keywords: MQTT protocol · IoT · Performance evaluation
Formal specification · Statisitcal model-checking

1 Introduction

Today, computers and the Internet are becoming increasingly a necessity of the modern life. Over time, the computer has been integrated in different objects of our daily life. Moreover, with the Internet, these objects can connect and communicate with each other, creating opportunities for more direct integration of the physical world into computer systems, and resulting in greater efficiency, accuracy and economic benefits, and moreover, human intervention will be reduced. The concept of linking objects to the Internet, known nowadays as the Internet of Things (IoT) [14]. The Internet of Things is considered as the current revolution of the Internet and Machine-to-Machine technologies. This can be seen in the connectivity of the physical objects or networked embedded devices of our world via the infrastructure systems such as smart houses, intelligent transportation, smart grid. Hence, the development of communication protocols is required to consider the IoT challenges such as the compatibility between the heterogeneous objects, the actions control, the autonomous reaction and the automatic identification, etc. In addition, The implementation of protocols for IoT has to deal with some problems related to the particular nature of the physical devices.

© Springer International Publishing AG, part of Springer Nature 2019
J. Mizera-Pietraszko et al. (Eds.): RTIS 2017, AISC 756, pp. 225–234, 2019.
https://doi.org/10.1007/978-3-319-91337-7_22

Among these problems, we find the problems of limited computational capacity, the low memory, and the low energy consumption. Such problems make the design of these protocols complicated. Hence, the implementation of IoT protocols must respect the standards and satisfy the expected good properties. The formal methods can ensure this last constraint.

In this paper, we present a formal study for one of the IoT application layer protocol Message Queue Telemetry Transport (MQTT 3.1.1) [1]. The originality of this paper is focusing on the formal modeling of the 3.1.1 version of the MQTT using timed and probabilistic timed automata, and on the other hand, the formal verification of some indispensable properties in networking such as liveness, safety and reachability and the performance analysis of some important metrics related to this protocol, using the statistical model checking provided in UPPAAL tool-set.

This paper is organized as follows: after introduction, Sect. 2 presents the description of the MQTT 3.1.1, its control packets, the structure of these packets and the quality of service levels. Section 3 proposes the modeling of MQTT 3.1.1 using timed and probabilistic automata and the verification of this protocol using UPPAAL SMC. Section 4 presents the related work overview, before concluding the work in Sect. 5.

2   MQTT 3.1.1 Informal Description

MQTT 3.1.1 is a message publishing and subscribing protocol. Clients do not communicate directly with each other, they publish messages on a broker, and the messages are composed of a content and a topic. The broker stores the last message for each topic. Clients who are interested in a subject's messages can retrieve them by signing in to the broker. This solution has the advantage to allow for several clients to communicate even if they are never connected at the same time to the broker. The communication using this protocol is performed by exchanging a set of MQTT control packets which are illustrated in the Table 1.

MQTT 3.1.1 follows an operating mode which is explained in the following subsections.

2.1   Connection/Disconnection

MQTT 3.1.1 uses persistent connections between clients and the broker, and it uses network protocols that guarantee a high level of reliability such as TCP. Before sending control commands, a client must first register with the broker, which is done with the CONNECT command. Various connection parameters can then be exchanged such as the client identifier or the desired persistence mode. The broker must confirm to the client that the registration has been taken into account, (i.e., it indicates that an error has occurred by returning a CONNACK accompanied by a return code). There is a PINGREQ command to let the broker know that the client is still active, the broker will respond with a PINGRESP to tell the client that the connection is still active. When the client wants to disconnect, it sends a

Table 1. MQTT 3.1.1 Control packets types

Control packet	Description
CONNECT	The first packet must be sent after the establishment of the network by the client
CONNACK	Acknowledgment packet of CONNECT must be sent by the server
PUBLISH	A packet for transporting application message which can be sent by both server and client
PUBACK	A packet of PUBLISH acknowledgment
PUBREC	The response to a PUBLISH packet
PUBREL	The response to a PUBREC packet
PUBCOMP	The response to a PUBREL packet
SUBSCRIBE	A packet can be sent by client to create one or more Subscriptions and registers a client's interest in one or more topics
SUBACK	A packet of SUBSCRIBE acknowledgment that confirms the reception and processing of a SUBSCRIBE packet
UNSUBSCRIBE	Can be sent by a client to unsubscribe from topics
UNSUBACK	A packet of UNSUBSCRIBE acknowledgment that confirms the reception of an UNSUBSCRIBE packet
PINGREQ	Can be sent by a client when there is no control packet to be sent to the server. This packet indicates to the server that the client is alive. It is also used for exercising the network to indicate that the network connection is active
PINGRESP	Can be sent by the server in response to a PINGREQ packet. It indicates that the server is alive
DISCONNECT	Can be sent by a client as the last control packet to the server in order to indicate the clean disconnection of the client

DISCONNECT command to the broker. Otherwise, the broker will consider the disconnection to be abnormal and will send accordingly the message of will *WILL* on behalf of the client to all the subscribers.

2.2   Subscriptions and Publications

Each published message is necessarily associated with a topic, which allows its distribution to the subscribers. Topics can be organized in tree hierarchy, so subscriptions can be based on filtering patterns. Subscription management is very simple and consists of three essential commands: (i) SUBSCRIBE: allows a subscriber to subscribe to a topic, once subscribed, it will then receive all the publications concerning this topic. A subscription, also, defines a quality of service level. The successful reception of this command is confirmed when the broker sends a SUBACK carrying the same packet identifier. (ii) UNSUBSCRIBE: gives the possibility to cancel a subscription, and thus, no longer receives subsequent publications. The successful reception of this command is confirmed when the broker sends a UNSUBACK carrying the same packet identifier. (iii) PUBLISH: initiated by a client, it allows to publish a message which will be transmitted by the broker to the possible subscribers. The same command will be sent by the broker to the subscribers to deliver the message.

3   Formal Modeling and Verification of MQTT 3.1.1

This section presents the formal modeling and the formal verification of the protocol MQTT 3.1.1 using UPPAAL SMC [11].

3.1   Formal Modeling Using Probabilistic Timed Automata (PTAs)

Probabilistic Timed Automata (PTAs) [18] are considered as a modeling formalism dedicated for the systems in which real-time and non-determinism are the main characteristics of their behavior. Probabilistic timed automata are an extension of classical timed automata. They allow a more effective representation of the probabilistic aspect of a timed system. The probabilistic timed automaton provided in UPPAAL SMC tool-set defines the stochastic interpretation by: (i) probabilistic choices (non-deterministic choices between multiple enabled transitions); (ii) probability distributions defined at the state level which are given either uniform distributions (in case of time-bounded delays) or exponential distributions (in case of unbounded delays).

Due to the limited number of pages, we opt to demonstrate the formal models using PTAs of the three principal entities that participate in the communication using MQTT 3.1.1. These models are shown in the Figs. 1a, b, 2, and 3.

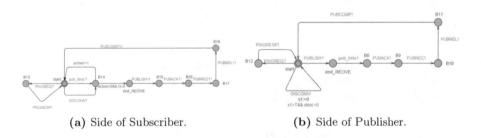

(a) Side of Subscriber.   (b) Side of Publisher.

Fig. 1. Model of the broker

3.2   Formal Verification of MQTT 3.1.1

We opt to verify some properties using the query language provided in the UPPAAL SMC. Before modeling the subscriber and the publisher using probabilistic automata, we try to verify some qualitative properties such as reachability, safety and liveness.

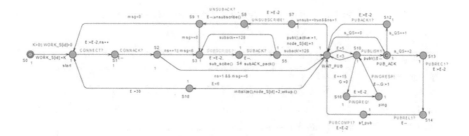

Fig. 2. Probabilistic model of the subscriber

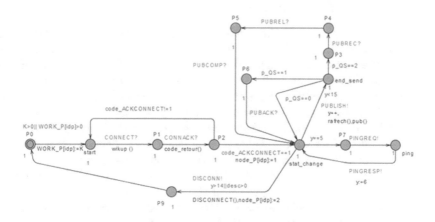

Fig. 3. Probabilistic model of the publisher

Verifying whether a subscriber and a publisher can arrive at an initialization state (S18 and P9 respectively) is formulated by the formulas: $E<>sub1.S18$ and $E<>pub1.P9$, respectively. Likewise, the Liveness property which is of the form "something will eventually happen" is verified using the following query: $A<>publish\_sub$.END_RECEIVE. In addition, the safety property is verified using the formula: $A\ [\ ]\ not\ deadlock$ which means that a deadlock must never occur in the model.

UPPAAL SMC gives us the possibility of adding temporal constraints to the formulas of the properties to be verified. Using this query language, we have examined whether the subscriber can receive a publication before that the waiting time exceeds or the value of X reaches a certain threshold where X is a clock added to the model of the broker in the publication phase at the publisher side, in order to present the "keep alive" property. We opt to verify this latter property with the thresholds T = 45000, 1000 and 100. The results obtained for the aforementioned properties are depicted in the Table 2.

Table 2. Satisfaction results of qualitative properties

Property	Query	Satisfaction
Reachability 1	$E<>sub1.S18$	Satisfied
Reachability 2	$E<>pub1.P9$	Satisfied
With T = 45000	$E<>(sub1.PUB\_ACK\ and\ pub\_publish.X1 < 45000)$	Satisfied
With T = 1000	$E<>(sub1.PUB\_ACK\ and\ pub-publish.X1 < 1000)$	Satisfied
With T = 100	$E<>(sub1.PUB\_ACK\ and\ pub\_publish.X1 < 100)$	Satisfied
Liveness	$A<>publish\_sub$	Satisfied
Safety	$A\ []\ not\ deadlock$	Undecidable

Discussion. In some cases, the verification of some qualitative properties such as safety (as seen in Table 1) and the reachability, in the case of 3 subscribers and 3 publishers, becomes undecidable due to the state space explosion in the model-checker. However the utilization of UPPAAL SMC and probabilistic timed automata solve this problem. Updating the subscriber and publisher models, to be probabilistic models, solves the undecidability problem. Thus the property examined before (with 3 subscribers and 3 publishers) becomes satisfied with probability interval as seen in Table 3. This latter table demonstrates also the subscriber probability of arriving at an initialization phase with a runtime lower or equal to 100 in the case of 9 publishers and 9 subscribers. UPPAAL SMC gives the possibility of drawing the probability functions. We opt to draw (as seen in Fig. 4a) the cumulative probability confidence interval of finishing the reception of a publication according to the number of nodes for a runtime $<= 10^3$ units of time. Likewise, the probability that a publisher disconnects and that its subscribers receives the *will* during a runtime $<= 10^3$ time units also studied and the results are depicted in Fig. 4b. As depicted in Fig. 4a, one can see that the protocol behaves well even when the number of publishers and subscribers increases. The probability of a successful publication is always similar when the number of entities develops in the system.

Table 3. Results of some quantitative properties.

Query	Satisfaction	Probability interval	Confidence
Pr[<=100]<>sub1.S18	Satisfied	[0.303567, 0.403432]	0.95
Pr[<=100]<>(NNI_P>1 and NNL_P>=2)	Satisfied	[0.902606, 1]	0.95

Figure 5a and b show the probabilities of having active publishers and active subscribers for a runtime $<= 10^4$ and 10^3 units of time, respectively. These two figures show that this probability, of course, increases with the time, hence having active entities for runtime 10^4 time units is more important than having this probability for 10^3 time units.

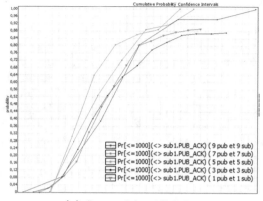

(a) Successful publication

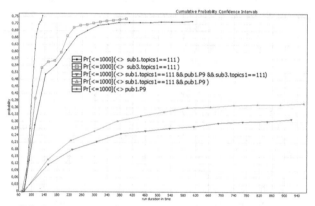

(b) Successful reception of the *will* from subscribers.

Fig. 4. Cumulative probability

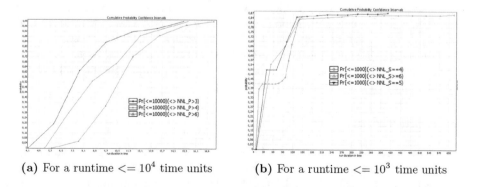

(a) For a runtime $<= 10^4$ time units   **(b)** For a runtime $<= 10^3$ time units

Fig. 5. Probability of having active publishers and subscribers

4   Related Work

Many researches have proposed and studied the performance of protocols for IoT [6,12,17,19]. The work reported in [6] proposed an enhanced version of Routing Protocol for Low-Power and Lossy Networks (RPL) for Internet of Multimedia Things. In [17], the authors proposed a new authentication protocol for Mobile IoT (MIoT) in order to construct a secure network. The proposed protocol was tested and simulated as well as modeled using Alloy language. The authors of [12] made another performance analysis of IoT security protocols (IPSec [21] and Datagram Transport Layer Security (DTLS) [20]). Indeed, the authors analyzed the impact of these protocols on the embedded devices resources. Hence, the authors proposed, in [19], a new protocol and they used OMNET++ simulator to test some properties such as end-to-end delay, energy efficient, etc., as well as they compared their proposal to traditional solution of RPL.

Formal methods are expected to ensure the complying with standards and to check the required properties. In the last few years, many works used formal methods in networking domain such as [9,15,16]. These latter works studied formally CSMA/CA protocols with three different ways; where in [15] the authors extended the CSMA/CA towards a new enhanced one and they discussed its performance by modeling its behavior using timed automata and they verified some qualitative properties such as liveness, safety and reachability. The work presented in [16] used the previous work [15] and modeled their proposal with probabilistic time automata and verified some qualitative and quantitative properties. With regard to the work of [9], the authors studied CSMA/CA of energy-harvesting WSNs using timed automata at the modeling level and the UPPAAL model-checker at the analysis level.

The use of formal methods in the analysis of IoT protocols still not yet developed. In fact, few works exist in the literature such as [4,7,8,10]. In [7], the authors developed two applications of test to study the performance of IoT web application. The test has also been taken into consideration in [10] where the Internet of things application layer protocols Constrained Application Protocol (CoAP) [22] and MQTT were tested and their latencies were compared for high packet loss by creating a middle-ware component. The work reported in [8] proposed a formal computational framework for publish/subscribe systems and studied the completeness and minimality properties. Talking also about publish/subscribe systems, the BonjourGrid [3] protocol which is based on these asynchronous systems was also studied in [4] by the formal modeling and verification using Colored Petri Nets (CPNs).

5   Conclusion

The application of formal methods in the study of distributed systems and communication protocols requires powerful tools for the modeling and verification of stochastic events. UPPAAL-SMC supplies these two requirements. In this paper, we have proposed an abstract model describing a dedicated application layer protocol for the Internet of Things. In fact, we have developed a formal model of the messaging protocol called MQTT. First, we have provided an exhaustive study of the protocol, then we have made a formal model with timed and probabilistic automata using UPPAAL tool-set. This UPPAAL is a suitable tool to verify this protocol using its SMC model-checker. The SMC model-checker allows a qualitative and a quantitative analysis of this protocol. In this paper, several important properties were verified (e.g., reachability, safety, vivacity). Besides these qualitative properties, some important metrics were analyzed (e.g., number of active and inactive nodes, success rate of transfer and reception of message, etc).

In our future work, we are planing to examine the MQTT protocol using another model-checker allowing the verification with a more important number of entities. On the other hand, we are opting for an automated approach to derive source code of protocols from their formal high level specifications.

References

1. Mqtt version 3.1.1 plus errata 01. http://docs.oasis-open.org/mqtt/mqtt/v3.1.1/ mqtt-v3.1.1.html. Accessed 09 July 2017
2. What is uml? http://www.uml.org/. Accessed 29 June 2017
3. Abbes, H., Cérin, C., Jemni, M.: Bonjourgrid as a decentralised job scheduler. In: 2008Asia-Pacific Services Computing Conference, APSCC 2008, pp. 89–94. IEEE (2008)
4. Abidi, L., Cérin, C., Evangelista, S.: A petri-net model for the publish-subscribe paradigm and its application for the verification of the BonjourGrid middleware. In: 2011 IEEE International Conference on Services Computing (SCC), pp. 496–503. IEEE (2011)
5. Al-Fuqaha, A., Guizani, M., Mohammadi, M., Aledhari, M., Ayyash, M.: Internet of things: a survey on enabling technologies, protocols, and applications. IEEE Commun. Surv. Tutorials 17(4), 2347–2376 (2015)
6. Alvi, S.A., Shah, G.A., Mahmood, W.: Energy efficient green routing protocol for internet of multimedia things. In: 2015 IEEE Tenth International Conference on Intelligent Sensors, Sensor Networks and Information Processing (ISSNIP), pp. 1–6. IEEE (2015)
7. Babovic, Z.B., Protic, J., Milutinovic, V.: Web performance evaluation for internet of things applications. IEEE Access 4, 6974–6992 (2016)
8. Baldoni, R., Contenti, M., Piergiovanni, S.T., Virgillito, A.: Modeling publish/subscribe communication systems: towards a formal approach. In: 2003 Proceedings of the Eighth International Workshop on Object-Oriented Real-Time Dependable Systems, WORDS 2003, pp. 304–311. IEEE (2003)
9. Chen, Z., Peng, Y., Yue, W.: Modeling and analyzing CSMA/CA protocol for energy-harvesting wireless sensor networks. Int. J. Distrib. Sens. Netw. 11(9), 257157 (2015)
10. Collina, M., Bartolucci, M., Vanelli-Coralli, A., Corazza, G.E.: Internet of things application layer protocol analysis over error and delay prone links. In: 2014 7th Advanced Satellite Multimedia Systems Conference and the 13th Signal Processing for Space Communications Workshop (ASMS/SPSC), pp. 398–404. IEEE (2014)
11. David, A., Larsen, K.G., Legay, A., Mikuăionis, M., Poulsen, D.B.: Uppaal SMC tutorial. Int. J. Softw. Tools Technol. Transf. 17(4), 397–415 (2015)

12. De Rubertis, A., Mainetti, L., Mighali, V., Patrono, L., Sergi, I., Stefanizzi, M.L., Pascali, S.: Performance evaluation of end-to-end security protocols in an internet of things. In: 2013 21st International Conference on Software, Telecommunications and Computer Networks (SoftCOM), pp. 1–6. IEEE (2013)
13. Granjal, J., Monteiro, E., Silva, J.S.: Security for the internet of things: a survey of existing protocols and open research issues. IEEE Commun. Surv. Tutorials 17(3), 1294–1312 (2015)
14. Gubbi, J., Buyya, R., Marusic, S., Palaniswami, M.: Internet of things (IoT): a vision, architectural elements, and future directions. Future Gener. Comput. Syst. 29(7), 1645–1660 (2013). Including Special sections: Cyber-enabled Distributed Computing for Ubiquitous Cloud and Network Services & Cloud Computing and Scientific Applications Big Data, Scalable Analytics, and Beyond
15. Hammal, Y., Ben-Othman, J., Mokdad, L., Abdelli, A.: Formal modeling and verification of an enhanced variant of the IEEE 802.11 CSMA/CA protocol. J. Commun. Netw. 16(4), 385–396 (2014)
16. Hmidi, Z., Kahloul, L., Saber, B., Othmane, C.: Statistical model checking of CSMA/CA in WSNs. In: Proceedings of the 10th Workshop on Verification and Evaluation of Computer and Communication System, VECoS 2016, Tunis, Tunisia, 6–7 October 2016, pp. 27–42 (2016)
17. Kumar, A., Gopal, K., Aggarwal, A.: Simulation and analysis of authentication protocols for mobile internet of things (MIoT). In: 2014 International Conference on Parallel, Distributed and Grid Computing (PDGC), pp. 423–428. IEEE (2014)
18. Kwiatkowska, M., Norman, G., Sproston, J.: Symbolic computation of maximal probabilisti reachability, pp. 169–183. Springer, Heidelberg (2001)
19. Le, Q., Ngo-Quynh, T., Magedanz, T.: RPL-based multipath routing protocols for internet of things on wireless sensor networks. In: 2014 International Conference on Advanced Technologies for Communications (ATC), pp. 424–429. IEEE (2014)
20. Rescorla, E., Modadugu, N.: Datagram Transport Layer Security Version 1.2 (2012)
21. Seo, K., Kent, S.: Security architecture for the internet protocol (2005)
22. Shelby, Z., Hartke, K., Bormann, C.: The constrained application protocol (CoAP) (2014)

Networking

Simulation Automation of Wireless Network on Opnet Modeler

Hafsa Ait Oulahyane[✉], Ayoub Bahnasse, Mohamed Talea,
Fatima Ezzahraa Louhab, and Adel Al Harbi

Laboratory of Information Processing (LTI), Faculty of Sciences Ben M'Sik,
Hassan II University of Casablanca, Casablanca, Morocco
{hafsa.aitoulahyane-etu,fatimaezzahraa.louhab-etu,
adel.harbi-etu}@univh2m.ma, a.bahnasse@gmail.com,
taleamohamed@yahoo.fr

Abstract. Today, wireless networks have experienced significant progress, which has given rise to several technologies that coexist with each other in the same environment forming heterogeneous environments. The problem posed by this architecture is the navigation between several networks in a transparent way, hence the appearance of a compromise between mobility, transparency and performance. The actual study of this architecture is difficult in terms of cost and equipment. The simulation under Opnet Modeler remains a good solution if the user masters the tool first and then the technologies to simulate.

This paper presents a new model for the simulations automation of handovers in homogeneous and heterogeneous networks under Opnet Modeler. This model allows users to generate different scenarios with a minimum knowledge of technologies based on a new Web interface and to launch these simulations under Opnet Modeler.

Keywords: Automation · Handover · Opnet modeler · Heterogeneous networks
Mobility · Simulation

1 Introduction

1.1 Mobile and Wireless Networks

Mobile and wireless networks are in full development due to the flexibility of their interface, which allows a user to change places while remaining connected. The communications between terminal equipment can be carried out directly or via base stations, called AP (Access Points) [1]. We will present a brief description of four mobile technologies namely Wi-Fi, WiMAX, UMTS, LTE.

Wi-Fi is one of the standards that allows deploying a wireless network by communicating several devices together, through the radio wave. For reasons of performance improvement (range, throughput, etc.), this standard has undergone several evolutions through the appearance of different versions [2]. It uses two different modes to connect

© Springer International Publishing AG, part of Springer Nature 2019
J. Mizera-Pietraszko et al. (Eds.): RTIS 2017, AISC 756, pp. 237–249, 2019.
https://doi.org/10.1007/978-3-319-91337-7_23

devices on a network: The ad hoc mode, which makes it possible to realize a peer-to-peer network, and the infrastructure mode, which requires access points to the wired local area network.

Worldwide Interoperability for Microwave Access (WiMAX) appeared to meet the Wi-Fi limitations [26]. With this large selection of frequencies, it is easier to interact with various types of products existing on the market [3]. It should be noted that the WiMAX standard introduces mobility and manages the Handover as well as introduces a new mechanism for quality management service.

Universal Mobile Telecommunications System (UMTS) is one of the third generation (3G) mobile telephony technologies that allows a wider frequency band to be used to transmit more data and therefore to obtain a higher throughput. The UMTS meets two clearly expressed user needs: On the one hand, the communication of data in all its forms, on the other hand the convenient communication between people or with machines in a mobility situation [4].

Long Term Evolution (LTE) is one of the fourth generation (4G) mobile telephony technologies that brings a real turning point in the abundance and disparity of existing solutions. The objective of this new generation is certainly to increase the bit rates and the applications supported by these networks but also to build a framework allowing their interoperability [5]. LTE also supports advanced VOIP since it employs a different form of radio interface, using OFDMA/SC-FDMA instead of CDMA.

1.2 Handover

Today the mobile user always seeks to remain connected anywhere and anytime, this becomes possible with the evolution of wireless networks, which offers heterogeneous networks where the user can change his point of attachment without service interruption.

This navigation between different types of networks is called Handover.

Two main types of handovers can be distinguished [24]:

- Horizontal Handover: When the mobile changes its point of attachment from one station to another, belonging to the same technology, and at the same level of network hierarchy (e.g. WiMAX to WiMAX).
- Vertical Handover: when the user moves between different network access technologies (example Wi-Fi to WiMAX) [25] (Fig. 1).

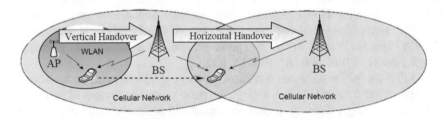

Fig. 1. Types of handovers

The rest of the paper will be organized as follows: In the Sect. 2, we will present the state of the art of the recent work related to the automation of wireless networks. In the Sect. 3, we will situate our contribution in relation to the work already done. In the Sect. 4, we'll talk about the motivation and the benefits of this model. We will present in the Sect. 5 the model "Simulation automation of wireless network" describing its architecture and its modules and its functioning. The Sect. 6 will be dedicated to the implementation of this model through a guided tour and demonstration. And finally, we end with a conclusion.

2   State of the Art

2.1   The Simulators

The realization of a research project on real-time networks remains difficult because of the expensive costs of implementing these components. Network simulators reduce this task in real time and accomplish it with less time and cost. There are many network simulators available on the market such as ns2, ns3, OMNET++, Opnet and Qualnet, We will focus subsequently on the simulators NS3 [12], OMNET++ [15], and OPNET [7].

- **NS3 (Network Simulateur version 3)**

The NS-3 simulator is an open-source discrete event and network simulator for research and specially designed for educational purposes [12]. It was developed in 2006. A large number of versions have been released so far after its first development. It is based on two programming languages: C++ and OTCL.

NS3 offers several advantages [12]: modularity, documentation, flexibility, large-scale network simulation, a good use of memory, and credibility. But it also has some limitations: the GUI limited and it does not support the use of IP addressing, design and alignment with Internet protocols, manipulation of multiple interfaces, details on 802.11 models, etc.

- **OMNET++ (Objective Modular Network Testbed in C++)**

OMNeT++ is an open source simulation environment with a solid graphical interface integrating a simulation kernel. It is a discrete object-oriented simulation tool, based on C++, free for academic uses. While there is a commercial version that is OMNEST. It was designed to simulate communication network systems, multi-processor systems and other distributed systems [15].

OMNET++ offers several advantages: Reusability and ease of learning. But it also has some limitations [12]: the number of protocols supported is insufficient, the lack of some features, and compatibility problem.

- **OPNET (Optimum Network Performance)**

OPNET Modeler is a very powerful simulation software, coded in C++ and designed in object oriented. It allows modeling, simulation of communication networks and performance analysis with its model libraries (routers, switches, workstations, servers) and protocols (TCP/IP, FTP, FDDI, Ethernet, ATM...). It is designed specifically for the development and analysis of communication networks, provides many details not available in simple resource-based simulation packages, and can be used in many areas of communication networks from diverse applications such as LANs and wireless networks [7]. OPNET can also be used in telecommunication network traffic modeling, protocol modeling, network queue modeling, validation of hardware architectures, and evaluation of the performance aspect of complex software systems.

Opnet offers several advantages: the possibility of modifying the models parameters, it provides a rich set of modules for each of the IEEE 802.11 protocol stacks, such as IEEE 802.15.4 and various routing protocols viz AODV and DSR, a graphical interface that facilitates simulations, also Opnet is the most modular and scalable engine for simulation engines.

Comparative study of the various simulators

Several works [11–13, 20] analyzed and compared the various simulators: NS3, OMNET++, OPNET and QUALNET. Table 1 is a general synthesis which summarizes the different characteristics of each simulator. Based on this comparison [21–23], we chose to work with Opnet Modeler because it is the best simulator.

Table 1. Comparison of different simulators

Characteristics	Simulators		
	NS3	OMNET++	OPNET
Language	C++	C++	C and C++
Licence	Open source	Academic/ Commercial	Academic/ Commercial
Availability	• Traditional model • Wireless • Ad-hoc	• Traditional model • Wireless • Ad-hoc	• Traditional model • Wireless • Ad-hoc • Advanced wireless sensors
Scalability	Limited	Limited	Large
Platform	Linux	Linux/MacOs, Unix	Microsoft Windows
Documentation	✓		✓
Communication with external modules			✓

2.2   Related Works

Currently, a great majority of the research carried out on the automation of the networks we will present some works in this section:

The study [14] put in place a test to evaluate the functionality of a new decentralized distribution automation architecture. The demonstration system consists of a distribution network simulation in the real-time simulation environment, including simulation control and monitoring devices as well as physical devices interfaced with the simulator as hardware in the test equipment loop. The system also involves automating substations units for real-time monitoring and control that are interfaced with the simulator and physical devices. The operating principle of the system is demonstrated by an example of a simulation case. The main objective of this system was to test the functionality of the decentralized distribution automation architecture and the way to any possibility of interfacing the automation system before implementing the demonstration fields concept.

This work [9] makes it possible to design a model for the dynamic simulations automation, multipoint and multi-architectural VPNs networks for Opnet Modeler and to create a tool allowing the customized management of scenarios based on a graphical WEB interface, some work has been done in the direction of automating network simulations for Opnet modeling and the translation of simulation scenarios into XML files intended for various simulators based on a WEB interface.

This thesis [15] proposes a generic approach to automate experiments on networks irrespective of the technology used or the type of evaluation platform. The proposed approach is based on the abstraction of the life cycle of the experiment into generic steps from which an experimental model and experimental primitives are derived. A generic experimental architecture is proposed, consisting of a generic experimental model, an experimental programming interface and an orchestration algorithm that can be adapted to simulators, emulators and network test benches. The feasibility of this approach is demonstrated by the implementation of a framework capable of automating experiments on any combination of these platforms.

The article [16] presents a comprehensive strategy to systematically determine, on the basis of the user's information (via mobile traces), why occurs the user disconnected. The automatic cause analysis project is characterized by a top-down model and provides an exhaustive classification of defects when caused by radio-related problems. The proposed method for identifying the origin of the radio has been applied both in a LTE simulator and a real LTE network, illustrating the usefulness of the approach.

3   Positioning of Our Contribution

The use of simulation becomes more and more frequent and necessary because of the complexity of the theoretical models and the impossibility in some cases of being able to solve them [6]. The simulation comes to meet these needs in order to satisfy, at the least cost, increasingly stringent requirements to improve network performance. But the simulation of handovers in heterogeneous networks and homogeneous under Opnet Modeler requires a good knowledge and understanding of the field of wireless networks to master the ins and outs of the simulator.

Currently, a great majority of the research carried out on networks under Opnet modeler, the article [17] studies heterogeneous networks and evaluates the scalability of vertical and horizontal handovers under Opnet modeling. Article [18] deals with the deployment of information and communication for the majority of rural areas to improve education and economic growth through the use of Opnet Modeler. The paper [19] proposes a new Zigbee compliant simulation model using the OPNET simulator. Based on the Zigbee MAC model in OPNET Modeler, it develops a network layer model and proposes an enhancement of the AODV routing algorithm to support node mobility, both of which are compatible with Zigbee protocols.

In this work, we propose a new model for the simulations automation of handovers in homogeneous and heterogeneous networks under Opnet Modeler. This model allows:

– Manage the different scenarios of the handovers based on a graphical WEB interface.
– Start the simulation under Opnet Modeler from the XML files generated by this model.

4   Motivation

The simulation is an essential tool in the performance study of any network. Several works [7, 9, 15, 17, 19] have dealt with simulations under Opnet Modeler and the automation of this tool, but unfortunately no work has dealt with the case of mobility in homogeneous and heterogeneous networks. This was a motivation to design this model of automation of the handovers in homogeneous and heterogeneous networks under Opnet Modeler.

With the rapid evolution of the telecom industry, mobile operators deploy new technologies (such as 4G, 5G) that will coexist with the old technologies. The major challenge posed by this new architecture is mobility. In order for a user to switch from one network to another in a transparent way, a whole mechanism is run in parallel. This transfer or what is called Handover will play a very important role in the future generations, and for this our model will be beneficial for the operators to facilitate the study of the different scenarios.

This model will not be beneficial only for operators but also for:

– The researchers: some simulations take a long time to be realized, this tool will allow the researchers to realize their different scenarios in short durations.
– Teachers: With the evolution of Opnet Modeler, the teacher must always master the new versions of the software, this is a waste of time and energy. Thanks to the modularity of our model, the teacher will no longer need to master each version, just indicate the version of Opnet, and the system upload the appropriate configuration.
– Students: Simulation was always an obstacle for students because they must have a prerequisite on the tool and master the technologies with which it will simulate. This is no longer necessary thanks to our model; the student will be able to perform hundreds of simulations with a minimum of knowledge and in a short time.

5   Wireless Network Automatic Simulation Model

5.1   Model Architecture

The proposed architecture is composed of two parts (Fig. 2):

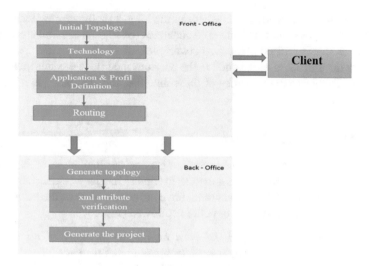

Fig. 2. Architecture of wireless network automatic simulation model

Front Office: It is the visible front by the user and in direct contact with him, thanks to this interface the user can set all the attributes of a handover.

This part is composed of four modules: Initial Topology, Technology, Application and Profile Definition, and Routing.

– Initial Topology: this module defines the network scale (world, campus, company or office), its size and the type of handover to simulate (horizontal or vertical).
– Technology: this module represents the technologies used in a Handover (Wifi, WiMAX, UMTS, LTE…), and defines the characteristics of each mobile station and each base station.
– Application and Profile Definition: this module is the responsible of the applications definition supported by the stations (video, voice, web…) as well as their class of quality of service (UGS, rtPS, nrtPS, Best Effort, ertPS). It also defines the application profile and its mode of operation.
– Routing: this module includes all parameters responsible for routing; it allows to choose the routing protocol (OSPF, EIGRP, RIP…) and addressing on the one hand, and to define the mobile nodes trajectory on the other side.

Back-Office: this part designates the internal processes to which the user has not access and allows processing the data collected in the first part in order to convert them into XML to simulate subsequently on Opnet Modeler. This part consists of three modules: Generate Topology, XML attributes verification and Generate the project.

– Generate Topology: this module creates the architecture shown in the previous modules; it generates mobile stations and base stations as well as their configurations (technology, routing protocols, path…).

– XML attributes verification: the proposed model is a modular model applicable to all versions of the Opnet. This module generates the appropriate xml attributes with the version that is located in the file network.dtd.
– Generate the project: after the audit in the previous module, the system generates the final file. Thanks to the sequence of these modules, the simulation is run in Opnet Modeler.

5.2 The Model Activity Diagram

The purpose of this model is the automation of generation simulations of the handovers in homogeneous and heterogeneous networks in Opnet Modeler by the scenario management based on a WEB GUI. The diagram above provides a view of the behavior of the model by describing the sequence of actions followed (Fig. 3).

1- *Once the user is connected, he must define the topology of his simulation, mentioning the network scale (world, campus, company, office...) and its area.*
2- *The user chooses the type of handover to be simulated, which can be either horizontal or vertical.*
3- *If the user chooses a vertical handover, a definition of the names of the two subnetworks is necessary.*
4- *The user must define the positioning of the adjacent networks according to their type (WiFi, WiMAX...) and the standards of each of them in order to avoid any break or packets loss.*
5- *If the user chooses the horizontal handover, he passes directly to the equipment configuration.*
6- *The user selects the technology to simulate (WiMAX, wifi, UMTS...) and indicates the number, model and the name for base stations and mobile stations.*
7- *In the case of existence of application, the user defines his name and the technology used (http, voice, database, email, video conferencing, FTP...).*
8- *The user must create a profile for this application, where he mentions the operation mode which can be either simultaneously or in series and finally chooses a quality level of service for this application.*
9- *The user must choose the Routing Protocol (RIP, IGRP, EIGRP, OSPF...).*
10- *The user defines the mobile's trajectory.*
11- *The system creates devices predefined by the user in the previous steps.*
12- *Creation of the XML attributes corresponding to the DTD schema to finalize the generation.*
13- *The generation of the final XML file containing all the equipment and their configurations, and the preparation of the simulation under Opnet Modeler.*

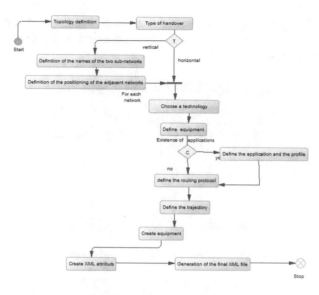

Fig. 3. Model activity diagram

6 Demonstration

After authentication, the model presents several menus to the user: Create a new configuration, open an existing configuration and the last tab help that provides information
about the application.

To create a new configuration, the user must complete the fields shown in Fig. 4,
indicating the type of network if it is a network of world order, of company, a campus
or simply an office. We will choose a network at the scale of a campus. He must also
indicate its size and the type of handover (horizontal or vertical).

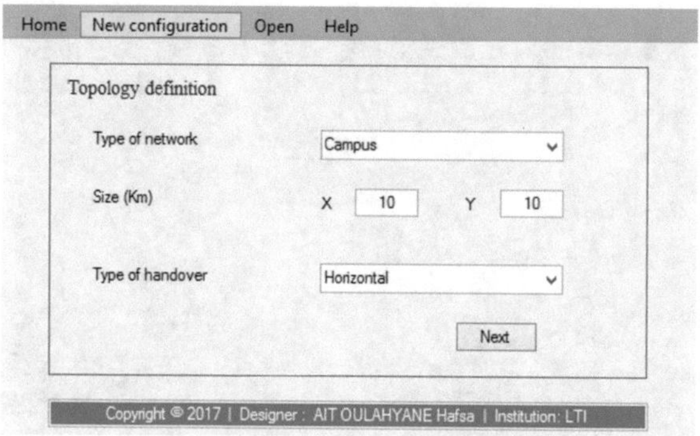

Home	New configuration	Open	Help

Topology definition

Type of network	Campus ⌄
Size (Km)	X 10 Y 10
Type of handover	Horizontal ⌄

Next

Copyright © 2017 | Designer : AIT OULAHYANE Hafsa | Institution: LTI

Fig. 4. New Configuration: Topology definition

After specifying the technology in Fig. 5, the user should mention the number of desired equipment, their model and name. The facilities are organized by type: Base station and mobile station (Fig. 6).

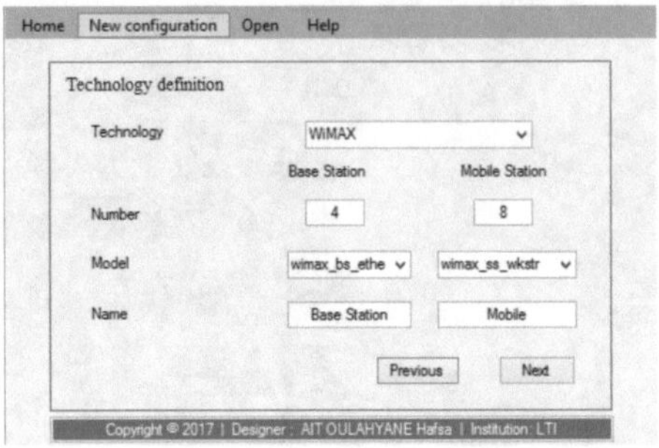

Fig. 5. Technology and equipment definition

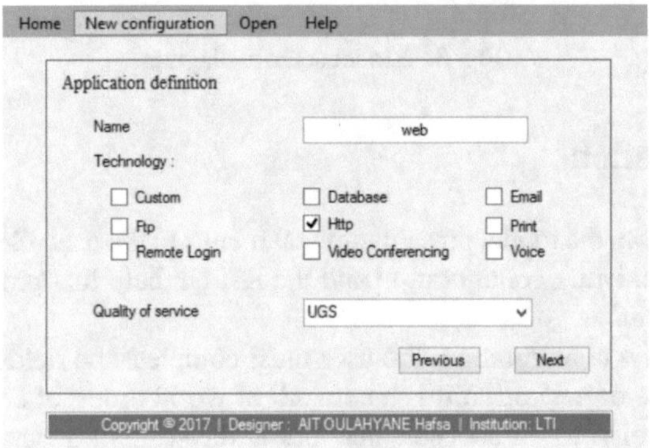

Fig. 6. Application definition

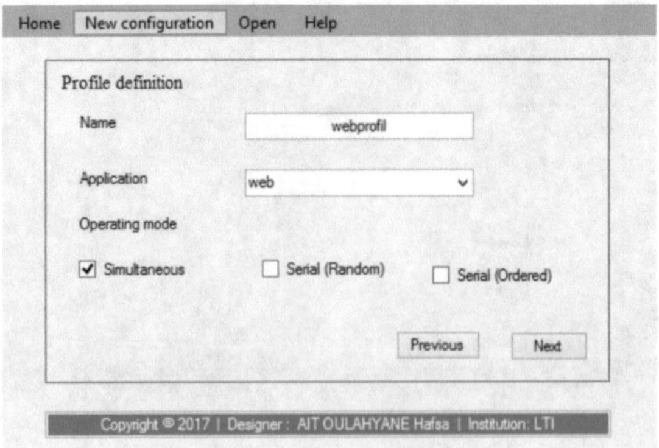

Fig. 7. Profile definition

If the user wants to set an application, he must first indicate its name, define the technology supported by this application, and finally choose a suitable class of quality of service.

In order to finalize the definition of applications and profile, the user must indicate the name of the profile, as well as the application and the operating mode (Fig. 7).

The last step of this configuration is to choose a routing protocol, then define the type of addressing and finally define the trajectory. In the case of the choice of a custom path, a menu will appear to indicate the path of this trajectory (Fig. 8).

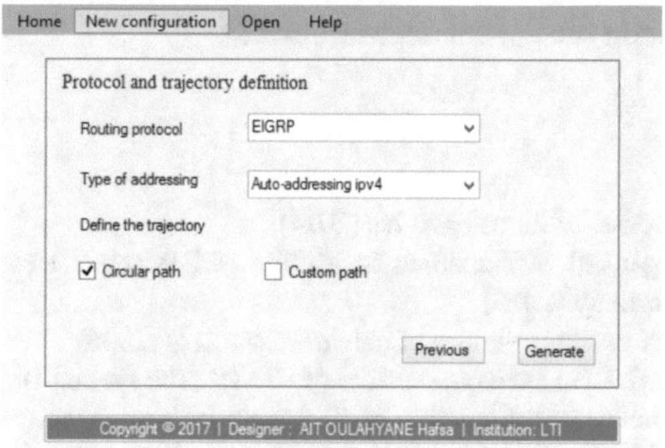

Fig. 8. Protocol and trajectory definition

After completing the configuration, an xml file is generated containing the architecture to simulate in Opnet Modeler. The simulation is run in Opnet Modeler (Fig. 9) showing the result of our Setup.

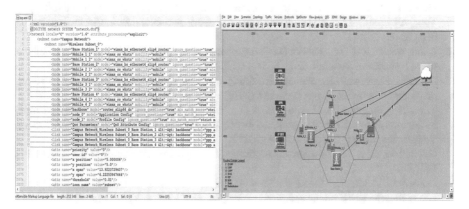

Fig. 9. Resulting XML file and configured architecture

7   Conclusion

The handovers simulation in the homogeneous and heterogeneous networks in Opnet Modeler is a solution, which allows studying certain conditions unrealizable in reality. However, this simulation requires a good mastery of the simulator and good skills in heterogeneous networks and mobile technologies.

In this paper, we proposed a model for the automation of handovers simulations in homogeneous and heterogeneous networks in Opnet Modeler, which allows generating the simulations through the tool implemented, based on a Web GUI. This model allows us to reduce simulation time and reduce error rates.

References

1. Pujolle, G.: Networks. Editions Eyrolles (2014)
2. Rahil, A.: Management of Handover in Mobile and Wireless Heterogeneous Networks. University of Burgundy (2015)
3. Mian, S.: WiMAX or the evolution of networks wireless (2006)
4. Didier, M., Lorenzi, J.H.: Economic issues of UMTS. The French Documentation (2002)
5. Salhani, M.: Modeling and simulation of 4th generation mobile networks (2008)
6. Ben Hamida, E.: Stochastic modeling and simulation of multisite wireless networks (2009)
7. Sidi Ykhlef, A., Kebir, K.: Modeling and simulation of a network using opnet modeler (2015)
8. Li, H., Lin, X.: An OPNET-based 3-tier network simulation architecture. In: Communications and Information Technology, IEEE International Symposium (2005)
9. Bahnasse, A., EL Kamoun, N.: Policy-Based Automation of Dynamique and Multipoint Virtual Private Network Simulation on OPNET Modeler (2014)
10. Habib, G.: Quality of service and quality control of a Discrete System Controlled in Wireless Network: proposal of a co-design approach applied to the IEEE 802.11 standard (2010)
11. Toor, A.S., Jain, A.K.: A survey on wireless network simulators. Bull. Electr. Eng. Inform. **6**(1), 62–69 (2017)
12. Chhimwal, P., Rai, D.S., Rawat, D.: Comparison between different wireless sensor simulation tools. J. Electron. Commun. Eng. **5**, 54–60 (2013)
13. Hafir, L., Mehaouad, K., Slimani, R.: Study and evaluation of the performance of routing protocols for wireless sensor networks (2016)
14. Tuominen, V., Reponen, H., Kulmala, A., Lu, S., Repo, S.: Real-time hardware-and software-in-the-loop simulation of decentralized distribution network control architecture. IET Gener. Transm. Distrib. **11**(12), 3057–3064 (2017)
15. Quereilhac, A.: A generic approach for the automation of experiments on computer networks (2015)
16. Gómez-Andrades, A., Muñoz, P., Serrano, I., Barco, R.: Automatic root cause analysis for LTE networks based on unsupervised techniques. IEEE Trans. Veh. Technol. **65**(4), 2369–2386 (2016)
17. Khiat, A., Bakkoury, J., El Khaili, M., Bahnasse, A.: Study and evaluation of vertical and horizontal handover's scalability using OPNET modeler. Int. J. Comput. Sci. Inf. Secur. **14**(11), 807 (2016)
18. Nathan, D., Chukwunonye, A., Henry, A.: Opnet based simulation for rural educational ICT connectivity. Int. J. Sci. Eng. Res. **7**(4), 1499–1504 (2016)

19. Xiaolong, L., Meiping, P., Jun, C., Changyan, Y., Hong, Z.: OPNET-based modeling and simulation of mobile Zigbee sensor networks (2015)
20. Babouri, K., Khellasi, L., Djebrouni, K., Besses, M.: The Simulators Networks: Network Technology. University of Science and Technology Houari Boumediene (2013)
21. Khiat, A., Bahnasse, A., Bakkoury, J., El Khaili, M.: Study, evaluation and measurement of IEEE 802.16 e secured by dynamic and multipoint VPN IPsec. Int. J. Comput. Sci. Inf. Secur. **15**(1), 276 (2017)
22. Khiat, A., Bahnasse, A., Khail, M.E., Bakkoury, J.: Study, evaluation and measurement of 802.11 e and 802.16 e quality of service mechanisms in the context of a vertical handover case of real time applications. Int. J. Comput. Sci. Netw. Secur. **17**(2), 119 (2017)
23. Khiat, A., Bahnasse, A., Khail, M.E., Bakkoury, J.: Impact of Qos mechanisms on the performance of dynamic web services in heterogeneous wireless networks (802.11 e and 802.16 e). Indian J. Sci. Technol. **10**(25) (2017)
24. Alsamhi, S.H., Rajput, N.S.: Neural network in intelligent handoff for QoS in HAP and terrestrial systems. Int. J. Mater. Sci. Eng. **2**, 141–146 (2014)
25. Zekri, M., Jouaber, B., Zeghlache, D.: A review on mobility management and vertical handover solutions over heterogeneous wireless networks. Comput. Commun. **35**(17), 2055–2068 (2012)
26. Lu, K., Qian, Y., Chen, H.H.: Wireless broadband access: Wimax and beyond-a secure and service-oriented network control framework for wimax networks. IEEE Commun. Mag. **45**(5), 124–130 (2007)

Smart SDN Policy Management
Based VPN Multipoint

Adel Alharbi[(✉)], Ayoub Bahnasse, Mohamed Talea, Hafsa Ait Oulahyane,
and Fatima Ezzahraa Louhab

Department of Physics, Faculty of Science Ben M'Sik,
University Hassan II, Casablanca, Casablanca, Morocco
{adel.alharbi-etu,hafsa.aitoulahyane-etu,
fatimaezzahraa.louhab-etu}@univh2m.ma, a.bahnasse@gmail.com,
taleamohamed@yahoo.fr

Abstract. SDN "Software Defined Networking" as a term does not refer to
anything specific, and specific is a modern approach led to the formation of a new
architecture in the world of networks by making the networks an open and free
world can be programmed and controlled smoothly and easily. Thanks to this
technology, many complexities of the traditional network have been solved which
were problematic in the implementation of services and technologies, and one of
these techniques, VPN technology. Even though VPNs have proved to be cost-
effective solutions, the lack of centralized network management capabilities of
current VPN deployment makes the management of growing networks time-
consuming and error-prone. Several studies have been proposed for centralized
policy management of intra/inter VPN domain, most solutions are only addressed
to only the static VPN (site-to-site) technology, according to our researches no
centralized management model based on a server for VPN Multipoint networks
over Lunix operation system was proposed. In this paper, we propose Smart SDN
Policy Management based VPN Multipoint (SSDPM based VPN Multipoint) for
centralizing managing security policies in dynamic networks using a new Java
interface.

Keywords: Smart SDN · VPN multipoint · Policy management · Security
Dynamic networks

1 Introduction

A LONG with rapid development of the Internet, the distributed network architecture
has increased network complexity and maintenance costs. The Software defined
Network (SDN) was widely implemented in the last year [1, 2]. Differing from the
traditional network, an SDN uses a centralized control architecture. The main concepts
of SDN include the separation of the control data plane [3, 4], and a programmable
network. In other words, an SDN facilitates flexible management and configuration of
a network[5] by pushing all control tasks to a centralized controller. Moreover, SDN
speeds the deployment of innovations or services and reduces optional costs through
programmable interfaces (e.g., Openflow [6]) in the controller. Due to the benefits of

© Springer International Publishing AG, part of Springer Nature 2019
J. Mizera-Pietraszko et al. (Eds.): RTIS 2017, AISC 756, pp. 250–263, 2019.
https://doi.org/10.1007/978-3-319-91337-7_24

SDN, network operators have recently realized that SDN is a promising solution for their networks.

VPN technology is a virtual private network built within the Internet public. It is recommended for a low cost compared to traditional solutions such as Frame Relay, AT [7, 8]. However, companies steadily increase the number of branches which poses a problem of scalability, while reconfiguring all sites is mandatory when a change is made VPN Multipoint technology solution is a software provide an economic and accessible alternative to hardware VPN solutions, and guarantee a full meshed connection between multiple sites with a dynamic, quick and automatic manner, it offers scalability, i.e. involves no extra configuration on already configured equipment. VPN Multipoint consists of two mainly deployment designs:

- HUB and SPOKEs are used to perform headquarters-to-branches interconnections,
- SPOKE -to- SPOKE are used to perform branch-to-branch interconnections

With this approaching the problem of scalability is resolved, in case we add a SPOKE, other equipment previously configured will undergo no further modifications, we had to configure the added SPOKE to register with the HUB and become a member of the current architecture.

On the other hand, if companies expand, and the number of spokes or endpoint devices increases, the network becomes too complex to manage and operate securely and efficiently. Furthermore, the manual configuration for each device also brings burdens for network operation and maintenance. For example, if an organization has 1,000 sites, you cannot afford to have one person at each site responsible for managing security policy. Thus, you want to have central management that allows for a handful of dedicated security specialists to manage the whole environment. Additionally, consistent policy throughout the entire network is more efficiently set through a centralized management interface. While the purpose that virtual private networks (VPNs) serve is fairly straightforward, the technology components that enable them can be rather complex. Fortunately, central management can go a long way in terms of alleviating concerns around securing remote access by making organizations more secure, efficient and productive.

People are currently working on applying SDN technology to centralized policy management of intra/inter VPN domain, most solutions are only addressed to only the static VPN (site-to-site) technology. By adapting SDN philosophy, this paper proposes a new model called Smart SDN Policy Management based VPN Multipoint (SSDPM based VPN Multipoint) for centralizing managing security policies in VPN multipoint technology using a new Java interface application to program the network elements. This model will have a global view of the network and can make smart decisions.

The remainder of the paper is organized as follows. In Sect. 2, we will present in an exhaustive way the recent state of the art of work related to the centralized management of security in secure networks. In Sect. 3, we will discuss positioning of our contribution. In Sect. 4, we will present our motivation to development a new model over open source Linux OS for this type of network. In Sect. 5, we will describe the design model "Smart SDN Policy Management Security System" and define various modules. In Sect. 6, we will thoroughly describe various steps required by the model to automatically configuration for validation our model. Finally, we will conclude in Sect. 7.

2   State of the Art

The centralized management of the network is an active field of research, using centralized management of the network reduces operational complexity, improves security, and lowers risk through consistent policy application.

2.1   Work on Management of Networks by SDN

The work presented in [9] proposes a three-tier framework for CNDP deployment. Under this framework, and expand the existing CNDP specification language, come out SDN traffic steering method and defense entity selection method. This paper complete the transformation from CDNPs to security configurations and finish rules deployment. Besides, it experiments the CNDP model in OpenStack cloud environment show that under the proposed framework, we can finish policy deployment automatically and correctly. Hence, the results can demonstrate the model well in real network.

The author in [10] proposes a dynamic resource management scheme, called the EnterpriseVisor engine that manages the distribution of network resources among slices. The EnterpriseVisor engine first slices an enterprise's network into virtual subnets, and then continues to monitor the resource utilization of each subnet. In the engine, dynamic resource scheduling of the subnet for each service uses the linear programming method. The engine then delivers the schedule as the resource allocation policy of the FlowVisor layer for data plane users. In order to find out the feasibility of the proposed EnterpriseVisor engine, an OpenFlow tested with four subnets is designed. Compared with the basic OpenFlow and OpenFlow/FlowVisor platforms, the EnterpriseVisor engine has higher network utilization, 25.7% and 13.4% higher, respectively, with only a minor sacrifice to control message latency.

2.2   Works on Building a Tunnel VPN

The article [11] introduced a method to implement VPN by using Free/SWAN software under Linux OS and setup configuration for different types of tunnelling technique such as the second tunnelling protocol (L2TP, PPTP and L2F), the third layer tunneling protocol (GRE and IPSec), and high-level tunnel technology (SOCKS) by edit the file configuration manually. This method could configure some high quality VPN products, which can achieve security and confidentiality of network data transmission, and meet the needs of most users.

This article [12] deals with the application of open source VPN technology and virtual computer software used to support Red Hat Enterprise Linux solution that uses OpenSwan and the IPSec sub-system to set up VPN IPSec configuration. This implementation relies on Gateway-to-Road Warrior connections, and is used to protect access to internal LAN that houses the classroom server, and provide LAN access to autonomous workstations from different networks. The VPN topology is diagrammed as follows:

Protected LAN—VPN Gateway INTERNET Road Warrior. The solution suffers from scalability and automatically management, namely the administrator must specify for each accounts access and modifier the file of configuration manually when add or delete connection. The article [13] discusses the aspect of deploying IPsec tunnels on Linux kernel 2.6 for the development of VPN security gateway. The solution proposed is to deploy different IPsec policies between a source and a destination.

The article [14] proposes a new method for provisioning of Internet VPN (Virtual Private Network), which will be used for ad-hoc personal activity and group work across users' networks over the Internet. In order to utilize Internet VPN in ad-hoc fashion, the users must configure manually the VPN routers and hosts in their networks correctly and rapidly. Because these configuration tasks require technical knowledge of the VPN, users often find it difficult and troublesome to accomplish the tasks, even though they are not new to the Internet. In this method is introduced two management servers placed at the ISP: the VPN management server handling VPN routers and the DHCP server providing hosts' configuration. With these servers, users can rapidly join their desired VPN by a simple and fairly small task, using the web-based GUI (Graphical User Interface), through which the VPN router configuration is automatically generated and provisioned from the VPN management server.

2.3   Work on Automation Management of VPN

The centralized management of VPN tunnels is an active field of research, and several contributions were made to negotiate and create VPN policies between devices from different areas [15] as well as to manage the control plane of IPsec protocol for multiple VPN [16].

Others work addressed the complexity of VPN management and deployment in the network. Authors in [17, 18] propose different solutions to simplify the L3VPN deployment and mitigate existing identical complexities.

The patent of invention [19] allows the automatic management of the VPN technology. This invention provides a way for managing partitioned VPNs on multiple subnets. By the present invention, the inventors proposed a system based on three elements: (1) authentication database, (2) Hardware Inventory Database and (3) VPN Management Module.

- The first element contains a list of registered users as well as the VPN type that they have access to (PPTP, SSL, and IPsec).
- The second element contains the list of sub-networks as well as the constituent devices.
- Thanks to the third element, the link between the two preceding on the device traversed, this module allows to change the VPN type for a given user, if it has the right of course to benefit.

In this work, the parameters of the security policy are not discussed. Even if we consider that maximum security is assured.

The article [20] proposes a method based on SDN network architecture for IPSec VPN automatic configuration management to solve the problem of manual configuration and commands that pose difficulty in manual faults removing, and the large number of client applications makes its efficiency even lower. This method is combined with SDN controllers and OpenFlow switches to improve its management configuration and user access management efficiency.

The article [21] presents a solution to the problem of security policy management for computer and networking resources by using IPsec and KeyNote which they are provided a mechanism to implement a granular security policy. The proposed solution presents an XML editor and a graphical user interface to create and manage a consistent and correct security policy. The interface has the simplicity of a simple menu-driven editor that not only provides KeyNote with a policy in the specified syntax but also integrates techniques to support administrative policy verification.

Previous works deal with the manual and central management of network and site-to-site VPN policies, our contribution is in relation with the multipoint aspect of VPN network over Linux Operation System, and it provides a model of centralized policy management through a software program based controller.

3   Positioning of Our Contribution

Several research studies have been proposed for centralized policy management of intra/inter VPN domain, most solutions are only addressed to VPN site-to-site technology but none of them have addressed the issue of VPN multipoint technology. Through this work, we propose an enhancement of the VPN multipoint networks in the form of an approach, that we have called "Smart SDN Policy Management System based VPN multipoint". Our approach allows:

Automatically management security policies and eliminating the need to manually configuring the network for user errors.

- Admins to manage services by abstracting higher-level functionality
- Cost Savings
- Provide a flexible way for network administrator to manage large networks as a single entity rather than hundreds or thousands of the discrete element
- Solve problems in existing technology represent limitations of reliability.

4   Motivation

Small and medium-sized enterprises: using Linux machines, the latter are not obliged to invest in a router with a security licence in order to secure the exchanges. Companies that use Open Source (OS) find that it offers the most flexibility of any third-party software alternative. You are, for example, never locked into a vendor, their costs, their buying structures, or their re-distribution terms. By using Open Source enables vendor independence. Our tool is based on an open source IOS and allows Administrator on VPN Multipoint Management on Linux machine to:

- Set up VPN connections with using a GUI man/machinery application designed for this type of networks.
- To help systems administrators verifiably enforce simple multi-layer network security policies
- Manage multiple sites simultaneously.

5   The Approach Description

Our approach Fig. 1 architecture is basically focused on allowing network administrators to manage and control the whole network through a software program based a controller. This goal is achieved through the separation of the data plane and control plane, which simplifies the networking services. The model is composed of three main components: "Management layer" "Control layer" and "Data layer". We will now briefly describe the model and its various components.

5.1   Management Layer

The management layer is the language of communication between User (network admin/ ops to be precise) and the devices. It used to convert user data from a graphical interface to specific device configuration. It deals with managing configuration files and definition the parameters and attributes of security and routing protocol-specific for enabling VPN multipoint technology. This layer has several management components: IPsec Policy management, VPN Tunnel management, Dynamic Routing and Key authentication management.

5.1.1   IPSec Policy Management

This component is control by a IPSec [22] that is a framework for providing a number of security services, as opposed to a single protocol or system, which is used to implement encryption function. IPSec provides a solution for Secured Tunnelling for ensuring Authenticated and Encrypted data flow. IPSec has two security protocols, IP Authentication Header (IPSec AH RFC2402) [23], which provides access control, connectionless message integrity, authentication, and anti-replay protection and IP Encapsulating Security Payload (IPSec ESP RFC2406) [24], which supports these same services, plus confidentiality. IPsec operates in two modes, tunnel and transport mode, transport mode does not change the initial header it sits between the network layer and transport. In order to prepare IPsec, it must prepare the Racoon file. Racoon is "an IKE daemon for automatically keying IPsec connections". The Racoon daemon was originally developed as part of the Kame project [The KAME project [25]] and is now part of the IPsec-Tools collection available on many Linux systems. Because Racoon supports fairly verbose logging of the parameters set at the start of the negotiations in debug mode, it is useful for troubleshooting. All the candidates successfully connected to Racoon first. Three files were used for the configuration:

- **racoon.conf** contains the encryption and hash algorithms for Phase 1 and Phase 2, and the PFS and lifetime settings for the keys. This file maps the Phase 1 parameters to a correspondent entity, which can be an IP address or a distinguished name from a certificate. In Phase 2, net blocks (optionally ports) that are valid for the Phase 2 algorithms and parameters are specified. Because the configuration uses fixed passwords (preshared keys or PSKs for short), we need a place to store this information. psk.txt maps the IP addresses of the correspondents to the passwords. Passwords can be specified in the clear or as hex strings.

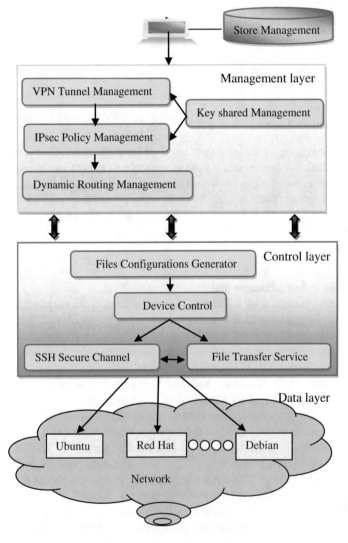

Fig. 1. Architecture smart SDN policy management

We need also to store the IPsec policy, which contains rules that define which packets should be sent down the IPsec tunnel negotiated by Racoon, or from which source packets are expected to exit the tunnel. We can do this quite easily at the command line with the setkey command, but it is more practical to store the policy in a setkey script. The rules contain source and target IP address ranges, protocols and port numbers, the direction (incoming or outgoing), and a statement stating whether ESP, AH, or both should be used. We can also tell Racoon to generate the policy automatically following the negotiation phase.

5.1.2 VPN Tunnel Management
Tunnel management functions are control by an OpenNHRP is an open source software that implements the Next Hop NBMA resolution protocol (as defined in RFC 2332). OpenNHRP can run on a standard Linux system to create multipoint VPN Linux route. OpenNHRP allows HUB and SPOKE to build a dynamic multipoint tunnel. Identification and authentication of tunnels will be made by Key file attributes.

5.1.3 Key Shared Management
This component is responsible for putting the identification key of the tunnel to pick the right tunnel before it can decapsulate the packet and look at the IP address. But better look at a scenario where we try to reach a destination behind the HUB, or behind SPOKEs. There, even theoretically, it could not be the IP address that is taken as a differentiator. It has to be the tunnel-key.

One more thing: If we extend VPN with IPsec, we typically also have to apply the tunnel protection in a different way. We need the "shared" keyword when there can be two tunnels between two devices and both have the same tunnel-source and tunnel-destination.

5.1.4 Dynamic Routing Management
This component has an open source software called Quagga Routing Suite [26]. Quagga is a routing software suite, providing implementations of Open Shortest Path First (OSPFv2, OSPFv3), Routing Information Version 1, 2 (RIP v1 and v2), Routing Information Protocol, Next Generation (RIPng) and Border Gateway Protocol version 4 (BGP-4) for Linux platforms. Quagga is a fork of GNU Zebra, which was developed by Kunihiro Ishiguro Quagga tree aims to build a more involved community around Quagga than the current centralized model of GNU Zebra.

The Quagga architecture consists of a core daemon, zebra, which acts as an abstraction layer to the underlying Lunix kernel and presents the Zserv API over a Lunix or TCP stream to Quagga clients. It is these Zserv clients, which typically implement a routing protocol and communicate routing updates to the zebra daemon.

- **Store Management:** the local store is used to store data and files configurations of each machine in the local administrator machine.

The management plane is sometimes considered as a subset of the control plane.

5.2 Control Layer

This layer is a logical entity, that completes the procedure of traffic steering and configurations are sending to related devices. This layer consists of two components:

5.2.1 Files Configurations Generator

It is an essential part of control layer. It uses to convert user data to specific device configuration.

5.2.2 Device Control

It communicates remotely with these devices using their user interfaces and delivers configurations to them using:

- SSH [27] Secure Channel: standardized by the secsh working groups was designed to support opens an encrypted SSH tunnel to the end devices.
- Files Transfer Service: This service is used to transfer configuration files through an SSH connection.

6 Data Layer

All the configurations are implemented in this layer finally, which contains a variety of Linux distributions that responsible for forwarding data.

7 Function of the VPN Multipoint Security Management System

In order to validate the developed module, we will present various required by the design "Smart SDN Policy Management System". The architecture is composed of one HUB and two Spokes. The operating-system users run is the Ubuntu. Ubuntu Linux was chosen for this research, as it is an operating system aimed at providing a secure, private access to the Internet. The graphs following show the steps to follow for centralized management.

Step1: Specify the type of the machine to configure
The user through the side configuration has a field Fig. 2 containing the type of machine (1), either to represent a HUB or SPOKE, and field to specify number of Spoke (2).
Step2: Define identity information
After specifying the type of machine to configuration, a window in Fig. 3 is available, the window is mainly composed of two parts: interface configuration (3), security and routing policies configuration (4).First part consists of two sections: HUB Configuration (5) and Spokes Configuration (6), the two sections are composed of the following fields: outside interface (7), public IP address (8), subnet mask of public address (9), tunnel IP address (10),subnet mask of the tunnel(11),inside interface(12), private IP address(13),subnet mask of private address(14), option (15) hostname of remote device and (16) SSH authentication data, (17) return to the main program

interface, option (18) is used to test the accessibility or inaccessibility of the remote device, the option (19) to go the next step 2/2.

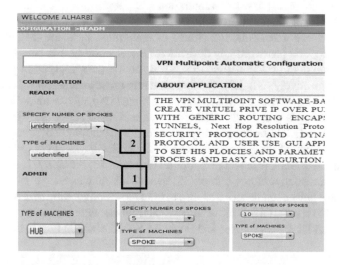

Fig. 2. Specifying machines type and number of spokes

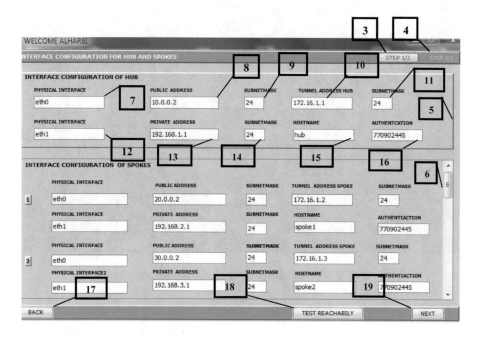

Fig. 3. Specifying machines data

Step 3: Define security policy and routing protocol

The second part, security and routing policies configuration Fig. 4 consists of four main sections: IPsec phase1 Configuration (20), IPsec phase 2 configuration (21). OpenNHRP configuration (22) and the choice of routing protocol (23). Section (20) is composed of three fields, the choice of encryption protocol (24), the integrity protocol (25), and password key derivation (26). Section (21) is composed of three fields, the mode of protocol IPsec to use ESP or AH (27), encryption protocols and integrity respectively (28) (29); the default mode is set to Transport. Section (22) is composed of two fields, OpenNHRP interface tunnel (30), mGRE tunnel key (31) used to separate tunnels and provide authentication and identifier of the NHRP network. The last section (23) allow the user to pick through a list the protocol to be implemented which can be one of these protocols RIP, OSPF or iBGP (32). In order to activate or not to activate routing protocol (33), option (34) return to the step 1/2, After we were completed the configuration, option (35) to save all policies and parameters in store management before sending, option (36) to send files configuration to network devices, option (37) to reset all fields all windows. After the network is running, we are doing ping between HUB –to- Spokes and Spoke-to-Spoke, show in Figs. 5 and 6.

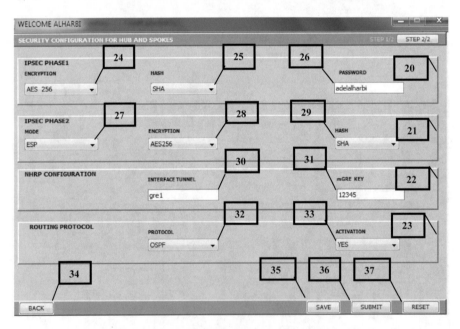

Fig. 4. Configuration of security and routing protocol

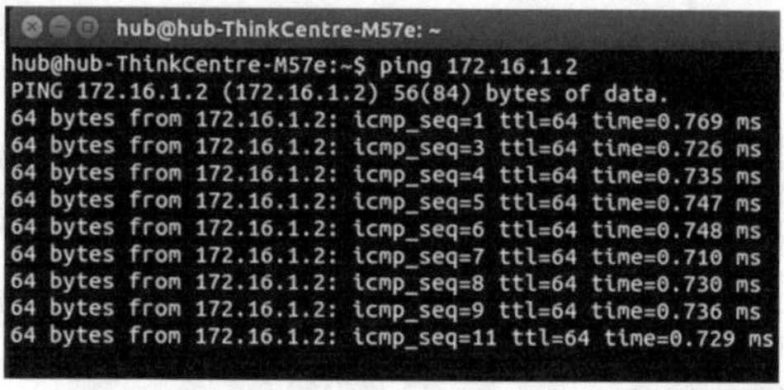

Fig. 5. Ping between HUB and Spoke1

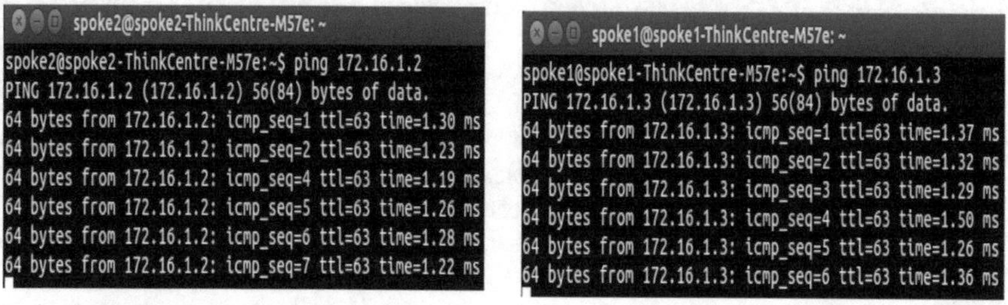

Fig. 6. Ping between Spoke1 and Spoke2

8 Conclusion

In this paper, we have presented a model Smart SDN Policy Management based VPN Multipoint, our model allowed Network administrators to centralize management of VPN Multipoint network with less complexity and simplifies large scale Policy security system deployment and place the configurations correctly into corresponding network devices using SDN philosophy. Our proposed model consists of three layers, Management layer that automates the deployment of VPN Multipoint instances using a new graphic interface to manage multiple components VPN tunnel Management, IPsec policy Management, Dynamic Routing Management and bypasses the error-prone tasks of VPN Multipoint configuration on provider edge device. The control layer includes Device Control for different device management. This layer completes the procedure of traffic steering and configurations are sending to related devices. Data layer supplies network elements that carries user traffic. By using our model these network elements can be used dynamically to the upper layers.

The model was implemented and tested on architecture consisting one Hub and two Spokes, the time require for an expert on VPN networks and Linux operation system for manual set up the devices is 45 min, we moved that to 1–2 min with the proposed model, in addition to time effectiveness the management error is null, and we have shown, how the user can interaction with the interface graphic to a manager the machine with his policies and attributes. We discussed the components of the model and its operation and presented the tool.

References

1. Shah, S.A., Faiz, J., Farooq, M., Shafi, A., Mehdi, S.A.: An architectural evaluation of SDN controllers. In: IEEE International Conference on Communications (ICC), pp. 3504–3508. IEEE (2013)
2. Sezer, S., Scott-Hayward, S., Chouhan, P.K., Fraser, B., Lake, D., Finnegan, J., Viljoen, N., Miller, M., Rao, N.: Are we ready for SDN? Implementation challenges for software-defined networks. IEEE Commun. Mag. **51**, 36–43 (2013)
3. Dixit, A., Hao, F., Mukherjee, S., Lakshman, T.V., Kompella, R.: Towards an elastic distributed SDN controller. In: ACM SIGCOMM Computer Communication Review, pp. 7–12. ACM (2013)
4. Ortiz, S.: Software-defined networking: on the verge of a breakthrough? Computer **46**, 10–12 (2013)
5. Xia, W., Wen, Y., Foh, C.H., Niyato, D., Xie, H.: A survey on software-defined networking. IEEE Commun. Surv. Tutor. **17**, 27–51 (2015)
6. Open Networking Foundation (ONF): Openflow specification version 1.4.0.ONFspecifications (2013).https://www.opennetworking.org/images/stories/downloads/sdn-resources/onf-specifica tions/openflow/openflow-specv1.4.0.pdf. Accessed Aug 2014
7. Matthews, A.R., Bhaskaran, S., Jou, L., Desai, S.: System and protocol for frame relay service over the internet. Google Patents (2007)
8. Chase, C.J., Holmgren, S.L., Medamana, J.B., Saksena, V.R.: Traffic management for frame relay switched data service. Google Patents (2001)
9. Gao, J., Xia, C., Wang, S., Zhang, H.: A SDN-based deployment framework for Computer Network Defense Policy. In: 2015 4th International Conference on Computer Science and Network Technology (ICCSNT), pp. 1253–1258. IEEE (2015)
10. Chen, J.-L., Ma, Y.-W., Kuo, H.-Y., Yang, C.-S., Hung, W.-C.: Software-defined network virtualization platform for enterprise network resource management. IEEE Trans. Emerg. Top. Comput. **4**, 179–186 (2016)
11. Zhang, J., Hu, W., Gao, F.: Construction of VPN gateway based on FreeS/WAN under Linux. In: 9th International Conference on Signal Processing, ICSP 2008, pp. 2876–2879. IEEE (2008)
12. Toderick, L.W., Lunsford II, P.J.: Using VPN technology to remove physical barriers in linux lab experiments. In: Proceedings of the 8th ACM SIGITE Conference on Information Technology Education, pp. 113–118. ACM (2007)
13. Li, Z.: Design and implementation of VPN security gateway based on Linux kernel 2.6. In: 4th International Conference on Computer Science & Education, ICCSE 2009, pp. 357–360. IEEE (2009)
14. Hori, K., Yoshihara, K., Horiuchi, H.: Provider provisioned internet VPN for personal communication environment. In: The 8th Asia-Pacific Network Operations and Management Symposium, pp. 190–201 (2005)
15. Baek, S.-J., Jeong, M.-S., Park, J.-T., Chung, T.-M.: Policy-based hybrid management architecture for IP-based VPN. In: NOMS, pp. 987–988 (2000)
16. Guo, X., Yang, K., Galis, A., Cheng, X., Yang, B., Liu, D.: A policy-based network management system for IP VPN. In: International Conference on Communication Technology Proceedings, ICCT 2003, pp. 1630–1633. IEEE (2003)
17. Lospoto, G., Rimondini, M., Vignoli, B.G., Di Battista, G.: Rethinking virtual private networks in the software-defined era. In: IFIP/IEEE International Symposium on Integrated Network Management (IM), pp. 379–387. IEEE (2015)

18. van der Pol, R., Gijsen, B., Zuraniewski, P., Romão, D.F.C., Kaat, M.: Assessment of SDN technology for an easy-to-use VPN service. Future Gener. Comput. Syst. **56**, 295–302 (2016)
19. Francis, W.J., McAloon, D.: Real-time automated virtual private network (VPN) access management. Google Patents (2017)
20. Li, Y., Mao, J.: Sdn-based access authentication and automatic configuration for IPsec. In: 2015 4th International Conference on Computer Science and Network Technology (ICCSNT), pp. 996–999. IEEE (2015)
21. Mohan, R., Levin, T.E., Irvine, C.E.: An editor for adaptive XML-based policy management of IPsec. In: 19th Annual Computer Security Applications Conference, Proceedings, pp. 276–285 (2003)
22. Atkinson, R., Kent, S.: Security architecture for the internet protocol (1998)
23. Atkinson, R., Kent, S.: IP authentication header (1998)
24. Atkinson, R., Kent, S.: IP encapsulating security payload (ESP) (1998)
25. The KAME project. http://www.kame.net/
26. Jakma, P., Lamparter, D.: Introduction to the quagga routing suite. IEEE Netw. **28**, 42–48 (2014)
27. Ylonen, T., Lonvick, C.: The secure shell (SSH) protocol architecture (2006)

Towards Smart Software Defined Wireless Network for Quality of Service Management

Azeddine Khiat[1(✉)], Ayoube Bahnasse[2], Mohamed El Khaili[1], Jamila Bakkoury[1], and Fatima Ezzahraa Louhab[2]

[1] Lab SSDIA, ENSET Mohammedia, University Hassan II, Casablanca, Morocco
azeddine.khiat@univh2c.ma
[2] Laboratory LTI, Department of Physics FS Ben M'SIK, University Hassan II, Casablanca, Morocco

Abstract. The quality of service (QoS) is one of the major concerns for most of the network administrators. However, QoS requires an implementation of policies, which are said to be static. This approach is not flexible with the diversity of users' services and their varying needs in terms of network resources.

Ensuring the end-to-end quality of service in existing network architectures is a permanent problem. The SDN paradigm appeared recently in response to the limitations of traditional network architectures. Its main advantages are to provide a centralized view of the network, programmability and separation of data and control plane.

As a result of this issue, in this paper, we propose a new extensible SSDWN (Smart Software Defined Wireless Network) architecture for the intelligent, dynamic, and adaptive management of QoS in a heterogeneous or homogeneous wireless network.

Keywords: SDN · Heterogeneous wireless networks · QoS · SDWN Smart SDN

1 Introduction

Nowadays, telecommunications technologies occupy a very important place in the daily lives of individuals, organizations and even research laboratories. This trend is justified by the many advantages that the telecommunications field has offered, particularly in terms of remote interconnection, file sharing, quad-play data transmission: (i) Internet access, (ii) landline telephony, (iii) online television and (iv) mobile telephony.

Like wired networks, wireless is an integral complement to the fixed broadband networks; It adds the necessary mobility element to the areas with a high throughput demand. Wireless networks are expected to replace wired broadband in some geographic areas, and should be expanded to other areas where wired infrastructure is not economically viable. Three competing mobile standards will accelerate the technical upgrade dynamics:

- The Universal Mobile Telecommunications System (UMTS) standard will evolve, first in High Speed Packet Access (HSPA) and then in the Long Term Evolution (LTE) standard;
- The Code Division Multiple Access Evolution-Data Optimized (CDMA EVDO) standard;
- The Worldwide Interoperability for Microwave Access (WiMax).

These mobile networks are currently a very active research area, given the limitations they represent mainly in terms of quality of service [1]. Users now route different flow categories (Real Time, Web-Based [2], Transactional, etc.) across heterogeneous networks. This constitutes a real problem concerning:

(i)   The user satisfaction (Quality of Experience - QoE).
(ii) The network resources management that are often restricted.
(iii) The scalability [3], because the increasing demand for bandwidth and the diversity of flows can lead to a saturation of the entire network.

1.1   Software Defined Networking

Since 2008, a new concept is emerging in the networks world, that of Software Defined Networking or SDN [4]. This concept, which emerges as a new paradigm, is very promising and is attracting much interest today.

The fundamental principle of SDN is the decoupling of the control plane (routing decision intelligence), management plane and the data plane (decisions execution) Fig. 1.

- Data plane: it is the part that manages the core of our equipment, its role is to route the packets from point A to point B.
- Control plane: this plane allows as its name suggests to control the data plane by establishing the rules that it must follow.
- Management plane: this plane deals with all matters relating to the equipment administration.

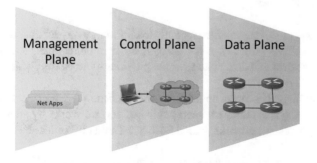

Fig. 1. The SDN architecture

SDN and Network Function Virtualization "NFV" [5] technology address these three challenges by providing flexibility in the management and configuration of such networks. These two technologies ensure a very high profitability of the wireless networks and extend the innovation field thanks to their modularity.

Indeed, SDN is a new approach in the creation and management of the network architecture (Wired or wireless) in which the management, control, and data layer are separated. The goal is to make the data layer's equipment transparent to the administrator so that the administrator can act across a unified console on the entire system.

With the SDN approach, a user can exploit several network infrastructures, depending on which one meets the contract between him and the operator (Service Level Agreement SLA [6]). This data control is therefore transferred to a software entity called a controller.

1.2 Positioning of Contribution

Quality of service management in wireless networks is generally difficult to ensure, for several reasons:

1. The QoS techniques change from one access network to another.
2. The resources reservation must be made at each attachment point.
3. The Scalability becomes a problem if the user profiles change or the number of intermediate equipment increases.

By adopting the SDN approach, we propose by this work a new model of the intelligent and adaptive management of the quality of service in a wireless network. This model is called Smart Adaptive QoS for Wireless Networks SSDWN.

The rest of the paper is organized as follows: in the second section, we will present the recent related works. In the third section, we propose the SSDWN model architecture. An evaluation of the model will be presented in Section four and finally, we will conclude.

2   Related Works

SDN technology [7–9] has emerged recently thanks to its many services such as network virtualization (NFV) [10, 11], self-management of architectures, and separation of management, control and data planes, which provides more flexibility and scalability to modern wireless networks.

This technology has been applied for a long time on wired networks [12, 13]. However, nowadays, researchers have published several works by adapting the SDN technology in wireless networks, called Software Defined Wireless Network "SDWN" [14–16].

Our contribution is in the context of the adaptation of the SDWN technology for the management of the QoS in wireless networks. Some work has been published in this field. The work [17] proposes a solution which allows to manage the QoS in a heterogeneous wireless network. However, this work does not deal with the case of users mobility, in other words, the QoS adaptation will be performed when the client is associated with an attachment point.

The work [18] proposes a solution to change the routing plane in a flexible and automatic manner to meet QoS needs of multimedia applications. However, this work does not deal with the management of the QoS policies for multimedia streams.

Taking into account our comments on previous work, we propose a new SDWN architecture for managing quality of service and mobility intelligently and dynamically. This model is called SSDWN acronym for Smart Software Defined Wireless Network.

3 SSDWN Model Architecture

We propose by this work a SSDWN model for the dynamic and smart management of QoS in wireless networks.

The different modules of SSDWN model are defined in Fig. 2.

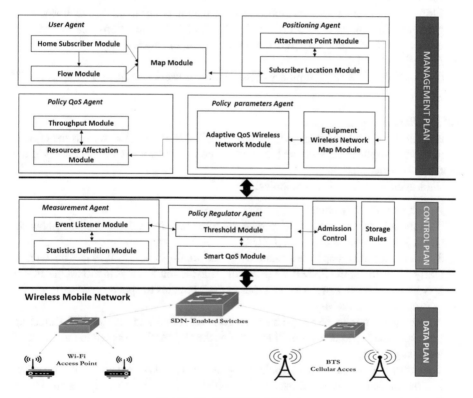

Fig. 2. The SSDWN architecture

The SSDWN model consists of three planes: Data plane, Control and Management.

3.1 Management Plane

The management plane consists of four agents: (i) User Agent, (ii) Positioning Agent, (iii) Policy QoS Agent, and (iv) Policy Parameters Agent.

(a) User Agent

This agent maintains a table of mobile users' identitiesm their roles and active network trafic. Then it performs mapping between applications and beneficiary users. This agent consists of three modules.

- Home Subscriber Module:
  This module allows users to be registered after the authentication phase. Radius, TACACS+ and Diameter can be used for this purpose.
  A user is identified by his MAC address or USIM. The Home Subscriber module allows static records as it can be dynamic.
- Flow Module:
  Flow Module, allows to maintain an information table about active flows on a network. This module is configured by standardized applications and is open for addition or modification when using proprietary applications.
- Classifier Module:
  This module allows a preliminary classification of users - traffic according to DSCP codes.

(b) Positioning Agent

This agent defines all the managed equipment (BS, AP, BTS ...) as well as the current attachment equipment and the next attachment equipment for each user defined in the Home Subscriber module.

This agent consists of two modules:

- Attachment Point Module:
  This module makes it possible to map the intermediate equipment of the network. (BSS, access point, routers, BTS, etc.).
  The devices can be populated manually by the administrator, as they can be dynamically detected via the SNMP V3 protocol. This module is very important because it is from this that we can know the types of QoS policies to apply.
- Subscriber Location Module:
  A user can be mobile, he can go from a wireless X network to a network Y. Detecting his exact position makes it possible to define the QoS policy adapted to the nature of his next gateway.
  The Subscriber Location (SL) module detects the current position of each subscriber. As, the distance between a user and its point of attachment can be calculated or detected by modern operating systems. Our application can predict the next point of attachment, and configure the next QoS policy.

(c) Policy QoS Agent

This agent allows you to define the different attributes of the QoS policies to be applied in a domain.

This agent consists of two modules.

- Throughput Module:
  Before proceeding to the final phase, it is first necessary to detect the uplink and downlink throughput of each Equipment defined in the module "Equipment Module". This module is indispensable because the allocation of the bandwidths must be made according to the available one. Without excess or waste.
- Resources affectation Module:
  After (i) authenticating and authorizing subscribers, (ii) detecting their active flows and customizing their new flows (iii) detecting attachment points (iv) and detecting their nature.
  The allocation resource module allows to associate for each class of the module a determined bandwidth.
  The bandwidth will be adapted dynamically, using the PB-SAQOS model proposed by BAHNASSE [19, 20].
  This module makes it possible to adapt the QoS policy according to: (i) Intermediate equipment, (ii) traffic class, (iii) appropriate QoS policies to the type of the network access.

(d) Policy Parameters Agent

This agent is the last gateway to generate an appropriate and exploitable QoS policy for a specific device.

In the case of a Wi-Fi network, this module allows a definition of the set of contention window durations for each class of traffic. Namely that by default the mode applied is HCF. The parameters on which action can be taken are, therefore, the class of service and the maximum and minimum window contention.

- Equipment wireless Map Module:
  This module allows you to assign a policy to an attachment point. This assignment must be done in real time.

3.2   Control Plane

The control plane consists of two agents: (in) Measurement Agent, and (ii) Policy Adapter Agent.

(a) Measurement Agent

This agent performs an active and passive metrology to: (i) Collect reports on the performance of the network and its transported applications and (ii) Detect further client to attachment point connections.

This agent consists of two modules:

- Event Listener Module:
  As soon as the first policies are delivered, it is necessary to carry out an active metrology on the performances of the applications in order to deduce if the policy is effective. Is that the allocated bandwidth is sufficient or the contention window needs to be further reduced.

This module performs metrology, it just collects the statistics, defined by the module that follows.

- Statistics Definition Module:
  This module is responsible for defining statistics to be collected. These statistics can be related to network, or applications.
  This module provides a list of evaluation criteria. The administration then specifies the criteria according to the context. That is, in the case of a conference, the administrator can rely on delay, jitter and loss rate only.

(b) Policy Regulator Agent

This agent plays the role of the QoS policy regulator. It makes it possible to compare the statistics of the data collected by the agent Metrology with a base of the recommended SLA thresholds. If a threshold is triggered, this module adapts the QoS policy in an autonomous way to meet the requirements of the network.

This adaptation can only be carried out by validation of the admission control. This agent consists of two modules.

- Threshold Module:
  This module allows to do two tasks: Define the standardized thresholds for each application and compare the statistics obtained by the module "Event listener" with this database. If a threshold is exceeded, processing will be triggered. This treatment will be detailed by the following module.
- Smart QoS Module:
  Through this module we will allocate additional bandwidths to certain services temporarily and remove their resources once the network is in critical condition and that priority applications or users are in need of bandwidth.

3.3   Data Plane

This layer is responsible for packets forwarding between the different devices in a domain. This layer may contain equipment of different access technologies.

4   SSDWN Model Validation

4.1   Network TestBed

In order to evaluate our SSDWN architecture, under OPNET Modeler, we performed a scenario in accordance with Fig. 3.

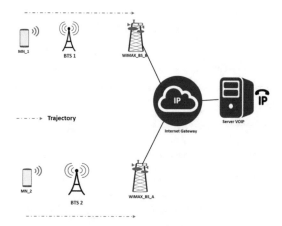

Fig. 3. The SSDWN Network Testbed

For the test traffic, we used VOIP traffic with the presence of other background traffic. The VOIP codec used is G729.

The metrics used to evaluate VOIP:

- The jitter: If two consecutive packets leave the source node with time stamps t1 & t2 and are played back at the destination node at time t3 & t4, then: jitter = (t4−t3) − (t2−t1). Negative jitter indicates that the time difference between the packets at the destination node was less than that at the source node.

- The latency: The total voice packet delay, called "analog-to-analog" or "mouth-to-ear" delay = network_delay + encoding_delay + decoding_delay + compression_delay + decompression_delay Network delay is the time at which the sender node gave the packet to RTP to the time the receiver got it from RTP.

In order to validate the SSDWN model, we created two scenarios: the first without any quality of service mechanisms and the second using SSDWN.

4.2 Obtained Results

4.2.1 VOIP Jitter

Figure 4 illustrates the jitter results, both scenarios offer a tolerable jitter, however our architecture offers even small delays. This can be justified by the fact that thanks to our SSDWN architecture, the bandwidth reserved for VOIP traffic is adjusted dynamically to the needs of subscribers.

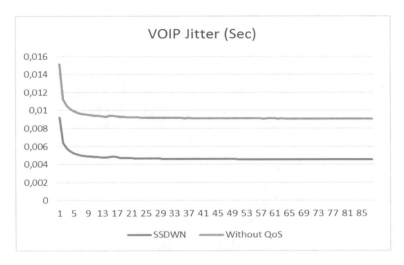

Fig. 4. VOIP Jitter

4.2.2 VOIP Latency

Figure 5 illustrates the latency results. It is clearly concluded that in the first scenario without QoS VOIP is extremely unusable as they far exceed acceptable thresholds. While our model offers a delay of 300 ms, offering an improvement of 142.857% compared to traditional QoS architectures.

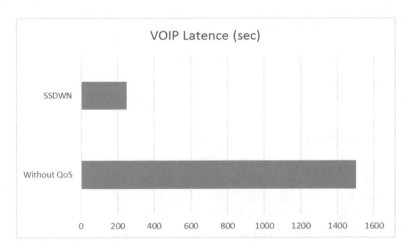

Fig. 5. VOIP Latency

5   Conclusion

The challenge of modern wireless networks is to integrate a certain flexibility in the configuration and guarantee the QoS of mobile users. SDN is a good choice for the integration of these new features.

Through this paper, we proposed a new SDN model for wireless radio and cellular networks called SSDWN. The proposed model guarantees a dynamic and intelligent management of the quality of service taking into account subscriber mobility.

The validation results of the model have shown encouraging results, minimizing VOIP jitter and latency delays.

In the next work, we will evaluate and measure the scalability of our model in large-scale networks.

References

1. Khiat, A., Bahnasse, A., Bakkoury, J., El Khaili, M.: Study, evaluation and measurement of 802.11e and 802.16e quality of service mechanisms in the context of a vertical handover case of real time applications. Int. J. Comput. Sci. Netw. Secur. **17**, 119 (2017)
2. Khiat, A., Bahnasse, A., El Khaili, M., Bakkoury, J.: Impact of Qos mechanisms on the performance of dynamic web services in heterogeneous wireless networks (802.11e and 802.16e). Indian J. Sci. Technol. **10**(25) (2017)
3. Khiat, A., Bahnasse, A., El Khaili, M., Bakkoury, J.: Study and evaluation of vertical and horizontal handover's scalability using OPNET Modeler. Int. J. Comput. Sci. Inf. Secur. **14**(11), 807 (2016)
4. Bailey, S., Bansal, D., Dunbar, L., et al.: SDN Architecture Overview. Open Networking Foundation, Ver, 2013, vol. 1 (2013)
5. Martins, J., Ahmed, M., Raiciu, C., et al.: ClickOS and the art of network function virtualization. In: Proceedings of the 11th USENIX Conference on Networked Systems Design and Implementation, pp. 459–473. USENIX Association (2014)
6. Yuanming, C., Wendong, W., Xiangyang, G., et al.: Initiator-domain-based SLA negotiation for inter-domain QoS-service provisioning. In: Fourth International Conference on Networking and Services, ICNS 2008, pp. 165–169. IEEE (2008)
7. Li, C.-S., Liao, W.: Software defined networks [Guest Editorial]. IEEE Commun. Mag. **2**(51), 113 (2013)
8. Monsanto, C., Reich, J., Foster, N., et al.: Composing software defined networks. In: NSDI, pp. 1–13 (2013)
9. Jammal, M., Singh, T., Shami, A., et al.: Software defined networking: state of the art and research challenges. Comput. Netw. **72**, 74–98 (2014)
10. Matias, J., Garay, J., Toledo, N., et al.: Toward an SDN-enabled NFV architecture. IEEE Commun. Mag. **53**(4), 187–193 (2015)
11. Hawilo, H., Shami, A., Mirahmadi, M., et al.: NFV: state of the art, challenges, and implementation in next generation mobile networks (vEPC). IEEE Netw. **28**(6), 18–26 (2014)
12. Velasco, L., Asensio, A., Berral, J., Castro, A., López, V.: Towards a carrier SDN: An example for elastic inter-datacenter connectivity. Opt. Express **22**(1), 55–61 (2014)
13. Dixit, A., Hao, F., Mukherjee, S., et al.: Towards an elastic distributed SDN controller. ACM SIGCOMM Comput. Commun. Rev. **43**, 7–12 (2013)

14. Hu, H., Chen, H.-H., Mueller, P., et al.: Software defined wireless networks (SDWN): part 1 [Guest Editorial]. IEEE Commun. Mag. **53**(11), 108–109 (2015)
15. Costanzo, S., Galluccio, L., Morabito, G., et al.: Software defined wireless network (SDWN): an evolvable architecture for W-PANs. In: 2015 IEEE 1st International Forum on Research and Technologies for Society and Industry Leveraging a Better Tomorrow (RTSI), pp. 23–28. IEEE (2015)
16. Derakhshani, M., Parsaeefard, S., Le-Ngoc, T., et al.: Leveraging synergy of 5G SDWN and multi-layer resource management for network optimization. arXiv preprint arXiv:1602.09104 (2016)
17. Lu, Z., Lei, T., Wen, X., et al.: SDN based user-centric framework for heterogeneous wireless networks. In: Mobile Information Systems, vol. 2016 (2016)
18. Tomovic, S., Prasad, N., Radusinovic, I.: SDN control framework for QoS provisioning. In: Telecommunications Forum Telfor (TELFOR), 2014 22nd, pp. 111–114. IEEE (2014)
19. Bahnasse, A., Elkamoun, N.: Policy-based smart adaptive quality of service for network convergence. Int. J. Comput. Sci. Inf. Secur. **13**(3), 21 (2015)
20. Bahnasse, A., El Kamoun, N.: A policy based management of a smart adaptive QoS for the dynamic and multipoint virtual private network. Int. J. Control Autom. **9**(5), 185–198 (2016)

Transformation of High Level Specification Towards nesC Code

Sara Houhou[1], Laid Kahloul[1(✉)], Saber Benharzallah[2],
and Roufaida Bettira[3]

[1] LINFI Laboratory, Computer Science Department,
Biskra University, Biskra, Algeria
Sara.houhou@gmail.com, kahloul2006@yahoo.fr
[2] LINFI Laboratory, Computer Science Department,
Batna University, Batna, Algeria
saber@yahoo.fr
[3] INSAT Laboratory, Computer Science Department,
Tunis El Manar University, Tunis, Tunisia
roufaida.bettira@yahoo.fr

Abstract. Automatic code generation is an important step in development of software especially for a particular domain application such as embedded system. WSN is a wireless network of a group of sensor nodes, connected with AdHoc method. The formal models is adopted in the embedded application to allow to the developer the design of their system at an abstract level, and guarantee to them an assure implementation. A lot of work proposed an approach for modelling and verifying WSN protocols based on formal models for the wireless sensor network [3, 17]. In this paper, we present an approach for automatically generate WSN implementations from Timed Automata models. The proposed approach generates a set of operations that are presented in the abstract model, which define each functionality provided by a the WSN application. To demonstrate the feasibility of the proposed approach, it will be applied to automatically generate the code for an example of WSNs application.

Keywords: Formal specification · Real time system · Transformation
Timed automata

1 Introduction

The innovative new solutions as Internet of Things (IoT), Cloud Computing, Smart Cities, etc. are based on the development of embedded systems. Embedded systems run on mobile and dynamic devices, thus they are restricted by the limited resources of these devices. These systems are exploited in many critical applications (military, health-surveillance, etc.). Hence, they must be reliable and efficient to protect and save the life of their users. WSNs are distributed systems typically composed of embedded devices. WSNs are composed of a number of connected distributed autonomous sensors, capable of measuring environmental parameters such as temperature, sound, pressure, etc. in order to accomplish a task. The sensors transmit their data through the network to

© Springer International Publishing AG, part of Springer Nature 2019
J. Mizera-Pietraszko et al. (Eds.): RTIS 2017, AISC 756, pp. 275–284, 2019.
https://doi.org/10.1007/978-3-319-91337-7_26

one or more base stations which are called sinks. These stations transmit collected data over the internet or via satellite to the central computer to analyze these data and make decisions [8]. On the software level, the wireless sensor network is seen as a set of deployed and connected components which act together following a set of protocols. Protocols define all the realization stages from the deployment to the exploitation and maintenance of the network. The design, specification and implementation of these protocols must be validated and verified to guarantee the well-functioning of the network. The use of formal methods in this domain attracted many researchers. Formal methods provide techniques allowing: the abstract description, the absence of confusion, the proof of properties, and the possibility of automatic refinement.

The presented work proposes an approach to automatically generated source code for wireless sensor networks systems from timed automata [2] specifications, which is a formalism extending the finite automata with variables of the clock type. A WSNs system is specified according to a set of defined constraints to ease the code generation, and verified using UPPAAL model checker. This specification is automatically translated to nesC code uses with the TinyOS operating system, which is a real-time and open source OS. This transformation is based on a set of rules, which define the mapping from timed automata to nesC code, and implement a Python code generator.

The remainder of this paper is structured as follows: Sect. 2 presents related work. Section 3 describes the provided approach. Section 4 defines the transformation rules. Section 5 presents the application of the proposed approach on a WSN system, Sect. 6 concludes.

2   Related Work

In the literature, many works have proposed approaches and techniques to derive detailed representations of a system by refining the high-level specifications. Some researchers addressed the issue of automating this refinement. In [7], the authors defined model-driven engineering (MDE) methodology for WSNs development. They defined three meta-models at different levels of abstraction, with automatic model transformations between these levels. The authors defined a domain-specific model for WSNs that is Domain-Specific Language (DSL) which is a meta-model of WSNs applications. Then, they used model-to-model transformation from this later into nesC meta-model. Finally, they proposed a mapping (model-to-text transformation) from NesC meta-model to nesC code. They included an intermediate abstraction level between WSNs-DSL and the nesC meta-model. This intermediate level used a subset of the UML meta-model describing a structure of the system. In [1] Ajih et al. presented a deployment testing of the DMAMAC protocol and validated its implementability and usability on real hardware. DMAMAC protocol is an energy efficient MAC protocol used for real-time process control. The authors used a traditional methodology to design it with Timed automata in [15], and implemented this later on Zolertia Z1 using nesC language, running in TinyOs platform in [6]. The advantage of their work is that the protocol can be implemented based on time synchronization, and it is evaluated in a real-environment. The work proposed in [7] defined two phases for generating nesC codes. Firstly, a semi-formal formalism (DSL) is used as input to generate an UML format which is also semi-formal. Then, using this output the second phase generates the nesC code. In our approach we propose to do it differently, where we begin with a formal input which is TA and transform this formal specification directly to a nesC code.

Besides the above works, several works have proposed generating code from Colored Petri Nets [4, 5, 9–13, 16]. In [12], a model-based approach is proposed to generate automatically software code from Pragmatic Annotated Colored Petri Nets (PA-CPN) models. PA-CPN is a syntactical restriction of Colored Petri Nets which describes protocol behaviors. They defined pragmatics formally in [13]. PA-CPN is used to extend the CPN modelling language with domain-specific elements to make it exploitable for code generation. The generated code is implemented in Groovy programming language as a target platform. The authors implemented the proposed approach in the software PetriCode which is presented in [9]. Also, They proposed an evaluation of this approach in [10] by using unidirectional framing protocol as example and generated code for different target languages and platforms (Java, Python, Clojure). In [11], the authors applied this approach on an industrial large-size protocol which is IETF WebSocket. The authors modelled the protocol and verified its proprieties (reachability, liveness and precedence) through state space exploration by using CPN Tools. Then, they generated automatically an implementation for the Groovy platform. In [5, 16], the authors proposed an approach based on Model Driven Software Engineering (MDSE) for protocol development for WSNs. The authors extended PetriCode tool to specific platform model to generate and simulate code source for WSNs protocols. This approach combined two tools CPN Tools and PetriCode tool [9]. The input of this approach is Colored Petri Nets models which specify the behaviour of the protocols. They modelled, verified, and simulated the given protocol via model checking and theorem proving using CPN Tools. Then, they translated the resulting CPN model to PetriCode which is transformed automatically to code for different platforms: MiXiM (a wireless network simulator) and TinyOS (an operating system for deployment). The authors generated C++ code for simulation, nesC code for MiXiM (or TinyOS) and they generated Timed automata for model checking using UPPAAL. They tested the generated code in real-world deployment on Zolertia Z1 motes. They used GinMac protocol presented in [14] as a case of study. In [4], another MDSE approach based on pragmatics is presented. This approach is semi-automatic and it consisted to transform platform-independent CPNs models to specific model platform (TinyOS). This approach exploited pragmatics (defined previously) to generate code for TinyOS platform. The authors used five manual refinement steps to construct the CPN model. In each step, they add pragmatics to refine the model at a suitable level for generating nesC code. After the fifth step, they used a software prototype written in java to transform automatically the refined model to TinyOs code. Finally, the RPL routing protocol is exploited as a case study.

The most of the previous works are based on Petri Nets. Indeed, often they used this formalism to model protocols then they implemented these protocols manually. However, our approach is defined to do that automatically by using a set of transformation rules. We assume that one of the most important aspect in modelling protocols is the timing, hence TA with their UPPAAL tool stay one of the most suitable formalism for such systems.

3   Approach Architecture

This section presents the global architecture of the proposed approach. The presented approach relies on model transformation as depicted in Fig. 1, p. 5. It is generic and easily extensible because the central part code generator is based on generic transformation rules. The Fig. 1 illustrates the architecture of the proposed approach where it shows that the approach is composed by two operational steps:

The First is the Modelling and Verification of the WSNs Application. In this step a designer models the protocols of WSN application in using UPPAAL Tool. During the modelling with TA, the designer of the systems might respect certain defined restrictions which represent functionalities linked with patterns of the target language. These restrictions are presented on timed automat formalism as a set of keywords. As presented in the Fig. 1, the target language of our transformation approach is nesC language which is used for TinyOS operating system. TinyOS follows standard patterns: the communication radio (sending and receiving packets), acquiring data from different sensors, sensing, time scheduling, system boot, and LEDs control. In order to present this functional information in the model, the following points represent the set of restrictions which must be used on the model to approach the abstract model towards the target language, facilitate the transformation, and make the generation of correct code.

- **Restriction 1:** use the procedure key PreparePacket () to define the packet structure.
- **Restriction 2:** use the procedures keys: LediToggle () to turn on and off LEDs, LediOn() to turn on LEDs, and LediOff() to turn off LEDs.
- **Restriction 3:** use the procedure key Ack() to enforce acknowledgements for the transmitted messages in the protocol.
- **Restriction 4:** use the procedure key TempRead () refer to the use of function that allows to collect and measure the temperature of the environment.
- **Restriction 7:** use the procedure key HumidRead () refer to the use of function that allows to collect and measure the humidity of the environment.
- **Restriction 8:** use the procedure key LightRead () refer to the use of function that allows to collect and measure the light of the environment.
- **Restriction 9:** use the procedure key VoltRead () refer to the use of function that allows to collect and measure the voltage.
- **Restriction 10:** use the procedure key AccelRead () refer to the use of function that allows to collect and measure the accelerometer.
- **Restriction 11:** use the variable is Succeed to check whether the previous sensing procedures are executed successfully.
- **Restriction 12:** use the key Comm as a channel to model the sending and receiving radio messages.

The model of a WSN application needs to be verified functionally before being implemented on the target platform. The verified TA model obtained in this step is then taken as input in the next step.

The Second Step is the Code Generation. It represents the transformation of the timed automata model resulted from the first step to nesC source code. This transformation is based on the defined rules, which are presented in Sect. 4. The code generation step takes as input the XML files of the TA model. This file is parsed in order to extract the functional information used in the transformation phase. Then, the code generator applies the transformation rules on, and generates a set of files with (.nc) extension.

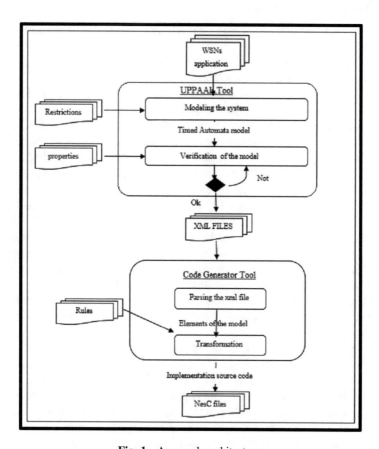

Fig. 1. Approach architecture.

4 Transformation from Timed Automata Model to Code

In this section, we present an intuitive transformation which consists of creating matching links between source model elements and target sample templates. Firstly, we consider a set of timed automata templates represent one or more WSN application according to the relation between the models. These templates have two types, templates corresponding to software hardware entities in the WSN, and templates used for only simulation purposes (i.e. humans, environment, preexisting systems, …etc.). The second type of templates must be precise during the generation of code from the part of user, and it is ignored in the transformation. Secondly, transitions, they represent the internal behavior of a component. Thirdly, for each template we generate a set of files (Module component file and configuration component file). finally, each module component uses the Boot interface and implements the Booted event which is the first point of execution in all WSNs systems. All the interfaces, used in the module, will be wired with their provided components in the configuration file. We can now describe the translation from TA to nesC code.

- **Rule 01.** Each variable declaration on timed automata model is mapped to variable declaration in the module file.
- **Rule 02.** The variables declared with type clock on the TA model is mapped to the use of the predefined component TimerC.
- **Rule 03.** Each guard on timed automata model is mapped to a standard conditional branching (IF-ELSE) bloc in the module file.
- **Rule 04.** The use of procedure(LedsiToggle() or LediOn() or LediOff() where i = 0..2) on transition in timed automata model are mapped to the use of the predefined component LedC. This component allows us to manipulate the Leds.
- **Rule 05.** The use of procedures (TempRead() or HumidRead()) on the transition in timed automata model are mapped to use of SensirionSht11C() component.
- **Rule 06.** The use of procedure LightRead() on the transition in timed automata model are mapped to use of the predefined component HamamatsuS10871TsrC().
- **Rule 07.** The use of sensing procedures mentioned previously in (R5, R6) leads to make a call of the Read command and the ReadDone event in the module file;
- **Rule 08.** The use of the variable is Succeeded is mapped to a particular If-Else bloc relates to the success of sensing events;
- **Rule 09.** The use of unicast or broadcast channel comm in the declaration part of TA model is mapped to the use of Radio. This later is provided by the predefined component ActiveMessageC.
- **Rule 10.** The use of the synchronization variable mentioned in R09 as a send channel comm! on the transition of timed automata is mapped to use of the pre-defined component AMSenderC(AM_RADIO). Note that the send channel must be preceded by the use of PreparePacket () procedure on the TA model, this procedure is mapped to use Packet interface in the module file provided by the AMSenderC component, and implement the structure of message to send in a header file, and declared a function preparePacket (typedef arg) in the module.
- **Rule 11.** The use of the synchronization variable mentioned in R09 as a receive channel comm? on the transition of Timed automata is mapped to the use of the predefined component AMReceiverC.

5   Illustrative Example

To illustrate the feasibility of our proposed approach, we detail the process of experimentation, from the TA model of the example and generate their code source implementation.

5.1   System Description

We describe an automatic watering system to keep the container garden of specialty peppers adequately watered. The system has a water barrel and pump, controlled by a sensor node that reads a soil humidity sensor in a pepper pot. When the soil humidity gets below a threshold, it turns on the pump until a humidity ceiling is reached. then it turns Off the pump.

5.2   Model of the Example

To model the system, we use UPPAAL Tool. This later is a model checker supporting the timed automata modelling. It is based on timed automata concepts, which are a set of clocks, channels for the synchronized systems (automata), variables and additional elements. The TA of the described system is presented in Fig. 2.

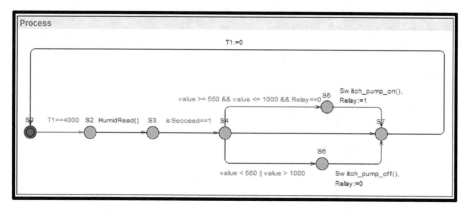

Fig. 2. TA of humidity soil sensing.

5.3   Implementation Code

In this subsection, we present the generation of the implementation written in nesC language. We obtained the latter by applying the rules of transformation defined by our approach on the model of experimentation presented shown in Fig. 2. We will try to explain each generated part. Firstly, the timed automata model is done as an XML file. Each XML file contains one or more template. In our case, the XML file has one template. For Each template; the model transformation creates two files (Module file, configuration file) and for each module file must be implement the booted event for the initialization of the system. The implementation of the booted event need the use of the interface Boot provided by the Main component of the system. Also, in the Configuration file we must wiring each used interface with its provided component as show the

Listings 1, 2. Secondly, the transition S0 to S1 used a guard (T1==4000) on the clock T1. This use of clock is transformed to the use of Timer in the system (application of the rule 02). This later consists to use of predefined interface Timer, which provided by the TimerC component. In addition, this guard related with the clock is used for precise the time cycle. In this case the model transformation takes the value of T1 as an argument in the started command of the clock (see Listings)

Listing 1. Skeleton of the Module file.

```
module ProcessM{
uses {
interface Boot ;
interface Timer<TMi l l i> as T1 ;
interface Read<uint16_t> as HumidRead ;} }
implementation {
// variables declaration uint16_t value ;
uint8_t Relay=0 ;
message_t _packet ;
event void Boot.booted() {
call T1.startPeriodic (4000) ;
}
event void T1.fired() {
      if (!( call HumidRead.read() == SUCCESS))
            ..............}
void Switch_pump_on (){..........}
void Switch_pump_off (){...........}
event void HumidRead.readDone(error_t result, uint16_t
val ){
value = -4.0 + 0.0405* val + (-2.8 * pow(10.0 , -
6))*(pow( val,2)) ;
      if (value >= 550 && value <= 1000 && Relay==0){
            Switch_pump_on ();
            Relay=1;
      ......
            }
      else {
            if (value < 550 || value > 1000) {
            Switch_pump_off ();
            Relay=0;                               }
            else {
            T1.stop() ;
                  }
      }
                                    }
}
```

Listing 2. Skeleton of the configuration file.

```
configuration SenseC{ }
implementation {
components MainC;
components TempC;
components new TimerMilliC () as Timer;
components new SensirionSht11C () as HumidRead;
TempC. Boot -> MainC;
TempC.T1 -> Timer ;
TempC.HumidRead -> HumidRead.Humidity;
}
```

6 Conclusion

Automatic code generation is an ambitious stage in the development of specific software like those of embedded systems. The paper proposes an approach to generate source code of WSNs systems. The paper shows that it is feasible to develop automatically a complete implementation of the WSNs system from formal specification models. The formal model chosen in this approach is timed automata. the choice of timed automata is that refer to the sensor network domain which is based on real-time constraints that provide sophisticated scheduling when many functionalities are in progress. The code generation is documented using a set of transformation rules. These rules are wired with the set of predefined restrictions using during the design step. Some improvements are already under realization. The first one consists of completing the rules to cover the code generation of all protocol kinds, and the tool implementation. Moreover, we will try to apply the proposed approach in complex case studies.

References

1. Somappa, A.A.K., Øvsthus, K., Kristensen, L.M.: Implementation and deployment evaluation of the DMAMAC protocol for wireless sensor actuator networks. Procedia Comput. Sci.
   83, 329–336 (2016)
2. Alur, R., Dill, D.L.: A theory of timed automata. Theor. Comput. Sci. **126**(2), 183–235 (1994)
3. Green, J., Bhattacharyya, S., Panja, B.: Real-time logic verification of a wireless sensor network. In: WRI World Congress on Computer Science and Information Engineering, vol. 3, pp. 269–273. IEEE (2009)
4. Herajy, M., Heiner, M.: A steering server for collaborative simulation of quantitative petri nets. In: International Conference on Applications and Theory of Petri Nets and Concurrency, pp. 374–384. Springer (2014)

5. Kumar, S.A., Simonsen, K.I.F.: Towards a model-based development approach for wireless sensor-actuator network protocols. In: 4th ACM SIGBED International Workshop on Design Proceedings, Modeling, and Evaluation of Cyber-Physical Systems, CyPhy 2014, pp. 35–39. ACM, New York (2014)
6. Kumar Somappa, A.A., Kristensen, L.M., Ovsthus, K.: Simulation-based evaluation of DMAMAC: a dual-mode adaptive mac protocol for process control. In: 8th International Conference on Simulation Tools and Techniques Proceedings, pp. 218–227. ICST (Institute for Computer Sciences, Social-Informatics and Telecommunications Engineering) (2015)
7. Losilla, F., Vicente-Chicote, C., Álvarez, B., Iborra, A., Sánchez, P.: Wireless sensor network application development: An architecture centric MDE approach. In: Proceedings of the First European Conference on Software Architecture, ECSA 2007, pp. 179–194. Springer, Heidelberg (2007)
8. Mottola, L., Picco, G.P.: Programming wireless sensor networks: fundamental concepts and state of the art. ACM Comput. Surv. (CSUR) 43(3), 19 (2011)
9. Simonsen, K.I.F.: PetriCode: a tool for template-based code generation from CPN models. In: International Conference on Software Engineering and Formal Methods, pp. 151–163. Springer, Cham (2013)
10. Simonsen, K.I.F.: An evaluation of automated code generation with the PetriCode approach. In: PNSE@ Petri Nets, pp. 289–306 (2014)
11. Simonsen, K.I.F., Kristensen, L.M.: Implementing the WebSocket protocol based on formal modelling and automated code generation, pp. 104–118. Springer (2014)
12. Simonsen, K.I.F., Kristensen, L.M., Kindler, E.: Generating protocol software from CPN models annotated with pragmatics. In: Brazilian Symposium on Formal Methods, pp. 227–242. Springer, Heidelberg (2013)
13. Simonsen, K.I.F., Kristensen, L.M., Kindler, E.: Pragmatics annotated coloured petri nets for protocol software generation and verification. In: Transactions on Petri Nets and Other Models of Concurrency XI, pp. 1–27. Springer, Heidelberg (2016)
14. Somappa, A.A.K., Kristensen, L.M., Ovsthus, K.: A formal executable specification of the GinMAC protocol for wireless sensor actuator networks. In: 2013 International Symposium on Wireless and Pervasive Computing (ISWPC), pp. 1–7. IEEE (2013)
15. Somappa, A.A.K., Øvsthus, K., Keistensen, L.M.: Towards a dual-mode adaptive MAC protocol (DMA-MAC) for feedback-based networked control systems. Procedia Comput. Sci. 34, 505–510 (2014)
16. Somappa, A.A.K., Simonsen, K.I.F.: Model-based development for MAC protocols in industrial wireless sensor networks. In: PNSE@ Petri Nets, vol. 2, pp. 193–212 (2016)
17. Zhang, F., Bu, L., Wang, L., et al.: Modeling and evaluation of wireless sensor network protocols by stochastic timed automata. Electron. Notes Theor. Comput. Sci. 296, 261–277 (2013)

Flexible Mobile Network Service Chaining in an NFV Environment: IMS Use Case

Youssef Seraoui[1(✉)], Mostafa Belmekki[1], Mostafa Bellafkih[1], and Brahim Raouyane[2]

[1] RAISS Team, CEDOC 2TI, National Institute of Posts and Telecommunications,
Rabat, Morocco
{seraoui,mbelmekki,bellafkih}@inpt.ac.ma
[2] N&DP Team, IT&NT Laboratory, Faculty of Sciences, Ain Chock, Casablanca, Morocco
raouyane_brahim@yahoo.fr

Abstract. With the emergence of the software defined network (SDN) and network functions virtualization (NFV) initiatives, mobile network operators (MNOs) look to make a gain from these leading technologies to mitigate hardware costs and reduce time to market. Especially with the use of NFV models, the MNO looks for scaling services up and down fast and also to reduce costs to align with network usage. Those procedures are conducted through flexible service chaining through implemented SDN controller functions along with NFV orchestration modules. In this paper, we give an introduction to a virtualized, flexible mobile network service chaining approach in an NFV environment, as an extended architecture of the 4G/LTE system defined by the 3GPP in order to draw a new design in favor of the 5G network, more especially from an Internet Protocol (IP) and Session Initiation Protocol (SIP) based network point of view. In this paper, we present our approach for dynamic and static service chaining. Moreover, as the IP Multimedia Subsystem (IMS) is one of the considerable use cases for NFV, we set up a testbed system with the goal of evaluating the behavior of a virtualized IMS core network, as a part of operator-based services, within a static service provisioning context. Experiment results show that the IMS system performance could be enhanced quickly in the case of unforeseen degradation by supporting a new VNF based service function path. In addition, SIP key performance indicators get enhanced in terms of registration delay and mean.

Keywords: Five generation (5G) · IP Multimedia Subsystem (IMS)
Network functions virtualization (NFV) · Software defined network (SDN)
Service function chaining (SFC) · Traffic steering

1 Introduction

Return on investment (ROI) is referred to as a measure of performance employed by organizations to evaluate the efficiency of an investment in infrastructures, software, or services. The concerns of these companies are therefore to boost benefits from such investment. The sector investing more in equipment is the telecommunications sector, that is, telecommunication operators (TOs) purchasing infrastructures for their data centers to meet the change of telecom technologies and to enhance the quality of service

© Springer International Publishing AG, part of Springer Nature 2019
J. Mizera-Pietraszko et al. (Eds.): RTIS 2017, AISC 756, pp. 285–299, 2019.
https://doi.org/10.1007/978-3-319-91337-7_27

(QoS) [1, 2] for their customers. With the concurrence of other TOs, the ROI could not achieve the expected results which are to increase incomes from that expenditure.

For TOs, reducing both capital expenditure (CapEx) and operational expenditure (OpEx) and thus increasing gains can be mainly achieved by virtualizing infrastructures, such as servers, telecom components, networks equipment, and other IT infrastructures.

Virtualization has played a considerable role in the IT sector. First, it has been applied to cloud computing, and then appeared in up-to-date concepts such as network functions virtualization (NFV) and software defined network (SDN) concepts.

An extended IMS framework with a high performance and scalable distributed storage was proposed in a previous study [3], as can been shown in Fig. 1, within a non-virtualized environment in conjunction with SDN technologies.

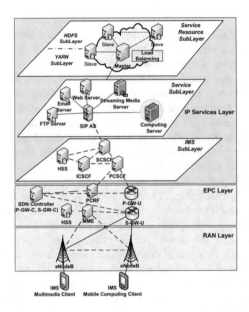

Fig. 1. Proposed, extended IMS framework within the 4G/LTE network

But, in this study, our approach concerns the abstraction of the control plane from the user plane and the dynamic implementation of traffic steering policy in the SGi-LAN and the packet data network (PDN) for flexible and comprehensive service function chaining in a virtualized environment through the use of new proposed controller functions and interfaces.

Many research studies and projects have been conducted to implement NFV based models in order to participate in the evolution of the four generation/long term evolution (4G/LTE) technology [3] toward a virtualized architecture planned to be termed five generation (5G). This architectural platform must meet the planned objectives concerning especially time-to-market, openness, infrastructure related costs, and the performance of telecom systems.

In this work, we propound new flexible mobile network service chaining approaches in an NFV milieu. Then, we shed light on two approaches; a static IP Multimedia

Subsystem (IMS) service provisioning approach and a dynamic IMS service provisioning approach. In the experiments section, we carry out benchmarks and assess the system performance [4] in a static context.

2   Background Information

In this section, we present, notable software components, technologies, and concepts we combined to propose a new approach for the telecoms sector in virtualized surroundings.

2.1   4G/LTE and IP Multimedia Subsystem (IMS)

In the 4G/LTE system, the core network is termed Evolved Packet Core (EPC) whose responsibility is authentication, control, charging, and so on. Mobility Management Element (MME), Serving Gateway (S-GW), PDN Gateway (P-GW), Policies and Charging Rules Function (PCRF), and Home Subscriber Server (HSS) are the core elements of the 4G/LTE architectural framework. The 4G core network handles both the control and user planes' traffic through their gateways; S-GW and P-GW.

The IMS [5] is an architectural framework that delivers multimedia services through an IP network, regardless of the access network. It implements Internet Engineering Task Force (IETF) protocols to establish signaling mechanisms by using the Session Initiation Protocol (SIP) [6] and invoking the Call Session Control Functions (CSCFs). Thus, the demand for multimedia services pass through various IP networks. On reports of the 3GPP, IMS is created to help access multimedia and voice services from wireless and wireline technologies. In general, it is used to achieve convergence between fixed and mobile networks.

2.2   Network Functions Virtualization (NFV) and Software Defined Network (SDN)

NFV and SDN are two widely popular initiatives and are implemented at the current time by a large number of hardware vendors and service providers. SDN consists in abstracting the control plane from the data plane and making networking more programmable via a networking operating system, the SDN controller. NFV is a technology aiming at virtualizing and deploying the network function upon commercial off-the-shelf (COTS) servers and at automating the life cycle of network services via a predefined operational framework.

In many cases, those two technologies are complementary but they can operate independently of each other.

Many efforts have been conducted to issue a unified model for NFV, called under the name European Telecommunications Standards Institute (ETSI) NFV framework. This model illustrates all the significant components to implement in an NFV based operational system. Five functional blocks compose this model, including operations and business support systems (OSS/BSS), Elemental Management System (EMS), the Virtual

Network Function (VNF) functional block, NFV Infrastructure (NFVI), and Management and Orchestration (MANO) that includes, in turn, NFV Orchestrator (NVFO), VNF Manager (VNFM), and Virtualized Infrastructure Manager (VIM).

2.3   Service Function Chaining (SFC)

Service function chaining (SFC) is a process for outweighing the basic destination based forwarding. It is typically thought of as an SDN approach. Indeed, SFC enables data packet traffic to route through an overlying network [7] via a defined path, named NFV Forwarding Graph (NFV FG) in the ETSI terminology [8] or Service Function Chain (SFC) in the IETF terminology [9], rather than the one that would be selected by routing table search on the packet's destination IP address in the case of traditional physical routing devices. It is often used in accordance with NFV when creating a collection of virtualized functions which would have usually implemented as physical networking devices connected in series by network cables.

2.4   OpenStack and OpenDaylight

As an operating system for cloud computing, OpenStack remains the leading software solution. It consists of many projects, but the most relevant are the networking project, known as neutron, and the compute project, named nova. The basic deployment of Openstack is composed of one controller and one or more compute nodes that host VMs, known also as instances, via the abstraction of the underlying resources by means of a hypervisor.

As a mature SDN controller, OpenDaylight (ODL) is more widely used by the research community that supports it as a reference controller. It opens into many virtualization initiatives, including OpenStack, through its southbound API for managing, for instance, SFCs in virtualized surroundings.

3   Service Function Chaining Approach

The following section describes the foundational concepts of the SFC architecture which we propose to be a part of the 5G network. The SFC concept enables dynamic ordering when constructing a service function chain and also topological independence in a way that it depends neither on the underlay network nor on the underlying hardware.

In a static way, the human intervention is primordial to build a network function forwarding graph (NFFG) also known as a service function chain (SFC). However, an automated fashion permits to scale up services dynamically providing a new dynamic service through a service chain being defined within a service descriptor (SD).

NFV is identical in nature to SDN in that both approaches include placing network management from the hardware level to the software level. This is why, it is crucial to consider both of the concepts when designing a new proposed architectural framework meeting 5G expectations in terms of virtualization.

The architecture and building blocks of our SFC approach are illustrated in Fig. 2. Obviously, this new virtualized environment could be controlled by two systems, an SDN controller and an NFV Management and Orchestration (MANO) system. The former manages centrally control plane considerations through different SDN applications, while the latter manages both the control plane and data plane components [10, 11] in a distributed way, and it mandates the SDN controller for SFC specifications.

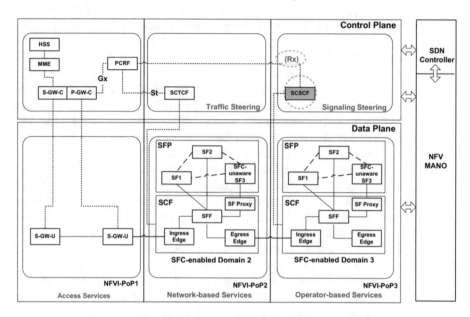

Fig. 2. Proposed SFC based mobile network solution for 5G networks

For access services, which are hosted in the 3GPP network within NFVI point of presence 1 (NFVI-PoP1), the concept of the abstraction of the control plane from the user plane is reflected by the separation of the control part of S-GW and P-GW from their data planes. Thus, the gateways processing data traffic are set in the data plane, that is, S-GW-U and P-GW-U, whereas the control components, like HSS, MME, PCRF, S-GW-C and P-GW-C are brought up to the control plane.

Concerning network-based services that are located within the NFV-PoP2, more especially in the SGi-LAN, we also apply the principle of the SDN. Therefore, in the control level we set a new function being defined by the 3GPP, the Service Chain Traffic Controller Function (SCTCF) for traffic steering. Moreover, 3GPP Technical Review [12] defines a new interface between the SCTCF and PCRF, termed St. This interface supplies the SCTCF with traffic steering rules for dynamic and coordinated implementation of SFC in the SGi-LAN. In the data plane, we define our SFC-enabled domain that consist of two SFC-enabled sub-domains, a service control function (SCF) sub-domain for data traffic classification and SFC encapsulation and a service function path (SFP) sub-domain for service chaining. For the SCF domain, it comprises two classifiers at the edge, Ingress Edge and Egress Edge whose roles are to classify and reclassify subscriber's data traffic before and after performing the SFC required, and a Service Function Forwarder (SFF) that forwards SFC-encapsulated packets directly to the required service functions (SFs) or through an SFC proxy that de-capsulate packets before routing them to SCF-unaware SFs. The SFP defines the set of virtual functions that are included in providing the requested service. It contains two kinds of functions, SFC-aware functions and SFC-unaware functions.

The third part represents our own approach toward the management of SIP systems in a way that is similar to the second network, that is, the SGi-LAN that hosts the network-based services. In the case of the IMS that is based on the SIP mechanism, the load balancing of SIP traffic is performed statically through the DNS server, more especially by the SRV records defined in zone files. However, in our approach, we intend to spread SIP traffic in a dynamic way based in the SDN and NFV perspectives. Consequently, this packed data network (PDN) hosts in turn operator-based services that rely on the SIP mechanism, and are located within NFV-PoP3. In the control level we propose a new function, the Service Chain Signaling Controller Function (SCSCF) for signaling steering. Moreover, 3GPP already defines an interface between the PCRF and AF, named Rx, but in the proposed approach this can be modified or remain unchanged because this reference point is placed rather between the PCRF and the new controller function, SCSCF. This interface, among other capabilities, would allow the PCRF to provide signaling steering policy to the SCSCF for more coordinated and comprehensive implementation of dynamic SFC in the signaling-related PDN. In the data plane, we define our SFC-enabled domain that consists of two SFC-enabled sub-domains, an SCF sub-domain for signaling traffic classification and SFC encapsulation and an SFP sub-domain for signaling service chaining. For the SCF area, it comprises two classifiers at the edge, Ingress Edge and Egress Edge whose roles are to categorize and reclassify signaling traffic before and after performing the SFC required, and an SFF that forward SFC-encapsulated packets directly to the required SFs or through an SFC proxy that de-capsulate packets before directing them to the SCF-unaware SFs. The SFP sub-domain encompasses the SFP that defines the set of SIP related virtual functions that are involved in providing an adequate signaling service. It contains two kinds of functions, SFC-aware and SFC-unaware functions.

In this architecture, an SDN controller is connected to all the mobile IP layers with the goal of controlling the traffic routing in an SDN conformant physical hardware or virtual switches using an SFC-encapsulation protocol, via its southbound API. The northbound API is used to communicate with NFV MANO. The SFC encapsulation is defined by the Network Service Header (NSH) [13] that is carried by the encapsulated packets so that they can be processed by the SCFs and then forwarded to the required SFs, according to the information comprised in the SFC encapsulation. As an example of the SDN controller, a leading solution being currently the most widely used by the SDN community code-named ODL.

NFV focuses on placing network functions in a virtualized background. With this migration, in reference to the ETSI NFV framework, the MANO permits to orchestrate both the service and VNF lifecycles. In fact VNFs could be available to the SDN controller platform as SDN applications, for that reason, the existence of both SDN and NFV is complementary. Additionally, MANO mandates the SDN controller for the implementation of dynamic SFC in the required network in proportion to the service description.

4   Service Provisioning Procedures: IMS Use Case

Providing telecom services can take place according to two approaches being present at the current time. The earliest approach is the static one where we can provide mobile services in a static way by chaining miscellaneous VNFs statically based upon the VMM that could manage directly VMs through their descriptor files or upon the VIM that manages instances through a given VMM. The next is the dynamic approach based on implementing an NFV model referred to as an ETSI NFV framework.

4.1   Dynamic IMS Service Provisioning (Service Chaining)

Dynamic IMS based service provisioning must be based, in addition to the VIM, on a specific model. The ETSI NFV model is the widely supported framework by the research community working on virtualization as it provides a complete paradigm to implement NFV solutions for dynamic service provisioning. An implementation of that model could cover a few parts, especially:

- NFVO: NFV orchestrator whose role is to manage network service chains or the lifecycle of the network service according to the service descriptor (SD).
- VNFM: VNF Manager that manages the lifecycle of VNFs in proportion to the VNF descriptor (VNFD) and VNF component (VNFC).

Two other components could be deployed using popular software solutions such as:

- NFVI: NFV Infrastructure that relies on a VMM that abstract the underlying hardware and uncouple the software layer from hardware. For networking connectivity, virtual appliances, including Open vSwitches (OVS), could be deployed in this virtual infrastructure functional block.
- VIM: Virtual Infrastructure Manager might depend on the widely used cloud manager known as OpenStack that provides different types of APIs to manage network, storage, and computing resources.

Figure 3 shows the proposed IMS based dynamic SFC architecture. The SFC procedure is triggered by NFV MANO and executed by the SDN controller through NSH that is added and carried by signaling packets for more dynamic steering of signaling traffic by the SCFs to SFP that are composed of basic IMS components, including Proxy-CSCF (P-CSCF), Interrogating-CSCF (I-CSCF), HSS, Serving-CSCF (S-CSCF), and Application Server (AS). The signaling steering policy is provided by PCRF to SCSCF via the Rx interface. One of the capabilities that might be provided by SCSCF, is dynamic load balancing in IMS VNFs for a high performance IMS system and for better performance of signaling session parameters.

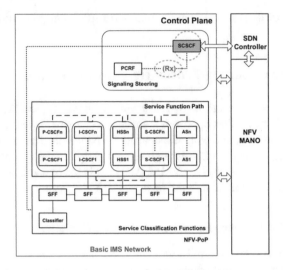

Fig. 3. Proposed IMS based dynamic SFC architecture

4.2 Static IMS Service Provisioning (Service Chaining)

Providing a mobile service along with the IMS is always preceded by a registration process and then a session setup related to a requested service. In this part we describe the mechanism of ensuring these two predefined actions along with service delivery. In a virtualized environment, the scenario remains the same as defined in the 3GPP recommendations, but only all the IMS components, including CSCFs, HSS, DNS, and the AS are virtualized considered as VNFs.

We consider the performance of all those elements. First step insists upon deploying and launching statically these IMS entities to provide the service required. Even though we hold a monitoring system that does not owe an orchestrator program, we can at any time according to the system performance indications add a new IMS component to the signaling service chain. For example, we could scale out one of the CSCFs functions when the VM's performance decreases either manually or using a script but the latter does not have an effect as dynamically conducted by an implemented solution based on NFV MANO.

In this approach, the main elements in providing such a service in a virtualized context is the VIM along with VMM since the former gives each VM the necessary system resources by the means of the latter that provides hardware abstractions.

Figure 4 proposes an IMS based SFC architecture. The IMS network is controlled by the VIM and the provisioning of the service is carried out statically, especially in proportion to IMS resource performance. In this case the load balancing mechanism is done statically by the DNS system according to what is configured in the zone file. Furthermore, the instantiation, scaling out, and termination of IMS VNFs are conducted manually through the VIM.

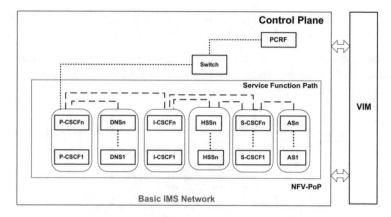

Fig. 4. IMS based static SFC architecture

5   Solution Evaluation

To show the effect of the virtualization technology and related approaches, in this section we outline the established simulation platform made up of different physical hardware, including COTS servers and switches, and of several software solutions containing OpenStack, Xen, and OVS as a cloud manager, a base VMM, and a virtual routing device respectively. Then, we evaluate the proposed approach in focusing on the static service provisioning, previously presented, by virtualizing the IP Services Layer including the IMS SubLayer. Next, we explore the different results obtained from this experiment.

5.1   Simulation Platform

Figure 5 displays the implementation of the framework testbed for system performance and IMS key performance indicators (KPI) benchmarks. Physical hardware incorporates 5 machines of 8 GB RAM, 4×3.2 GHz CPU, and Ubuntu 16.04 as a host operating system (OS).

A detailed scheme of the testbed setting is plotted in Fig. 5. The description is given as follows:

- The first machine hosts emulated IPTV clients installed based on UCT IMS Clients.
- The second machine comprises the Mininet platform for creating a virtual network topology by employing python scripts.
- The third machine is an OpenStack compute node dedicated to deploy two virtual IMS core settings with the help of the OpenIMSCore package [14] for a simulated IMS core network, Xen as a VMM, and OVS for bridging different IMS entities including two other virtual servers, a DNS server and a IPTV AS [15].
- The fourth machine is a compute node which incorporates virtual media servers.
- The fifth machine is a place for Hadoop based data center to host service resources and also it represents an object node for storing VNF images.

- The sixth machine considered as a monitoring system since it hosts the SDN/Open-Stack controller along with a QoS and performance system.

In this virtualized background, OpenStack was used in our simulation platform for controlling and managing VNFs and all the virtual network components – virtual switches – for both NFV and SDN considerations.

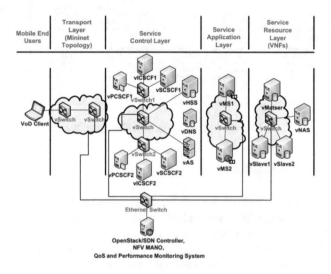

Fig. 5. Detailed testbed system

5.2   Benchmark Results

Through this study, we aim to assess the impact of virtualization on the behavior of the system, especially from the control plane side, so all measurements were performed in reference to SIP messages.

Two kinds of measurements were made. The first type is from the client's viewpoint for evaluating accessibility performance issued by IMS and network performance and the second measurement is commensurate with resource performance, especially for the integrating server, I-CSCF.

We run an experiment gradually for 100 emulated IPTV clients in two contexts; with one virtual IMS (vIMS) core network and then with two virtual IMS core networks. In the second vIMS core network, two main entities were present and virtualized, vP-CSCF2 and vI-CSCF2. Obviously, during the IMS registration scenario, the most requested IMS entity is I-CSCF, as it is the intermediate point between the visited network and the home network in the case of a roaming access, and also it communicates during that process with three entities, P-CSCF, HSS, and lastly S-CSCF. Another consideration is that before selecting an adequate I-CSCF, P-CSCF contacts the relevant I-CSCF according to the answer of the DNS server. Normally, a static load balancing is defined in the zone file of the DNS server provided by the SRV records so as to select the required I-CSCF in contrast to a dynamic load balancing being based on the load of the server. Accordingly, we have limited the measurements of the system performance to the I-CSCFs, especially to the vI-CSCF1 before and after adding another helping server, vI-CSCF2, to ensure load balancing and availability.

As can be noticed from the Figs. 6, 7, and 8, measurements were captured in terms of memory usage, throughput, and CPU usage.

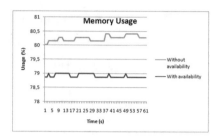

Fig. 6. Memory usage

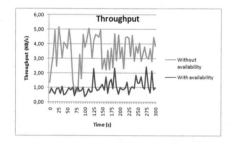

Fig. 7. Throughput

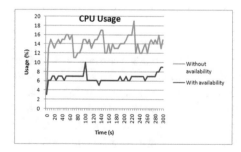

Fig. 8. CPU usage

Figures 9 and 10 depict the benchmark measurements from the client's side for measuring the delay of IMS KPI related parameters, including registration delay [16] and session setup time known also as mean [17].

Fig. 9. Registration delay

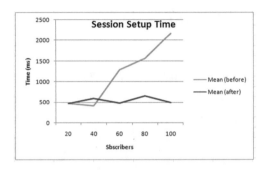

Fig. 10. Session setup time

Figure 9 marks the registration delay, a parameter used to evaluate how much the system is faster during the registration process, while Fig. 10 graphs the session setup time to measure the latency when establishing an IMS session before receiving the video on demand (VoD) service being used in this simulation.

5.3 Discussion

Experiment results were taken in virtualized surroundings within two contexts. The first context with one virtualized IMS core and the second one with two virtualized IMS cores. Indeed, we added the second context in order to show how important it is to keep the telecom system available by using the virtualization technology and also to have good performance in the case of growing demands from mobile users.

Scaling out has been performed statically as an introduction to a dynamic implementation of the ETSI NFV model along with NGOSS based standard telecom models in our future works.

Measurements were made in conjunction to the system usage in the vI-CSCF1 before and after scaling out from one vI-CSCF, i.e. vI-CSCF1, to two vI-CSCFs including vI-CSCF2. The reason for choosing these IMS entities is that they participate most during the registration operation for requests/answers, but the same considerations are applied to the other IMS servers for increasing the availability of services.

For the system performance attendant upon SIP IMS servers, the memory usage before and after setting up a new static service chain decreased slightly with almost an average rate of 2%. Moreover, the memory fluctuation is not important enough compared to the other parameters. For throughput, after the new static service, it significantly decreases resulting in substantially decreased CPU usage, which means that the static load balancing, ensuring by the SRV records in the DNS zone file, is achieved between interrogating servers – vI-CSCF1 and vI-CSCF2 – in lieu of the case where the SIP traffic is routed only via one I-CSCF.

The enhanced system performance confirms the availability of IMS related resources with the presence of static IMS service chaining.

As can be noticed from the measurement graphs related to SIP KPI, two main metrics have been measured including register delay and session setup time. Accordingly, replicating the VNFs has permitted to have two service provisioning graphs with each one its own IMS VNFs with a shared hard disk location to allow also live migration and facilitate maintenance as well. Typically, when the number of connected mobile users increases, the number of requests for registration and session set up systematically augments and therefore system performance could deteriorate if there are no necessary measures to make the system resources available. With the benchmark results, the IMS based telecom system can keep the performance of all the VNFs based VMs participating in the session establishment by creating a new service chain to respond more to the growing number of mobile subscribers.

6   Related Works

Few publications can be found in the literature that deal with the subject of the mobile SFC, particularly in the framework of the IMS ecosystem.

A 3GPP earlier study [18] explored diverse use cases for backing network traffic classification and network service chain selection capabilities per MNO's policies so as to realize flexible and efficient mobile service steering in the (S)Gi-LAN network. However, the main downside of this study is that no solution evaluations were performed. On the contrary, our study proposed an SFC based network solution for the entire 5G networks including the IMS use case. Also, a solution evaluation was conducted to validate our approach.

Another 3GPP study [12] was an enhancement of the earlier study [18] but for examining and evaluating architectural enhancements to supply mobile service steering policies per the 3GPP service requirements presented in TS 22.101 [19]. Moreover, this technical report was restricted to delivering policies for traffic steering via interfaces defined by the 3GPP, including the St interface. Yet, our work tackled the mobile service steering in a virtualized environment shedding light over the NFV and SDN technologies. Furthermore, we proposed a new steering function for the signaling traffic with the possibility of keeping or modifying the name and role of the 3GPP Rx interface for meeting these new signaling steering requirements defined in our paper.

Authors [20] from Intel provided in their paper technical details of the evaluation and testing methods of dynamic SFC using the Intel Open Network Platform (ONP) reference architecture. The evaluation platform included OpenStack, ODL SDN Controller, and OVS along with the Intel Data Plane Development Kit (Intel DPDK). Concerning our paper, we validated our approach by leveraging almost the same open source solutions except Intel specific tools, but we need to develop more scripts to have efficient dynamic service chain provisioning with the help of the ODL controller.

For evaluating SGi-LAN services using virtualization solutions, the paper in [21] exposed the implementation of a proposed solution for virtualizing the services on the SGi-LAN employing the SDN and NFV initiatives. Unfortunately, they showed their results with no graphs in the case of a growing number of subscribers in this virtualized network. In contrast to this work, our implementation was carried out in relation to SIP related services in the IMS network as a first implementation. Also, results showed good SIP related performance, especially for the session time setup and registration delay, within this new virtualized, scalable ecosystem.

Another paper [22] presented dynamic load balancing in the IMS network utilizing NFV. In comparison with this work that remains specific to load balancing for IMS networks, our generic approach supports large functionalities for SIP based systems including the IMS for, for instance, load balancing and QoS considerations through the proposed SCSCF.

7   Conclusion and Perspectives

This study shed light on a new approach for flexible mobile network service chaining in an NFV environment as a step to collaborating on the evolution of the 4G/LTE technology toward the 5G network.

Many concepts and technologies were included with the goal of enhancing traffic steering and the performance of the telecom system in a virtualized context. Resource performance in relation to the control plane was assessed, and thus registration delay and session setup time diminished in this context.

New implementations with dynamic service chaining will be the subject of our future works.

References

1. de Gouveia, F.C., Magedanz, T.: Quality of service in telecommunication networks. Telecommun. Syst. Technol. 2(5), 77–97 (2009)
2. Huston, G.: Next steps for the IP QoS architecture. IETF RFC 2990 (2000)
3. Seraoui, Y., Raouyane, B., Bellafkih, M.: An extended IMS framework with a high-performance and scalable distributed storage and computing system. In: International Symposium on Networks, Computers and Communications (ISNCC 2017). IEEE, Marrakech (2017)
4. Seraoui, Y., Bellafkih, M., Raouyane, B.: A high-performance and scalable distributed storage and computing system for IMS services. In: 2016 2nd International Conference on Cloud Computing Technologies and Applications (CloudTech), pp. 335–342. IEEE, Marrakech (2016)

5. 3rd Generation Partnership Project: IP multimedia subsystem (IMS); stage 2 (release 14). 3GPP TS 23.228 V14.2.0 (2016)
6. Rosenberg, J., et al.: SIP: session initiation protocol. IETF RFC 3261 (2002)
7. Duan, Z., Zhang, Z.L., Hou, Y.T.: Service overlay networks: SLAs, QoS, and bandwidth provisioning. IEEE/ACM Trans. Netw. (TON) **11**(6), 870–883 (2003)
8. European Telecommunications Standards Institute: Network functions virtualisation (NFV); use cases. ETSI GR NFV 001 V1.2.1 (2017)
9. Halpern, J., Pignataro, C.: Service function chaining (SFC) architecture. IETF RFC 7665 (2015)
10. Boucadair, M.: Service function chaining (SFC) control plane components & require-ments. IETF Internet-Draft (2016)
11. Gramaglia, M., Haeffner, W.: Service function chaining dataplane elements in mobile networks. IETF Internet-Draft (2017)
12. 3rd Generation Partnership Project: Architecture enhancement for flexible mobile service steering (release 13). 3GPP TR 23.718 V13.0.0 (2015)
13. Quinn, P., Elzur, U.: Network service header. IETF Internet-Draft (2017)
14. The Open Source IMS Core Project. http://www.openimscore.org
15. Spiers, R., Marston, R., Good, R., Ventura, N.: The UCT IMS iptv initiative. In: 2009 Third International Conference on Next Generation Mobile Applications, Services and Technologies, pp. 503–508. IEEE, Cardiff (2009)
16. Almes, G., Kalidindi, S., Zekauskas, M., Morton, A.: A one-way delay metric for IP performance metrics (IPPM). IETF RFC 7679 (2016)
17. 3rd Generation Partnership Project: Key performance indicators (KPI) for the IP multimedia subsystem (IMS); definitions (release 14). 3GPP TS 32.454 V14.0.0 (2017)
18. 3rd Generation Partnership Project: Study on flexible mobile service steering (FMSS) (release 14). 3GPP TR 22.808 V14.1.0 (2015)
19. 3rd Generation Partnership Project: Service aspects; service principles (release 15). 3GPP TS 22.101 V15.1.0 (2017)
20. Gasparakis, J., et al.: Evaluating dynamic service function chaining for the Gi-LAN. White paper (2016)
21. Grønsund, P., et al.: A solution for SGi-LAN services virtualization using NFV and SDN. In: 2015 European Conference on Networks and Communications (EuCNC), pp. 408–412. IEEE, Paris (2015)
22. Dandin, K., Hokelek, I., Kurt, G.K.: Dynamic load management for IMS networks using network function virtualization. In: NOMS 2016 - 2016 IEEE/IFIP Network Operations and Management Symposium, pp. 1011–1012. IEEE, Istanbul (2016)

Performance Analysis of the Vertical Handover Across Wifi/3G Networks Based on IEEE 802.21

Mohamed Lahby[1(✉)] [iD] and Abderrahim Sekkaki[2]

[1] Laboratory of Mathematics and Applications,
Ecole Normale Superieure of Casablanca, University Hassan II, Casablanca, Morocco
mlahby@gmail.com
[2] Laboratory of Research and Computer Innovation,
Faculty of Sciences Ain Chock of Casablanca,
University Hassan II, Casablanca, Morocco
sekkabd@gmail.com

Abstract. Recently, the fast evolution of wireless technologies has enabled the development of multi-mode terminals which can support more radio interfaces. In this context, the users can enjoy various mobile applications by using network at anytime and anywhere with the best quality of service (QoS). The IEEE 802.21 is a recent standard that aims at enabling seamless vertical handover between heterogeneous technologies.

In this paper we propose the performance analysis of the vertical handover decision between two networks WiFi (IEEE 802.11) and 3G (UMTS) by using IEEE 802.21 for two traffic flows: FTP and video streaming. The QoS metrics used for this evaluation include throughput, end-to-end delay and jitter. The Testbed for performance evaluation is done by using the ODTONE which is an open source platform implementing the IEEE 802.21 protocol.

Keywords: Heterogeneous wireless networks · Vertical handover
IEEE 802.21 · ODTONE · Performance analysis

1 Introduction

The fourth generation (4G) system is considered as a heterogeneous environment which integrate various wireless technologies such as 3G (UTMS, IEEE 802.11a, IEEE 802.11b, etc), and 4G (IEEE 802.16, LTE, LTE-A). This heterogeneous environment, can ensure diversity for multimedia applications and provide to motabtobile user the ability to be connected by using a new range of the connexion referred to a new paradigm namely mobile Internet. This new paradigm of connectivity allows total freedom to the user, freedom to enjoy various applications anywhere and anytime, freedom to be connected at any access network, freedom to choose any service, etc.

© Springer International Publishing AG, part of Springer Nature 2019
J. Mizera-Pietraszko et al. (Eds.): RTIS 2017, AISC 756, pp. 300–310, 2019.
https://doi.org/10.1007/978-3-319-91337-7_28

In parallel, telecommunications companies have already proposed and developed intelligent mobile terminal such as mobile phones, smart-phones, IPAD, which can support the service of mobile Internet. These a new equipment have integrates multiple interfaces which offer the possibilities for the user to maintain a seamless service continuity in this heterogeneous wireless networks. Usually the terminal mobile can select at the same time several wireless technologies taking into account QoS requirements of applications.

However, the selection of the most suitable access network by satisfying Always Best Connected (ABC) Paradigm [1], becomes a significant challenge in this heterogeneous environment. To cope with this issue, IEEE 802.21 standard [2] aims to choose and to maintain a continuous connection during a transition from one network to another. Handover algorithms based on IEEE 802.21 have been widely studied in the literature in order to solve and to optimize the handover vertical problem. These algorithms can be classified into five categories [3]: received signal strength (RSS) based, cost function based, utility function based, artificial intelligence algorithms based and multi-attribute decision making (MADM) based.

In the RSS-Based algorithms [3,4], the received signal strength criterion is used to made the network selection decision. For that, the network which have the high value of the signal is selected among available networks. In the cost function based algorithms [3,5], the goal is to find the network which minimize the output of the cost function. This one is defined according various metrics such as bandwidth, delay, jitter, security, etc. The utility based functions class [6,7], are commonly used to describe the level of satisfaction of the mobile terminal according a set of services offered by the network access. In the artificial intelligence algorithms [3,8], based some algorithms such as fuzzy logic, genetic algorithms and neural networks techniques are used to interpret imprecise information, and to perform an accurate decision for network selection Finally, the MADM-based class represent mathematical algorithms which can be used to evaluate the different alternatives based on their attributes. Many studies have been done in this area [9,10,17,18]. Lahby et al. [9] have proposed an algorithm based on M-AHP and GRA, firstly the M-AHP method is used to weigh different criteria, n the second step the GRA method is applied to rank the alternatives.

By using the IEEE framework, the aim of this paper is to evaluate the performance of the vertical handover between two networks WiFi (IEEE 802.11) and 3G (UMTS) for two traffic flows: FTP and Streaming. The vertical handover decision algorithm implemented in IEEE 802.21 is based on the received signal strength criterion. The remainder of this paper is organized as follows. Section 2 describe the IEEE 802.21 architecture. Section 3 gives an overview of vertical handover decision based IEEE 802.21. Section 4 includes our Testbed and the experimental results. In the final section we conclude this paper.

2   IEEE 802.21 Architecture

IEEE 802.21 standard [4] is also namely Media Independent Handover (MIH) aims to maintain a continuous connection during a transition from one network to another. The standard provides intelligence at the data link layer and other network information adapted in the upper layers to optimize handover between heterogeneous networks. The transition from one network to another can be useful when the environment changes or a new network appears more attractive and more efficient QoS conditions. The Fig. 1 illustrates the IEEE 802.21 architecture.

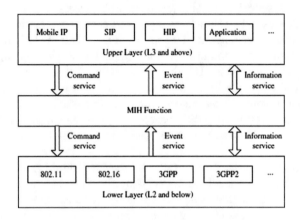

Fig. 1. Illustration of 802.21 architecture

The MIH defines this passage regardless of media independent handover function (MIHF). The MIHF function is a thin layer within the stack of mobility management protocols of the mobile node and network elements. This allows the association to the handover between heterogeneous networks. This function provides services to the upper layers through an interface regardless of technology and provision of services of the lower layers through a variety of interfaces and this type of service access point (SAPs).

The MIHF provides three types of services: Media Independent Event Service (MIES), Media Independent Command Service (MICS), and Media Independent Information Service (MIIS). The MIES aims to provide and to predict link changes. The MIES is divided into two categories, link events and MIH events. Link events are generated from the lower layer and transmitted to MIH layer. The MIH events are the events forwarded from MIH to upper layers. MICS refers to the commands, such as initiate handover and complete handover, sent from higher layers to lower layers. It allows enabling handover mechanism. MICS includes MIH command and Link command. MIIS provides a framework by which MIHF can discover different network information existing within a geographical area to facilitate seamless handover when roaming across these networks. The MIIS provides a bidirectional way for the two layers to share information such as current QoS, performance information and availability of service.

3   A Brief Overview of Vertical Handover Decision Based IEEE 802.21

The IEEE 802.21 is a recent standard of IEEE that aims at enabling seamless service continuity among heterogeneous networks including the IEEE 802 standard family, 3GPP and 3GPP2. In the literature review, variety of vertical handover strategies based IEEE 802.21 have been proposed and developed. In [11], the authors have provided an ample overview of IEEE 802.21 specification. The authors [12], have proposed the use of the IEEE 802.21 standard framework to achieve mobility of the terminal mobile by enabling the handover of IP sessions from one data-link layer access technology to another. The simulation results showed that Media Independent Handover framework is limited only to the initiation and the preparation phases of the handoff process, but it does not address the issue of authenticating and authorizing when a mobile node wants to use another BS/Aps/cell.

In [13] the emerging IEEE 802.21 standard have used to ensure seamless mobility between two radio access technologies IEEE 802.16m for 100 Mb/s high mobility connections and 802.11 VHT for 1 Gb/s at lower mobility. In [14], the authors have proposed the support of a hierarchical structure of Media Independent Information Service (MIIS) based IEEE 802.21 architecture, pointing out how it can enhance the mobile nodes (MNs) experience in different handover scenarios. In [15] the multiple attributes decision making-based terminal-controlled vertical handover decision scheme using MIH services is proposed in the integrated WiFi and WiMAX networks. The simulation results show that the proposed scheme provides smaller handover times and lower dropping rate than the RSS-based and cost function-based vertical handover schemes.

In [16], the authors have proposed handover performance between 3G and Wireless LAN access networks. The mobile devices are based on the IEEE 802.21 cross layer architecture. In addition handover algorithm is based on signal thresholds. It relies on the information provided by the Media Dependent layers and the Mobile IP Layer. References [17,18] the authors have provided an ample overview of handover vertical algorithms based on multi attribute decision making (MADM). The IEEE 802.21 framework is used to implement and to evaluate the performance concerning the MADM algorithms through simulations.

However, the major drawback of above references, lie in the performance results which are given by using simulation based on NS 2 simulator or Matlab simulator. To deal with issue, we have founded only two works [19,20] in literature review. In reference [19], the authors have developed framework for the implementation of the IEEE 802.21 Media Independent Handover (MIH) standard. This framework is applied to evaluate the handover performance between two heterogeneous networks Wifi (IEEE 802.11) and Wimax (IEEE 802.16e). The experimental results showed that the MIH services can be used to reduce packet losses and the network discovery. Finally the authors [20] have deployed IEEE 802.21 protocol by developing a novel open source solution namely ODTONE.

4   Experimental Results and Discussion

In this section, we present the testbed environment used to analyze the performance between 3G/WiFi, then we describe the network topology and performance metrics used in this testbed. Finally we give the experimental results according to QoS metrics.

4.1   The Testbed Environment

The experimental results presented in this paper are obtained using odtone platform [20]. This framework is used to define an appropriate topology that allows vertical handover between Wifi and 3G. Both Wifi parameters and 3G parameters used during our testbed are shown in Table 1.

Table 1. IEEE 802.11 and 3G parameters

Parameter	Values of IEEE 802.11	Value of 3G
Frequency	2.462 GHz	2.437 GHz
Bandwidth	7 Mb/s	1,5 Mb/s
Transmission power	200 dBm	200 dBm
Coverage radius	80 m	1,5 Km

4.2   Our Proposed Scenario

Figure 2 shows the network topology of the testbed deployed to evaluate the handover performance between 3G/Wifi. This topology consists of two access points (AP): AP for the 802.11 network, AP for 3G network and the mobile terminal (MT) is a laptop computer with two access interfaces that provide network connectivity by using 3G or Wifi. The mobile terminal is situated with regard to both APs of a distance of 1m, then it moves away from 5 m during the experiments.

We notice that ODTONE framework is implemented in mobile terminal in order to support seamless handover. We performed two experiments to measure the qos metrics for two flows as such as FTP traffic and streaming traffic. The time of Testbed has been fixed at 300 s. The traffic captures, and the average of the Qos metrics representing the extracted results are performed in each 15 s. Moreover, the circulation of flow starts just after the first 15 s.

4.3   The QoS Metrics

In this paper, we use three performance evaluation metrics to evaluate the vertical handover which are:

- Throughput: is the average rate of successful data delivery over a communication channel per unit time. It is measured in Kbits per second (Kb/s).

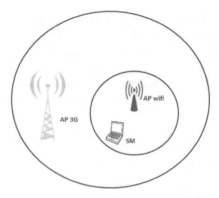

Fig. 2. Network topology to evaluate the performance of vertical handover

- End to end delay: refers to the time taken by data packet to be transmitted across a network from source to destination.
- Jitter: is defined as a variation in the delay of received packets.

4.4   The Experiment 1

This first experiment consists to analyze the performance of Qos metrics according to FTP traffic. For that, we calculate the average value of throughput, the average delay and the average jitter. The average throughput of FTP traffic is illustrated in Fig. 3. When the terminal mobile distance is equal to 1 m, Wi network provide best throughput than 3G network. With terminal mobile distance fixed to >5 m, also WiFi network provides better throughput than 3G network. In both distance from WiFi and 3G, for the first seconds the both APs are equal and competitor but after several seconds WiFi is the better despite decreasing but still remains the most favorable to the platform. We notice that the mobile terminal distance affects the average of throughput value at vertical handover, because decrease with distance with regard to both APs.

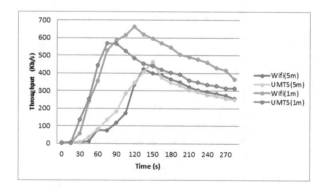

Fig. 3. Throughput (Kb/s) for FTP traffic

The average delay for FTP traffic is shown in Fig. 4. The curves obtained with FTP traffic during the handover process degrade very fast after the launch of the service and begin to stabilize later 100 s. We notice that the distance does not have any effect on the value of delay. Indeed the calculation of delay also depends on the varied size of the packages transmitted on the network. But, with terminal mobile distance fixed to >5 m, 3G provides better average delay than WiFi network.

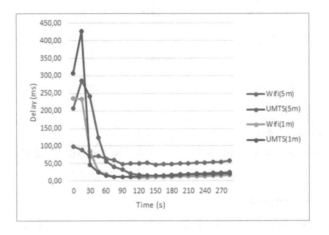

Fig. 4. The average delay (ms) for FTP traffic

The average jitter for FTP traffic is illustrated in Fig. 5. In interval [0..90 s] the Wifi network is showing highest average jitter in comparison of 3G network according to both distances 1 m and 5 m because the Wifi network supports only the low coverage network. Moreover, from 90 s, the average jitter is null for Wifi and 3G.

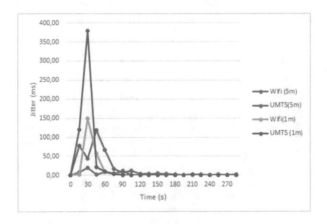

Fig. 5. The average jitter (ms) for FTP traffic

4.5   The Experiment 2

In this experiment, we analyze the performance of QoS metrics according to streaming Video traffic. So, we calculate the average value of throughput, delay and jitter. We start with throughput average in streaming traffic is illustrated in Fig. 6. In the beginning of service provisioning and for any type of network, the throughput increases with distinguished manner. The AP WIFI show as the rest away the best in terms of throughput compared to 3G and also during the progression in time.

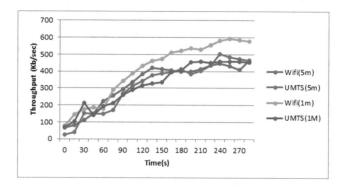

Fig. 6. Throughput (Kb/s) for streaming traffic

The average delay for streaming Video traffic is shown in Fig. 7. We notice that, the service starts with a significant peak. WIFI period is best compared to 3G, however the last timeout value (10 ms) is above the threshold (5 ms). The delay has a very important impact on the quality video by apparition frozen zones and loss of frame I in RTP packets (Real Transport Protocol).

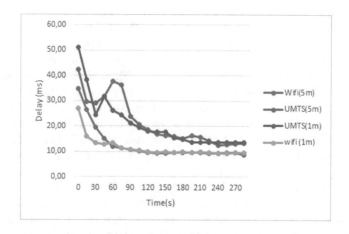

Fig. 7. The average delay (ms) for streaming traffic

In addition for the both distances 1 m and 5 m, the curves obtained during the handover process show that for both distances 1 m and 5 m, the Wifi network present low average delay in comparison of 3G network.

The Fig. 8 show the curves of average jitter (ms) for streaming traffic according to different distances. From the curves it is clearly indicated that when the terminal mobile distance is equal to 1 m, in the interval [0..60], the average value of jitter is >15 ms concerning both networks WiFi and 3G. Then from 60 s, this average value this becomes constant, 10 ms for WiFi and 15 ms for 3G. So, the WiFi network provides better jitter than 3G network.

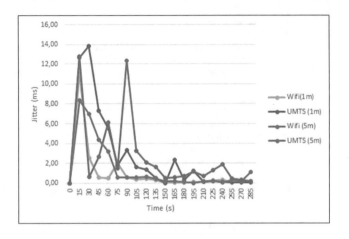

Fig. 8. The average jitter (ms) for streaming traffic

4.6   Discussion

Our experimental results shows the reliability of WiFi in terms of throughput, delay and jitter. Measures is made between WIFI and 3G services with two consumers of bandwidth (FTP, VoD) taking into consideration that the mobility is made a signal power based or other words the distance from the AP. The results shows a competition between 3G and WiFi, both types of networks show their capacity for service provision, however a large delay value in VoD; WIFI is always better seen several constraints but to provide 3G service continuity is required. Also, in both experiments, it is remarkable that the platform performs the handover in which case the QoS in the current network is better; for example, when approaching the 3G network you are leaving the WIFI because of high signal power but to poor QoS so on.

Finally, from the above analysis concerning FTP traffic, it can be concluded for both distances 1 m and 5 m, that Wifi network shows hight average throughput, low average delay, and low average jitter rate in comparison of 3G network. However, when the MS distance is equal to 1 m the network selection decision is favorable because the MS is connected to Wifi, but when the distance is 5 m the decision is not favorable because, the Ms is connected to 3G network. So, we confirm that the vertical handover algorithm based on Received Signal Strength(RSS) not efficient to choose the best access network.

5   Conclusion

In this work, we proposed an evaluation performance of vertical handover between WiFi and 3G by using 802.21 for FTP and video streaming. We notice that the implementation of an ODTONE open source solution which can support 802.21 standard is done. Our experimental results showed that for each traffic class, the Wifi network shows better Qos in comparison of 3G network. However, when the mobile terminal distance is equal to 1m the network selection decision is favorable because the mobile terminal is connected to Wifi, but when the distance is 5 m the decision is not favorable because, the mobile terminal is connected to 3G network.

Finally, we can conclude that the vertical handover decision based on RSS not efficient to choose the best access network. For this reason, in our future work, we investigate to introduce multi-attribute decision making (MADM) algorithms based IEEE 802.21, in order take into account multiple criteria.

References

1. Gustafsson, E., Jonsson, A.: Always best connected. IEEE Wirel. Commun. Magaz. **10**(1), 49–55 (2003)
2. IEEE 802.21: IEEE standard for local and metropolitan area networks, part 21: Media independent handover services (2009)
3. Yan, X., et al.: A survey of vertical handover decision algorithms in fourth generation heterogeneous wireless networks. Comput. Netw. **54**(11), 1848–1863 (2010)
4. Yan, X., Mani, N., Sekercioglu, Y.A.: A traveling distance prediction based method to minimize unnecessary handovers from cellular networks to WLANs. IEEE Commun. Lett. **12**(1), 14–16 (2008)
5. Zhu, F., McNair, J.: Optimizations for vertical handoff decision algorithms. In: Proceedings of the 2004 IEEE Wireless Communications and Networking Conference (WCNC04), Atlanta, Georgia, USA, pp. 867–872, March 2004
6. Vuong, N., Thinh, Q., Doudane, G., Nazim, A.: On utility models for access network selection in wireless heterogeneous networks. In: IEEE Network Operations and Management Symposium (NOMS), pp. 144–151 (2008)
7. Xiaoyan, W., Qinghe, D.: Utility-function-based radio-access-technology selection for heterogeneous wireless networks. Comput. Electr. Eng. **52**, 1–12 (2015)
8. Nasser, N., Guizani, S., Al-Masri, E.: Middleware vertical handoff manager: a neural network-based solution. In: Proceedings of the 2007 IEEE International Conference on Communications (ICC07), Glasgow, Scotland, pp. 5671–5676 (2007)
9. Lahby, M., et al.: Network selection mechanism by using M-AHP/GRA for heterogeneous networks. In: The Sixth Joint IFIP Wireless and Mobile Networking Conference (WMNC), pp. 1–6 (2013)
10. Lahby, M., et al.: A hybrid approach for network selection in heterogeneous multi-access environments. In: 4th IFIP International Conference on New Technologies, Mobility and Security (NTMS), pp. 1–5 (2011)

11. De La Oliva, A., et al.: An overview of IEEE 802.21: media-independent handover services. IEEE Wirel. Commun. **15**(4), 96–103 (2008)
12. Taniuchi, K., et al.: IEEE 802.21: media independent handover: features, applicability, and realization. IEEE Commun. Magaz. **47**(1), 112–120 (2009)
13. Eastwood, L., Migaldi, S., Xie, Q., Gupta, V.: Mobility using IEEE 802.21 in a heterogeneous IEEE 802.16/802.11-based IMT-advanced (4G) network. IEEE Wirel. Commun. **15**(2), 26–34 (2008)
14. Buiati, F., Villalba, L.J.G., Corujo, D., Sargento, S., Aguiar, R.L.: IEEE 802.21 information services deployment for heterogeneous mobile environments. IET Commun. **5**(18), 2721–2729 (2011)
15. Wu, J.S., Yang, S.F., Hwang, B.J.: A terminal-controlled vertical handover decision scheme in IEEE 802.21-enabled heterogeneous wireless networks. Int. J. Commun. Syst. **22**, 819–834 (2009)
16. De la Oliva, A., et al.: IEEE 802.21 enabled mobile terminals for optimized WLAN/3G handovers: a case study. SIGMOBILE Mob. Comput. Commun. Rev. **11**(2), 29–40 (2007)
17. Lahby, M., Silki, B., Sekkaki, A.: Survey and comparison of MADM methods for network selection access in heterogeneous networks. In: 7th IFIP International Conference on New Technologies, Mobility and Security (NTMS), pp. 1–6 (2015)
18. Obayiuwana, E., et al.: Network selection in heterogeneous wireless networks using multi-criteria decision-making algorithms: a review. Wirel. Netw. **23**, 1–33 (2016)
19. Lim, W.-S., et al.: Implementation and performance study of IEEE 802.21 in integrated IEEE 802.11/802.16e networks. Comput. Commun. **32**(1), 134–143 (2009)
20. Corujo, D., Guimaraes, C., Santos, B., Aguiar, R.: Using an open source IEEE 802.21 implementation for network-based localized mobility management. IEEE Commun. Magaz. **49**(9), 114–123 (2011)

Privacy and Security in Intelligence

A New Encryption Scheme to Perform Smart Computations on Encrypted Cloud Big Data

Ahmed El-Yahyaoui$^{(\boxtimes)}$ and Mohamed Dafir Ech-Cherif El Kettani

Mohammed V University in Rabat, Rabat, Morocco
{ahmed_elyahyaoui, dafir.elkettani}@um5.ac.ma

Abstract. Cloud computing is a new paradigm of information technology and communication. Performing smart computations in a context of cloud computing and big data is highly appreciated today. Fully homomorphic encryption (FHE) is a smart category of encryption schemes that allows working with the data in its encrypted form. It permits us to preserve confidentiality of our sensible data and to benefit from cloud computing powers. Currently, it has been demonstrated by many existing schemes that the theory is feasible but the efficiency needs to be dramatically improved in order to make it usable for real applications. One subtle difficulty is how to efficiently handle the noise. This paper aims to introduce an efficient FHE based on a new mathematic structure that is noise free.

Keywords: Fully homomorphic encryption · Lipschitz integers
Scheme · Cloud · Security · Smart computations

1 Introduction

Companies are increasingly challenged to provide better services to their customers, stay ahead of competition, lower costs and increase profits, and work more efficiently. Cloud computing (Fig. 1) has manifested as a powerful computing model in the last decade, with numerous advantages both to clients and providers. The concept of cloud computing refers to the use of memory and storage capacities and calculation of shared computers and servers, interconnected through the Internet, obeying the principle of grid computing. One of the obvious huge advantage is that clients can delegate their complex computations and benefit from the best technologies and computation powers at low costs. The cost benefits presented by cloud technologies is one of the major arguments that justify the spreading of cloud computing in many industries. During the last few years, enterprises culture of accepting cloud computing was developed and many companies had shown their ready to adhere cloud and benefit from its capacities, but businesses are now finding that there is a number of security issues that have to be treated when venturing into the cloud.

Privacy of sensible data is one of most important security issues. Leakage of some data can cause huge damages to its owners. In general, to save privacy of our data it is advised to encrypt it before storing it on a remote cloud server. Using classical encryption schemes as RSA, AES, 3DES… allows clients to preserve data privacy during transmission to the cloud, but if a client requests the cloud to perform a complex

© Springer International Publishing AG, part of Springer Nature 2019
J. Mizera-Pietraszko et al. (Eds.): RTIS 2017, AISC 756, pp. 313–320, 2019.
https://doi.org/10.1007/978-3-319-91337-7_29

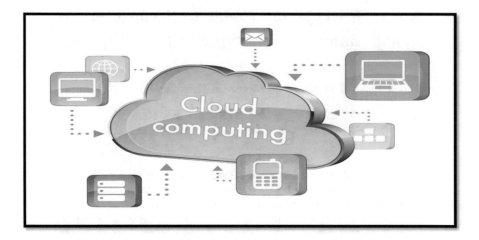

Fig. 1. Cloud computing oversimplification

treatment on its data, he should share his private key with the remote cloud server. This traditional use of cryptography may not be the best solution in terms of privacy, especially if we consider the cloud as an untrusted part.

One solution to this problematic is doing smart computations on encrypted data, this idea was early introduced by Rivest, Adleman and Dertozous in 1978 [1], authors conjectured the existence of a privacy homomorphism. Today we are using the notion of Fully Homomorphic Encryption (FHE) rather than privacy homomorphism.

FHE schemes (Fig. 2) are considered as the next generation algorithms for cryptography, it is a type of smart encryption cryptosystems that support arbitrary computations on ciphertexts without ever needing to decrypt or reveal it. In a context of cloud computing and distributed computation this is a highly precious power. In fact, a significant application of fully homomorphic encryption is to big data and cloud computing. Generally, FHE is used in outsourcing complex computations on sensitive

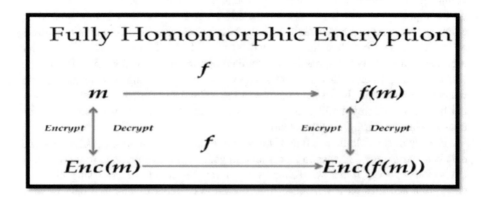

Fig. 2. Fully Homomorphic Encryption diagram

data stored in a cloud as it can be employed in specific applications for big data like secure search on encrypted big data and private information retrieval. It was an open problem until the revolutionary work of Gentry in 2009 [2]. In his thesis, Gentry proposed the first adequate fully homomorphic encryption scheme by exploiting properties of ideal lattices.

Gentry's construction is based on his bootstrapping theorem which provides that given a somewhat homomorphic encryption scheme (SWHE) that can evaluate homomorphically its own decryption circuit and an additional NAND gate, we can pass to a "levelled" fully homomorphic encryption scheme and so obtain a FHE scheme by assuming circular security. The purpose of using bootstrapping technique is to allow refreshment of ciphertexts and reduce noise after its growth.

Gentry's construction is not a single algorithm but it considered as a framework that inspires cryptologists to build new fully homomorphic encryption schemes [3–6] …. A FHE cryptosystem that uses Gentry's bootstrapping technique can be classified in the category of noise-based fully homomorphic encryption schemes [7]. If this class of cryptosystems has the advantage to be robust and more secure, it has the drawback to be not efficient in terms of runtime and ciphertext size. In several works followed Gentry's one, many techniques of noise management are invented to improve runtime efficiency and to minimise ciphertext and key size's [8,9,10…], but the problematic of designing a practical and efficient fully homomorphic encryption scheme remains the same until now.

In the literature we can locate a second category called free-noise fully homo-morphic encryption schemes which do not need a technique of noise management to refresh ciphertexts. In a free-noise fully homomorphic encryption scheme one can do infinity of operations on the same ciphertext without noise growing. This class of encryption schemes is known as faster than the previous one, involves simple opera-tions to evaluate circuits on ciphertexts and do not require a noise management tech-nique, but it suffers from security problems because the majority of designed schemes are cryptanalyzed today.

2 Our Techniques and Results

We propose an efficient noise-free fully homomorphic encryption scheme that uses the ring of Lipschitz's quaternions and permits computations over encrypted data under a symmetric key. We exploit properties of non-commutativity of Lipschitz integers to build our efficient fully homomorphic encryption scheme.

3 Mathematical Background

3.1 Quaternionique Field $\mathbb{H}$

A quaternion is a number in a general sense. Quaternions encompass real and complex numbers in a number system where multiplication is no longer a commutative law.

The quaternions were introduced by the Irish mathematician William Rowan Hamilton in 1843. They now find applications in mathematics, physics, computer science and engineering.

Mathematically, the set of quaternions $\mathbb{H}$ is a non-commutative associative algebra on the field of real numbers $\mathbb{R}$ generated by three elements i, j and k satisfying relations: $i^2 = j^2 = k^2 = i.j.k = -1$. Concretely, any quaternion q is written uniquely in the form: $q = a + bi + cj + dk$ where a, b, c and d are real numbers.

The operations of addition and multiplication by a real scalar are trivially done term to term, whereas the multiplication between two quaternions is termed by respecting the non-commutativity and the rules proper to i, j and k. For example, given $q = a + bi + cj + dk$ and $q' = a' + b'i + c'j + d'k$ we have $qq' = a_0 + b_0 i + c_0 j + d_0 k$ such that: $a_0 = aa' - (bb' + cc' + dd')$, $b_0 = ab' + a'b + cd' - c'd$, $c_0 = ac' - bd' + ca' + db'$ and $d_0 = ad' + bc' - cb' + a'd$.

The quaternion $\bar{q} = a - bi - cj - dk$ is the conjugate of q.

$|q| = \sqrt{q\bar{q}} = \sqrt{a^2 + b^2 + c^2 + d^2}$ is the module of q. The real part of q is $Re(q) = \frac{q+\bar{q}}{2} = a$ and the imaginary part is $Im(q) = \frac{q-\bar{q}}{2} = bi + cj + dk$.

A quaternion q is invertible if and only if its modulus is non-zero, and we have $q^{-1} = \frac{1}{|q|^2}\bar{q}$.

3.2 Reduced Form of Quaternion

Quaternion can be represented in a more economical way, which considerably alleviates the calculations and highlights interesting results. Indeed, it is easy to see that $\mathbb{H}$ is a $\mathbb{R}$-vectorial space of dimension 4, of which $(1, i, j, k)$ constitutes a direct orthonormal basis. We can thus separate the real component of the pure components, and we have for $q \in \mathbb{H}$, $q = (a, \boldsymbol{u})$ such that $\boldsymbol{u}$ is a vector of $\mathbb{R}^3$. So for $q = (a, \boldsymbol{u}), q' = (a', \boldsymbol{v}) \in \mathbb{H}$ and $\lambda \in \mathbb{R}$ we obtain:

1. $q + q' = (a + a', \boldsymbol{u} + \boldsymbol{v})$ and $\lambda q = (\lambda a, \lambda \boldsymbol{u})$
2. $qq' = (aa' - \boldsymbol{u}.\boldsymbol{v}, a\boldsymbol{v} + a'\boldsymbol{u} + \boldsymbol{u} \wedge \boldsymbol{v})$ Where $\wedge$ is the cross product of $\mathbb{R}^3$.
3. $\bar{q} = (a, -\boldsymbol{u})$ and $|q|^2 = a^2 + \boldsymbol{u}^2$.

3.3 Ring of Lipschitz Integers

The set of quaternions defined as follows:
$\mathbb{H}(\mathbb{Z}) = \{q = a + bi + cj + dk / a, b, c, d \in \mathbb{Z}\}$ Has a ring structure called the ring of Lipschitz integers. $\mathbb{H}(\mathbb{Z})$ is trivially non-commutative.

For r $n \in \mathbb{N}^*$, the set of quaternions:
$\mathbb{H}(\mathbb{Z}/n\mathbb{Z}) = \{q = a + bi + cj + dk / a, b, c, d \in \mathbb{Z}/n\mathbb{Z}\}$ has the structure of a non-commutative ring.

A modular quaternion of Lipschitz q $\in \mathbb{H}(\mathbb{Z}/n\mathbb{Z})$ is invertible if and only if its module and the integer n are coprime numbers, i.e. $|q|^2 \wedge n = 1$.

3.4   Quaternionique Matrices $\mathbb{M}_2(\mathbb{H}(\mathbb{Z}/n\mathbb{Z}))$

The set of matrices $\mathbb{M}_2(\mathbb{H}(\mathbb{Z}/n\mathbb{Z}))$ describes the matrices with four inputs (two rows and two columns) which are quaternions of $\mathbb{H}(\mathbb{Z}/n\mathbb{Z})$. This set has a non-commutative ring structure.

There are two ways of multiplying the quaternion matrices: the Hamiltonian product, which respects the order of the factors, and the octonionique product, which does not respect it.

The Hamiltonian product is defined as for all matrices with coefficients in a ring (not necessarily commutative). For example:

$$U = \begin{pmatrix} u_{11} & u_{12} \\ u_{21} & u_{22} \end{pmatrix}, \quad V = \begin{pmatrix} v_{11} & v_{12} \\ v_{21} & v_{22} \end{pmatrix} \Rightarrow$$

$$UV = \begin{pmatrix} u_{11}v_{11} + u_{12}v_{21} & u_{11}v_{12} + u_{12}v_{22} \\ u_{21}v_{11} + u_{22}v_{21} & u_{21}v_{12} + u_{22}v_{22} \end{pmatrix}$$

The octonionique product does not respect the order of the factors: on the main diagonal, there is commutativity of the second products and on the second diagonal there is commutativity of the first products.

$$U = \begin{pmatrix} u_{11} & u_{12} \\ u_{21} & u_{22} \end{pmatrix}, \quad V = \begin{pmatrix} v_{11} & v_{12} \\ v_{21} & v_{22} \end{pmatrix} \Rightarrow$$

$$UV = \begin{pmatrix} u_{11}v_{11} + v_{21}u_{12} & v_{12}u_{11} + u_{12}v_{22} \\ v_{11}u_{21} + u_{22}v_{21} & u_{21}v_{12} + v_{22}u_{22} \end{pmatrix}$$

In our article we will adopt the Hamiltonian product as an operation of multiplication of the quaternionique matrices.

3.5   Schur Complement and Inversibility of Quaternionique Matrices

Let $\mathcal{R}$ be an arbitrary associative ring, a matrix $M \in \mathcal{R}^{n \times n}$ is supposed to be invertible if $\exists N \in \mathcal{R}^{n \times n}$ such that $MN = NM = I_n$ where N is necessarily unique.

The Schur complement method is a very powerful tool for calculating inverse of matrices in rings. Let $M \in \mathcal{R}^{n \times n}$ be a matrix per block satisfying: $M = \begin{pmatrix} A & B \\ C & D \end{pmatrix}$ such that $A \in \mathcal{R}^{k \times k}$.

Suppose that A is invertible, we have: $M = \begin{pmatrix} I_k & 0 \\ CA^{-1} & I_{n-k} \end{pmatrix} \begin{pmatrix} A & 0 \\ 0 & A_s \end{pmatrix}$ $\begin{pmatrix} I_k & A^{-1}B \\ 0 & I_{n-k} \end{pmatrix}$ where $A_s = D - CA^{-1}B$ is the Schur complement of A in M.

The inversibility of A ensures that the matrix M is invertible if and only if A_s is invertible.   The   inverse   of M is: $M^{-1} = \begin{pmatrix} I_k & -A^{-1}B \\ 0 & I_{n-k} \end{pmatrix} \begin{pmatrix} A^{-1} & 0 \\ 0 & A_s^{-1} \end{pmatrix}$ $\begin{pmatrix} I_k & 0 \\ -CA^{-1} & I_{n-k} \end{pmatrix} = \begin{pmatrix} A^{-1} + A^{-1}BA_s^{-1}CA^{-1} & -A^{-1}BA_s^{-1} \\ -A_s^{-1}CA^{-1} & A_s^{-1} \end{pmatrix}.$

For a quaternionique matrix $M = \begin{pmatrix} a & b \\ c & d \end{pmatrix} \in \mathcal{R}^{2\times 2} = \mathbb{M}_2(\mathbb{H}(\mathbb{Z}/n\mathbb{Z}))$ where the quaternion a is invertible as well as its Schur complement $a_s = d - ca^{-1}b$ we have M is invertible and:

$$M^{-1} = \begin{pmatrix} a^{-1} + a^{-1}ba_s^{-1}ca^{-1} & -a^{-1}ba_s^{-1} \\ -a_s^{-1}ca^{-1} & a_s^{-1} \end{pmatrix}.$$

Therefore, to randomly generate an invertible quaternionique matrix, it suffices to:

- Choose randomly three quaternions a, b and c for which a is invertible.
- Select randomly the fourth quaternion d such that the Schur complement $a_s = d - ca^{-1}b$ of a in M is invertible.

4 A New Fully Homomorphic Encryption Scheme

We place ourselves in a context where Bob wants to store confidential data in a very powerful but non-confident cloud. Bob will later need to execute complex processing on his data, of which he does not have the necessary computing powers to perform it. At this level he thinks for, at first, the encryption of his sensitive data to avoid any fraudulent action. But the ordinary encryption, which he knows, does not allow the cloud to process his calculation requests without having decrypted the data stored beforehand, which impairs their confidentiality. Bob asks if there is a convenient and efficient type of encryption to process his data without revealing it to the cloud. The answer to Bob's question is favorable, in fact since 2009 there exist so-called fully homomorphic encryption, the principle of which is quite simple: doing computations on encrypted data without thinking of any previous decryption.

In order to profitably benefit from the technological advance of the cloud computing and to outsource its heavy calculations comfortably, Bob needs a robust highly secure fully homomorphic encryption scheme whose operations, addition and multiplication, are done in a judicious time and whose the generated noise during a treatment is manageable.

To help Bob take full advantage of the powers of the cloud, we introduce a probabilistic symmetric fully homomorphic encryption scheme without noise. The addition and multiplication operations generate no noise. We can describe our cryptosystem as follows:

Key Generation

- Bob generates randomly two big prime numbers p and q.
- Then, he calculates $N = p.q$.
- Bob generates randomly an invertible matrix $K = \begin{pmatrix} a_{11} & a_{12} & a_{13} \\ a_{21} & a_{22} & a_{23} \\ a_{31} & a_{32} & a_{33} \end{pmatrix}$
  $\in \mathbb{M}_3(\mathbb{H}(\mathbb{Z}/N^2\mathbb{Z}))$.
- Bob calculates the inverse of K, Which will be denoted K^{-1}.

- The secrete key is (K, K^{-1}).

Encryption

Lets $\sigma \in \mathbb{Z}/N^2\mathbb{Z}$ be a clear text. To encrypt σ Bob proceed as follows:

- Bob transforms σ into a quaternion: $m = \sigma + \alpha Ni + \beta Nj + \gamma Nk \in \mathbb{H}(\mathbb{Z}/N^2\mathbb{Z})$ such that $\alpha, \beta, \gamma \in \mathbb{Z}/N\mathbb{Z}$

- Bob generates a matrix $M = \begin{pmatrix} m & r_1 & r_2 \\ 0 & r_3 & r_4 \\ 0 & 0 & r_5 \end{pmatrix} \in \mathbb{M}_3\big(\mathbb{H}(\mathbb{Z}/N^2\mathbb{Z})\big)$ such that $r_i \in$

  $\mathbb{H}(\mathbb{Z}/N^2\mathbb{Z}) \forall i \in [\![1, 5]\!]$ are randomly generated.
- The ciphertext of σ is $C = Enc(\sigma) = KMK^{-1} \in \mathbb{M}_3\big(\mathbb{H}(\mathbb{Z}/N^2\mathbb{Z})\big)$.

Decryption

Lets $C \in \mathbb{M}_3\big(\mathbb{H}(\mathbb{Z}/N^2\mathbb{Z})\big)$ be a ciphertext. To decrypt C, using his secrete key (K, K^{-1}), Bob proceeds as follows:

- He computes $M = K^{-1}CK$.
- Then he takes the first input of the resulting matrix $m = (M)_{1,1}$
- Finally, he recovers his clear message computing $\sigma = m \bmod N$.

Addition and Multiplication

Let σ_1 and σ_2 be two clear texts and $C_1 = Enc(\sigma_1)$ and $C_2 = Enc(\sigma_2)$ be their ciphertexts respectively.

It is easy to verify that:

(1) $C_{mult} = C_1 \times C_2 \bmod N^2 = Enc(\sigma_1) \times Enc(\sigma_2) \bmod N^2 = Enc(\sigma_1 \times \sigma_2)$.
(2) $C_{add} = C_1 + C_2 \bmod N^2 = Enc(\sigma_1) + Enc(\sigma_2) \bmod N^2 = Enc(\sigma_1 + \sigma_2)$.

5   Comparison with Other Schemes

As it is shown in Table 1, our cryptosystem presents good performances compared to other existing schemes. Its ciphertext and key sizes depend linearly to cleartext space dimension. The other schemes use a small cleartext space which influences the runtime

Table 1. Comparison of the Performances of FHE Schemes

Algorithm	Cleartext space	Secret key	Public key	Ciphertext
Gentry [2]	$\{0,1\}$	n^7	n^3	$n^{1.5}$
Smart-Verc [3]	$\{0,1\}$	$O(n^3)$	n^3	$O(n^{1.5})$
DGHV [4]	$\{0,1\}$	$\tilde{O}(\lambda^{10})$	$\tilde{O}(\lambda^2)$	$\tilde{O}(\lambda^5)$
CMNT [8]	$\{0,1\}$	$\tilde{O}(\lambda^7)$	$\tilde{O}(\lambda^2)$	$\tilde{O}(\lambda^5)$
Batch DGHV [9]	$\{0,1\}^l$	$\tilde{O}(\lambda^7)$	$l.\tilde{O}(\lambda^2)$	$l.\tilde{O}(\lambda^5)$
LI-WANG [10]	$\mathbb{Z}/N\mathbb{Z}$	$O(3N)$	NA	$O(3N)$
Our scheme	$\mathbb{Z}/N^2\mathbb{Z}$	$O(9N^2)$	NA	$O(9N^2)$

of the algorithm. In our case we are using a large cleartext space which allows us to encrypt big messages and perform computations directly on ciphertexts. We can observe that the complexity of Li-Wang's scheme is smaller than ours, but this scheme uses a smaller cleartext space.

6 Conclusion and Perspectives

In this paper, we presented a new fully homomorphic encryption scheme. It is symmetric, noise free and probabilistic cryptosystem, for which the ciphertext space is a non-commutative ring quaternionic based. The cleartext can be a large number $\sigma \in \mathbb{Z}/N^2\mathbb{Z}$. Our encryption scheme finds its effective applications in the domain of smart computations on encrypted data in cloud computing as it can be applied also to big data security. It is an efficient and practical scheme whose security is based on the problem of solving an over-defined system of quadratic multivariate polynomial equations in a non-commutative ring. In the next work we will implement this cryptosystem and proof its security.

References

1. Rivest, R., Adleman, L., Dertouzos, M.: On data banks and privacy homomorphisms. Foundations of Secure Computation, pp. 169–180 (1978)
2. Gentry, C.: A fully homomorphic encryption scheme, Ph.D. thesis, Stanford University (2009) http://crypto.stanford.edu/craig
3. Smart, N., Vercauteren, F.: Fully Homomorphic Encryption with Relatively Small Key and Ciphertext Sizes, Cryptology ePrint Archive, Report 2009/571 (2009). http://eprint.iacr.org/
4. van Dijk, M., Gentry, C., Halevi, S., Vaikuntanathan, V.: Fully homomorphic encryption over the integers, Cryptology ePrint Archive, Report 2009/616 (2009). http://eprint.iacr.org/
5. Chunsheng, G.: Fully Homomorphic Encryption Based on Approximate Matrix GCD. eprint.iacr.org/2011/645
6. Vikuntanathan, V., Brakerski, Z.: Efficient Fully Homomorphic Encryption from (Standard) LWE. http://eprint.iacr.org/2011/344
7. EL-Yahyaoui, A., EL Kettani, M: Fully homomorphic encryption: state of art and comparison. https://www.academia.edu/25106824/Fully_Homomorphic_Encryption_State_of_Art_and_Comparison
8. Coron, J., Mandal, J., Naccache, D., Tibouchi, M.: Fully Homomorphic Encryption over the Integers with Shorter Public Keys. http://eprint.iacr.org/2011/441
9. Coron, J., Lepoint, J., Tibouchi, M.: Batch fully homomorphic encryption over the integers (2012). eprint.iacr.org/2013/36
10. Li, J., Wang, L.: Noise-free Symmetric Fully Homomorphic Encryption based on noncommutative rings, Cryptology ePrint Archive, Report 2015/641 (2015)
11. Yagisawa, M.: Improved fully homomorphic encryption with composite number modulus. Cryptology ePrint Archive, Report 2016/50 (2016). http://eprint.iacr.org/

New Real Time Method for Air Traffic Control Based on the Blocking Area

Sallami Chougdali[1(✉)], Khalifa Mansouri[1], Mohamed Youssfi[1], and Youssef Balouki[2]

[1] Signals, Distributed Systems and Artificial Intelligence Laboratory, ENSET,
Hassan II University, Casablanca, Morocco
s.chougdali@gmx.fr, khmansouri@hotmail.com, med@youssefi.net
[2] Department of Computer Science, FSTS, Hassan I University, Settat, Morocco
balouki.youssef@gmail.com

Abstract. In this paper we present a new real time method for Air Traffic Control inside Terminal Management Advisor (TMA) space based on the Earliest Deadline First algorithm (EDF). The proposed method consists to; (i) modulate the landing aircraft process on the real time task with period Ti, deadline Di and execution time Ci, (ii) evaluate and manage the aircraft landing priority using EDF specifications and (iii) execute the aircraft landing process. The main goal of this research is optimizes the global Air Traffic Delay ATD in order to minimize the aircraft delays, to preserve the environment and to offers quality passenger service. To evaluate the performances of the new real time method, we compare it with the FCFS (First Come First Served) method most used actually by air traffic controller to manage the air traffic inside the TMA space.

Keywords: Air Traffic Control · Real time scheduling algorithm · FCFS method
Earliest Deadline First · Terminal Management Advisor

1 Introduction

Air Traffic Management consists by tree processes:

- Air Traffic Control process (ATC process): consist to manage and separate the aircrafts with safety and optimal manner in the sky as they fly and at the airports where they land and take off again. This activity is provided by the air traffic controller in the airport Tower and in the regional control center. ATC process composed by tree parts; (i) Regional Air Traffic Control (RATC), that allow to separate and manage the aircrafts in the reginal airspace, (iii) Approach Air Traffic Control (AATC) consists to manage and separate the aircrafts in the Terminal Management Advisor (TMA), that the airspace around the airport, (iii) Aerodrome Air Traffic Control (ArATC) consists to manage and control the aircrafts in the ground, taxiway and in the aircrafts parking.
- Air Traffic Flow Management (ATFM): consists by the all activities that are done before flights take place, such as, flight plan traitement, aircrafts take-off planning.

© Springer International Publishing AG, part of Springer Nature 2019
J. Mizera-Pietraszko et al. (Eds.): RTIS 2017, AISC 756, pp. 321–331, 2019.
https://doi.org/10.1007/978-3-319-91337-7_30

For safety reasons, air traffic controllers cannot handle too many flights at once so the number of flights they control at any one time is limited. Sophisticated computers used by Air Traffic Flow Management calculate exactly where an aircraft will be at any given moment and check that the controllers in that airspace can safely cope with the flight. If they cannot, the aircraft has to wait on the ground until it is safe to take off.

- Aeronautical Information Services (ATS): These services are responsible for the compilation and distribution of all aeronautical information necessary to airspace users, which include safety, technical, navigation informations.

The air traffic continues to grow in the world, so the air transportation becomes a highly competitive area and it forecasted to face strong challenges. Therefore, the Air Traffic Management processes are considered us complex tasks. In this contribution we are focused on the Approach Air Traffic Control process, specially, on the aircraft landing scheduling process. Aircraft Landing Management consists to manage the aircrafts landing with the safety and optimal manner, in order to eliminate the aircraft delay and congestion areas into TMA space.

Currently, the widely used method for scheduling the landing of a set of aircraft is FCFS – First Come First Served -, where landing aircraft is done in the order of their arrival on the Terminal Management Advisor (TMA) of the airport. Cert the FCFS presents some advantages such as; it is easy to implement and it also minimizes the number of aircraft deviations, but it is a sequential method, it can give rise to congestion areas into TMA space, which affect the air traffic safety and does not preserve the environment. Also, using FCFS method the aircraft of low speed can affect the landing time of others of high speed, which can generate the aircraft delay and affect the service quality. These limitations have motivated a large number of scientists to study the Aircraft Landing Management problem. So, there are many advanced methods and algorithms based on linear programming methods and meta-heuristic approaches to solve this problem.

In this work, we consider the aircraft landing management as a real-time system and we proposed a new real time method based on the Earliest Deadline First algorithm (EDF).

After this introduction, the paper is organized as follows; Sect. 2 describe the actual method used by the air traffic controller, the proposed method is presented in Sect. 3, the new system architecture is discussed in the Sect. 4, Sect. 5 is reserved to experiments and results discussion. Finally, Sect. 6 presents the conclusions and the direction of future studies.

2   Air Traffic Management in TMA Space Based on FCFS Method

The Terminal Management Advisor is a space around the airport with a cylindrical shape; in general the runway is placed in the center of its base. Each TMA offers one entry point, also it has one holding point which defined by VOR equipment. Every aircraft enter to TMA space from the entry point, flay until the holding point and wait the landing authorization or landing clearance from the control tower when the runway becomes free.

In general the air traffic controllers use the FCFS method; the first arrived on the holding point is the first receiving the landing clearance when the runway becomes free and other constraints required by the national and international norms are satisfied such as the minimum separation distance between two aircrafts imposed by ICAO (International Civil Aviation Organization) (Fig. 1).

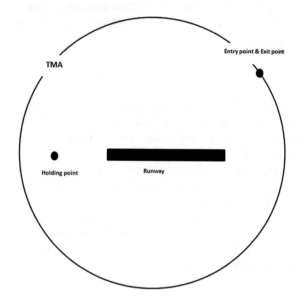

Fig. 1. Terminal Management Advisor architecture for FCFS method

2.1 Inference Process

Actually, the main process used by Air Traffic Controller to manage the air traffic in the TMA space based on the FCFS method:

Basic consideration:

- TMA has a cylindrical shape with a radius equals 40 NM (Nautical Mile);
- Successive landing are spaced at a minimum time separation;
- The aircraft landing have absolute priority over aircraft departure;
- The runway can be used by one aircraft at a time;
- Aircraft are served on a FCFS (First Come First Served);
- The TMA offers one entry point and one exit point;
- The entry point and the exit point is the same;

Algorithm:

1. The aircrafts enter to TMA space from the only entry point, respecting the minimum time separation between two successive arrivals.

2. All aircrafts flaying in the TMA space until holding point which defined by VOR equipment.

3. When the runway is free, the Air Traffic Controller (ATC) gives the landing authorization of the waiting aircrafts in order of these arrivals at the holding point. (First Come First Served), respecting the minimum time separation between two successive landing.

4. The ATC can give the take-off authorization, when any aircraft waiting in the holding point. Respecting the minimum time separation between two successive departures.

5. Aircraft with special conditions such as, aircraft in distress situation have a priority over the others.

2.2 Discussion

Since the FCFS method in the TMA space offers some advantages; it's easy to implement and it also minimizes the number of aircraft deviations, so it's offers the high air traffic safety level. But this method has limitations; the aircraft of low speed can affect the landing time of others with high speed, subsequently, the global cost of landing. Also, FCFS don't offer more flexibility to air traffic controllers to manage the air traffic in this critical step. These limitations have motivated a large number of scientists to study the Air traffic management problems in the TMA space. So, there are many advanced methods and algorithms based on linear programming methods and meta-heuristic approaches to solve these problems. In the following, we propose a new real time method for aircraft landing scheduling in order to manage the landing operations with an optimal manner and ensure the air traffic safety.

3   New Real Time Method for Air Traffic Management in the TMA Space

The new method consist to schedule the aircrafts landing and taking-off operations in the TMA space based on the on the Earliest Deadline First algorithm (EDF).

Earliest Deadline First (EDF): it is preemptive, dynamic and variable priority. The priority is granted to the task which has the term or deadline time is the closest. EDF is characterized by two advantages; (i) the rate of use of CPU can reach 100% for the set of tasks deadlines on requests (ii) it offers less dead-time compared to RM and DM, but EDF presents difficulties to implement and priority inversion problem can be generated.

The proposed method allows of giving an optimal solution of air traffic management inside TMA space, so the main objectives are:

- Solve the Aircraft Landing scheduling with optimal way.
- Give enough flexibility to air traffic controller.
- Minimize the remaining fuel costs of all aircrafts to be landed by meeting their most economic target landing times at preferred speed.
- Ensure the air traffic safety.

In this case, we consider the TMA space with multiple entry points and exit points, also, we eliminate the holding point (Fig. 2).

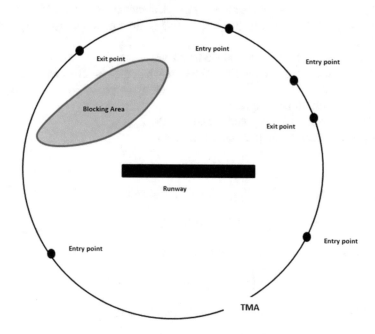

Fig. 2. Terminal Management Advisor illustration for the proposed method

3.1 System Modelization

1. The TMA is modulating by two agents; the scheduling agent S-Agent and the Air Traffic Controller agent ATC-Agent. The S-agent consists of scheduling the aircraft landing and taking-off inside the TMA space. The ATC-Agent allows executing and managing the aircraft landing and taking-off.
2. Each aircraft is characterized by:
   - V_i: The aircraft landing speed, it is considered constant during the landing process.
3. Aircraft departure process:
   - $T_{departure\_contacti}$: The time when the aircraft asked departure authorization.
   - $T_{real\_departurei}$: The time when the aircraft tacking off.
   - **DD:** Aircraft Departure delay

$$DD = T_{real\_departurei} - T_{departure\_contacti} \qquad (1)$$

4. Aircraft landing process is considered as a real-time task.
   - $D_{E\text{-}L}$: The distance between the entry point and the runway threshold, it's the distance flaying by aircraft during the landing process.

- $T_{Landing\_contacti}$: The time when the aircraft arrives in entry point and it's asked landing authorization.
- T_{Entryi}: The time when the aircraft take clearance to enter the TMA space.
- $T_{Th\_Landingi}$: The aircraft theoretical landing time or the aircraft target landing time.
- $T_{Real\_Landingi}$: The aircraft real landing time.
- R_i: The aircraft activation time, the time when the aircrafts arrives at the TMA entry point.
- T_i:: The aircraft landing period.
- D_i: The aircraft landing deadline equals $T_{Th\_Landingi}$.
- C_i: The aircraft landing execution time, so

$$C_i = D_{E-L/Vi} \qquad (2)$$

- CT (t_i): The aircraft landing Critical time at a time t_i:

$$CT(ti) = D_i - C_i(ti) \qquad (3)$$

- DLI: Aircraft Landing delay inside TMA space:

$$DLI = \left(T_{Real\_Landing} - T_{Th\_Landing}\right) - DLO \qquad (4)$$

- DLO: Aircraft Landing delay outside TMA space:

$$DLO = T_{Landing\_contacti} - T_{Entryi} \qquad (5)$$

The aircraft landing task is:

- **Ready:** when the aircraft flying at the entry point and waiting the authorization from the S-agent to enter inside the TMA space.
- **Executed:** when the aircraft flying inside the TMA space and it's in the landing phase.
- **Blocked:** when the aircraft landing process is blocked, the aircraft flying in the blocking area.
- **Terminated:** the aircraft landing process is terminated when the aircraft landed and runway become free.

3.2 Global Air Traffic Delay Definition

In this research, the global Air Traffic Delay inside the TMA space between t_i and t_{i+1} is defined by:

$$ATD\,(ti, ti+1) = \sum_{1}^{n} (DLI * Wli) + (DLO * Wlo) + \sum_{1}^{m} (DD * Wd)$$

- **n:** the aircraft number that landing between t_i and t_{i+1}.
- **m:** the aircraft number that tacking-off between t_i and t_{i+1}.

- **Wli:** delay landing weight coefficient inside the TMA space equals **0.6**.
- **Wlo:** delay landing weight coefficient outside the TMA space equals **0.2**.
- **Wd:** delay departure weight coefficient inside the TMA space equals **0.2**.

This contribution consists to minimize the global air traffic delay, under constraints:

$$\textbf{Di} \leq \textbf{Ci}$$
$$\sum_{i=1}^{n} \left(\frac{Di}{Ti} \right) \leq 1$$

3.3 Inference Process

Basic consideration:

- TMA space has a cylindrical shape and it's contains one runway.
- TMA space offers multiple entry points.
- TMA space offers multiple exit points.
- TMA space contains one blocking area.
- Successive landing are spaced at a minimum time separation; ST_{min} (the minimal separation time).
- The aircraft landing have absolute priority over aircraft departure.
- The runway can be used by one aircraft at a time.
- The landing priority is grant to aircraft with the minimal critical time **CT** at a given time t_j. (According to EDF algorithm).
- The aircraft landing with the specifics conditions have absolute priority.

Inference process:

- Each time equals ST_{min} the S-Agent does:
  1. Updates the ready aircrafts list, which arrived in the one of the TMA entry point.
  2. Classifies aircrafts in the ascending order of their minimal critical time CT.
  3. Gives the landing authorization to the aircraft with low critical time CT.
  4. Updates the aircrafts data in data base.
  5. Coordinates the aircraft landing process with ATC-Agent.
- Each time equals STmin the ATC-Agent does:
  1. Check the aircrafts CT that flaying inside TMA space.
  2. Give the authorization to aircraft with low critical time CT to continue its landing.
  3. Orient other aircrafts to blocking area.
  4. If there is no aircraft in landing process, the ATC-Agent manage the tacking-off process.
  5. Update the aircrafts data in data base.
  6. Coordinate the aircraft tacking-off with S-Agent.

4 Implementation and Infrastructure

The proposed method is developed as virtual computing system; the system is integrated in order to deliver the optimal solution over all the Air Traffic Management inside the

TMA space. This method is implemented with Java using Real-Time Specification for Java (RTSJ) and they run on a Computer of 2.3 GHz CPU.

The Fig. 3 shows the basic components of our new real time method architecture and the relationship among the components. The application layer consists of two components S-Agent allows to schedule the air traffic operations inside the TMA space and ATC-Agent permits to execute these operations, the application layer considers the Real-Time Java Specifications and uses the JamaicaVM as Virtual Machine. The second layer is the data layer; it contains aircraft data and air traffic informations.

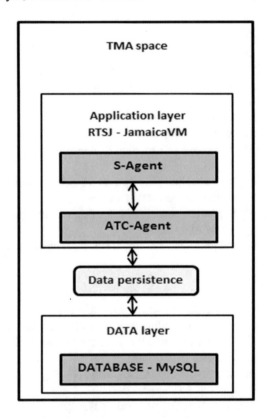

Fig. 3. The implementation architecture of the proposed method

5   Experiments and Results Discussion

To evaluate the feasibility and efficiency of the proposed method, we compared it with FCFS method (First Come First Served); this is the most used method to solve the Air traffic control in the TME space. Two comparative experiments have been carried out on the benchmark set to illustrate the performance of the proposed method. The Comparative study between our approach and existing algorithms such as FCFS method is based on calculation of the Global Air Traffic Delay in the TMA space between tow different instants. The results are presented in the Tables 1 and 2.

Table 1. Comparison the global Air Traffic Delay if the aircraft number inside the TMA space change.

Aircraft number	Global Air Traffic Delay ADT (min)	
	Proposed method	Existing method (FCFS)
5	0	5
10	1	13
50	1	27
100	3	35
200	5	43

Table 2. Comparison the global Air Traffic Delay if the entry and exit point number change.

Entry point number	Exit point number	Global Air Traffic Delay ADT (min)	
		Proposed method	Existing method (FCFS)
1	1	3	35
2	2	2	37
3	2	1	42
4	2	1	45
5	3	0	51

The proposed method performances are tested through two experiments; the first one consists to investigate the effects of the aircraft number end the second experiment allows to evaluate the proposed method behaviors with the different entry point number.

5.1 First Experiment

This step consists to execute the proposed method with the different aircrafts sequences, each time, we change the aircraft number and we consider the TMA space with one entry points and one exit point. The comparative results were obtained by executing the proposed method and the existing method (FCFS) for 10 independent executions, and averaging over the achieved results.

The global Air Traffic Delay is calculates for a period equals 3 h and we are used the real data of the arrivals and departures aircrafts at the international airport of Casablanca.

The aircraft number presents the total aircraft operations (landing and tacking-off) inside the TMA space.

When the aircraft operations number inside the TMA space increases, the proposed method presents high performances than the FCFS method.

5.2 Second Experiment

In this part we consider the TMA space with multiple entry and exit points and we evaluate the effects of the entry and exit points' number on the proposed method performances. The simulation is operated with 100 aircraft operations (landing and tacking-off) inside TMA space.

Also the experiment consists to change the entry and exit points' number and evaluate the performances of the proposed method. The comparative results were obtained by executing the proposed method and the existing method (FCFS) for 10 independent executions, and averaging over the achieved results. The global Air Traffic Delay is calculates for a period equals 3 h and we are used the real data of the arrivals and departures aircrafts at the international airport of Casablanca.

The comparative result shows the benefits of the proposed method than the existing method if the entry and exit points increases.

5.3 Discussion

To show the benefits of the proposed method, we are compered this method with the FCFS method and we choose tow selection parameters; (i) aircraft number, (ii) the entry and exit point number.

The proposed method does not requires holding the aircraft in the holding point, due to the air traffic operations scheduling based on the real time method and to the high synchronization of both landing and departure operation. The economy of 43 min of flight time, in 3 h of flight analysis, is a strong indicator of the benefits of this method. Also, this approach eliminates the holding point so it economizes the VOR equipment cost.

Other important benefits of our approach; it offers more flexibility to air traffic controller and ensure the air traffic safety.

6 Conclusion and the Direction of Future Studies

This work presents the new real method for air traffic control inside the TMA space. The proposed method presents enough benefits than the exiting method such as FCFS method; it's eliminate the aircraft holding problem inside the TMA space, so it preserves the environments, ensures the air traffic safety and offers the high passenger service quality. Also, our approach gives to the air traffic controller more the flexibility to manage the air traffic inside the TMA space with the optimal manner.

The researches in the Air Traffic Management are still in developing stage, so the future work consists to:

– Integrate the real radar data and manage the air traffic information in real time.
– Extend this approach to manage the air traffic form the departure airport, in the transit space, to the arrival airport.

References

1. Weigang, L.I., Marcos, V.P.D., Daniela, P.A., Antonio, M.F.C.: Intelligent computing methods in air traffic flow management. J. Transp. Res. Part C **18**, 781–793 (2010)
2. Yu, S., Cao X., Zhang, J.: A real-time schedule method for aircraft landing scheduling problem based on cellular automation. J. Appl. Soft Comput. **9**, 175–187 (2011)
3. Ciesielski, V., Scerri, P.: Real time genetic scheduling of aircraft landing time. In: Proceedings of the 1998 IEEE International Conference on Evolutionary Computation, New York, USA, pp. 360–364 (1998)
4. Boysen, N., Fliedner, M.: Scheduling aircraft landings to balance workload of ground staff. Comput. Ind. Eng. **60**, 206–217 (2011)
5. Soomer, M.J, Franx, G.J.: Scheduling aircrafts landings using airlines preferences. Eur. J. Oper. Res. **190**, 277–291 (2008)
6. Chougdali, S., Roudane, A., Mansouri, K., Youssfi, M., Qbadou, M.: New model for aircraft landing scheduling using real-time algorithms scheduling. In: Proceedings of International Conference on Intelligent Systems and Computer Vision, Fez (2015)
7. Hu, X.B., Paolo, E.D.: Binary-representation-based genetic algorithm for aircraft arrival sequencing and scheduling. J. IEEE Trans. Intell. Transp. Syst. **9**, 301–310 (2008)
8. Hu, X.B., Chen, W.H.: Genetic algorithm based on receding horizon control for arrival sequencing and scheduling. J. Eng. Appl. Artif. Intell. **18**, 633–642 (2005)
9. Tavakkoli-Moghaddan, R., Yaghoubi-Panah, M., Radmehr, F.: Scheduling the sequence of aircraft landings for a single runway using a ling the sequence of fuzzy programming approach. J. Air Transp. Manage. **25**, 15–18 (2012)
10. Fahmy, M.M.M.: A fuzzy algorithm for scheduling non-periodic jobs on soft real-time single processor system. J. Ain Shams Eng. J. **1**, 31–38 (2010)

The Data Mining: A Solution for Credit Card Fraud Detection in Banking

Hafsa El-kaime[✉], Mostafa Hanoune, and Ahmed Eddaoui

Laboratory of Information Technology and Modeling,
Faculty of Science ben M'sik, Hassan II University, Casablanca, Morocco
hafsa.elkaime@gmail.com, mhanoune@gmail.com,
ahmed.eddaoui@gmail.com

Abstract. The banking sector has recognized a radical transformation in its services, play an indispensable role in the development and feasibility of electronic commerce, which is coming to help the purchase to its customers. However, it becomes a major target for fraudsters through internet transactions that have become the cause of majority fraud. So, the fight against this fraud is an obligation on banks to ensure the safety of payment. We present our fraud detection approach based on data mining techniques: classification, clustering and research association.

Keywords: Association · Data-mining · E-commerce · Fraud
Classification · Clustering

1 Introduction

1.1 General Context

Every day the banking systems collect enormous quantities of data, whether it is the information of the customers, the details of the deals, the details of credit card, the risk profiles of data of financing of trade… Thousands of decisions are taken in a bank daily, however, this process, subject to the errors and exposed to anomalies, because of the enormous volume of the transaction and historical data.

So among the essential activities of banks in the development and the feasibility is the e-commerce, indeed by means of an electronic board, simple to treat, several people chosen to purchase by internet view that it is simple and fast on one hand, and on the other hand retail websites make appeal to hosts which put into service a banking solution supplied by the partner bank of the retail website. So, the financial transaction is insured by a banking body which ensures both parties (trading and customer) the good progress of the payment.

1.2 Problematic

Because of the diversity of the services presented by banks, and as any activity the banking sector is displayed to frauds. Indeed there are two types of fraud: Remote fraud and nearby card, now the rate of remote fraud is growing fast especially after the

© Springer International Publishing AG, part of Springer Nature 2019
J. Mizera-Pietraszko et al. (Eds.): RTIS 2017, AISC 756, pp. 332–341, 2019.
https://doi.org/10.1007/978-3-319-91337-7_31

appearance and the development of the e-commerce: sales on the Internet so reached 45 billion euros in 2012 [1]. And afterward, the rate of fraud of the remote payments is growing fast [2], of 0.24% in 2007 in 0.32 in 2011, that is a 36% increase.

Thus, then the credit card fraud is a massive phenomenon, that the bankers absolutely have to fight against, and we identified our fields of study, to work on the class of strong importance which is the remote fraud.

1.3   Solutions Existing

The fraud detection is generally based on methods of data mining especially the networks of neurons, as the system, Falcon developed from 1995 [3], then varied statistical techniques [4], also the extraction of the rules of networks of neurons to detect the fraud [5], and recently, analyzes it of social networks.

1.4   Proposal and Structure

In this article one remedy this problem by the most powerful methodology of the forecasts which is the data mining, through its techniques which are the classification, the segmentation and the search for association allowing fighting against the fraud at the same time for the detection, the treatment, and the correction.

The present article is presented as follows: in the first place we shall define more exactly the fraud on the Internet; then we shall present the most recent results of the searches using the techniques of the data mining for the fraud detection, then the brief definition of the data mining, so its techniques; then, we shall rise how our proposal helps the bankers decrease the rate of the fraud. We end with a conclusion and present suggestions for the future searches.

2   Literature Review

2.1   The Credit Card Fraud on the Internet

The development of the e-commerce has a very strong impact on the increase of the credit card fraud on the Internet. The transactions on the Internet so became the majority cause of credit card fraud and the trend increases at a sustained pace.

At the end of 2014, the number of cards in circulation is 10.9 million increasing by 11.2% compared with 2013. The e-commerce activity generated 2 million operations for 1.2 billion DH with 592 sites instead of 554 in 2013 [6].

According to the report of the Monitoring center of the safety of payment cards published in July 2013, evoked by Zineb TAMTAOUI, Product Manager Electronic Banking and Technological Services at BMCE Bank during the Forum of the currency and the electronic payment in French-speaking Africa held in Casablanca, "The rate of fraud by bank card progresses for the fifth consecutive year. He reached 0.080% of the total of the transactions in 2012 against 0.077% in 2011. The amount of the frauds, increasing by 9% in 450.7 million euros, evolved of 4% more with regard to that of the transactions by cards.

The evolution of the rate of fraud all around the world, the Fig. 1 summarizes this evolution in France according to the statistics of the annual report of the monitoring center of the safety of payment cards [7].

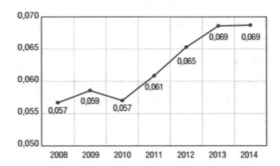

Fig. 1. Evolution of the rate of fraud of French cards

2.2   The Evaluation of Fraud Detection

The algorithms of fraud detection of the credit card consist of identifying the transactions with a probability raised to be fraudulent, to based on models history of fraud. The use of machine learning in the fraud detection was a subject interesting these last years. Different systems which are based on the learning machine were successfully used to remedy this problem of fraud, in particular, the networks of neurons [8], Bayesian study, artificial immune systems [9], settle of association [10], hybrid social models [11], discriminating analysis [12] and analysis of network [13].

Most of these studies compare their algorithms proposed in an algorithm of mark and then make the comparison by using measures which are extracted by means of a matrix of confusion by calculating the cover and the relevance, however these measures can not be the most appropriate criteria of evaluation by estimating models of fraud detection, because they suppose tacitly that the errors of false classification carry the same cost, at the same time with transaction correct.

This Hypothesis does not hold in the practice when incorrectly the forecast of a fraudulent transaction as door legitimises an appreciably different financial cost that the inverse case. Besides the distribution of class among transactions is constant and balanced and typically the distributions of a set of data of fraud detection are biased.

For the analysis of the fraud by means of the social networks, he uses a type of bipartite social network. It is a network in which knots are two different types: cards and traders, and where he can have of link there that between different knots of type. Then, the projection of the bipartite network in two simple networks.

However, this simple model using initial variables describing the transactions has very low performances in relevance and on the cover, thus it was necessary to implement additional techniques to try to reach big performances.

Thus, we shall try to obtain a system of detection having a good cover; not to allow passing too much case of fraud, and good relevance; to raise alerts only advisedly. Thus, through the most powerful methodology of the forecasts which is the data mining, and our system of fraud detection which bases itself on the presence of two bases that through his techniques which are the classification, the segmentation and the search for association to allow fighting against the fraud at the same time for the classification of the base given of the clientele of the bank and the constitution of a system of signature of fraud. These two bases constitute the skeleton of our proposal.

3 Ours Approach of Fraud Detection

The review of the literature shows that the fraud detection requires results of the factors of performance very high and that the techniques of data mining can supply a precision improved with regard to the methods cities before. Our approach for the fraud detection is based on the techniques of data mining.

3.1 Technique of the Data Mining

The data mining: excavation of data, or drilling of data, is a process of exploration and analysis of big volumes of data with the aim of a part to return them more under-standable and on the other hand to discover significant correlations, that is rules for classification and prediction, the most current ultimate purpose of which is the decision-making support. We distinguish big generally two sectors of techniques of data-mining, descriptive techniques not overseen as ACP, AFC, automatic classification... And techniques of predictions, for example, decision trees, networks of neurons (Fig. 2).

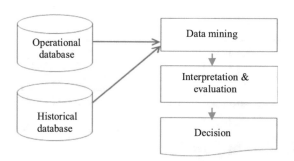

Fig. 2. The decisions made by data mining

3.2 The Data Mining in Banks

The banking sector is at the head of all other industrial domains thanks to the use of the techniques of data mining in its big databases clienteles. Although banks used statistical tools of analysis with a little of success for several years and maintaining banks manage to understand clearly the behavior of certain customers which was invisible by means of the new tools of data mining.

Some applications of the data mining in this domain are:

- Predict the reaction of the customers to the changes of the interest rates.
- Identify quickly the customers who will be the most susceptible in the new offers of p products.
- Identify the regular customers.
- Predict the customers who may change their cards of membership during the next quarter.
- Detect the fraudulent activities in the transactions by credit cards.

3.3   Process of Fraud Detection

The process of fraud detection aims at generating an alert in case of:

- In real time: block a transaction at the time of the authorization request if its code is similar in signature-fraud
- In time postponed: notice that a taken place transaction (the day before) was deceitful to avoid the fraud future on the same card. And the same principle takes place, by comparing the code of this transaction with the signature - fraud.

In all our research, we shall handle the case of detection in batch mode, the problem of real time detection is similar, except the additional constraints of calculation time (Table 1).

Table 1. History of transactions

Fraud	Model	Result
0	0	
0	0	
0	0	
1	0	
1	1	VP
0	0	
0	0	
1	0	FN
0	1	FP
1	1	VP

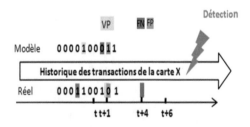

Fig. 3. The fraud detection process

After a fraudulent transaction in t, the transaction in the date t + 1 is planned fraudulent, what is confused with the real case (true positive TP), in t + 4 the transaction is fraudulent but it is not detected (false negative FN), and in t + 6 the transaction is to detect fraudulently and the reality also proves it (Fig. 3).

4 Methodology

As this explained at the top among the very strong applications of the data mining in banks it is to detect and to plan the fraud.

Then our methodology of fraud detection articulates around three pillars:

- Firstly: a classification of the base given of the clientele of the bank "BDC".
- Secondly: the constitution of a system which he is named "SSF" System signature swindler, of whom we are going to gather all the frauds which were existed really to model their behavior.
- Thirdly the functioning of our system of the cart banking fraud detection on the Internet.

So that our system of detection works correctly, it is necessary that these entrances which are BDC and SSF are ready of being to inject in our system of fraud detection.

4.1 Classification of the BDC

Given that the bank contains hundreds in thousands of the users, then the first thing to do it is to include the behavior of every customer, what is not obvious. Thus, we call on to the technique of the data mining which is the classification to train classes of the customers who have similar behavior, the objective of this classification is to understand very well the purchasing behavior of the carriers of the bank card.

We used the mixed automatic classification, because the hierarchical method is very expensive in calculations, even for powerful computers. Furthermore, when two individuals are allocated to the same group, they will not be separate anymore, contrary to the mobile averages. On the other hand, the big advantage is that we can choose the number of groups posteriorly by measuring the qualities of representation and possibly to compare several partitions. The less greedy, mobile averages in the calculation, present the inconvenience that it is necessary to choose a prior the number of desired groups.

We can thus envisage a mixed method, allowing to gain in times of calculation and to choose then the number of groups. For that purpose, we start by the method of the mobile averages for a rather high number of groups (example 20 groups) and we chain from these groups a hierarchical method, which will be practicable by any computer.

The result of this algorithm it is classes of customers that have common points, for example: Class A contains all the customers who have a fixed wage vary between [10000 DH, 20000 DH], who make for 90% of their purchase via the internet and that the average of the transactions follows a law exponential. Whereas class B consists of customers who have an amount average a month =5000 DH, by which the average of the transactions follows a normal law and who have never purchased by the internet.

Then through this classification, we can model the behavior of every class of customer of the bank. This part is going to be detailed and will be the objective of our next article, where we are going to explain in detail the way of functioning of this database of clientele.

4.2   Constitution of the SSF

The second element which is fundamental for the functioning of our system of fraud detection it is the constitution of an SSF (system of signature of fraud), its role is to detect the fraud. This system of detection consists of a database which contains frauds, which we name them after "signature-fraud", every fraud has its own signature, determined from cards having really cheated.

This signature contains all the information on the frauds (cards having cheated) collected from the old fraudulent transactions which really existed: place and the hour of the fraud, nature, rising, the website, the swindler (if he is arrested by the police) the number of the card cheated.

4.3   Fonctionnement Du SDF

After the constitution of the BDC and the SSF, our system is going to be ready to be established.

The functioning of our system is the following one: after the arrival of the trans-action, we shall cross to handle it, then we plan it in our BDC, then we apply the method of search research for association, later we calculate the distance between the established center of the classes of the BDC and the transaction, then we allocate the transaction to the class the distance of which is minimal (Fig. 4).

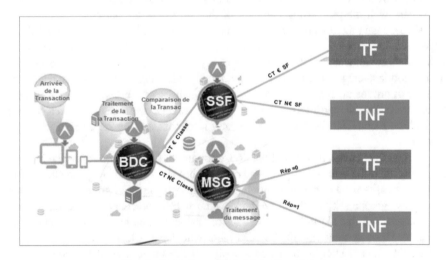

Fig. 4. Operation of fraud detection system

We distinguish here two cases:

- Code of transaction belongs to a class If the transaction is similar to the behavior of a customer of a class, we shall cross to compare this transaction with the signatures of the SSF, always we shall use the algorithm of research for the association, so we shall distinguish two cases: If the code of transaction is similar to a signature of fraud, the system throws an alert that the transaction is fraudulent: TF, otherwise the transaction is not fraudulent: TNF.

- Code of transaction belongs to no class: thus it is a doubtful transaction, it is possible that it is a new behavior of a customer, as it is possible that is a fraudulent transaction, then we send an immediate message to the owner of the card, he is asked a question - that its answer was determined by the agent of the bank at the time of the opening of the account - as the date of birth, the profession of the father... - a personal question that only the real customer of the card know its answer.
  Then if the answer is positive, thus it is well about the real carrier of the card and afterward TNF otherwise the transaction is fraudulent TF.

When the system indicates a warning message of fraud, we distinguish generally two cases of fraud, that is a fraudulent transaction 100%, when the code transaction is confused with the signature-fraud, or a doubtful transaction, when the code of transaction belongs to no class, is it is explained that it is possible that it is a new behavior of the real carrier of the card the system repeats the processing to ask a set of questions (bound to the given of the carrier of the card) in person who is making the transaction, to know if she is good the carrier of the bank card or a swindler.

4.4   Evaluation of the Performances of Detection

We shall use as the measure of performances of detection two indicators: the cover, it is the rate of the identified case of fraud and the relevance the rate of really deceitful alerts. We define the matrix of confusion:

Table 2. Matrix of confusion

		Provided		
		P	N	
Réel	P	TP	FN	F
	N	FP	TN	N F
		A	N A	

The "couverture" and the "pertinence" are defined in the following way:

$$Couv = TP/F = TP/TP + FN$$
$$Pert = TP/A = TP/TP + FN$$

Then the desired objective is to maximize the coverture, thus it is necessary to minimize FN, not to allow passing too much case of fraud, and maximize the relevance, thus not much fraud positive to generate the alerts only after the real fraud detection [15] (Table 2).

5   Conclusion

In this paper, we presented our approach to fraud detection through the method of data mining resting on three techniques: the classification, the segmentation, and the research for the association. This system of fraud detection bases itself mainly on the classification of the databases of the carriers of banking card then the constitution of the base of signature of the frauds from the history of cards having cheated. Also this paper handles the subject of the banking fraud in a global way, through our system of fraud detection, remains to detail the mixed automatic classification of the base given of the clientele of the bank, a subject which is going to be detailed in our next article.

References

1. Fevad: Bilan e-commerce 2012 (2013). http://www.fevad.com/espacepresse/bilan-e-commerce-45-milliards-d-euros-en-2012
2. OSCP: Rapport annuel. Observatoire de la sécurité des cartes de paiements (2011a). http://www.banque-france.fr/observatoire/telechar/2-statistiques-pour-fraude-2011-rapport-annuel-2011-observatoire-securite-cartes-paiement.pdf
3. Hassibi, K.: Detecting payment card fraud with neural networks. In: Lis-boa, P.J.G., Edisbury, B., Vellido, A. (eds.) Business Applications of Neural Networks: The State-of-the-Art of Real-World Applications. Progress in Neural Processing, vol. 13. Chap. 9, pp. 141–158. World Scientific Publishing (2000)
4. Hand, D.J., Weston, D.J.: Statistical techniques for fraud detection, prevention and assessment in mining massive data sets for security. In: Fogelman-Soulié, F., Perrotta, D., Pikorski, J., Steinberger, R. (eds.) Advances in Data Mining, Search, Social Networks and Text Mining and Their Applications to Security. NATO ASI Series, pp. 257–270. IOS Press (2008). http://videolectures.net/mmdss07_hand_stf/
5. Neural Network Rule Extraction to Detect Credit Card Fraud
6. Cartes bancaires, le taux de fraude en progression écrit par Sara BAR-RHOUT, Catégorie: Économie. Publication: 24 mars 2015. Mis à jour: 1 avril 2015. Affichages: 328
7. Rapport annuel de l'observatoire de la sécurité des cartes de paiement, p. 16 (2014)
8. Maes, S., Tuyls, K., Vanschoenwinkel, B., Manderick, B.: Credit card fraud detection using Bayesian and neural networks. In: Proceedings of NF 2002 (2002)
9. Bachmayer, S.: Artificial Immune Systems, vol. 5132, pp. 119–131 (2008)
10. Sánchez, D., Vila, M., Cerda, L., Serrano, J.: Association rules applied to credit card fraud detection. Exp. Syst. Appl. (2009). http://linkinghub.elsevier.com/retrieve/pii/S0957417440 8001176, http://www.springerlink.com/index/rq58w1v614933838pdf
11. Krivko, M.: A hybrid model for plastic card fraud detection systems. Exp. Syst. Appl. (2010). http://linkinghub.elsevier.com/retrieve/pii/S0957417410001582
12. Mahmoudi, N., Duman, E.: Detecting credit card fraud by modified fisher discriminant analysis. Exp. Syst. Appl. (2015). http://linkinghub.elsevier.com/retrieve/pii/S09574174140 06617

13. Van Vlasselaer, V., Bravo, C., Caelen, O., Eliassi-Rad, T., Akoglu, L., Snoeck, M., Baesens, B.: APATE: A Novel Approach for Automated Credit Card Transaction Fraud Detection using Network-Based Extensions. Decision Support Systems (2015). http://linkinghub. elsevier.com/retrieve/pii/S0167923615000846
14. Kantardzic (2011)
15. Mehboob, B., Liaqat, R.M. Abbas, N.: Student performance prediction and risk analysis by using data mining approach. J. Intell. Comput. **8**(2) 49–57 (2017)

Internet of Breath (IoB): Integrative Indoor Gas Sensor Applications for Emergency Control and Occupancy Detection

Simon Fong[1(✉)], Chintan Bhatt[2], Dmitry Korzun[3], Shuang-Hua Yang[4], and Lili Yang[5]

[1] Department of Computer Information Science, University of Macau, Macau SAR, China
ccfong@umac.mo
[2] Patel Institute of Technology, Charotar University of Science and Technology,
Anand, Gujarat, India
chintanbhatt.ce@charusat.ac.in
[3] Department of Computer Science, Petrozavodsk State University,
Petrozavodsk, Republic of Karelia, Russia
dkorzun@cs.karelia.ru
[4] Department of Computer Science and Engineering,
Southern University of Science and Technology, Shenzhen, China
yangsh@sustc.edu.cn
[5] Department of Mathematics, Southern University of Science and Technology, Shenzhen, China
yangll@sustc.edu.cn

Abstract. A design framework is proposed in this paper for a new type of the Internet-of-Things technology. We specifically use the term Internet-of-breaths (IoB) which is a wireless sensor network (WSN) for enhancing fire-and-rescue (FRS) operations with real-time information of the air quality and human presence at the proximity of the fire-site. Though IoB can be useful in emergency scenario, IoB would be a useful tool in detecting human occupancy in a confined area. It can estimate approximate head-counts by inferring from the size of the room and the measured CO_2 concentration level. Based on the previous experiences of SafetyNET with is a successful project by Loughborough University UK IoB extends the capacity of the SafetyNET sensor with capability of CO_2 concentration measurement as an integrative approach. With this new sensing capability, new information such as human occupancy can be measured. By knowing whether and how many people exist in a certain location and its nearby locations, extra intelligence can be obtained in addition to toxic gas, fire severity, etc.; this will great help FRS operations. With the human occupancy information, the proposed research here is to focus on designing and building prototypes of applications that can benefit from knowing the human occupancy in real-time, through the IoB wireless sensing network. Integration of this new information on existing WSN data and systems is investigated. Possibilities of new applications based on human occupancy detection is discussed in the aspects of WSN technology.

Keywords: Internet-of-Things · Internet-of-Breath · Wireless sensor network
Human occupancy detection

© Springer International Publishing AG, part of Springer Nature 2019
J. Mizera-Pietraszko et al. (Eds.): RTIS 2017, AISC 756, pp. 342–359, 2019.
https://doi.org/10.1007/978-3-319-91337-7_32

1   Introduction

Fire disasters are on the rise worldwide. Out of the many rescue missions, some firemen sacrificed their lives in fire-and-rescue (FRS) operations, due to the inability to better know the actual and real-time situations at the fire sites. In urban areas, this critical decision making based on real-time fire information is important as many new buildings and mega malls are being rapidly developed. The interiors of those mega buildings are complex, and maybe packed with people too. On the other hand, old buildings in some cities are crowdedly clustered and their floorplans may not be even available or outdated. This problem is especially grave in the world's most overcrowded cities. They are packed with large residential population as well as the number of visitors that are ever increasing, adopting a new generation of fire control using IoT technology and real-time analytics is imperative for being a smart city. In this paper, a conceptual framework of a novel type of Internet-of-things technology, which is called Internet-of-breaths (IoB) which is a wireless sensor network (WSN), for enhancing fire-and-rescue (FRS) operations with real-time information of the air quality and human presence at the proximity of the fire-site is proposed and explored.

The significance of this Internet of Breath (IoB) project is to deploy an improved type of sensors together with low-energy consumption wireless sensor network over (or in supplementary to) the existing Automatic fire detection system (SADI) in an urban city. With the new system in place, real-time information about the fire sites can be known instantly, can be monitored and even predicted using machine learning (e.g. severity of fire, estimated damage, spreading direction of fire) through the constantly collected data from the sensors. In addition, which is very important, the human occupancy rate can be detected (up to certain accuracy) in each room or location that has the installed sensors over the fire site and nearby proximity – this will greatly help plan for the best rescue and evacuation tasks in fire control. The early generation of smoke alarms are individual, they only sound off and sprinkle water when the heat-fuse is broken. The current generation of smoke alarms which is known as SADI wires up the sensors, alarms into a central control, as a closed-loop system. Please see diagram below (Fig. 1).

Currently real-time information is unavailable; SADI detects fire only when the temperature or smoke concentration exceeds certain preset threshold. That is, fire is detected only AFTER fire has happened. With the new IoB system, which could be used to complement the existing SADI, real-time information can be collected, interpreted and analyzed as shown in the diagram below. As such, valuable real-time information that are being collected, include: temperature, humidity, toxic gas, radiation, light intensity, particulate matter (PM) and carbon dioxide (CO_2). In particular, our proposed new sensor is comprised of CO_2 sensor which measures the ambient background of CO_2 concentration and fluctuations of CO_2 concentration – from there we can infer approximately how many people there are at the fire scene and nearby locations. Human occupancy is a mature technology and it has been widely adopted recently. Since the data are collected in real-time and stored up, we can infer new information which was previously unavailable, such as: (1) the development of fire for early detection and continuous monitoring; (2) estimate using machine learning the intensity of the fire and predict its spreading/expansion when multiple spatial sensors are used; (3) by knowing and/or

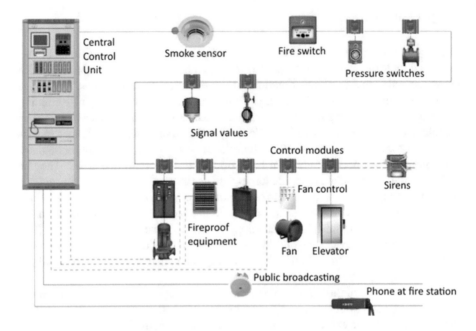

Fig. 1. Current fire alarm system that wires up the components

predicting approximately how many people there are at the fire scene, given the previous readings before the fire broke out, and (4) by using big data for referring to historical records and using potentially real-time weather data, the characteristics of the fire can be analysed and its damage can be predicted. The conceptual diagram about how the fire/gas sensors and the human occupancy detection sensors being wired up, with real-time information polling is shown in Fig. 2.

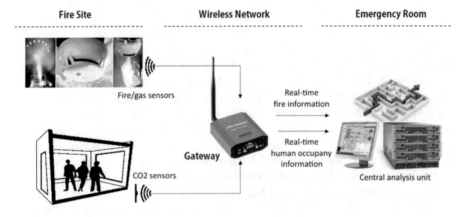

Fig. 2. IoB framework to integrate fire/gas sensors and CO2 sensors for real-time data collection

The reminder of this paper is structured as follows. Section 2 presents the research motivation, explaining why IoB is necessary in our daily lives. It also shows about related work and latest research progress similar to the proposed system. The IoB design framework is described in detail in Sect. 3. Feasibility of WSN design and implementation for IoB is discussed in Sect. 4. Section 5 concludes the paper.

2 Motivation and Related Work

Mainly IoB is about using IoT wireless sensor network techniques for gathering real-time intelligence at fire scenes for fire control and people rescue. The motivation is to use affordable resources (not-too-expensive IoT technologies) for useful outcomes, such as saving lives in case of emergency. The other motivation is human occupancy detection – it works like automatic means in counting humans. Although the use is limited to indoor, it has significant usefulness in addition to emergency situations. When the new sensors are deployed, as a side benefit, the capabilities of measuring CO_2 concentration in real-time enable guessing head counts which otherwise would be a tedious and costly process if it were to be done manually – e.g. bus drivers who record the number of passengers in a bus using pen and paper.

2.1 Features

In addition to fire control, the following features of the IoB sensors could be made possible tapping on the CO_2 measurement and machine learning computation.

IoB sensors installed at static locations, e.g., buildings, rooms, halls, stations, airports, seaports, etc.

- Auto human presence counting for business intelligence, e.g. knowing how many people have attended a seminar, a concert, an exhibition, etc.
- Keep tracking of human traffic flows, e.g. lobbies, offices, hotels, immigration control, can have an approximate figure about number of visits per hour, at different times, days, dates, locations.
- Early warning for preventing stampedes tragedy when the CO_2 concentration and the size of the locations and the rate of the continuing "built up" of human counts rising are known by the IoB sensors. Please note that this would work for cases that are indoor. For outdoor cases, such as the recent Shanghai stampedes due to too many people swamping to a particular place for new year count-down, the ultimate head count in an outdoor place would have to be inferred by summing up the CO_2 concentration levels from all public transports (e.g. buses) that are head to the same direction. Or if there are any entrance lobby, it might work though it is difficult.
- Normally, when no extreme cases happen, the IoB sensors can be used for measuring the presence of attendees, crowdness and overcrowdings – this is significant to tourism and resort operators who care about whether an event, a room, a hall has

many visitors/attendees, and whether these attendees are comfortable, being over-crowded or not. As an extra feature, when the sensitivity of the IoB sensor is fine-tuned and adjusted properly, it can be used for security – detecting whether human presence exists in a restricted confined area.

IoB sensors can be installed at mobile locations, e.g. public buses, casino shuttle buses, turbo jet ferries, taxis, etc.

- The number of passengers can be known in real-time. This would be useful business intelligence in optimizing routes, bus frequency, traffic management etc. However, it is acknowledged that extensive tests are required because on the road there would be plenty of CO_2 emissions; other factors such as whether and how many vehicle windows are opened, the air quality at different sections of the roads etc. At least there would be approximate estimate of passenger numbers being detected in real-time on the move.
- As another level of business intelligence in e-governance of smart city, the "direc-tions" of human flows in approximate volumes and velocities can be estimated in real-time; it aids an emerging smart city initiative known as crowd-sensing which would be useful for traffic planning. In some countries, such information can be used to predict whether a riot may take place.

This IoB design features a set of not-too-complex techniques for profound results – it makes use of available and affordable state-of-the-art sensors and IoT components. The current SADI system can integrate with gas sensor capabilities with CO_2 concentration and real-time remote data transmission technology. With CO_2 concentration detection we can have new possibilities, in addition to the existing emergency disaster monitoring: (a) human occupancy detection, for preventing indoor stampede, for government/organization knowing information about "estimated" head counts of people visit, in buildings scenarios such as exhibitions, theatres, indoor stadiums, classrooms, showrooms, etc. and in dynamic scenarios such as buses, trains, trams, boats where real-time passenger occupancies can be collected via wireless sensor networks.

So for Smart City managements, government can know about the human occupancies (crowd-sensing) as well as indoor air quality for business intelligence, for facility planning and optimization. The CO_2 measurements could be used as a forewarning for fire detec-tion (sudden surge within a short time frame) for emergency management. For example, as a possibility, we can extend the current WiFi and ZigBee with LoRaWSN, powered by ThingPark China[1] - it can cover radius in kilometers, communicating with end devise support by common battery, making this project very feasible for deployment in urban or metropolitan city.

Another main feature of this project is: it potentially can integrate well with other Smart City projects. E.g. IoT energy consumption control can be the next step of indoor crowd-sensing, by air-conditioning; human presence detection can integrate with security systems; human occupancy measurements and the results should integrate with smart city transportation system, and air quality and health monitoring system etc. The useful data

---

[1] http://www.cww.net.cn/article?id=409382.

collected, can feed into some organizational and/or government big data systems for analysis.

2.2 Other Works

In China, automatic fire alarm system is getting very popular. However, most of them are standalone systems, lack of connectivity, real-time data collection and analysis capabilities.

In South East Asia, smart cities projects especially for Fire and Safety are moving on. Please see recent news:

http://www.thestar.com.my/metro/community/2017/08/17/putting-the-spotlight-on-cybersecurity-industry-experts-to-showcase-products-at-international-fire-a/

On 8/9/2017, India just announced about their plans in using IoT to improve fire and safety measures.

http://realty.economictimes.indiatimes.com/news/industry/govt-authorities-taking-proactive-measures-to-transform-chennai/60423261

In other developed countries, the co-author of this paper, Professor Yang's from Loughborough University, UK, have R&D and implemented IoT enabled WSN called SafetyNet[2], and now it is in trial use. However, SafetyNet lacks CO_2 detection hence human occupancy detection capabilities.

In Singapore, there is a research funded by the Singapore's National Research Foundation through a research grant to the Berkeley Education Alliance for Research in Singapore (BEARS) for the Singapore-Berkeley Building Efficiency and Sustainability in the Tropics (SinBerBEST[3]) Program. They proved that counting human by CO_2 concentration is technically possible [1].

In Northern Europe and Northwestern part of Russia, the co-author of this paper, Dr. Korzun from Petrozavodsk State University, emphasized the role on data processing challenge in the Industrial Internet [2]. In particular, we face with the Big Data problem when large-scale WSN networks are used in surrounding physical environment.

3 Design of Internet-of-Breath

The IoB which is abbreviation for "Internet of Breath" is a research project of implementing a wireless sensor network with improved smoke sensor with CO_2 concentration measuring capability. It is designed to apply for (1) enhancing Fire and Rescue Services (FRS) in emergency, providing real-time situational information, analytics, prediction for critical decision making; and for (2) indoor human occupancy counting tool for business intelligence and other purposes. IoB could be installed on fixed locations such as buildings, air/sea ports, or on mobile objects such as firefighters, vehicles, and public transports etc.

---

[2] http://www.lboro.ac.uk/microsites/enterprise/safetynet/index.html.
[3] http://bayen.eecs.berkeley.edu/sites/default/files/journals/sensing_by_proxy.pdf.

3.1   Auto-sensing for Real-Time Dynamic On-site Environment Information

Running on low-cost battery power, IoB wireless gas sensors that are installed in buildings gather real-time information about temperature, humidity, existence of hazardous gas (for example, carbon monoxide, flammable gas, etc.), smoke intensity, flame intensity, CO_2 concentration etc. This information would be relayed through the network to a control centre. During FRS, the control centre base on the real-time information to decide and support front-line firefighters.

By sampling the air instantly and constantly, IoB can early detect a potential fire from the abnormal readings of environment variables (sudden rise of temperature, drop in humidity, initial emerging of thick smoke, flame, toxic gas) even before a fire is visually detected or felt by a human and called the emergency hotlines 999/911 etc. At present, fire alarms are triggered off only when a fire has already been burning at high intensity. Often, burning can be detected slightly earlier using IoB and information is monitored in real-time via a computer. We plan to conduct a series of tests in testing the sensitivity of the popular smoke detectors on the market and IoB sensors which take into account of a combination of various parameters, especially in scenarios where a small burning (that goes undetected) escalate into a serious fire.

With the information by IoB sensors available on the radar, fire control commanders can make better decisions: How severe is the fire? How and where and when did it arise and evolve? The proximity information? What are the readings from the nearby IoB sensors? Are there any chemical or hazard gas present? Currently on-site fire information is limited. The existing fire alarm system only raised an alarm, the air/fire detection and analysis would have to be done manually by the firefighters. Officers at the control room have to rely on their experience and decide with uncertainty. With IoB, the officers can better analyze the situation based on real-time information. (Just as a side-line: if any IoB sensor or part of the network segment goes offline that implies the fire is raging very seriously – they are swallowed up by fire).

As IoB continuously feeding in various environmental parameters, hopefully more efficient and effective decisions could be made at the beginning of the fire. During the FRS, officers who are working on-site or at the command centre can benefit from the dynamic timely information.

3.2   Challenges in Sensor Design and Network Connectivity

Transmitting fire site information reliably to incoming fire engines and to the control centre in real time is essential for speeding up the FRS operation. Earliness is very crucial in FRS. Data and system integration is a challenging part of IoB project, where the new data feeds and GUI would have to integrate smoothly into the existing ICT network currently used by Macau Fire department. For the sensors, robust wireless mote technology as well as tough encasing is needed for low-power consumption but capable in functioning in harsh conditions like extreme heat, explosion, while being able to maintain the communication link for as long as possible even under direct fire. For the network design, it is anticipated that the remaining network will still function though some parts of it is damaged in fire.

The co-author, Professor Yang's from SUST (Southern University of Science and Technology, Shenzhen) and adjunct professor from Loughborough University UK, has built a similar sensor for fire control, as shown below. The work to be done in the proposed project is to improve the sensor by in-cooperating CO_2 concentration probe in the sensing block; and test the whole sensor block together with cost-effective fire-proof casing under harsh conditions (Fig. 3).

Fig. 3. Gas sensor used in SafetyNET (Image courtesy by Professor S.H. Yang)

The IoB wireless network may be installed on new buildings to be built or added on existing buildings. Therefore no physical wiring cable would be needed. The sensing network would be designed and implemented in mesh network topology that supports communication protocols that are self-organizing, resilient and easy to deploy. It is required that even some sensors may be damaged, the rest of the sensors and the network still function for as long as possible, transferring the minute information to the center. We would have to test the coverage of each sensor, and simulate different scenarios where the network is partially connected for performance evaluation.

In the preliminary design, a self-healing ad-hoc network connecting hundreds of IoB sensors is used covering rooms to floors and to buildings in a city of size of Macau. The underlying communication by ZigBee is proposed that supports a robust wireless sensor network installed around a building at specific locations where motes are installed. Each mote is identified with a special code that maps to a pre-registered location prior to installation. In normal situations, only ZigBee is operating. However, in emergency situations, another layer of communication utilizing WiFi and Digital Audio Broad-casting (DAB) is turned on. This provides some communication channels among the fire-fighting crews, on-side or mobile in fire engines to communicate.

This middle communication layer also should work as an information uplink feeding real time info into some existing emergency control data network that is to be used by emergency personnel. A typical network architecture is shown in Fig. 4.

However, the scope of IoB is focused on the wireless sensor network layer, testing the coverage of ZigBee versus LoraLan which is recently emerged as a new WSN protocol, claiming to have coverage of kilometres. In this project, extensive testing will be conducted over the prototype that we build for the resilience of such WSN, and test the reliability and synchronization of the live data feeds from many gas sensors.

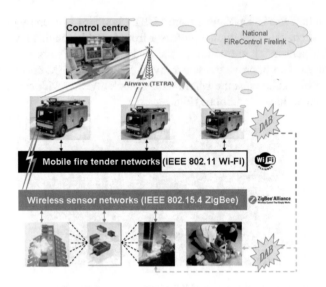

Fig. 4. Architecture of SafetyNET (Image courtesy by Professor S.H. Yang)

3.3 Integration of CO2 Probe in the IoB Sensor

A novel feature of this project is integrating CO2 probe in the sensor, thereby enabling human occupancy and counting. Currently many CO2 concentration probes are available on the market. Our proposed work is to test and find the most suitable CO2 probe in integrating into the IoB probe, in consideration of size, power consumption, sensitivity and durability. In the case of FRS, as shown in the two pictures below, with the information of human occupancy detection from the CO2 concentration, we can know from some data mining, whether and roughly how many people are trapped in certain rooms. So the task and the route of checking the rooms could be prioritized in the hope of saving more lives in time. This feature is especially useful for buildings with many rooms, such as large hotels where knowing human occupancy is important.

The following diagrams in Fig. 5 depict the current work done so far by co-PI, Professor Yang of SUST. The diagram beneath depicts about the extension of the WSN to IoB where CO2 sensors can detect human presence in addition to detecting fire.

Of course, when a fire broke out, the CO2 concentration would rise drastically, the readings after the start of the fire would not be so reliable. However, the human occupancy readings just before the start of fire should be taken for reference – so that firefighters would have some rough ideas of where the locations possibly would have had people presence. Extensive testing is needed for CO2 sensors simulating smoke environment to see how effective human occupancy detection and counting would still be in the presence of different thickness of smoke in a confined space. Also, different sizes of rooms should be tested per sensor, to learn about its maximum sensing capability.

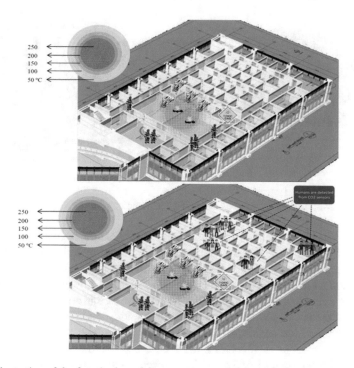

Fig. 5. Illustration of the functioning of the gas sensor and the new IoB sensor (Image courtesy by Professor S.H. Yang)

By the same logics and IoB sensors, the CO_2 concentration measuring probes can be installed on mobile objects such as public buses, boats, cruise ships and ferries, just for example. Extra real-time information can be provided telling not only where the buses are, but how many passengers approximately are inside (by using some algorithms to estimate the head counts from the CO_2 readings).

In this example, the real-time occupancy information of a public bus from IoB can integrate with the existing public bus App. Passengers can get the latest update on the approximate number of passengers in buses. The diagram is illustrating a scenario where an incoming bus is almost full. But the next bus that is coming behind is quite empty. So the user can decide to wait a little longer for the less crowded bus coming on the way, or to squeeze up the arriving bus (Fig. 6).

Of course, one challenge is the calibration of the CO_2 sensor. Lots of tests by trial-and-error in adjusting the sensitivity as well as developing some intelligent computer program algorithms are needed to generate the best accurate results. The other challenge is the communication link. Currently most public buses in Macau are equipped with Wi-Fi and GPS capability. The mobile CO_2 sensors and the whole IoB network needs to integrate and operate smoothly with the existing data communication facility at the buses. The sensors are easy to obtain; nevertheless, careful calibration of sensitivity and coding the estimation algorithms for inferring head counts from CO_2 concentration which fluctuates when the buses move, remain a challenge.

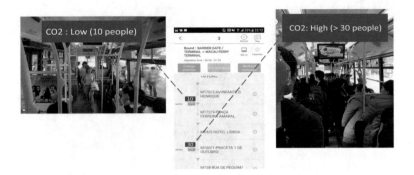

Fig. 6. Mobile App shows the head counts estimated by IoB sensor in a public bus

Overall, this project is mainly about designing applications, extending the currently working SafetyNET WSN that was successfully developed by a co-author of this paper, Professor Shuang-Hua Yang few years ago. A lot of testing and calibration, from the sensors to the ad-hoc WSN, are anticipated. When achieved, which is likely because it builds upon a previous successful similar project, the results are profound – it will be an important piece of the whole smart city strategy. Fire control and human occupancy detection using IoT are useful part of a smart city.

However, the scope of this IoB project is focused on the WSN layer, testing the coverage of ZigBee versus LoraLan which is recently emerged as a new WSN protocol, claiming to have coverage of kilometres. In this project, extensive testing will be conducted over the prototype that we build for the resilience of such WSN, and test the reliability and synchronization of the live data feeds from many gas sensors.

4   Considerations of Wireless Communications Technologies

IoB is a kind of ubiquitous but remote sensing system for measuring the CO_2 concentration as well as other air-quality readings in a confined area, for two objectives: one is to early detect fire, feeding real-time information about severity of fire and hazard of the current air quality in different locations to a central server for analysis; for another objective, together with suitable software and machine learning algorithm, it could be used for estimating head-counts in a room.

IoB requires installing remote air sensing devices physically at convenient places. Such motes are linked through gateways and to relay over to central servers. Recently, similar wireless monitoring systems are implemented [3]. Bluetooth and ZigBee are the main wireless transmission protocols used in the IoB wireless monitoring systems.

The development of remote IoB Monitoring Systems (MSs) through telecommunication networks has become a popular approach for WSN. Such systems facilitate remote monitoring of air quality by local devices that are equipped with ZigBee radios in Wireless Sensor Networks (WSNs) for large area monitoring [4] or as part of the Smart Home concept [5]. The air quality readings in the format of IoB signals are transmitted to a local hub and then sent to an IoB server that can be used for further analysis offering long-distance indoor air quality monitoring. New developments in mobile devices, air sensors, and wireless equipment further create application possibilities using the online MSs. Furthermore, MSs can provide automatic alerts to the users and emergency control centre based on the efficient IoB data processing, preset thresholds and logical rules at the analytics parts. When working hand in hand between the sensing parts and the analytics side, IoB can effectively detect anomalies early. Typically, the performance of ZigBee and Bluetooth are relatively similar. ZigBee is often found in environmental control systems and home monitoring systems. ZigBee is also currently in numerous applications including patients' monitoring in healthcare and other real-time smart home monitoring systems. While there are many gas sensor devices that use either Bluetooth or WiFi, an early plan is to first focus primarily on IoB monitoring into home control networks (with few rooms), and thus focusses on ZigBee based estate or town indoor air monitoring. Eventually we want to move to metropolitan scale indoor air monitoring using Lora wireless techniques that can cover up to 1.5 km radius. These three scales of WSNs would be tested, from small to large, in terms of feasibility that includes power consumption, sensitivity, accuracy and manageability.

Wireless communication enabled monitoring systems that offer light weight and ubiquitous sensing at low cost, ideal for huge mass market penetration. Recently, progression in air quality sensors, networking, and data processing is enabling new applications and services. Ubiquitous air monitoring is becoming emergent to support real-time detection in emergency control such as fire, un-known hazardous gas attack (by either natural cause or deliberate attack), and human head-counts.

4.1 Air Monitoring System Architecture

For air quality monitoring and human occupancy detection, remote sensing device that is installed on a room ceiling or as a portable device that can be attached to a wall or indoor decoration should consists of wireless enabled processor, such as ZigBee to transmit the IoB signal from the room. Typically, for any detected abnormalities, action is taken by giving an alert message to the user or some security authority. To enable communication out of the house or a building, to a remote server, 3G and 2G cellular, or an Internet connection can be used to assist the communication relay [6]. In general, a cloud based architecture is proposed and to be simulated in this project such as the one shown in the following diagram which has the general feature of IoT gateway. It has three domains, application domain, networking domain and sensing domain. Through the cloud service, data analytics and decision making can be made at the remote server. The data feeds are being delivered at the sensing later. The preliminary architectural design is shown in Fig. 7.

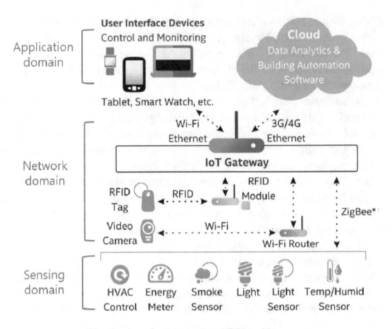

Fig. 7. A typical three-tiered IoT architecture.

4.2 IoB Signals and ZigBee-Based IoB Monitoring

Typically, the IoB sensors' characteristics can be mathematically modelled using similar equation that is found in [7]:

$$G(x_i, \eta_i) = s(x_i, \eta_i) \tag{1}$$

where G is the sensor's output, x_i is the sensing position, and η_i represents the representative parameter. Furthermore, one of the applied methods is to model the IoB signals, in the form of air quality readings is time-series. In temporal domain the time-series that are the pre-processed data feeds collected by IoB sensors could be modelled using the Fourier transform, as shown in the following equation.

$$AQ(t,f) = \frac{1}{2\pi} \left| \int e^{-j2\pi f z} b(\tau)\, f(\tau - t) d\tau \right|^2 \pi r^2 \tag{2}$$

where $AQ(t, f)$ is the output sensed signal, and $f(\tau - t)$ is the sliding window function along $b(\tau)$ that captures the data segment by segment. The sampling could be adjusted depending on how high the resolution of early fire detection or human head counting are needed. Each signal can be measured by each type of individual probe. It depends on the application settings. In case of fire control, there would be various probes that measure heat, lumen, humidity, CO2 as well as a collection of toxic gas indices. In this case, the signals could be a composite of individual data feeds that come from various air sensing probes from the proximity of the fire scene – direct at the fire site and also the nearby areas potentially would be propagated to. A special sensing device is needed to be designed and tested, by using and combing available probes on the market. The processed IoB output signal is then fed to the house or building hub via ZigBee for further processing. Possibly the signals may have to propagate through a hierarchical design of networks, like concentric rings of sensing circles. A typical hierarchy, from outer to inner, in the sense of data processing and analytics, should have levels consists of rooms, floors, buildings, estates, towns, cities, provinces, and country etc.

Typically, the foremost objective of the ZigBee network is then to be able the individual hub to have high coverage throughout a building, directly or via mesh networking. Therefore, the effectiveness of the ZigBee operation and the power characteristics models in the sensor battery would be one of the essential research tasks in this project.

ZigBee is a mature, popular and open technology that can handle low-power hence low-cost demands of the wireless networks using short-range radio communication. It is based on mainly on Radio Frequency (RF) so that applications built on it can have a long battery life. It is featured with secure networking, but data are transmitted at relatively low data rate. A feasible communication networking topology suitable for IoB could be ZigBee Mesh or similar. It supports high reliability and extensive data range, because data would have to be propagated, may be aggregated, pre-processed through multiple room and building hubs.

The ZigBee Alliance published the ZigBee specifications for similar IoT applications using wireless sensor networks; they provide control for low cost connectivity and low power consumption by the sensor devices that can be functioned on consumer-grade batteries. The ZigBee implementation at the IoB networks are designed according to the IEEE 802.15.4 standards [8]. Similar to any sensing device that is powered by home-use batteries, ZigBee could be configured to toggle between sleep and active modes in the communication. In the sensing nodes with ZigBee, using battery has benefits of, device operating flexibility, affordable, and scalable for deploying the sensors in large number throughout the hierarchical network, that may cover far and wide. A small amount of power is only consumed at the ZigBee devices during the reception or transmission of data. In IoB a very low duty cycle can be set by the IEEE 802.15.4 standard. Therefore, it is crucial that in the IoB networks, connected by ZigBee, the sensing device would have to be in sleep mode most of the time for assuring a long battery life. The sensors would have to be programmed that, only when sampled data becomes different from the previous samples, it turns into active mode for transiting the new data [9]. That is, when some exception happens in the air-quality, the anomaly is detected for perhaps at successive sampling cycles, data then would be sent. By this design the data analytics would have to assume that during long silent period, the sampled data do not change though in fact no data might have been received. Some intelligent logics would be needed to be programmed for this. Moreover, the IoB network should be configured that takes care of prioritizing the delivery of real-time signals according to the severity level of the detected air abnormal quality or normal conditions. In a common sense, in case of fire, or sudden surge of $CO2$ rate, the data needs to be prioritized for reaching the central data analytic centre as soon as possible.

In this way, we can estimate the power consumption of an IoB node using ZigBee to be an approximate function of the battery duty cycle, processing, reception windows, electronic components and systems of the network, battery energy of the ZigBee network. The estimated parts of battery energy consumption are simply stated in the following expression [10] as below:

$$P_C = P_t + P_r + P_s + P_{id},$$ (3)

where, P_C is the total power consumption, power consumption owing to the transmitted signal and the received signal are P_t and P_r respectively. During the sleeping mode, the power consumption is represented by P_s while P_{id} represents the power consumption due to the idle/sleep state at which zero packets are received/transmitted. The battery lifetime can be calculated in hours based on the following expression:

$$BT = L_c/L^n,$$ (4)

where, BT represents the battery life time in hours, the battery capacity is L_c in mAH, and the load current is L in mA, where, n is the Peukert's exponent ranging from 1 to 1.3. These simple expressions can be used to estimate the battery life time of the ZigBee device [11].

Employing soft computing techniques [12, 13] to develop an integrated ZigBee platform in IoB monitoring can be considered as a novel and active domain. This monitoring system infers patterns from different locations and time domain for suspicious signs which we should pay attention to [14]. This analytics platform can be supplemented by using a fuzzy logic based diagnosis system [15]. Research and development has created intelligent smart homes offering a reliable connection to the Internet and cloud services. Research and development has also created IoT based health monitoring systems enabled by sensors and ZigBee [16], but in this project, we are taking a level further because air-quality does fluctuate faster than health; they are more dynamic and sometimes quite difficult to detect and analyse due to many possible influential factors externally. For indoor air quality monitoring to be truly ubiquitous and cost-effective, this project calls for a number of hardware/software designs, implementations and tests to be considered as an integral part of the overall smart city architecture, rather than separate or as added feature. Such integration will maximize the performance, the reliability and security of the indoor air-quality monitoring system through the existing home hubs.

Tentatively, a number of research tasks are listed below, ranging from design, implementation, and testing intensively of the following for feasibility studies:

A. Different brands and models of probes/sensors which are to collect air quality samples for indoor monitoring. The challenge is that we need to find out which type of sensor is most appropriate for IoB sensor operating in different conditions.
B. Different communication devices, topology and protocols are to be modelled and simulated. The challenge is to find out which communication facility is most appropriate for integration, for coverage, for robustness and for use in emergency situations.

C. Different analytics including data processing, data mining, machine learning and real-time predictions would be tried. The challenge is to find the most suitable data analytics models and techniques for different application scenarios.

In addition to technical challenges mentioned above, user requirements are recommended to be conducted as early as possible as a part of the Smart City initiatives; relevant government department in-charges and may be the tentative end-users of the IoB system for smooth project execution and mutual understandings of the transparent project progress.

For future extension, when the IoB system is to be deployed as a real-life fully functioning IoT based smart city project in full scale covering most of a city, there would be a heterogeneous architecture and data sensitivity issues. Our research prototype could potentially grow into full scale where the ownership of the sensing data would be publicly owned by the government, as well as privately owned by organizations, e.g. hotels who want to keep the human occupancy data commercially confidential. However, certain sharing policy will need to be set up for emergency control, e.g. fire with the fire department, public safety and upon request by the police force enforcement when necessary. Such heterogeneous network could be made possible, as a future extension of the prototype that is to be built in this proposed IoB prototyping project. The heterogeneous network features about distributed data processing in a fog-computing environment which help prevent bottleneck of data processing/analysis at the central command centre. Data processing and may be some simple analytics should be delegated to the edge of the IoB network as much as possible for spreading out the data processing loads as well as controlling the data traffic flows.

5   Conclusion

In this paper, we proposed and studied the feasibility of a novel Internet-of-thing model called Internet of Breath (IoB): Integrative Indoor Gas Sensor Applications for Emergency Control and Occupancy Detection. A WSN architecture is adopted for designing and constructing a prototype similar to SafetyNET. The sensing layer and the network layer of SafetyNET would be very similar to IoB WSN. At the data layer, measurements of CO_2 concertation levels are sampled though they do fluctuate from time to time as human move in and out of the rooms. Special software program by machine learning should be coded for inferring the CO_2 readings to head counts based on the room characteristics and the environmental variables. The new intelligence harvested would be fed to the application layers enabling new smart city applications; the possibilities will be studied, modelled, simulated, and tested scientifically. At the same time, the robustness and efficacy of the new CO_2 sensing probes would be tested extensively with the consideration of operating under very harsh conditions such as extreme heat at fire scenes. A number of technical challenges are discussed, and further issues are anticipated, ranging from the bottom sensing layer, hardware, software, application designs, user acceptance testing, and user requirement studies. This paper serves as a blue-print and directions for future works aim at intensively modelling, implementing, fine-tuning this IoB model into a useful reality.

Acknowledgement. The authors are thankful to the financial support from the research grants, (1) MYRG2015-00128-FST, titled 'Temporal Data Stream Mining by Using Incrementally Optimized Very Fast Decision Forest (iOVFDF)' offered by RDAO/FST, University of Macau and Macau SAR government. (2) FDCT/126/2014/A3, titled 'A Scalable Data Stream Mining Methodology: Stream-based Holistic Analytics and Reasoning in Parallel' offered by FDCT of Macau SAR government. The work of D. Korzun is supported by the Ministry of Education and Science of Russia within project # 2.5124.2017/8.9 of the basic part of state research assignment for 2017–2019.

References

1. Jin, M., Bekiaris-Liberis, N., Weekly, K., Spanos, C., Bayen, A.: Sensing by proxy: occupancy detection based on indoor CO2 concentration. In: The Ninth International Conference on Mobile Ubiquitous Computing, Systems, Services and Technologies, UBICOMM 2015, 24 July 2015, Nice, France (2015)
2. Gurtov, A., Liyanage, M., Korzun, D.: Secure communication and data processing challenges in the industrial internet. BJMC **4**(4), 1058–1073 (2016)
3. Ebrahim, N., Deen, M.J., Mondal, T.: A wireless wearable IoB sensor for long-term applications. IEEE Commun. Mag. **50**(1), 36–43 (2012)
4. Cano-Garcia, J.M., Gonzalez-Parada, E., Alarcon-Collantes, V., Casilari-Perez, E.: A PDA-based portable wireless IoB monitor for medical personal area networks. In: Proceedings of MELECON, pp. 713–716 (2006)
5. Hui, T.K.L., Sherratt, R.S., Diaz Sanchez, D.: Major requirements for building smart homes in smart cities based on Internet of Things technologies. Future Gener. Comput. Syst. **76**, 358–369 (2017)
6. Pantelopoulos, A., Bourbakis, N.G.: A survey on wearable sensor-based systems for health monitoring and prognosis. IEEE Trans. Syst. Man, Cybern. B. **40**(1), 1–12 (2010)
7. Sarma Dhulipala, V.R., Kanagachidambaresan, G.R.: Cardiac care assistance using self configured sensor network—a remote patient monitoring system. J. Inst. Eng. India Ser. B. **95**(2), 101–106 (2014)
8. Sinem, C.E., Qadri, S.F., Awan, S.A., Amjad, M., Anwar, M., Shehzad, S.: Applications, challenges, security of wireless body area networks (WBANs) and functionality of IEEE 802.15.4/ZIGBEE. Sci. Int. (Lahore) **25**(4), 697–702 (2013)
9. Qadri, S.F., Awan, S.A., Amjad, M., Anwar, M., Shehzad, S.: Applications, challenges, security of wireless body area networks (WBANs) and functionality of IEEE 802.15.4/ZIGBEE. Sci. Int. (Lahore) **25**(4), 697–702 (2013)
10. Hussein, A., Samara, G.: Mathematical modeling and analysis of zigbee node battery characteristics and operation. MAGNT Res. Rep. **3**(6), 99–106 (2015)
11. Srinivasan, R., Turker, D.Z., Park, S.W., Sánchez-Sinencio, E.: A low-power frequency synthesizer with quadrature signal generation for 2.4 GHz zigbee transceiver applications. In: Proceedings of ISCAS 2007, pp. 429–432 (2007)
12. Kuo, W.-H., Chen, Y.-S., Jen, G.-T., Lu, T.-W.: An intelligent positioning approach: RSSI-based indoor and outdoor localization scheme in ZigBee networks. In: Proceedings of International Conference on Machine Learning and Cybernetics (ICMLC), pp. 2754–2759 (2010)
13. Pinto, A.: Machine learning methods and technologies for ubiquitous computing. In: Proceedings of AI*IA 2012 Doctoral Consortium, pp. 38–42 (2012)

14. Fong, S., Yan, Z.: A security model for detecting suspicious patterns in physical environment. In: Third International Symposium on Information Assurance and Security, pp. 221–226 (2007)
15. Jiang, M., Cui, P., Faloutsos, C.: Suspicious behavior detection: current trends and future directions. IEEE Intell. Syst. **31**(1), 31–39 (2016)
16. Talpur, M.S.H., Liu, W., Wang, G.: A ZigBee-based elderly health monitoring system: design and implementation. Int. J. Auton. Adaptive Commun. Syst. **7**(4), 393–411 (2014)

Implementation of an Hierarchical Hybrid Intrusion Detection Mechanism in Wireless Sensor Network Based on Energy Management

Lamyaa Moulad[1]($\boxtimes$), Hicham Belhadaoui[2], and Mounir Rifi[2]

[1] ENSEM/EST, University Hassan II, Casablanca, Morocco
Lamyaa.moulad@gmail.com
[2] EST, University Hassan II, Casablanca, Morocco

Abstract. In the last few years, Wireless Sensor Networks (WSN) have attracted considerable attention within the scientific community. The applications based on Wireless Sensor Networks, whose areas include agriculture, military, hospitality management… etc, are growing swiftly. Yet, they are vulnerable to various security threats like Denial Of Service (DOS) attacks. Such issues can affect and absolutely degrade the performances and cause a dysfunction of the network and its components.

However, key management, authentication and secure routing protocols aren't able to offer the required security for WSNs. In fact, all they can offer is a first line of defense especially against outside attacks. Therefore, the implementation of a second line of defense, which is the Intrusion Detection System (IDS), is deemed necessary as part of an integrated approach, to secure the network against malicious and abnormal behaviors of intruders, hence the goal of this paper. This allows to improve security and protect all resources related to a WSN.

Different detection methods have been proposed in recent years for the development of intrusion detection system, In this regard, we propose an integral mechanism which is in fact a hybrid Intrusion Detection approach based Anomaly, Detection using support vector machine (SVM), specifications based technique and clustering algorithm to decrease the consumption of resources, by reducing the amount of information forwarded. So, our aim is to protect WSN, without disturbing networks' performances through a good management of their resources, especially energy.

Keywords: WSN · IDS · Misuse detection · Anomalies
Specification-based detection · DOS attacks · Hybrid intrusion detection system
Support vector machine (SVM) · False alarm · Detection rate

1 Introduction

Sensors nodes are low power electronic devices, that cooperate to form a network called wireless sensor network (WSN), often deployed in hostile areas, difficult to access. They are equipped with small batteries with limited energy which makes it very expensive and difficult to replace or charge these sensors' batteries.

© Springer International Publishing AG, part of Springer Nature 2019
J. Mizera-Pietraszko et al. (Eds.): RTIS 2017, AISC 756, pp. 360–377, 2019.
https://doi.org/10.1007/978-3-319-91337-7_33

Recently, the demand of wireless sensor networks (WSN) [1–3] have become a promising future to many new real applications, where data is communicated insecurely to critical destination, such as emergency evacuations security, health monitoring, soldiers in battlefield, biometric application in airport, etc.. Thus, WSN are exposed to various malicious attacks, which can generate an overconsumption of energy. Therefore, controlling energy consumption is important to secure a WSN, which means that during the implementation, communication protocols dedicated to WSNs must consider the level of power consumption to provide optimal management of this vital resource.

The goal of this work is to implement an integral mechanism, a new hybrid intrusion detection system [4] for WSN using the clustering algorithm, to reduce the information forwarded and decrease the consumption of resources, especially energy. In general we have combined two main techniques, anomaly-based detection, that class data into normal and abnormal (binary classification), to detect malicious behaviors. We have also, applied misuse detection technique called also (signature) to determine known attack patterns, specifications based technique, and other techniques. Therefore, the combination of those techniques, the benefit from the advantages of the two detection techniques, can absolutely offer a high detection rate and low false positive. This mechanism can make a better decision in order to detect new kinds of intrusions.

The paper is organized as follows: In Sect. 2, we provide a background information about IDS in WSNs and related works. Section 3 elaborates on the proposed scheme and architecture of our proposed Hybrid Intrusion Detection System. Section 4 contains The simulation results with analysis of the proposed scheme are discussed. In Sect. 5, We conclude our work with a further discussion of research directions

1.1 Background of Ids Security in WSNs

This paper examines one of the most important axes of Wireless Sensor Networks, which is security and particularly Intrusion Detection Systems (IDS) [14]. As already stated, Intrusion detection systems are defined as the second lines of defense; However, Key management and authentication represent just a first line of defense against just external attacks. Therefore, IDSs, allows detection and prevention from both internal and external attacks and all kinds of intrusions (Intrusion is defined as an unauthorized activity in a network.) (Fig. 1).

Fig. 1. Intrusion detection architecture

Each IDS [27] contains 3 modules:

(a) Data Collection modules: collect the information sent, received and forwarded by the sensors.
(b) Intrusion detection module: it depends on the intrusion detection technique used (Signature, Anomaly or Specification-based detection), IDS agent sends an alarm message mentioning the suspect node, to all network.
(c) Intrusion detection module: In case of abnormal behavior the ids send an alarm to the rest of components, and remove the intruder.

IDSs are classified into 3 main techniques: Anomaly based, Signature based and Specification-based detection (Fig. 2).

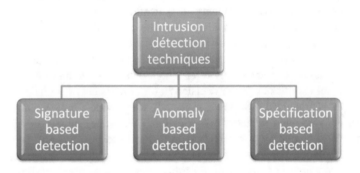

Fig. 2. Intrusion detection techniques

Misuse detection (Signature): Misuse detection based IDS have a predefined set of rules that are designed on the basis of previously known security attacks, so the behavior of nodes is compared with well-known attack patterns already existing in database. The disadvantages are that this technique needs knowledge of attacks' patterns and can't detect new attacks, so we always have to update attack signatures database.

Anomaly detection: this technique works on the basis of threshold, it compares the behavior of observed nodes with normal behavior. This model first describes normal behaviors which are established by automated training (as SVM..) and then flags as intrusions any activities varying from these behaviors. it has the ability to detect new intrusions, but, it has a major disadvantage of missing out on well known attacks. The anomaly based model has a high detection rate, but it has also a high false positive rate.

Specification-based detection: This model is based on deviations from normal behaviors which are defined by neither machine learning techniques not by training data. Yet, specifications are defined manually and describe what normal behavior is and monitor any action with respect to these specifications.

However, to improve the level of detection, we can use another solution called the hybrid Intrusion Detection model. Which is a combination of detection techniques already mentioned. Therefore, this combination allows the system to benefit from theirs advantages. This mechanism can make a better decision, which might detect new kinds of intrusions with higher detection rate, and lower false alarm.

2 Related Works

In previous works, and as we consider proposing hybrid HIDS system, there are some proposed hybrid schemes integrated for clustered sensor networks.

In [16] a detection system is proposed for WSN. To get an hybrid model, the combined version of Cluster-based and Rule-based intrusion detection techniques is used and eventually evaluated the performance of intrusion detection using hybrid technique and detection graph shows ratings like attack rating, data rating and detection net rating with the attack name and performs better in terms of energy, but the model proposed still weak and it can not detect new intrusions.

In [15], Su et al. proposed energy efficient hybrid intrusion prohibition system for CWSNs. They use intrusion detection and intrusion prevention techniques to get an hybrid security system. Their system contains collaboration-based intrusion detection subsystem which uses cluster head monitoring and member node monitoring. In this scheme, member nodes monitor the cluster heads and the cluster heads monitor their own cluster members by using alarm table and HMAC. This scheme can detects the intruder in case of member nodes are monitors, but when cluster nodes are monitors, the scheme fail because of using the only shared key between cluster head and member node.

Abduvaliyev et al. [14, 25] proposed a hybrid IDS (HIDS) based on both anomaly and misuse detection techniques in a cluster WSN (CWSN) topology. The results showed that the proposed scheme allows a high detection rate with low level of energy consumption. However, this model does not detect most network attacks.

Yan et al. proposed hierarchical IDS (CHIDS) based on clusters. The authors took advantage of this approach and install on each cluster-head an IDS agent (core defense). This agent contains three modules: a supervised learning module, an anomaly detection module based on the rules and decision-making module. The simulation results showed that this model has a high detection rate and lower false positive rate. But, his main disadvantages of this scheme is: The IDS node is static (runs only in the cluster-head), in this case the intruder uses all his strength to attack this hot element and subsequently disrupts the network. The implementation of this detection mechanism requires many calculations in cluster-heads, and that can decrease the network lifetime.

Hai et al. 4 proposed a hybrid, lightweight intrusion detection system integrated for sensor networks (SN), using the scheme of Roman et al. [5]. Intrusion detection scheme takes advantage of cluster-based protocol to form a hierarchical network (HN) to give an intrusion framework based on anomaly and misuse techniques. In their proposition, IDS agent consists of two detection modules, local agent and global agent. The authors apply their model in a process of cooperation between the two agents to detect attacks with greater accuracy (both agents are in the same node). The disadvantage of this scheme is the sharp increase in signatures, which can lead to an overload of the node memory.

In recent work, Coppolino et al. [6] presented a hybrid, lightweight, distributed IDS (HDIDS) for WSN This IDS uses both misuse-based and anomaly-based detection techniques. It is composed of a Central Agent (CA), which performs highly accurate intrusion detection by using data mining techniques, and a number of Local Agents (LA) running lighter anomaly-based detection techniques on the motes.

Sedjelmaci et al. implemented a lightweight Framework for securing WSN that combines the advantages of cryptography and IDS technology in order to detect the most dangerous network attacks, and provide a trust environment based on clusters. The results show that the model performs well in terms of detection rate, and generates high overhead and energy consumption.

Yassine Maleh et al. implemented a hybrid, lightweight intrusion detection model integrated for sensor networks, intrusion using cluster-based architecture. This model uses anomaly detection based on support vector machine (SVM) algorithm and some of signature rules. the proposed hybrid model give efficiency in terms of detecting attacks and false positives rates compared to previous schemes, however the charge of CH can cause an early dysfunction of this element.

3   Proposed Hybrid IDS

The proposed model contains specification based technique, signatures based technique using some fixed rules representing most dangerous attacks in Wireless Sensor Network, and anomaly detection based on SVM technique, which is designed to confirm

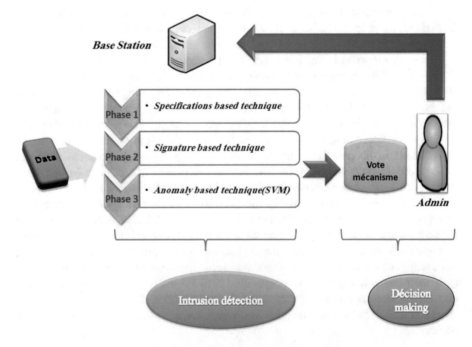

Fig. 3. Architecture of proposed hybrid IDS.

the malicious behavior of a target identified by behavior detection technique, and analyze data for classification (Fig. 3).

3.1 Intrusion Detection Used Techniques

Phase1: Behavior Based Detection (Specification-Based)

This technique adopts the same principle as the detection based anomalies that, any deviation of normal behavior is considered as intrusion. This technique fit a statistical model (usually normal behavior) to the data provided. Then, It applies a statistical inference test to determine if an instance belongs to this model or not. The bodies that have a low probability of being generated from the learned model are reported as anomalies.

However, the definition of the behavior model is performed in a manual way and not automatically using a learning algorithm, because it uses thresholds defined by the user to identify areas of abnormal data. It is similar to a Non parametric learning (statistical) the techniques that offer greater flexibility with respect to parametric learning techniques because they require no prior knowledge of the data distribution. This simplifies the detection system, and significantly reduces the rate of false negative detections. Compared to the detection based on anomalies, this technique seems to be best suited to the limitations of sensor networks.

Phase2: Anomaly Detection Using SVM

In this section a description of SVM and feature selection are presented:

Support Vector Machines

Support vector machines are a set of supervised learning techniques used for classification of network behavior. The aim of SVM classifier is to determine a set of vectors called support vectors to construct a hyperplane in the feature spaces. In our context, a distributed binary classifier to normal and abnormal, which permits detection of every malicious act.

$$\sum_{i=1}^{n} \alpha_i y_i x_i$$
$$\min\left\{ \frac{\|w\|^2}{2} + C \sum_{i=1}^{n} \varepsilon_i \right\} \tag{1}$$

$\sum_{i=1}^{n} \varepsilon_i$ is the constraints on the learning vectors, and C is a constant that controls the trade off between number of misclassifications and the margin maximization (Fig. 4).

The Eq. (1) can be deal by using the Lagrange multiplier [17]:

Classification hyperplane Given the training datasets,

$$(x_i, y_i) \ i = 1, \dots n \ y_i \ \varepsilon \ \{-1, +1\}, x_i \ \varepsilon \ R^d$$

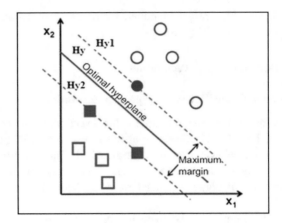

Fig. 4. Hyperplane

We want to find the hyperplane that have a maximum margin:

$$W.x = b$$

Where w is a normal vector and the parameter b is offset. In order to find the optimal hyperplane, we must solve the following convex optimization problem:

$$maximise \ l(\alpha) = \sum_{i=1}^{n} \alpha_i - \frac{1}{2}\sum_{i=1}^{n}\sum_{j=1}^{n} \alpha_i\alpha_j y_i y_j k(x_j, x_i)$$

$$subject \ to \sum_{i=1}^{n} y_i\alpha_i = 0, \ and \ 0 \leq \alpha_i \leq C \ for \ all \ 1 \leq i \leq n \tag{2}$$

$K(x_j,x_i)$ is the kernel function and α_i are the Lagrange multipliers. Referring to the condition of Kuhn-Tucker (KKT), the x_is that corresponding to $\alpha_i > 0$ are called support vectors (SVs).

Once the solution to Eq. (2) is found, we get [17]:

$$y_i(w.x_i + b) \geq 1 - \varepsilon_i, \varepsilon_i \geq 0, 1 \leq i \leq n \tag{3}$$

Thus the decision function is written as:

$$f(x, a, b) = \{\pm 1\} = sgn\left(\sum_{i=1}^{n} y_i\alpha_i k(X_j X_i) + b\right) \tag{4}$$

SVM is more suitable for intrusion detection in case where new signature is detected. Also, SVM, provide low false positive, and satisfied results with low training time compared to neural networks [18].

Phase 3: Misuse Based Detection (Signature)

Misuse or signature based detection is used to prevent network against malicious behavior using a set of rules. There is five main rules for each attack, rule to detect an excessive demand of energy (E(d)>E, The rule to detect the Selective forwarding attack, represented by the number of packets dropped (PDR). The rule to detect the Hello flood attack is the received signal strength (ISSR) at the IDS agent, The rule to detect the Black hole attack is defined by the number of RDP (greater than threshold δissrbh). Finally, the rule to detect the wormholes attack is the power signal (above the threshold δissrhwh).

Phase 4: Cooperative Decision Making Approach (Voting Mechanism)

In this approach, each node participates in the detection and management of intrusion decision.

The goal of the decision making model is to analyze the results of all detection techniques used which are the behavior's specification, anomaly and misuse detection models and validate when an intrusion occurs or not. Then, it reports the results to the administrator of network, to help them handle the state of the system, update the database of signatures, make further countermeasures, and prevent the system by sending an alarm if an intrusion occurs.

3.2   Network Structure and IDS Agents Location Process

(A) Structure of the network:

As mentioned before, the detection approach uses cluster-based topology to decrease the quantity of packets forwarded through the network while increasing the network lifetime. by designating a leader of the group called cluster-head (CH) - via a cluster election - that collect data received from member nodes to prepare it for the mobile sink (MS) use, then and while moving trough CHs, the MS aggregate data (collected by CHs), instead of sending it to the base station (BS), in order to reduce the charge and also support the CH (Fig. 5).

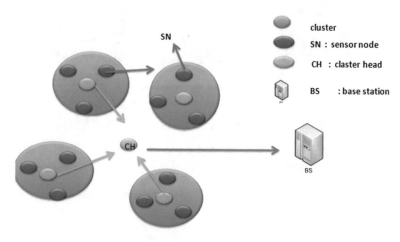

Fig. 5. Network Structure

The base station starts the process of CH election, CHs calculate residual energy using the equation $V_i(t) = [Initial - E_i(t)]/r$, where Initial is the initial energy, r is the current round of CH selection and $E_i(t)$ is the residual energy. According to obtained value, Base station calculates the average value and average deviation. Then CH is elected dynamically according to his residual energy.

CH starts the CH election procedure for nodes. Old CH broadcasts a message about the withdrawal of authority. New CH sends alert messages to the member nodes. CH is responsible for authentication of the other members of the cluster, and the base station is responsible for CH authentication. Because of limited battery life and resources, each agent is only active when needed [24].

(B) IDS location process

In this proposed scheme, an IDS [26] agent is located in every sensor node. Each cluster contains two kinds of agents: local IDS agent and global IDS agent. Because of the limited battery life and resources, each agent is only active when needed, To avoid the above issues, we place a sensor node called mobile sink which act as an intermediate between the cluster-head and the base station. The mobile sink (MS) is kept in moving state so that the intruder may not find the location of the node easily. The proposed cluster-based wireless sensor networks topology is shown in the (Fig. 6). The MS gathers the data from each of the cluster-head when it moves near to the corresponding clusters. The mobile sink reduces the work load of the cluster-head. When the cluster-head transmits the data to the mobile sink, the energy of the cluster-head is reduced [11, 12].

Figure 6 below describes the process of IDS agents location in network.

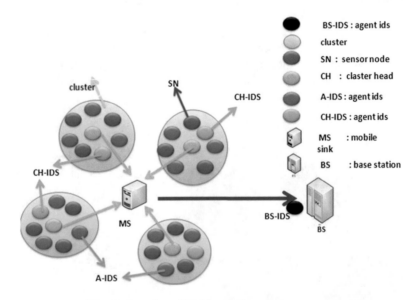

Fig. 6. Location of IDS in wireless sensor network

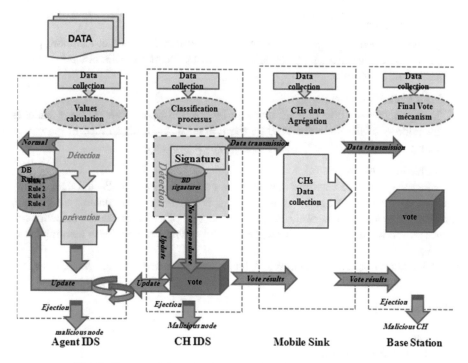

Fig. 7. Process of detection between WSN IDS agents components

In this hybrid IDS architecture, and by using hierarchical architecture, Our aim is to utilize cluster-based protocols in energy saving, to reduce computational resources and data transmission redundancy. In this context, we propose an intrusion framework based information sharing (Fig. 7).

- Intrusion detection at Member nodes:

Data Collection modules, and intrusion detection are in general, the principal components in this type of agent.

1. Data Collection Module:
   Is responsible to collect the information sent, received and forwarded by sensor. This node stores in his database id of the node analyzed and compute values of some parameters such Energy, NPD, NPS, RSSI, NRM, JITTER... in every node.
2. Intrusion Detection Module:
   This module apply a mechanism that the cluster have a special behavior, so any deviation of the normal values fixed for parameters mentioned, represent an abnormally that need to be fixed immediately, by alarming CH of the cluster. This IDS can supervise even the CH when needed.

- Intrusion detection at CHs:

Proposed clustering algorithm chose for each cluster, the CH that has more power resources to manage and aggregate data from cluster members. This powerful node is composed of 3 modules:

1. Data Collection Module:
   Is responsible of collecting packets sent by the IDS agent. This message includes the address of the node analyzed by the IDS agent then, transmitted to the abnormality detection module for intrusion detection process.

   Behavior classifier:
   Then the Behavior classifier classifies the node behavior of collected data already transmitted by the ids agent, as trustworthy if no match with database signature, attacker if rule signature is confirmed, and suspect if not an attack but the behavior still shows an abnormality in this case we need to apply detection module for learning based on SVM.
   After computation and analysis of the values collected and the fixed rules, the behaviour is classified into:

Classification {

If (packet is Normal)
{ Launch of voting process }

Elseif (packet matches a signature)
        {Declare the intruder node with exclusion and   classification of the attack)
}

Else   { (calculate SVM)
Launching voting processes}

}

2. Intrusion Detection Module: (Signature + SVM)

This kind of IDS uses discovery protocol based on the fixed rules signatures representing most dangerous attacks in Wireless Sensor Network (Sect. 3. phase 3), then transmitted to the abnormality detection module for learning and classification process.

3. Voting mechanism:

Regarding collaborative process, the cluster-head uses the voting mechanism. if there is no correspondence between the intrusion detected by predefined signatures attackers and the anomaly detection, IDS agent sends a message to the CH, this one use voting to make a final sure decision on the suspect node. If more than ½ of IDS nodes located in the same cluster voted for malicious suspected target, the CH rejects that node and calculates the appropriate rule of this new intrusion detected. CH sends an update message to all IDSs that are in the same cluster and CHs neighbors. This message contains the ID of the malicious node and this new rule (and signatures). When IDS agent receives this message it is an update of its signature table.

Mobile sink:

Each mobile sink gathers the data from each of the cluster-head in the same radio coverage area when it moves near to the corresponding clusters to reduce the work load

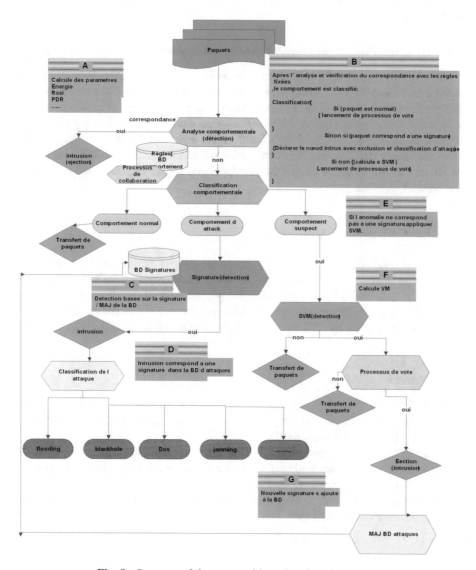

Fig. 8. Structure of the proposed intrusion detection model

of the cluster-head. When the cluster-head transmits the data to the mobile sink, the energy of the cluster-head is reduced, this information will be transmitted to the base station for a monitoring process.

- Intrusion detection at Base station:

The CH monitoring sends to the base station a report of intrusion, includes the CH suspect, if exist, and the type of attack detected. The base station performs a polling mechanism to identify malicious nodes. In the case where more than ½ of the votes are in favor of the attack, the CH is excluded from the sensors network and a new CH is elected (Fig. 8).

3.3   Dynamic Process for Intrusion Detection System

In the suggested approach, if (1/2) of IDS nodes within the cluster have consumed more than 25%, 50% and 75% (in tree level) of their energy; new IDSs are elected and receive the actual set of intrusion signature from the cluster head. The older ones are designated as ordinary. Then new IDSs election depends on the residual energy and the placement strategies suggested by Khalil et al. new IDS nodes are elected, they compute locally the SVN and the distributed algorithm for training SVMs is performed as alluded above. This can protect the network from energy depletion and prolonging the network lifetime.

4   Experimental Evaluation

To evaluate the performance of the proposed hybrid IDSs. we have used the KDDcup'99 dataset [10] as the sample to verify the efficient of the hybrid detection mechanism and valid it by compare with one proposed by Abduvaliyev et al. [14] and Su and Chang [15]. [13] according to the false positive rate (false alarm), detection rate and energy generated by IDS agents, in order to determine the effectiveness of our scheme.

4.1   Dataset

The KDD 99 intrusion detection dataset is developed by MIT Lincoln Lab in 1998, each connection in the dataset has 41 features and it's categorized into five classes: normal and four attack behaviors (Dos, Probe, U2r, R2 l).

Our analysis is performed on the "KDD" intrusion detection benchmark by using its samples as training and testing dataset. We focus on all categories of attacks and specially Dos attacks, which are defined as anomalies behavior.

The training data used at each IDS comprises of 50 normal and 50 anomalous samples include Dos attacks [17].

To determine the effectiveness of our proposed hybrid intrusion detection system we tried to analyze some important metrics, which are: detection rate (DR), the false positive rate (FP) and energy, according to the formulas:

$$Detection\ Rate = \frac{Number\ of\ detected\ attacks}{Number\ of\ attacks} \times 100\%$$

$$False\ Positive\ Rate = \frac{Number\ of\ misclassified\ connections}{Number\ of\ normal\ connections} \times 100\%$$

Total energy consumption $E_t = E_A + E_M$

1 - Detection Rate: is the percentage of attacks detected on the total number of attacks;
2 - False positive rate (false alarms): is the ratio between the number classified as an anomaly on the total number of normal connections;
3 - Total energy consumption: it calculate the total amount of energy consumed in all nodes in the network.

4.2 Simulation Results

The sensor nodes are deployed in a randomized grid fashion, The network is composed of 10 clusters that contains 1–7 nodes over all the nodes are static. distributed in a field of 100×100. An interference model for radio simulations. The rest of the specifications of a sensor node for detection module are defined in the Table 1 below (Fig. 9).

Table 1. Simulation parameters

Parameter	Value
Simulation time	900 s
Simulation area	100 *100 m
Number of nodes	100
Radio Model	Lossy
Number of cluster	10
IDS agents per cluster	1–7
Routing Protocol	HEED modifier
MAC	TDMA
Radio range	20 m
Initial energy	5 J
Power consumption for transmission	1.6 W
Power consumption for reception	1.2 W
Power consumption in idle state	1.15 W

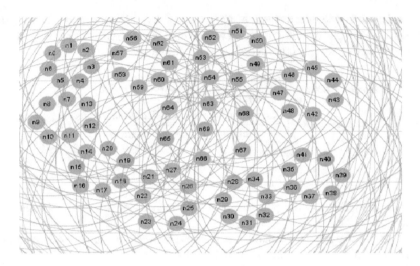

Fig. 9. Senario of 10 clusters

- *Detection rate:*

The proposed scheme in Fig. 10, is effective when the number of member nodes are increased. In addition, the probability of a missed detection affects the efficiency of our scheme. However, the proposed model performs better in term of detection rate, exceeding over 98.5% comparing to schemes proposed by Abduvaliyev et al., W.T. Su and K.M. Chang.

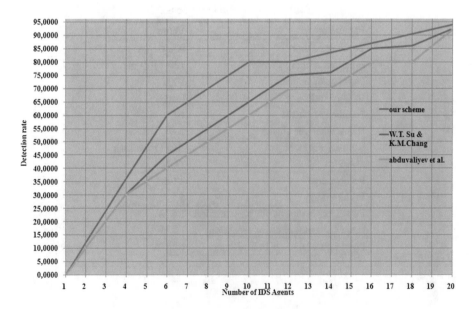

Fig. 10. Detection rate

• *false positive rate*

The probability of false positive detection is shown in Fig. 11. It indicates that the increasing number of nodes results in an increase in the probability of a collision. So, Fig. 11 shows a low false alarms (1.8%) and a short detection time, compared to the scheme proposed by Abduvaliyev et al. and W.T. Su, K.M. Chang.

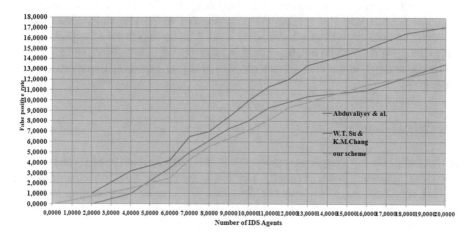

Fig. 11. False positive rate

• **Energy Consumption**

Figure (12) illustrates the total amount of energy consumed in the network. It is clear that our model is the less energy consuming scheme comparing to the other schemes proposed by Abduvaliyev et al. and W.T. Su, K.M. Chang.

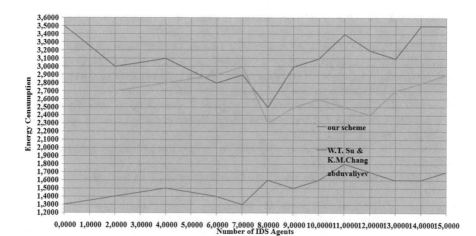

Fig. 12. Energy Consumption (j)

Detection and false positive rates were respectively of the order of 98.5% and 1.8%. As shown in Figs. (10) and (11) the two diagrams show a high detection rate with low false alarms and a short detection time, compared to the scheme proposed in the reference.

Furthermore, our detection model requires less energy to detect these attacks, compared to the approach used by the authors mentioned. This improvement was achieved through our use of a cluster-based topology that aims to select a single node in a cluster (cluster-head) to transmit data aggregated at Mobile sink, which allows grouping packets from cluster-heads, then send it to the base station, especially that each IDS agent is based on a policy that minimizes packet transmission, which, in turn, will save energy. In conclusion, we can say that our approach improves network lifetime.

5   Conclusion

In this paper, we have implemented a security mechanism which is a hybrid Intrusion Detection approach based Anomaly Detection, based on support vector machine (SVM), specifications, and the Misuse Detection WSN, using the clustering algorithm to decrease the consumption of resources specially the energy by reducing the amount of information forwarded, so, our aim was to a safe WSN without damaging the network, by the good management of resources specially the energy. All results show that all attacks are detected with low false alarm and high detection rate.

As the future research directions, we will analyze, evaluate and implement our model with various attacks in a real environment; also a soft hybrid model will be proposed and compared to this present model.

References

1. Houngbadji, T.: Réseaux ad hoc: système d'adressage et méthodes d'accessibilité aux donnée, Thesis 2009, école polytechnique de Montreal (2009)
2. Akyildiz, I.F., et al.: Wireless sensor networks: a survey. Comput. Netw. **38**, 393–422 (2002)
3. Karl, H., Willig, A.: A short survey of wireless sensor networks. IJCT (2004)
4. Strikos, A.A.: A full approach for intrusion detection in wireless sensor networks. School of Information and Communication Technology, March 2007
5. Mitchell, R., Chen, I.-R.: Department of Computer Science, Virginia Tech, Falls Church, VA 20191, United States 'A survey of intrusion detection in wireless network applications'. Comput. Commun. **42**, 1–23 (2014)
6. Masri, W.: Dérivation d'exigences de Qualité de Service dans les Réseaux de Capteurs Sans Fil sur TDMA, Thesis (2009)
7. Haboub, R., Ouzzif, M.: Secure and reliable routing in mobile Ad hoc networks. (IJCSES) (2012)
8. Moulad, L., Belhadaoui, H., Rifi, M.: Estc/Ensem UH2C Implementation of a security mechanism of WSN based on energy management. IJEAT (2013)

9. Prasanna Venkatesan, T.: An effective intrusion detection system for manets. Int. J. Comput. Appl. (IJCA) (0975–8887). International Conference on Advances in Computer Engineering and Applications (ICACEA-2014) at IMSEC, GZB

10. Sedjelmaci, H., Feham, M.: Novel hybrid intrusion detection system for clustered wireless sensor network. (IJNSA) **3**(4), July 2011

11. Huh, E.-N., Hai, T.H.: Lightweight intrusion detection for wireless sensor networks, Thesis (2009)

12. Maleh, Y., Iaeng, M., Ezzati, A.: Lightweight intrusion detection scheme for wireless sensor networks. IAENG, IJCS

13. Madhumathi, C.S.: Efficient cluster head selection and mobile sinks for cluster-based wireless sensor networks. Int. J. Sci. Eng. Res. (IJSER)

14. Abduvaliyev, A., Lee, S., Lee, Y.-K.: Energy efficient hybrid intrusion detection system for wireless sensor networks. In: 2010 International Conference on Electronics and Information Engineering, ICEIE 2010, Department of Computer Engineering, Kyung Hee University, Suwon, Korea (2010)

15. Su, W.T., Chang, K.M., Kuo, Y.H.: eHIP: an energy efficient hybrid intrusion prohibition system for cluster-based wireless sensor network. J. Comput. Netw. **51**, 1151–1168 (2007)

16. Deshmukh, R.: An intrusion detection using hybrid technique in cluster based wireless sensor network. J. Eng. Res. Appl. (IJERA) **3**(4), 2153–2161 (2013). ISSN: 2248–9622

17. Yan, K.Q., Wang, S.C., Wang, S.S., Liu, C.W.: Hybrid intrusion detection system for enhancing the security of a cluster-based wireless sensor network. In: Proceedings of 3rd IEEE International Conference on Computer Science and Information Technology, China, pp. 114–118 (2010)

18. KDD Cup 1999 Data (1999). http://kdd.ics.uci.edu/databases/kddcup99/task.html

19. Nurtanio, I., Astuti, E.R., Purnama, I.K., Hariadi, M.: Classifying cyst and tumor lesion using support vector machine based on dental panoramic images texture features. IAENG Int. J. Comput. Sci. **40**(1), 29–37 (2013)

20. Yuan, L., Parker, L.E.: Intruder detection using a wireless sensor network with an intelligent mobile robot response. IEEE Southeastcon **1**, 37–42 (2008)

21. Patel, M., Aggrwal, A.: Security attacks in wireless sensor networks: a survey. In: International Conference on Intelligent Systems and Signal Processing, March 2013

22. Meena Kowshalya, A., Sukanya, A.: Cluster in algorithms for heterogeneous wireless sensor networks - a brief survey. Int. J. Ad Hoc Sensor Ubiquitous Comput. **2**(3), 57–69 (2011)

23. Hai, H., Khan, F., Huh, E.: Hybrid intrusion detection system for wireless sensor networks. LNCS, vol. 4706, pp. 383–396, August 2007

24. Maleh, Y., Ezzati, A.: Contributions to Security in Wireless Sensor Networks and Constrained Networks in Internet of Things, Thesis (2017)

25. Abduvaliyev, A., Pathan, A.K., Zhou, J., Roman, R., Wong, W.: On the vital areas of intrusion detection systems in wireless sensor networks. IEEE Commun. Surv. Tutor. **15**(3) (2013)

26. Sedjelmaci, H., Senouci, S.M.: A lightweight hybrid security framework for wireless sensor networks. In: IEEE International Conference on Communications (ICC), vol. 1, pp. 3636–3641, June 2014

27. Krontiris, I., Benenson, Z., Giannetsos, T., Freiling, F., Dimitriou, T.: Cooperative intrusion detection in wireless sensor networks. LNCS, vol. 5432, pp. 263–278, February 2009

A New Medical Image Processing Approach for the Security of Cloud Services

Mbarek Marwan[(✉)], Ali Kartit, and Hassan Ouahmane

Laboratory LTI, Department TRI, ENSAJ, University Chouaïb Doukkali,
Avenue Jabran Khalil Jabran, BP 299, El Jadida, Morocco
marwan.mbarek@gmail.com, alikartit@gmail.com,
hassan.ouahmane@yahoo.fr

Abstract. Implementing cloud computing in medical fields would undoubtedly help achieving the best health outcomes. Obviously, this model simultaneously improves the quality of clinical decisions through advanced IT services, and lowers operating expenses. Indeed, cloud services are usually characterized by remarkable features such as cost-efficient, availability and easy exploitation. In particular, image processing using cloud has presently gained an expanding interest. Since cloud is an evolving technology, the usage of this new paradigm in such a sensitive domain requires filling the potential gaps related particularly to data privacy and security. In order to maintain data privacy, several security measures, which are based on different techniques and countermeasures, are developed, especially Service-Oriented Architecture (SOA), Secure Multi-party Computation (SMC), homomorphic cryptosystems and Secret Share Scheme (SSS). Although these existing methods are generally a promising approach, applying them to process medical data has negative effects on performance and privacy. In fact, they are inadequate to deal with a very high volume data effectively because they are originally designed for individual pixel values and text data. The main contribution of this paper is to provide a novel solution based on three-level architecture and clustering technique to secure Software-as-a-Service (SaaS) model. In this case, we use K-means clustering method to break up the secret image into a fixed number of regions, thereby processing each portion in a distinct node. This approach is meant to eliminate or reduce the risk of the potential disclosure of sensitive data. Further, we use a trusted component acting as an interface between consumers and cloud providers for minimizing security risks and cloud security threats. Specifically, this architecture is a highly efficient solution to mitigate anonymity and unlinkability issues in cloud environment. Simulation results have demonstrated the utility of the proposed methodology in ensuring the safety of medical data when using cloud services.

Keywords: Cloud computing · Security · K-means · Image processing

1 Introduction

It is commonly agreed that recent developments in medical imaging have revolutionized the accuracy of diagnosis process, thereby improving treatment outcomes. Obviously,

© Springer International Publishing AG, part of Springer Nature 2019
J. Mizera-Pietraszko et al. (Eds.): RTIS 2017, AISC 756, pp. 378–392, 2019.
https://doi.org/10.1007/978-3-319-91337-7_34

using on-premise applications requires significant effort and financial investment compared to cloud-based software. Therefore, the cloud adoption in healthcare domain seems more promising to boost the utilization of Health Information Technology (HIT). With this approach, clients typically rent the desired cloud resources and pay only for the services they have procured [1]. Additionally, delegating some tasks to a cloud hosting provider, such as computation operations and data management capability is the main significant extra value offered by cloud computing. Even though there are enormous advantages to be gained from a move to cloud, various challenges facing this model are not sufficiently addressed. Specifically, various factors are the root cause of the most common cloud security threats and vulnerabilities, particularly virtualization security risks [2], data storage location challenges [3], web technology threats [4], systems interoperability issues [5, 6] legal and regulatory constraint [7, 8]. Conventionally, clients initially encrypt their digital data before uploading them to the cloud computing for security and privacy. We therefore believe that well-established techniques provide necessary and sufficient mechanisms to process data in accordance with the legal data protection regulations. In the literature, various methods are suggested for securing cloud-based services, namely homomorphic cryptosystems, Service-Oriented Architecture (SOA), Secure Multi-party Computation (SMC) and Secret Share Schemes (SSS). However, in a sensitive field such as the healthcare, using theses existing techniques poses several challenges due to many constraints and limitations. Mostly, these methods are unfortunately inadequate to process large volume medical data as they need huge computational costs. Besides, medical records require appropriate data protection measures since they are present in diagnosis processes to avoid degradation of image quality. Alternatively, we propose a simple and efficient methodology to use cloud services safely. The first strategy is multi-region segmentation which is based on clustering techniques such as K-means. The second is three-level architecture to avoid direct connection between healthcare organizations and cloud providers, thereby reducing potential cloud vulnerabilities. This is principally accomplished through data classification along with distributed data processing system. Consequently, the proposal is an appropriate solution to secure cloud services context where users process their data remotely. Essentially, this solution provides mechanisms to mitigate security issues, especially confidentiality, anonymity and unlinkability.

The rest of this paper is organized as follows. Sections 2 and 3 examine the related work in this field and discuss the limitations of existing approaches. In Sect. 4, we describe the fundamentals of the proposed solution to address security problems in cloud-based services. Sections 5 and 6 report experimental results and discuss the security analysis of our proposal. Conclusion and perspectives are given in Sect. 7.

2 Related Work

In general, the security problems of cloud computing have been the interest of many researchers. In this context, classical cryptographic techniques, like AES, RSA DES, 3DES, ECC, ECDH, are commonly used to store data securely. However, these techniques do not allow operations over encrypted data. Currently, there are various methods to process outsourced data safely. This section, therefore, provides real-world applications and a classification of existing security methods, found in the literature, to ensure the security of data processing in a cloud environment.

2.1 Service-Oriented Architecture (SOA)

Basically, SOA is an efficient method because it facilitates interoperability among diverse environments. Hence, this approach has been used intensively in cloud-based services. In [9], the authors developed a framework, which is based on XML standard to process medical data remotely. In this framework, additional security mechanisms are implemented in utility components to deal with privacy and data security problems. This includes a number of methods that ensure access control, traceability and authentication. The major drawback of this type of solution is that it processes only raw images, and hence, leads to the exposure to internal risks. In the same line, a cloud application is used in [10] to process medical images easily and quickly. In this solution, Chiang et al. use typically SOA and ImageJ tools to provide a cost-efficient way to analyze digital records. In this respect, the proposal encompasses three different major modules to ease the task of processing medical images: Presentation layer, Service layer, Business Logic and ImageJ plugin framework. Although such features are useful, this application suffers from many security issues caused by various technologies involved in data processing. In [11], Moulick et al. suggest a distributed image processing system named DIPE framework to help healthcare professionals handle their health records efficiently. Functionally, the proposed solution, which offers affordable image processing services, consists of three components, i.e., Messaging, Programming and Services. Unfortunately, this framework does not guarantee that privacy of patients cannot be compromised, especially malicious local threats.

2.2 Homomorphic Encryption

The central idea behind using homomorphic encryption as a security mechanism is to have a possibility to perform operations on encrypted data. Because of these advantages, the usage of this technique in cloud computing has grown enormously. In [12], a cloud application is developed to process heath records accurately and consistently. The proposed method belongs to a new generation of homomorphic encryption techniques, which rely on learning with error (LWE) approach. First, we apply this scheme on medical images. We then perform basic operations on encrypted data without decryption. In particular, a real-world implementation case is presented, which sustains both addition and multiplication on encrypted digital records. In this system, healthcare organizations encrypt all digital records before being uploaded on cloud computing for a secure medical analysis. However, this type of encryption algorithm, in general, involves huge computational costs, and hence, has a negative effect on cloud system's performance. In [13], Gomathisankaran et al. propose a new cryptosystem to securely process encrypted digital data stored off-site. In fact, cloud-based applications still suffer from various security risks and threats. In light of these facts, it is necessary to take measures to ensure that confidentiality is maintained during image processing. In this respect, the authors use homomorphic technique, especially Residue Number System (RNS) scheme. An important feature of this proposal is that it allows computations to be carried out on encrypted images. For instance, the approach consists of executing basic image processing functions, such as edge detection through the Sobel filter. Consequently, this solution is a useful technique to delegate specific tasks to non-trusted or semi-trusted cloud providers.

2.3 Secret Share Scheme (SSS)

Another potential approach to addressing privacy considerations in cloud services is Secret Share Scheme (SSS) technique. In part, this is because this method is completely secure against insider attacks. Additionally, it supports direct computation on ciphered data thanks to its homomorphic properties. In [14], the authors describe an application to carry out data visualization safely using cloud computing. In this scheme, SSS method is used together with pre-classification volume ray-casting technique. The main objective is to strengthen the security and privacy protections in cloud environment by splitting the secret data into many small portions. In this case, each node handles specific parts of the secret data, thereby avoiding data disclosure. In the decryption process, clients combine only some shares to recover the processed image. In that same context, Lathey et al. [15] present a novel way to process encrypted image in off-site servers. To deal with security problems in cloud environment, the authors adopt SSS method to enable distributed data processing. Furthermore, this cryptosystem is designed to enhance data protection when outsourcing medical image analysis. This technique is typically used to provide some basic digital image processing in encrypted domain, such as noise removal, contrast and edge enhancement and antialiasing. Interestingly, the proposal provides an efficient mechanism to mitigate division homomorphic property. As a consequence, this solution allows healthcare organizations to use cloud services safely for image processing. In [16], the authors use SSS method to carry out volume ray-casting process on off-site servers. In such an approach, various nodes are involved in data processing, thereby working together effectively to achieve this goal. More precisely, this technique supports distributed data processing to enhance data protection and performance. This is achieved through a framework of three main functional modules, including the Server, the Interpolation module, the Compositor and the Client Interface. For instance, the proposal permits to carry out interpolation and to view medical images, thereby offering ray-casting tools as a service.

2.4 Secure Multiparty Computation (SMC)

Another way of solving security challenges in cloud-based services is to use the Secure Multiparty Computation (SMC) method. This technique has become widely used for effective collaborative working. This situation usually involves beneficial ownership and a third party which processes medical data on behalf of its clients. Consequently, it allows clients to secure image processing at an external party safely. In [17], the authors develop a protocol to secure distributed linear image filtering. In this case, the proposed method basically encompasses concepts such as rank reduction technique and random permutation approach. Besides, the proposal applies secure inner product protocol to enforce data protection when processing data. Similarly, Avidan et al. [18] suggested another technique for a secure digital processing through the SMC solution. Formally, each significant portion is represented by feature vectors. Accordingly, SMC technique is normally applied on these generated vectors instead of the row input image. Practically, this permits one to carry out face detection task remotely. The major disadvantage of this method is the inability to reconstruct the original image from the generated vectors.

3 Comparative Analysis of Existing Approaches

Obviously, image processing through remote cloud tools would dramatically cut costs. However, even though most clients are aware of the benefits of this new concept, they may prefer on-premise solutions because of security challenges. In this regard, comprehensive effort was made to deal with threats and vulnerabilities in cloud computing. Therefore, various techniques are suggested to help clients processing their data remotely and securely, namely SOA, homomorphic encryption, SSS and SMC. These existing approaches still have many drawbacks which hamper its usefulness in healthcare domain. More precisely, SOA technique does not provide any mechanism to protect data against insider security risks. In fact, this method usually processes data in the traditional row format, thereby exposing them to security problems. To this aim, homomorphic encryption encodes digital data to avoid disclosure of confidential information. Normally, homomorphic approach is not adequate for large volume of image data because it requires heavy computations. To address this challenge, SSS method is introduced for parallel data processing, thereby speeding up system performance. Since this technique creates encrypted shares, it is impossible to process these small portions directly. In addition, these necessary adjustments are, in general, extremely confusing and hard to accomplish. Essentially, this is due to the considerable modifications and changes occurring particularly during the encryption stage. In this context, SMC protocol is another way to secure Software as a Service (SaaS) model. More technically, this protocol offers the possibility of potential cooperation between two or more parties to enables two parties to collaborate for evaluating a joint computation with private inputs. Unfortunately, this technique is extremely computationally expensive to be run in practical imaging tools, which manipulate individual pixel values. Even worse, a high security level requires complex SMC protocols, which are usually useless and hard to implement.

In summary, the results of this study show that existing solutions are not sufficiently mature to meet cloud-based services requirement in terms of security and privacy. In this regard, we suggest a novel method to mitigate security issues in this new concept. The originality of our proposed approach is twofold. First, it is a simple and efficient solution, since it relies only on segmentation to guarantee data confidentiality. Second, it uses a three-level architecture model to reduce security problems facing traditional architectures.

4   The Proposed Framework

The proposed solution combines multiple components to minimize the impact of risks associated with cloud vulnerabilities and threats. In order to design a security-efficient framework, the proposal is founded upon three main modules namely CloudServ, CloudSec and Clients, as presented in Fig. 1.

Fig. 1. The proposed architecture for image processing over the cloud

In this section, we provide a comprehensive description and purpose of each module involved in the security process. In this regard, our methodology consists of the following key components to secure patients' data in an untrusted cloud environment.

4.1   Clients

Basically, it refers to a data owner of an electronic health record (EHR), therefore it may be either patients or healthcare professionals. It is commonly considered the simplest element of the proposed architecture. Typically, this module offers an easy way that enables the data owner to access and process their personal medical data, as well as share them with authorized users. Functionally, consumers use a web interface provided by the CloudSec to upload their sensitive data into the cloud safely. Technically, CloudSec offers necessary mechanisms to establish a secure channel between healthcare organizations and cloud service providers (CSPs). In this respect, we suggest Secure Transport Layer (STL) in which the certificates are exchanged and checked to scramble data in transit, thereby preventing data disclosure.

4.2   CloudSec

Normally, this module is the central element in the proposed architecture because it is a reliable and trustable component. Essentially, it offers a web interface that helps consumers to use remote cloud imaging tools safely. Consequently, the primary role of CloudSec is to ensure that cloud services comply with the security and privacy requirements, especially in healthcare domain. Besides, it assists healthcare organizations in analyzing medical data using cloud computing. In this case, we suggest clustering technique, in particular K-means algorithm, to split an image into small portions for data scrambling, thereby guaranteeing data confidentiality.

Under these assumptions, we use a distributed system to analyze each region at a distinct node. Theoretically, this approach ensures that data privacy cannot be compromised, since malicious users are unable to reveal a secret using a single region. The proposed privacy-protection mechanism to secure processing of medical data in the cloud is presented in Fig. 2.

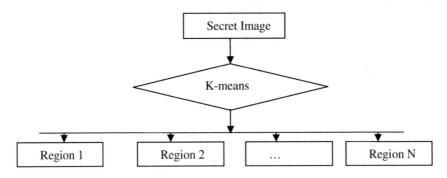

Fig. 2. Data security using K-means algorithm

Since the cloud service provider (CSP) is semi-trusted, CloudSec hosts a database to store users' identifying information locally, thereby mapping real information. In this case, it generates a pseudonym identifier from patient's ID through a one-way hash function, such as SHA 256. This technique guarantees that each digital data is associated with a pseudonym of specific user. In doing so, the proposal provides protection of the user's identity, ensures anonymity-preserving and maintains unlinkability. Besides these considerations, the CloudSec module offers the possibility to directly access and downloads the processed data easily. In practice, this module is often implemented on an on-premise resource, such as cloud-based hardware security modules (HSM). To sum up, CloudSec acts as a proxy that provides measures and mechanisms to consumers to secure cloud utilization, and hence, reduce risks resulting from direct connections with insecure public cloud.

4.3 CloudServ

In general, there are two reasons that provide a solid foundation to account for leveraging a multi-cloud approach to process medical data. First, the choice of this environment is usually motivated by its capability to deliver endless computational resources. Additionally, it is an ideal computational architecture for a distributed data processing model, and hence, it is the fastest way to analyze data. This would dramatically improve the performance of our proposed solution. Second, this approach is meant to enhance the confidentiality of outsourced data. In fact, this technique spreads data across multiple virtual nodes to prevent accidental data disclosure. We also bear in mind the case where one or more cloud becomes temporary unavailable. Indeed, this architecture ensures business continuity and high fault tolerance, since it is composed of multiple cloud servers. To summarize, the proposed approach for securing cloud-based image processing consists of three principal components. The general overview and main components of the proposed framework is illustrated in Fig. 3.

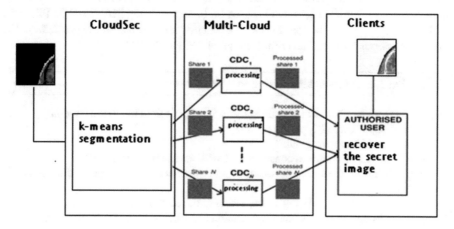

Fig. 3. The principle of the proposed framework

5   Simulation and Results

First, we present, in this section, a general description of the clustering technique used in this work, which is involved in data protection procedure. Second, the results are illustrated by a simulation study using some medical images to prove the correctness of our proposed approach.

5.1   An Overview of K-means Algorithm

In computer vision, image segmentation using clustering approach refers to the classification procedure of an image into multiple groups, namely clusters. Recently, a wide range of methods have been suggested to achieve this objective, including K-means clustering, Fuzzy C-means clustering, subtractive clustering method, etc. Typically, our proposal relies on the K–means algorithm because it is the most common clustering method and is widely used in practice. Mainly, the choice of this technique is due to its advantages in deployment, efficiency, simplicity and rapidity, which are considered the distinctive features of this proposed framework [19]. Basically, this type of algorithms is developed based on squared Euclidean distance to completely determine objects of interest, and then, to separate them from the background areas for identifying groups of similar objects. To this aim, a partial stretching enhancement is usually applied on the digital image to improve image quality efficiently. Meanwhile, the subtractive clustering algorithm is used to obtain the required centroids. The latter is typically based on the density of surrounding data points. Interestingly, K-means algorithm uses these centers during segmentation process. Often, it is useful to apply a median filter on the segmented image to eliminate any unwanted small parts from the final image.

5.2   Implementation Results

To demonstrate the correctness and show the advantages of the proposed approach, we implement the Algorithm 1, which is based on K-means, using MATLAB. In this case, each medical image will be split into four distinct regions to ensure data protection. The main steps of our proposed technique are given below [20, 21]:

Algorithm 1. SecureImageProcessing

Inputs: I (x×y), k, where I is host image, and k is the desired number of regions
Outputs: < M_p>, where M_p processed image
   Do until no object move group
      Step 1: Create the centroid coordinate (It will be randomly assigned)
      Step 2: Measure the euclidean distance d between each object pixel and the nearest centroids
$$d = \|p(x, y) - c_k\| \quad //c_k \text{ refers to the cluster centers}$$
      Step 3: Group the object based on minimum distance with the centroid, and then update the centroids.
$$c_k = \frac{1}{k} \sum_{y \in c_k} \sum_{x \in c_k} p(x, y)$$
   End do
      Step 3: Reshape the cluster pixel into images to generate k regions $R_1, R_2,..., R_k$
      Step 4: Create, for each region $Reg_i \subset I$, an image R_i using a bijective function Fpos
         /*In this case, we use Fpos function to save exactly the coordinates of all the pixels within a specific region in the image and generated image R_i using a logical link between two images.

$$Reg_i \rightarrow R_i$$
Fpos:
$$(x, y) \rightarrow (z, t)$$

Where (x, y) represents the initial position of a pixel i in the region Reg_i of the input image I. In contrast, (z, t) refers to the position pixel in the image R_i.

      Step 5: Process generated regions separately $R_1, R_2,..., R_k$
      Step 6: Use $Fpos^{-1}$, which is the reciprocal function of Fpos to reconstruct the final processed image M_p using different R_i.

Return < M_p>

Therefore, the process of image analysis requires a parallel image processing of created regions instead of the original image. Based on this approach, we apply separately to each generated region some basic image processing operations. Particularly, in this study, we aim at improving safety as well as the quality of the health records. Towards this objective, we use segmentation technique to split the secret image into four distinct portions. In this situation, we process only these generated shares to get the final result in order to maintain privacy. In this situation we reconstruct the final image, after the post-processing step, using the same procedure and a predefined reciprocal function.

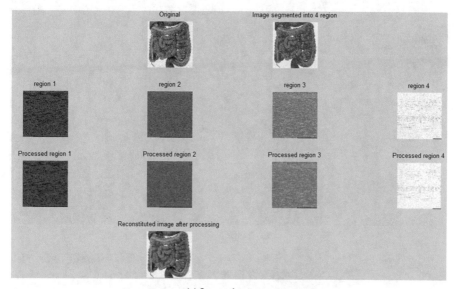

(a) Image 1

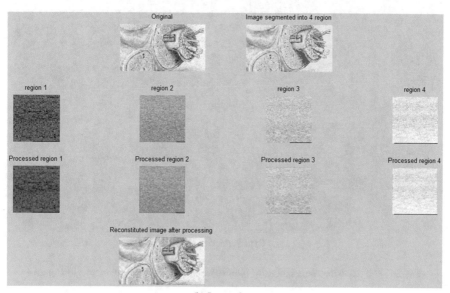

(b) Image 2

Fig. 4. The experimental results using Gaussian filter. (a) Image 1. (b) Image 2

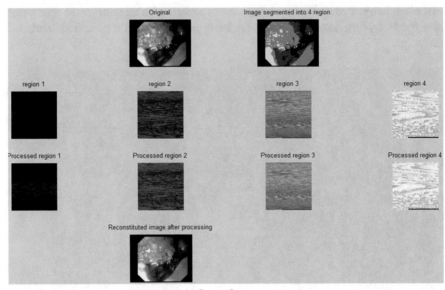

(a) Image 3

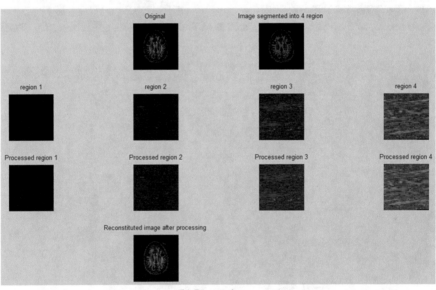

(b) Image 4

Fig. 5. The experimental results for intensity enhancement. (a) Image 3. (b) Image 4

In this section, we present some simulation results, as an illustration, to underpin the feasibility of our proposed solution. In this context, we use four colored medical images to evaluate the effectiveness of the proposed method for protecting data confidentiality. First, we apply the Gaussian filter to reduce the level of noise in two different medical images, as illustrated in Fig. 4.

Similarly, we perform image enhancement by increasing the intensity of pixels to improve the quality of medical images. The second type of simulation is illustrated in Fig. 5.

The results of simulation prove that we can analyze the secret image by processing only created regions. Hence, segmentation approach based on K-means ensures the security and privacy when using cloud services to process digital health records.

6   Security Analysis of the Proposed Framework

The practical results show that our proposal is an efficient scrambling method to conceal and alter the visual information, thereby confirming theoretical approaches. As shown in the simulation results, a single region does not provide enough information provide to make fairly accurate predictions about the secret image. Typically, K-means algorithm creates several portions to prevent inside attacks that are often conducted by an untrusted cloud provider. In this section, we analyze histogram parameters to measure the robustness of the proposed solution, especially the degree of encryption quality. Figure 6 represents the histogram of the secret image 3 and the histogram for each region.

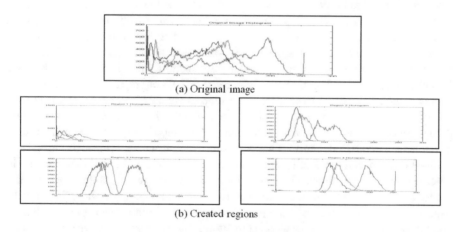

(a) Original image

(b) Created regions

Fig. 6. Histogram of different images. (a) Original image. (b) Created regions

Furthermore, we use a multi-cloud architecture to enable distributed data processing to improve Quality-of-Service (QoS), particularly running time and availability. Meanwhile, CloudSec uses pseudonym method to assure anonymity and unlinkability. To sum up, the proposed framework is a simple and efficient approach to meet privacy preserving requirements in cloud-based services. The comparison between the previous approaches and proposed one is presented in Table 1. In this regard, a $\sqrt{}$ is used to denote that a specific subject is covered in the proposed solution.

Table 1. Comparison of the related works with the proposed approach

References	Techniques	Confidentiality	Performance	Simplicity	Anonymity	Unlinkability
[9–11]	SOA		√	√		
[12, 13]	Homomorphic	√				
[14–16]	SSS	√				
[17, 18]	SMC	√				
Our proposal	K-means	√	√	√	√	√

7 Conclusion

The usage of cloud computing in healthcare has been growing recently because it offers an easy way to implement IT services. In particular, this approach delivers affordable, reliable online imaging tools on the detriment of the expensive on-site solutions. In parallel, using this model would help healthcare organizations improve efficiency and patient outcomes through the externalization of IT functions. Even though cloud is an extremely useful tool in healthcare domain, consumers have many factors to consider when choosing cloud services. In fact, data protection is the most challenging problem on the road to successful and consistent implementation of this model. Currently, a variety of security mechanisms have been developed to secure data processing over cloud computing. The adopted approaches involve several techniques, including homomorphic cryptosystems, Service-Oriented Architecture (SOA), Secure Multi-party Computation (SMC) and Secret Share Schemes (SSS). As outlined above, these methods are not sufficiently mature to support cloud adoption in healthcare and provide needed security tools. In this respect, a novel solution, which is based on K-means algorithm, is suggested to secure digital data processing. Most importantly, we use an additional component, named CloudSec, to enhance protection against unauthorized disclosure and unlinkability, as well as ensuring anonymity. We also use a multi-cloud architecture in order to effectively minimize cloud security issues. Experimental results prove that the proposal can achieve the data confidentiality in cloud using simple and efficient techniques. In summary, we expect the proposed approach, in this paper, to be useful for promoting data processing over cloud computing. Our principal objective is to facilitate effective communication among healthcare professionals and to boost telemedicine services. This is achieved by using cloud services to cut down the costs of implementing electronic health records systems (EHR). Our future work includes the application of complex image processing algorithms. In addition, we plan to use a fully distributed version of the K-means algorithm [22], and then, implement this method in Hadoop framework via the MapReduce functions as illustrated in [23]. This would improve computation time by executing multiple processes in parallel. We also intend to use watermarking techniques to mitigate issues of authentication and integrity.

References

1. Mell, P., Grance, T.: The NIST definition of cloud computing. Technical report, National Institute of Standards and Technology, vol. 15, pp. 1–3 (2009)
2. Mazhar, A., Samee, U.K., Athanasios, V.: Security in cloud computing: opportunities and challenges. Inf. Sci. **305**, 357–383 (2015)
3. Fernandes, D.A.B., Soares, L.F.B., Gomes, J.V., Freire, M.M., Inácio, P.R.M.: Security issues in cloud environments: a survey. Int. J. Inf. Secur. **13**(2), 113–170 (2013)
4. Marwan, M., Kartit, A., Ouahmane, H.: Cloud-based medical image issues. Int. J. Appl. Eng. Res. **11**, 3713–3719 (2016)
5. Petcu, D.: Portability and interoperability between clouds: challenges and case study. In: Abramowicz, W., Llorente, I.M., Surridge, M., Zisman, A., Vayssière, J. (eds.) ServiceWave 2011. LNCS, vol. 6994, pp. 62–74. Springer, Heidelberg (2011)
6. Ahuja, S.P., Maniand, S., Zambrano, J.: A survey of the state of cloud computing in healthcare. Netw. Commun. Technol. **1**(2), 12–19 (2012)
7. Pearson, S., Benameur, A.: Privacy, security and trust issues arising from cloud computing. In: Proceedings of the IEEE Second International Conference on Cloud Computing Technology and Science (CLOUDCOM), pp. 693–702. IEEE Computer Society, Washington, DC (2010)
8. Al Nuaimi, N., AlShamsi, A., Mohamed, N., Al-Jaroodi, J.: E-health cloud implementation issues and efforts. In: Proceedings of the International Conference on industrial Engineering and Operations Management (IEOM), pp. 1–10 (2015)
9. Todica, V., Vaida, M.F.: SOA-based medical image processing platform. In: Proceedings of the of IEEE International Conference on Automation, Quality and Testing, Robotics (AQTR), pp. 398–403 (2008). https://doi.org/10.1109/AQTR.2008.4588775
10. Chiang, W., Lin, H., Wu, T., Chen, C.: Building a cloud service for medical image processing based on service-orient architecture. In: Proceedings of the 4th International Conference on Biomedical Engineering and Informatics (BMEI), pp. 1459–1465 (2011). https://doi.org/10.1109/BMEI.2011.6098638
11. Moulick, H.N., Ghosh, M.: Medical image processing using a service oriented architecture and distributed environment. Am. J. Eng. Res. (AJER) **02**(10), 52–62 (2013)
12. Challa, R.K., Kakinada, J., Vijaya Kumari, G., Sunny, B.: Secure image processing using LWE based homomorphic encryption. In: Proceedings of the IEEE International Conference on Electrical, Computer and Communication Technologies (ICECCT), pp. 1–6 (2015)
13. Gomathisankaran, M., Yuan, X., Kamongi, P.: Ensure privacy and security in the process of medical image analysis. In: Proceedings of the IEEE International Conference on Granular Computing (GrC), pp. 120–125 (2013)
14. Mohanty, M., Atrey, P.K., Ooi, W.-T.: Secure cloud-based medical data visualization. In: Proceedings of the ACM Conference on Multimedia (ACMMM 2012), Japan, pp. 1105–1108 (2012)
15. Lathey, A., Atrey, P.K.: Image enhancement in encrypted domain over cloud. ACM Tran. Multimed. Comput. Commun. **11**(3), 38 (2015). https://doi.org/10.1145/2656205
16. Mohanty, M., Ooi, W.T., Atrey, P.K.: Secure cloud-based volume ray-casting. In: Proceedings of the International Conference on Cloud Computing Technology and Services (CloudCom 2013), pp. 531–538. IEEE (2013). https://doi.org/10.1109/CloudCom.2013.77
17. Hu, N., Cheung, S.C.: Secure image filtering. In: Proceedings of the of IEEE International Conference on Image Processing (ICIP 2006), pp. 1553–1556 (2006)

18. Avidan, S., Butman, M.: Blind Vision. In: Leonardis, A., Bischof, H., Pinz, A. (eds.) Computer Vision – ECCV 2006. LNCS, vol. 3953, pp. 1–13 (2006). Springer, Heidelberg (2006). https://doi.org/10.1007/11744078_1

19. Dhanachandra, N., Manglem, K., Chanu, Y.J.: Image segmentation using K-means clustering algorithm and subtractive clustering algorithm. Proc. Comput. Sci., 764–771 (2015). https://doi.org/10.1016/j.procs.2015.06.090

20. Abdul-Nasir, A.S., Mashor, M.Y., Mohamed, Z.: Colour image segmentation approach for detection of malaria parasiter using various colour models and K-means clustering. J. WSEAS Trans. Biology Biomed. **10**(1), 41–55 (2013)

21. Gulhane, A., Paikrao, P.L., Chaudhari, D.S.: A review of image data clustering techniques. Int. J. Soft Comput. Eng. **2**(1), 212–215 (2012)

22. Oliva, G., Setola, R., Hadjicostis, C.: Distributed K-means. Submitted to IEEE Transactions on Mobile Computing. http://arxiv.org/abs/1312.4176

23. Mao, Y., Xu, Z., Li, X., Ping, P.: An optimal distributed K-Means clustering algorithm based on cloudstack. In: Proceedings of the IEEE International Conference on Information and Automation, China, pp. 3149–3156 (2015). https://doi.org/10.1109/ICInfA.2015.7279830

Application Areas of Real-Time Intelligence

Introduction to Sociology of Moroccan Online Social Networks: Evolution Analysis of the Moroccan Community Activity on Facebook

Jaafar Idrais$^{(\boxtimes)}$ (iD), Yassine El Moudene (iD), and Abderrahim Sabour (iD)

Modelisation, Systems and Technologies of Information (MSTI),
High School of Technology, Ibn Zohr University, Agadir, Morocco
{jaafar.idrais, yassine.elmoudene}@edu.uiz.ac.ma,
ab.sabour@uiz.ac.ma

Abstract. Within the context to conduct studies of sociology on the Moroccan community on the Online Social Networks (OSNs), using a rich dataset we will analyze the evolution of the interactions and we will extract the different characteristics related to the behaviors of Moroccan users on Facebook. First, we present our dataset, then we project the interaction's evolution over time on some possible temporal axis (years, months, days and hours), to display the regular models reflecting the behavior of the users, two resolutions have been adapted. After we will study the different time slots, a big interactions observed during the evening which caused a time lag and a reduction of the hours of sleep among the users, which means an important attachment of the Moroccan community to Facebook.

Keywords: Social networks · Data mining · Graph mining · Networks analysis
Graphs theory · Community detection · Response-time · Community interaction

1 Introduction

With the revolution of computer technologies, social networks have exceeded their classical unfolding, we are now involved in indirect relationships with people and individuals imposed rather than chosen, in other words we are in front of a very large world of complex and virtual relationships. The first grouped communication system (Lee 2013) is the CBBS (Computerized Bulletin Board System) developed by W. Chrisensen and R. Seuss in 1978 which connects users using a connection server, messages are published on a dashboard accessible on the server [1]. In the 90s, there was the development of real-time IRC (Internet Chat Relay) communication systems and the appearance of "chat rooms" that bring a set of users with the ability to chat and share information together, where does the idea of today's social networking groups come in: Twitter, Facebook, WhatsApp, etc.

© Springer International Publishing AG, part of Springer Nature 2019
J. Mizera-Pietraszko et al. (Eds.): RTIS 2017, AISC 756, pp. 395–408, 2019.
https://doi.org/10.1007/978-3-319-91337-7_35

The complexity of humanitarian and associative questions coupled with the inherent necessity of a hierarchy based on strong central control means that this type of organization is able to take full advantage of the findings of these laws by organizing networks Flexible and adaptable. Although the indicators may appear insufficient to conduct a study in much formal sociology of the Moroccan community, the approach used is to avoid the complexity of the process applied in the sociology polls by exploiting the visible part of the iceberg via analysis the acquired data. In other words, the objective of this work is the extraction of knowledge related to the evolution of the behavior of the Moroccan community existing on Facebook via accessible data analysis.

In this study we focus on Facebook as our OSN for many reasons: the data access is a big challenge, the social networks like WhatsApp does not offer this possibility, on the other hand, Facebook is characterized by an easy access to public data, especially by offering APIs for the developer community like the Social Graph API.

2   Measures and Methodology

In this section we will present the definitions related to the context and describe the data used and the work process. One of the initial steps was to find the right dataset for our problematic, data that can help to analyze the behavior of the Moroccan community on OSNs, which resulted in no concrete results gives a start to create an appropriate dataset to our problem.

2.1   Definitions

Social Network: According to (Acquisti and Gross) [3] Social Networks are defined as web services giving users the environment that offers the possibility of building a public profile, having a list of friends (other users) With which they share a connection (Backstrom et al.) [2]. this system also offers the ability to view and browse the list of connections and those made by other people in the system [4].

Profile: A profile in the context of an information retrieval activity is defined as the set of dimensions that enable one to write and/or infer one's intentions and perception of relevance (Tamine et al. 2007) [5]. A user profile can help the system to locate information that best meets the needs of each user and adapt to its characteristics in a centralized context (Zayani 2008) [6] or in the distributed context (Rebai et al. 2013) [7]. A profile was defined as the entity that represents the user in the system, it is created by filling in information fields that are typically descriptors, such as age, location, interests, etc., and profile may also have a profile picture loaded by the account owner. The visibility of a profile depends on the system and the choice of the user. Some environments give the accounts a public status by default and a simple search on Google can lead to it, other systems give the owner the ability to make their profiles undetectable by search engines or only visible to friends.

2.2 Dataset

In the fact that we are trying to profile the behavior of the Moroccan community based on the analysis of interactions and reactions as well as the dataset exposed to the public do not meet our needs, this development we decided to create a data set adequate for this purpose.

Presentation

Referring to the traffic management websites (socialbakers.com, 2016) a list of the pages and groups most consulted by the Moroccan community was the starting point for the creation process of the dataset: the vocation was to recover all public information Offered by the Facebook Graph API in an incremental way, other pages and groups arrive on the queue according to a semi-supervised selection process. We have collected our own database by performing the Crawling via the API Graph tool of Facebook by referring to some recommendations of the work of (Gjoka et al. 2011) [8] (Table 1).

Table 1. Representation and description of the dataset

Years	2007-2017	Interactions	
Groups and Pages number's	892	Total	37 301 327
Feed's number	2 161 471	comments	10 998 543
Interaction's number	37 301 327	tags	457 956
User's number	10 437 154	reactions	25 844 828
Users with action visible	228 782		

Limitations

During this study, we are limited by the privacy settings of Facebook. Our dataset includes only public groups and we only collect public interactions. On Facebook the public interactions include the "LIKE" mention between the users, comments, tags, etc. Another limitation related to the security policy constraints of identifying "user-agents" for restricting robotic use, the extraction process does not offer the possibility of extracting the entire list of members Group there is a loss margin of between 2% and 5%. Only posts and comments have a crated time, the other types: likes and reactions does not offer this information which leads us to focus only on comments in this study.

2.3 Work Process

Behavior covers various social activities and represents what users do online, activities are at multiple scales: browsing feeds, chatting with friends, making comments., etc. Our aim is to understand how these reactions evolve over time, and deduct hours that experience high activity rates. To know the impact of the post's time on the reaction density, we have followed the process below: after the first phase which consists of collecting data from the social network studied and the creation of the dataset, the data filtering stage appears essential in this step, we will filter data by type, whether they are comments or maybe posts, then comes the next step to standardize data and calculate the cumulative distribution function (CDF) will be done to study the evolution of the interactions according to the different time axes (years, months and days of the week), this will make it possible to notice the important temporal periods and which have a rate high activity. The process will be applied separately for the posts then for the comments and after the interaction in a general way.

In order to have better visualization, we will adopt two resolutions: a representation by the hours of the day with a step of 5 min and a step of 15 min in order to make superimpositions between years. The handling performed complement each other and represent projections of the same dynamic system over different phase spaces (Fig. 1).

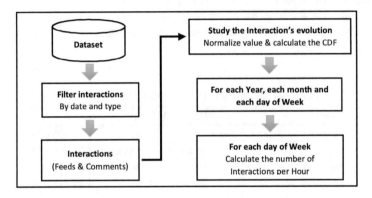

Fig. 1. Work Process

3   The Evolution of Interactions Over Time

Multiple and diversified uses (chat, contact, sharing links, photos and videos, etc.) have made social networks a favorite destination for Moroccan Internet users, the percentage of participation has changed by 38% between 2009 and 2012 [9]. This can be explained by increasing the daily connection time between 1 and 2 hours per day by using access via a desktop computer and then connecting via laptops and mobile phones.

Figure 2 shows a strong growth in the number of publications in recent years, with a number exceeding 300 thousand, unlike the years before 2013 where this number did not exceed 150 thousand, this can be explained by the simplicity of use of the equipment and web content, as well as the Internet represents a means of entertainment for people who benefit from the ease of access especially with the development of mobile technology, which most noticed by exceeding the number of publications by comments number ranging from double in 2013 to 7 times after 3 years.

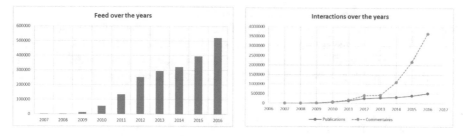

Fig. 2. The evolution of interactions over the years between 2007 and 2017

3.1 Evolution of Interactions Per Month

After drawing up an annual representation of the interactions, we plotted the monthly evolution graph between 2007 and 2016 in order to visualize the evolution of interactions and publications over the course of the months.

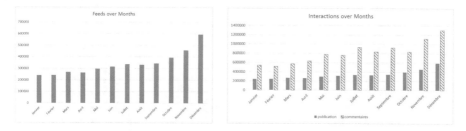

Fig. 3. Evolution of interactions over the months between 2007 and 2017

From Fig. 3: the months: October, November and December are characterized by a large number of feeds ranging from 400 thousand to 600 thousand, whereas the first four months of the beginning of the year are slightly (Between 200 and 300 thousand), making the most active year-end of other periods.

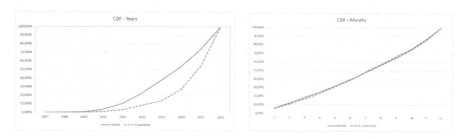

Fig. 4. Cumulative distribution function CDF of interactions

Analyzes used histogram in order to represent the frequency of the various measured variables as well as the distribution of the associated values, but does not give the gap's percentage between two values, that is why we chose to plot the CDF graphs or as say the cumulative distribution function. Figure 4 shows that there is an exponential growth of comments number, while the growth of publications follows an almost linear evolution.

3.2   Evolution Per Hour of Day

Years and months gives a more general view about publications and comments evolution, whereas hours give more visibility on how the interactions evolve during day. Figure 5 gives a view of the number of interactions per hour, the morning is characterized by a growth of activity of commenting from 7 am until 3 pm ranging from 100 thousand to 470 thousand, then the number of comments decreases between 4 pm and 5 pm (almost 400 thousand). At 6 pm a resumption of activity which takes a maximum value of 650 thousand comment at 10 pm, amortization from continuing until the first hours of the following day.

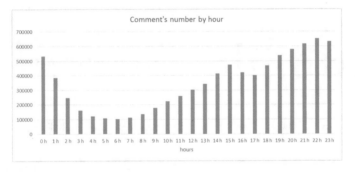

Fig. 5. Comment's evolution per Hour

For each day of the week and for each year we calculated the average of the interactions as a kind of aggregation, Fig. 6 shows that there is a high growth in the mean value of the interactions between 2009 and 2016, from 110 to 4100, and a clear deformation that indicates the fall hours, before there was a kind of continuity of use indicating that people do not stop interacting which really reflects a great attachment of users to social networks. The most active periods before 2013 are between 6 pm and 8 pm, but after this year there was a lag around 9 pm until 11 pm.

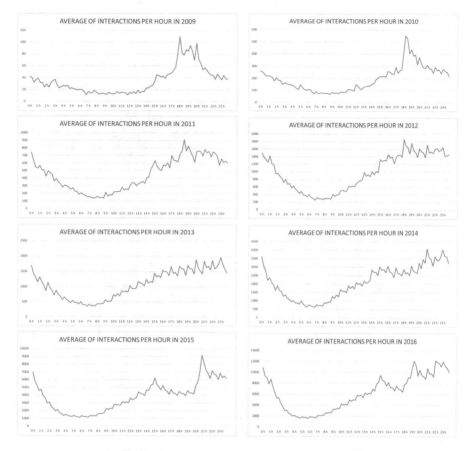

Fig. 6. Average of Interactions per Hour by Years

3.3 Representation by Hours with a Step of 5 Min

To zoom on the distribution of information over days with a 5 min step, Fig. 6 shows a change of the regular pattern that characterizes the days of the week, the presence on the social network can be observed during the years 2009, 2010, etc. where the change of the model begins with the disappearance of the only period of connection and the significant daily presence on Facebook. Figure 7 shows the evolution of visible reactions over years of the Moroccan community on Facebook, we notice the behavior transition from a semi-periodic towards a quasi-periodic behavior with intervals of inactivity increasingly shorter. Before 2009: the data acquired represents only 0.03% of the total reactions. It can be seen that during the years 2009 and 2011 the beginning and the end of week represent a low reactivity and that the maximum agitation is associated to mid-week days, with a slight decrease of the thresholds of the inactivity intervals.

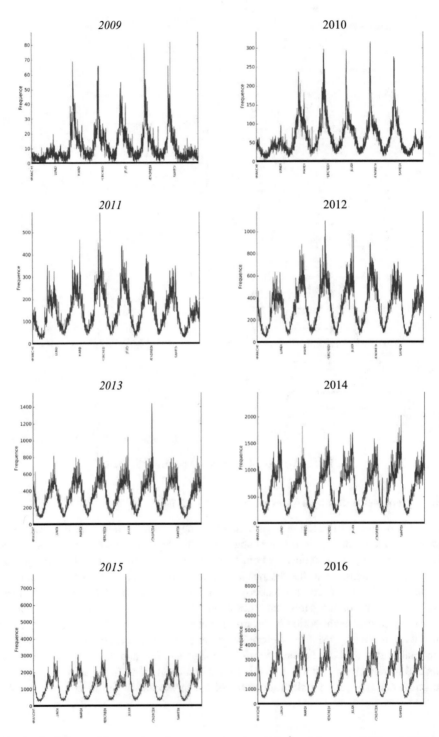

Fig. 7. Evolutions of the number of comments per day of the week between 2009 and 2017 by the 5-min step

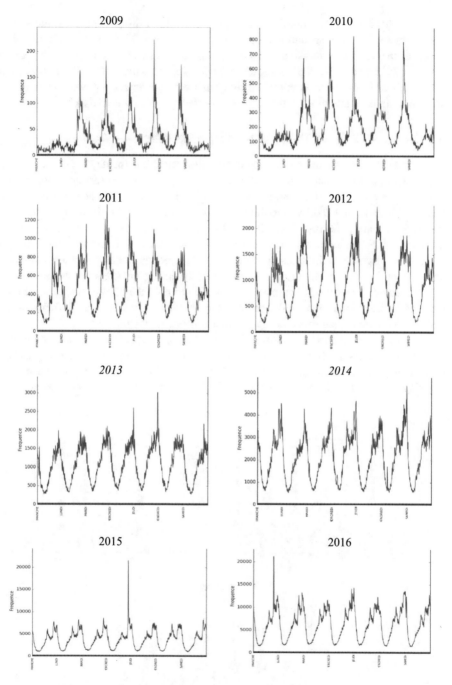

Fig. 8. Evolutions of the number of comments per day of the week between 2009 and 2017 by the 15-min step

From 2012 to 2014: a quasi-periodic behavior with a disappearance of the spikes that characterizes the distribution associated with the years 2009 to 2011. Giving birth to one of the periods of maximum reactivity and periods of acute inactivity. The period between 2015 and 2016: the quasi-periodic behavior is maintained while displaying a better regularity on the maximum reactivity phase characterized by two spikes of the same one finds a flattening of the spike associated with the inactivity phase.

3.4   Representation by Hours with a Step of 15 Min

This model is based on the information distribution relative to the days of the week by a step of 15 min, Fig. 8 shows the existence of a distortion on the users behavior related to the duration of connection and access to social networks, this can be noticed through years 2010, 2011 and 2012 where the model changes begins with the disappearance of the peak notion or the rush hour. In 2009 peak hours are very apparent and a weak presence on Facebook, on other hand one we can talk about intervals indicating a significant presence on the social network in 2017, which is explained by the amplitude and distribution of values during days and exactly during the second half-day.

The figures superimposition makes it possible to note the differences related to the changes in the graphs shape at the end of years. Figure 9 shows the graph deformation and the decrease of activities during the first 8 h of the day over years.

This can be explained as follows: during the early years, Facebook was impressed the internet surfers in such a way that they substitute their sleep for connecting, but by dint of time the users begin to adapt with the new intruder. People connect during working hours (8 am–4 pm), in addition to the above-mentioned remark concerning the

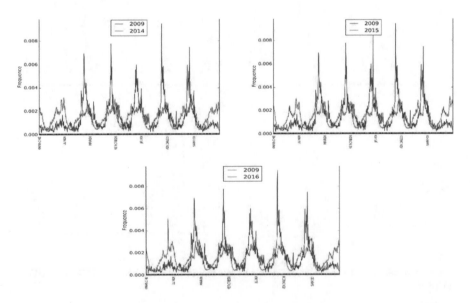

Fig. 9. Superposition of the days of the week between 2009 and 2014, 2015 and 2016

absence of a single peak hour, two connection periods with an almost uniform characteristic appeared (4 pm–7 pm) and (8 pm–11 pm).

Through the various reported remarks, it can be said that the behavior of Moroccan community on Facebook has undergone a significant change during the last ten years, the presence duration on Facebook have changed so much, users spans the whole day on communication instead of a few special moments like the evening. This shows that Moroccans users have managed to adapt with the new intruder in their daily lives.

4 Discussion

According to this study based on our dataset, we have shown the evolution of the interactions numbers between 2007 and 2017 regardless of type, whether they are publications or comments. For reasons of weaknesses of the quantities of information before 2009 we have limited our study interval beyond 2009 (Fig. 10).

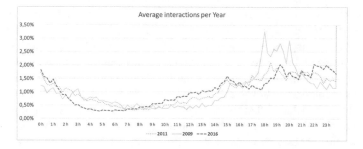

Fig. 10. Interaction model over the years 2009, 2011 and 2016 by hour

One of the asked questions during this study is how the reactions are distributed over months, days of week and during hours of the day. Figure 2 showed a big Facebook activity in terms of feeds, which grew from 0.6% in 2009 to 26.2% in 2016 (almost 43 times) with an annual average of 12.5%. The number of comments has grown exponentially between 2010 and 2016 with an average of 14.3% per year (45.60% in 2016), whereas before 2012 the value did not exceed 3% of total number of comments.

In January 2017, statistics proved that Facebook is the most popular network in Morocco with an average interaction number of 84311. Then, Twitter with 501 interactions, in terms of response rate Facebook still always the first with 87% and 16% for Twitter (socialbakers.com 2017). With the spread of Facebook in North Africa, Fig. 6 also showed a change of daily presence on website between 2009 and 2016. The cyber cafes was played a very important role and some limited ADSL access for families to the internet access, social networks impressed people, which involved a lag time of sleep up to 3 but with the evolution of 2G, 3G and 4G mobile technology, the number of users reached almost 12 million Moroccan users on Facebook (Internet World Stats, June 2016) with a presence on the site during the hours of (Between 8 am and 4 pm) at most of the periods of remarkable connections during the evening (4 pm–7 pm) and (8 pm–11 pm). In recent years (from 2013 to 2015) Facebook has become the most popular and preferred social network followed by WhatsApp and Twitter [10], in terms of usage, chat is the major element followed by the access for taking news and then sharing or post comments. The attention devoted to social networks such as Facebook and WhatsApp is due to the free and ease of use.

5 Related Work

Several studies have been carried out to understand the structure and evolution of large-scale online social networks [11, 12] through the extraction of topological features, but are not interested in how social networking links are used to interact. The study of the activity network (Chun et al. 2008) [13] based on the comments published by users in the guest books of others on the online social network of "Cyworld" showed that the structures of the activity network and the social network were similar and that interactions between users seem bidirectional. On the other hand using Facebook, (Wilson et al. 2009) [14] was demonstrated the inverse trend, that there is a significant difference between the structure of the activity network and the social network by examining the evolution of interactions number per user. They showed that the activity network has fewer small global properties compared to the social network. The study of the evolution of activity among Facebook users, specifically the difference between the components of the activity network of social network, (Viswanath et al. 2009) [15] was concluded that the strength of social links can vary rapidly over time without changing the majority of the theoretical properties of activity network.

Another line of research is taken by (Wilson et al. 2012) [14] who are trying to see if social bonds are valid indicators of real interactions between users. They studied in detail the user's interactions on Facebook, they analyzed the interaction graphs which have fewer nodes to higher degrees, and the remark made is that the overall diameter of the graph increases by one significantly. By concluding that to achieve realistic and accurate results, the finding is that research and studies on social applications must use real indicators of user interactions instead of social graphs.

6 Future Work

In future work, we will study and follow certain phenomena that appear within this community in order to see how the evolution of the reactions is influenced, also to make a comparative study of the different states that will initiate the work towards a new axis concerning content analysis.

7 Conclusion

During this work we were able to extract some characteristics related to the behavior of the Moroccan community on Facebook, a large number of people was very attracted by the new technologies, Social networks are among the main interest of Moroccan Internet users, every day they become more attached to Facebook, this is explained by the exponential evolution of the numbers of comments in the last years. Facebook is a free expression space, several social and political movements consider it as a free working space and a better environment for the propagation of information. An important change has infected the behavior of the Moroccan community on Facebook, users are influenced in such a way that they substitute their sleep only to interact with their friends on social networks.

References

1. Lee, J.: How We Talk Online: A History of Online Forums, From Cavemen Days to the Present. Accessed 20 Aug 2015
2. Backstrom, L., Huttenlocher, D., Kleinberg, J., Lan, X.: Group formation in large social networks: membership, growth, and evolution. In: Proceedings of 12th International Conference on Knowledge Discovery in Data Mining, pp. 44–54. ACM Press, New York (2006)
3. Acquisti, A., Gross, R.: Imagined communities: awareness, information sharing, and privacy on the Facebook. In: Golle, P., Danezis, G. (eds.) Proceedings of 6th Workshop on Privacy Enhancing Technologies, pp. 36–58. Robinson College, Cambridge (2006)
4. Mislove, A., Marcon, M., Gummadi, K.P., Druschel, P., Bhattacharjee, B.: Measurement and analysis of online social networks. In: Proceedings of IMC (2007)
5. Tamine, L., Calabretto, S.: Search for Contextual Information and Web, Book on Web Information Research, Hermes Edition, forthcoming. IRIT, France (2007)
6. Zayani, C.A.: Contribution to the definition and implementation of mechanisms for adapting semi-structured documents. Doctoral Thesis. University of Toulouse (2008)
7. Rebai, R.Z., Zayani, C.A., Amous, I.: An adaptive navigation method for semi-structured data. In: Morzy, T., Hrder, T., Wrembel, R. (eds.) Advances in Databases and Information Systems. AISC, vol. 186, pp. 207–215. Springer, Heidelberg (2013)
8. Gjoka, M., Kurant, M., Carter Butts, T., Markopoulou, A.: Practical recommendations on crawling online social networks. IEEE J. Selected Areas Commun. 29(9), October 2011
9. National agency for Telecoms Regulation, Study about the using of ICT in Morocco, access and use (2012)
10. National agency for Telecoms Regulation, Annual rapport of 2015, pp. 10–11, June 2015
11. Ahn, Y-Y., Han, S., Kwak, H., Moon, S., Jeong, H.: Analysis of topological characteristics of huge online social networking services. In: Proceedings of WWW (2007)
12. Kumar, R., Novak, J., Tomkins, A.: Structure and evolution of online social networks. In: Proceedings of SIGKDD (2006)

13. Chun, H., Kwak, H., Eom, Y.-H., Ahn, Y.-Y., Moon, S., Jeong, H.: Comparison of online social relations in volume vs interaction: a case study of cyworld. In: Proceedings of the 8th ACM SIGCOMM Conference on Internet Measurement, pp. 57–70. ACM (2008)
14. Wilson, C., Boe, B., Sala, A., Puttaswamy, K.P.N., Zhao, B.Y.: User interactions in social networks and their implications. In: Proceedings of Eurosys (2009)
15. Viswanath, B., Mislove, A., Cha, M., Gummadi, K.P.: On the evolution of user interaction in Facebook. In: WOSN 2009, pp. 37–42 (2009)

Introduction to Sociology of Online Social Networks in Morocco. Data Acquisition Process: Results and Connectivity Analysis

Yassine El Moudene$^{(\boxtimes)}$, Jaafar Idrais$^{(\boxtimes)}$, and Abderrahim Sabour$^{(\boxtimes)}$

High School of Technology, Ibn Zohr University, BP:33/S, 80000 Agadir, Morocco
yassine.elmoudene@gmail.com, jaafar.idrais@gmail.com, ab.sabour@uiz.ac.ma

Abstract. The aim of this paper is to study the Moroccan active community behavior on online social networks, firstly the choice of Facebook as OSN is justified by the statistics that ranks it in the first position compared to other OSNs like Twitter, LinkedIn,... The lack of a specific database dedicated to Moroccan community necessite the implementation of a data acquisition process. In second part, this paper presents macro data connectivity analysis, a visualization was made by applying ForceAtlas2 layout algorithm.

Keywords: Online social networks · Facebook · Data extraction
Community detection · Intra-group connectivity · Layout algorithm
ForceAtlas2

1 Introduction

Sociology as quite other science saw its axles of influence as well as its objectives evolved or even transferred in the course of this 20 last years, the analysis of the new communication technologies impact NTIC has create a sub-industry devoted, the arriving of web 2.0 [1] had a prodigious impact since he has allows to people to pass from a simple receiver to a more important role or he participates in the contents production, what allows the generation of huge data quantity in a period of time. The interlocking of the user in the process of generation of contents also allows to measure importance of one networks social online in comparison with other one.

2 Online Social Networks (OSN)

2.1 Definition

An online social network is a service [2] that allows individuals to:

- Build a public or semi-public profile in a delimited system.

© Springer International Publishing AG, part of Springer Nature 2019
J. Mizera-Pietraszko et al. (Eds.): RTIS 2017, AISC 756, pp. 409–418, 2019.
https://doi.org/10.1007/978-3-319-91337-7_36

– Articulate other users list with whom they share a connection.
– View and browse their connections list and those made by other users in the system.

So an online social network is an Internet community where individuals interact, often through profiles that represent their personalities and their networks.

2.2 OSN Evolution

SixDegrees.com [3] is the first recognizable social networking site launched in 1997. The next wave of OSN began with Ryze.com in 2001. Friendster was launched in 2002 as a social complement to Ryze [4]. As of 2003, many new OSNs have been launched, while MySpace [5] has attracted the attention of the majority of the media in the US and abroad, OSNs growing in popularity worldwide. Friendsk won the attraction in the Pacific Islands, Orkut became the first OSN in Brazil before growing rapidly in India, Mixi adopted a widespread adoption in Japan, LunarStorm took off in Sweden, Dutch users have embraced Hyves, Grono Has captured Poland, Hi5 was adopted in the small countries of Latin America, South America and Europe, and Bebo has become very popular in the United Kingdom, New Zealand and Australia. Facebook [6] debuted in early 2004 as a Harvard OSN only, starting in September 2005, Facebook has grown to include high school students, professionals within the corporate networks and eventually everyone.

2.3 OSNs Popularity Comparison

The world map[1] showing the most popular OSNs by country in 2017, according to Alexa traffic data[2] & Similar Web. Facebook is the main social network in 119 of the 149 countries analyzed, but was arrested in 9 territories by Odnoklassniki, Vkontakte and Linkedin. Interestingly, in some countries, such as Botwana, Mozambique, Namibia, Iran and Indonesia, Instagram wins and some African territories prefer LinkedIn. Overall, LinkedIn is in 9 countries, Instagram 7, while VKontakte and Odnoklassniki grow in Russian territories. In China, QZone still dominates the Asian landscape and Japan is the only country where Twitter is the leader.

3   Semi-supervised Extraction Process

Online social networks are part of the Web, but their data representations are very different from the general web pages. Web pages that describe an individual, a page, a group, in an OSN are generally well structured, as they are usually generated automatically, unlike general web pages that could be written by a person. The lack of fact that we seek to profile the behavior of the Moroccan community from interactions and reactions analysis has necessitated the creation of a data set adequate for this purpose.

---

[1] http://www.vincos.it/world-map-of-social-networks/.

[2] http://www.alexa.com/.

3.1 Inaccessible and Inadequate Dataset

Before starting data retrieval, a search for Existing Datasets, for a such study, was initially done, although the existing Datasets are inaccessible [7,8], they are not adequate to the studied problem, thus comes the need for Defines a semi-supervised data acquisition process for the collection of a specific database meeting the requested requirements.

3.2 Semi-supervised Extraction Process

To extract this data, a queued data structure to store the groups list and pending pages to be crawled is set up. Initially, By referring to the social media analysis Websites as SocialBakers [9] a list of most consulted pages and groups by the Moroccan community feeds the starting list of the process. Then, at each stage, the first entry in the queue is finalized: all publications, comments on publications, sub-comments (comment comments), reactions (towards publications and comments) for this entry (Group or page) are scanned to extract the different types of information. The Fig. 1 display the semi-supervised collection process followed.

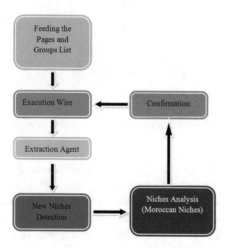

Fig. 1. Semi-supervised data acquisition process

3.3 Extracted Data and Its Structure

Facebook Data Structure. An user can stick to a group, Like A Page, Publish a publication, Comment on a publication or on a comment, React to a publication or to a comment, Share a publication, or Tag (identify) another user in a publication or a comment. So, by going through these different types of information, different connections graphs between all users can be constructed.

To assure the coherence and the data structure evolution, a MySQL database is used for the agents. It allows to unfold new agents without being subjected of time to stop or to shape, while supporting a proportional increase of performances. It is important to note that the failure of one agent does not affect the operation of others. The schema of database is illustrated on the Fig. 2.

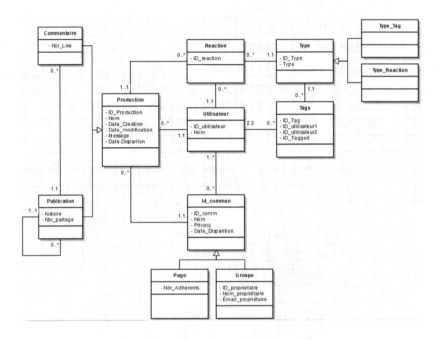

Fig. 2. Class diagram used for the backup of extracted data

Extraction Speed. However, in order to guarantee a browsing speed, we have set up a set of machines called Extraction Agent, each of these agents takes care of the extraction of the various information related to specific groups, at the end of the extraction of all types of information from of a group (page), This latter is marked as exploring, to avoid a re-run of the group (page).

3.4   Limit

Accessible Data. Online social networks are inherited from complex dynamic systems, the faculty of making accessible only a visible part and hiding a large part of it like an iceberg. During the data extraction, some limitations are imposed by Facebook:

- responses dimension returned by Facebook: to minimize traffic through its network, Facebook responds to requests with a limit of no more than 100 records, and at the end of these 100 records provides a paging next link that contains the page with the rest of information.

– Lasted of validity of the access token: to avoid accesses by robot, this length is restricted at 2 h. This problem is resolved by implementing the Access token validity surveillance Process illustrated in the Fig. 3.

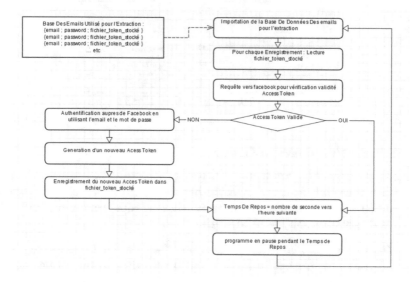

Fig. 3. Access token validity surveillance Process

Information Dissipation. Despite that a serious study has not yet been started on the extracted data, the system has triggered the disappearance of a set of nodes (group, pages, publications,...).

4   Results and Preliminary Analysis of Connectivity

4.1   Data Statistics Extracted

4.2   Global Statistics

Since the launch of the data extraction script on 01 January 2017 at the level of the various agents, up to the time of the writing of these lines, 2000 Moroccan groups and pages have been explored, this operation has made it possible to recover different Types of information cited above, the database is a size of 15 Gigabyte and contains about 100 Millions records. Below is a detailed table for the different information extracted (Table 1):

Table 1. Global extracted information statistics

Table	Number records	Sizes (in MB)
Users	13096530	1900
Groups or pages	2864	1,2
Members	4662253	212
Publications	5733444	3600
Comments	21652626	5400
Reactions	49551978	4000
Tags	1061392	144
Total	98761087	15256,2

4.3   Relationship Between Group

Extracted data are the product of Moroccan's users behaviour within facebook. With those informations, relation between groups (pages) are measured (via counting among common user between those groups) then visualized (with ForceAtlas2 spatialisation layout algorithm).

$$Put \begin{cases} U_{act}(A) & \text{Users who realized the action "act" in the group or page A} \\ R_{act}(A) & \text{The relation between A and B according to the action "act"} \end{cases}$$

Then the relation R between A and B varies according to the type of action studied, the following measures are defined:

$$R_{publication}(A, B) = E_{publication}(A) \cap E_{publication}(B)$$

$$R_{comment}(A, B) = E_{comment}(A) \cap E_{comment}(B)$$

$$R_{reaction\_Pub}(A, B) = E_{reaction\_Pub}(A) \cap E_{reaction\_Pub}(B)$$

$$R_{reaction\_Com}(A, B) = E_{reaction\_Com}(A) \cap E_{reaction\_Com}(B)$$

The results displayed in the Table 2, the nodes having a weak relation are hidden and keep only the nodes having R (A, B) > 1000.

Table 2. Detailed description of group connectivity results

ACT type	Nodes	Edges	Degree	Modularity	Community
Comments	37	225	6.081	0.175	3
Publications	67	107	1.597	0.621	11
Like comments	20	71	3.55	0.136	3
Like publications	22	119	5.409	0.133	2

Visualization with ForceAtlas2: ForceAtlas2 [10] is a spatialization layout algorithm implemented in Gephi software which allows to have a disposition aimed by force. nodes regimented as charged particles, while edges attract their nodes, as springs. This force creates a movement which converges on a balanced state. This final shape should help in the data interpretation (Fig. 4).

Algorithm 1. ForceAtlas2

Input: Undirected graph $G = (V, E)$, *iterations*, *gravitational* and *repulsive* force scalars f_g and f_r

Output: A position $p_v \in R^2$ for each $v \in V$

1 $globalspeed \leftarrow 1.0$
2 **forall the** $v \in V$ **do**
3 $\quad$ $p_v \leftarrow \text{random}()$
4 $\quad$ $f_v \leftarrow (0.0, 0.0)^T$ $\hspace{4cm}$ // Net force on node v
5 $\quad$ $f_v \leftarrow (0.0, 0.0)^T$ $\hspace{3.5cm}$ // f_v(preceeding iteration)
6 **end**
7 **for** $i \leftarrow 1, iterations$ **do**
8 $\quad$ BH.rebuild() $\hspace{3.8cm}$ // (Re)build Barnes-Hut tree
9 $\quad$ **forall the** $a \in V$ **do**
10 $\quad\quad$ $f_v \leftarrow f_v - p_v$ $\hspace{3.8cm}$ // (Strong) Gravity
11 $\quad\quad$ $f_v \leftarrow f_v + k_r.BH.force_{at}(p_v)$ $\hspace{1.8cm}$ // Repulsion
12 $\quad\quad$ **forall the** $w \in neighbors(v)$ **do**
13 $\quad\quad\quad$ $f_v \leftarrow f_v + \frac{p_v - p_w}{|p_v - p_w|}$ $\hspace{2.5cm}$ // Attraction
14 $\quad\quad$ **end**
15 $\quad$ **end**
16 $\quad$ UpdateGlobalSpeed()
17 $\quad$ **forall the** $v \in V$ **do**
18 $\quad\quad$ $p_v \leftarrow localspeed(v) * f_v$ $\hspace{2.5cm}$ // Displacement
19 $\quad\quad$ $f_v \leftarrow f_v$
20 $\quad\quad$ $f_v \leftarrow (0.0, 0.0)$
21 $\quad$ **end**
22 **end**
23 **Function** $LOCALSPEED$ (v) $\hspace{3cm}$ // For a node v
24 return $\dfrac{globalspeed}{1.0 + \sqrt{(globalspeed + swing(v))}}$
25 **Function** $SWING(v)$ $\hspace{3.8cm}$ // For a node v
26 return $|f_v - f_v|$

For each Type studied, we notice the variation of the number of community extracted, the publications, being the source of the other information, generated 11 community to us.

(a) Connection between groups by users who have published within these groups

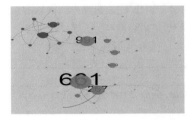

(b) Connection between groups by users who have commented within these groups

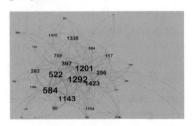

(c) Connection between groups by users who have reacted to publications

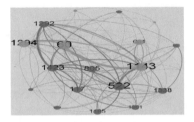

(d) Connection between groups by users who have reacted to comments

Fig. 4. Connectivity between groups by action type

4.4 Reactions Statistics

Since Wednesday 24th February 2016, Facebook offers reactions in order to offer the much-claimed alternative to the simple button I like. The Table 3 shows a clear view on the extracted reactions distribution.

Table 3. Extracted reactions per type

Reaction type	Before 24_02_2017	After 24_02_2017
Angry	1	235
HAHA	543	581638
LIKE	6965483	27594050
LOVE	1091	780799
SAD	0	532
THANKFUL	0	0
WOW	50	106603
Total	6967168	29063325

The button "LIKE" has received a large use percentage(95 %), which is normal because before the appearance of these new reaction buttons (before

24/02/2016), All the reactions were a simple "LIKE", hence the interest of neglecting the comparison of this button with the different reaction types. So by neglecting the "LIKE" button we notice that two buttons are the most used, namely "LOVE" (53%) and "HAHA" (39.5%), Reflecting a social trend of the Moroccan population, to be emotional and humorous. This observation is validated by the negligible number of the reaction type "Angry" (235 reactions) and of the reaction type "SAD" (532 reactions).

4.5   Publications Statistics

To make the publication more visible, the Facebook user enriched it with content: Image, video, event, Link or just leave text. Among all the types of contents cited, the photos are the visual cues that gave rise to the More feedback from users (Table 4).

Table 4. Type of publication and statistics

Publication type	Publication	Reactions	Comments	Shares
Photos	2020197	12779890	7374995	138346468
Videos	701269	4282558	1334805	23976143
Links	1595325	2767789	1530594	18206035
Events	74833	290521	806053	111450
Normal	1441698	7034893	4158864	52453299
Total	5833322	27155651	15205311	233093395

5   Conclusion

OSNs are among the most intriguing phenomena of recent years. In this work a description of data extraction from Facebook was made via the semi-supervised process, after general statistics for of Moroccan community Facebook use was given, a macro intra-group connectivity analysis was Carried out using graph theory concepts. The visualization is done by applying the Spatialization algorithm ForceAtlas2. A general macro-description of the extracted data was developed in the second part, which requires in-depth studies and analyzes in order to extract indicators on the social behavior of the studied population as well as to profile the Moroccan users, those studies Are the future axes to be undertaken.

References

1. Stenger, T., Coutant, A.: Les réseaux sociaux numériques: des discours de promotion à la défnition d'un objet et d'une méthodologie de recherche (2010)
2. Boyd, D.M., Ellison, N.B.: Social Network Sites: Definition, History, and Scholarship (2008)

3. Ahmed, I., Qazi, T.F.: A look out for academic impacts of Social networking sites (SNSs): a student based perspective (2011)
4. Danias, K., Kavoura, A.: The role of social media as a tool of a company's innovative communication activities (2013)
5. Urista, M.A., Dong, Q., Day, K.D.: Explaining why young adults use MySpace and Facebook through uses and gratifications theory (2009)
6. Ellison, N.B., Steinfield, C., Lampe, C.: The Benefits of Facebook Friends: Social Capital and College Students' Use of Online Social Network Sites (2007)
7. De Meo, P., Ferrara, E., Fiumara, G., Provetti, A.: On Facebook, most ties are weak (2014)
8. Laboratory for Web Algorithmics Datasets. http://law.di.unimi.it/Datasets.php
9. Social Bakers. Facebook Statistics by Country (2017). https://www.socialbakers.com/statistics/facebook/pages/total/morocco/
10. Jacomy, M., Venturini, T., Heymann, S., Bastian, M.: ForceAtlas2, a Continuous Graph Layout Algorithm for Handy Network Visualization Designed for the Gephi Software (2014)

Designing Middleware over Real Time Operating System for Mobile Robot

M. A. Rabbah[1(✉)], N. Rabbah[2], H. Belhadaoui[1], and M. Rifi[1]

[1] RITM Laboratory ESTC, Hassan II University,
BP. 8012, Casablanca, Morocco
mrabbah@ieee.org
[2] Laboratory of Structural Engineering, Intelligent Systems
and Electrical Energy, ENSAM, Hassan II University,
BP. 20000, Casablanca, Morocco

Abstract. Commanding some part of mobile robots or connected objects requires real time capability, especially for industrial need and applications where security is paramount. Real time operating system play here a big role by providing satisfactory level of safety. Our study will focus on comparing the general-purpose operating system Raspbian with PREEMPT_RT real time operating system, both running on the embedded system Raspberry PI 3. We will study those OS behaviors and performances under load and while receiving a critical event from outside sensors or connected object. At the same time, we will raise the impact that some scheduling policies have on a designed mobile robot.

Keywords: RTOS · Mobile robot · Embedded system · Scheduling policies

1 Introduction

In real application, the need of controlling the robot behavior is paramount; the major areas of robots application like surgery and industrial doesn't accept any fault tolerance. The mobile robots are equipped with embedded computer system that control them; most of those embedded cards, run operating system (OS) that is responsible of: managing device I/O, Communication, Memory, Task and Timer, and finally Interrupt and event handling. We can classify OS to two main categories: a general purpose OS (GPOS) like Debian, or Windows... and RTOS like VxWorks or Xenomai...

Before developing any real robot application, most developers choose a middleware that will provide the application the necessary independence from the heterogeneity of hardware and the OS. In our last work [1], we present the challenges facing those middleware, and one of them was the real time capability. In this contribution, we will detail more this criterion, by studding a robot behavior while processing a Pulse With Modulation (PWM) signal. This paper is organized as follows: In Sect. 2 briefly surveys the related work. Section 3 presents the concepts to master for developing Real time mobile robot application. Section 4 concludes the paper and outlines some possible future work.

2 Related Work

In [2] a self driving miniature vehicle was developed based on commercial-off-the-shelf design, sensors and actors are interfaced with STM32F4 discovery board witch run ChibiOS/RT real-time OS, and the OpenDaVINCI middleware was used; however OpenDaVINCI doesn't support real time tasks. In this work [3] the authors have used ATmega1280 board, which run FreeRTOS. The results show that a higher priority task could interrupt the execution of a low priority task and resume its own execution.

In this study [4] authors design a car driven by a Raspberry PI. Both response time of the ultrasonic sensor and time taken by the robot to move full rotation were measured, and the results show an improvement in the reaction with the patched OS. Authors in [5] present RT-Est a RTOS for semi-fixed-priority scheduling algorithms; experimental evaluations show that this scheduling is well suited for mobile robots.

In this study [6], authors explain the long integration time problem encountered every new release of hardware while using Xenomai RTOS; authors propose a solution by migration to RT-Preempt OS. Results show that the proposed solution is good enough in terms of determinism and solve RT-IRQ conflicts. The authors here [7] developed a Humanoid Robot using XBotCore middleware. The RTOS used was Xenomai. The middleware implements four tasks in real time behavior, however it does not offer any abstraction layer for running real time process.

In this contribution [8], a benchmarking was done between Xenomai and a RT patched Linux. The processing time at low load and overload scenarios show that RT Linux patch has better performance than Xenomai. In the following work [9], authors compare scheduling latency between PREEMPT-RT Linux and LITMUS RT. Since LITMUS RT is based mainly on Linux, there wasn't a big difference between those two OS.

3 Concepts to Master for Developing Real Time Mobile Robot Application

When we want to build a robotic application that must satisfy real time constraints, many concepts must be mastered, we think that we must capitalize on all researchers already done, and package all this knowledge in a middleware layer. So we will allow beginners as experts to develop effective application in an easy way, to illustrate our goal we will explain the built in robot for demonstration, after that we will address the test environment. By later, we will expose the results and we will interpret them.

3.1 Hardware Architecture

We built a differential robot having the ability to navigate independently, and avoiding obstacles (Fig. 1), the components of the robot are as follows:

- HC-SR04 Ultrasonic sensor for detecting obstacle: working on 5 V DC its range detection 2 cm to 0.5 m, resolution of 0.3 cm and frequency of 40 kHz
- Servomotor: instead of having several ultrasonic sensors, we opt to use a servomotor that will allow the Ultrasonic scan an area of 180°.

- Two motors: powered by 12 V
- L298 N H-bridge Dual Motor Controller
- Lipo Battery: produce 12 V and power directly the H-Bridge
- DC-to-DC Converter adjustable allows step down voltage from 12 V to 5.1 V.
- Raspberry Pi 3 embedded card, it is connected to the ultrasonic sensor throw two digital general purpose input output, one for the trigger and other for the Echo, also it control the servomotor throw a PWM signal, and motors via the H-bridge, by two digital output for direction and one PWM signal for the velocity (Fig. 2).

Fig. 1. Obstacles avoidance mobile robot.

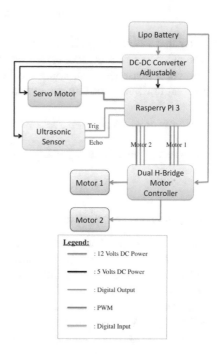

Fig. 2. Schematic diagram of the hardware architecture including the connection plan.

3.2 Test Environment

Firstly, we want to show the benefit behind building a middleware on the top of a RTOS by comparing a RTOS with a GPOS behaviour. Raspberry Pi 3 came Raspbian GPOS. Most RTOS like (FreeRTOS, Xenomai, ChibiOS/RT…) that was supported in earlier version of this board doesn't work any more. So one important thing about choosing a RTOS is the community behind it, we opted for PREEMPT_RT RTOS that patch the default one.

To study each OS behavior, we proceed to a latency test by following those steps:

- We overloaded the four CPU cores using "stress" utility command (1):

$$\$ > \quad stress\ \text{-}cpu\ 4 \tag{1}$$

- After that we simulate a thousands of external interrupt request (IRQ) by running from a remote host the following command (2):

$$\$ > \quad sudo\ ping\ \text{-}i0.01\ raspberrypi\_host \tag{2}$$

- After that we used "cyclictest" utility to create a configurable number of threads. The processor affinity mask is also set. After execution, each thread starts a periodic execution phase [9]. The command bellow (3) run 20k loops with 200 µs cycle interval lunching four threads, one in each CPU with priority 99:

$$\$ > \quad sudo\ cyclictest\ -l20000\ \text{-}m\ \text{-}Sp99\ -i200 \tag{3}$$

- We specify the scheduling policy and priority by using "chart" command (4):

$$\$ > \quad sudo\ chrt\ \text{-}f\ \text{-}p\ cyclictest\text{-}pid\text{-}number \tag{4}$$

We can choose between many scheduling policies among:

- SCHED_FIFO: First in, first out, real time processes.
- SCHED_RR: Round robin real time processes.
- SCHED_OTHER: Normal time/schedule sharing.
- SCHED_BATCH: Almost the same as the SCHED_OTHER, but the process is considered always the most CPU consuming.
- SCHED_DEADLINE: implements the Constant Bandwidth Server algorithm.
- SCHED_DEADLINE: uses the Greedy Reclamation of Unused Bandwidth algorithm.

After launching the test, we got the result showing in Figs. 3, 4 and 5:

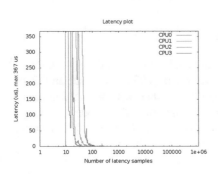

Fig. 3. Cyclictest result for Raspbian GPOS using SCHED_OTHER policy.

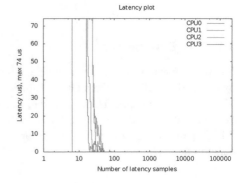

Fig. 4. Cyclictest result for PREEMPT_RT using SCHED_OTHER policy.

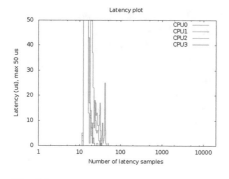

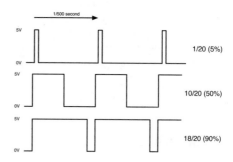

Fig. 5. Cyclictest result for PREEMPT_RT using SCHED_FIFO.

Fig. 6. PWM signal sent to Raspbian GPIO.

Figure 3 represents the cyclyctest result for Raspbian GPOS using SCHED_OTHER policy; the latency was about 367 μs. Figure 4 represents the cyclyctest result for PREEMPT_RT using SCHED_OTHER policy; the latency was about 74 μs. Figure 5 represents the cyclyctest result for PREEMPT_RT using SCHED_FIFO policy; the latency was about 50 μs, which represent a frequency of 20 kHz.

To better understand the impact of the latency in real robot application, we will choose the PWM signal and compare the GPOS and the RTOS behavior. The servo angle of mechanical rotation is determined by the width of the electrical pulse that is applied to the control wire (Fig. 6), the servomotor expects to see a pulse every 20 ms, for example, a 1.5 ms pulse will make the motor turn to the 90° position.

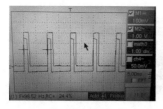

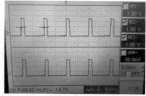

Fig. 7. PWM signal sent by the GPOS (Red) and the one sent by the RTOS (Green) using SCHED_OTHER policy.

Fig. 8. PWM signal sent by RTOS using SCHED_FIFO policy (Blue) compared to GPOS (RED).

Fig. 9. Both SCHED_FIFO (Blue) and SCHED_OTHER policies (Green) in RTOS, have the same precision.

The problem here if the program that sends the signal is pre-empted the pulse width change and the angle change also, this nose is well illustrated in this video [10].

Under the same test conditions as explained in Sect. 3.2, we execute a program that send a PWM signal to servomotor with a hight priority, for both GPOS and RTOS, and we change each time the scheduling policy. Using an Oscilloscope attached to Raspberry PI pin we measured in each case the output signal, Figs. 7, 8 and 9.

Figure 7 show the PWM signal sent by the GPOS (Red) and the one sent by the RTOS (Green) both using SCHED_OTHER policy for the same angle (90°), it is clear that the RTOS is more accurate, also in Fig. 8 we show a better result for the RTOS (Blue) using SCHED_FIFO policy compared to GPOS. In Fig. 9 we remark that both SCHED_FIFO (Blue) and SCHED_OTHER policies (Green) in RTOS, have the same precision but SCHED_OTHER policy can cause an uncontrolled behavior when it is pre-empted; those videos [11, 12] show exactly the behavior of each one.

4   Conclusion

In this work, we demonstrate the important of using Real Time Operating System (RTOS) over General Purpose Operating System, a mobile robot was built to illustrate the behavior of Raspbian and PREEMPT_RT operating systems, with PREEMPT_RT OS and FIFO scheduling policy we were able to satisfy the constraint of 20 kHz. Also in this contribution, we show how we can test effectively any operating system based on Linux, by using utilities like "stress", "chrt" and "cyclictest". At the end, we show the problem encountered during controlling the robot servomotor while sending a PWM signal, and we propose a solution that combine a SCHED_FIFO policy and a RTOS.

When we want to build a robotic application that must satisfy real time constraints over OS, many concepts must be mastered, and we think instead of each researcher or developer does this task in an individual way, we must capitalized all researchers efforts, and package all this knowledge in a middleware layer in future work. Therefore, we will allow beginners as experts develop effective application in an easy way.

References

1. Rabbah, M.A., Rabbah, N., Belhadaoui, H., Rifi, M.: Challenges facing middleware for mobile robots in smart environment. Int. J. Sci. Eng. Res. 7(11), 33–40 (2016)
2. Berger, C., Hansson, J.: COTS-architecture with a real-time OS for a self-driving miniature vehicle. In: SAFECOMP 2013-Workshop ASCoMS (Architecting Safety in Collaborative Mobile Systems) of the 32nd International Conference on Computer Safety, Reliability and Security (2013)
3. Moster, J.E: Integration of a real-time operating system into automatically generated embedded control system code (2014)
4. Murikipudi, A., Prakash, V., Vigneswaran, T.: Performance analysis of real time operating system with general purpose operating system for mobile robotic system. Indian J. Sci. Technol. 8(19), 1–6 (2015)
5. Chishiro, H., Yamasaki, N.: RT-Est: real-time operating system for semi-fixed-priority scheduling algorithms. In: 2011 IFIP 9th International Conference on Embedded and Ubiquitous Computing (EUC). IEEE (2011)
6. Gosewehr, F., Wermann, M., Colombo, A.W.: From RTAI to RT-Preempt a quantitative approach in replacing Linux based dual kernel real-time operating systems with Linux RT-Preempt in distributed real-time networks for educational ICT systems. In: IECON 2016-42nd Annual Conference of the IEEE Industrial Electronics Society. IEEE (2016)

7. Muratore, L., Laurenzi, A., Hoffman, E.M., Rocchi, A., Caldwell, D.G., Tsagarakis, N.G.: XBotCore: a real-time cross-robot software platform. In: IEEE International Conference on Robotic Computing (IRC). IEEE (2017)
8. Marieska, M.D., Kistijantoro, A.I., Subair, M.: Analysis and benchmarking performance of real time patch Linux and Xenomai in serving a real time application. In: 2011 International Conference on Electrical Engineering and Informatics (ICEEI). IEEE (2011)
9. Cerqueira, F., Brandenburg, B.: A comparison of scheduling latency in Linux, PREEMPT-RT, and LITMUS RT. In: 9th Annual Workshop on Operating Systems Platforms for Embedded Real-Time Applications. SYSGO AG (2013)
10. Rabbah, M.A.: General Purpose Operating System PWM Servo Motor Behavior 1. https://www.youtube.com/watch?v=6GLR42VCaGE. Accessed 08 Nov 2017
11. Rabbah, M.A.: Real time operating system: PWM servo motor behaviour using SCHED_OTHER policy. https://www.youtube.com/watch?v=XRVKavq-7xI. Accessed 08 Nov 2017
12. Rabbah, M.A.: Real time operating system: PWM servo motor behaviour using FIFO policy. https://www.youtube.com/watch?v=sk0OI3NdQf8. Accessed 08 Nov 2017

Sensor Fault Detection and Isolation for a Robot Manipulator Based on High-Gain Observers

Khaoula Oulidi Omali[✉], M. Nabil Kabbaj, and Mohammed Benbrahim

Integrations of Systems and Advanced Technologies Laboratory(LISTA), Sidi Mohammed Ben Abdellah University, Faculty of Sciences of Fez, Fez, Morocco
{khaoula.oulidiomali,n.kabbaj,mohammed.benbrahim}@usmba.ac.ma

Abstract. This paper deals with the analysis and design of a fault detection and isolation scheme (FDI) for a class of affine nonlinear systems using high gain observers. The observers are applied for robot manipulator especially robot named Articulated Nimble Adaptable Trunk "ANAT" in order to detected and isoleted sensor fault. The simulations results prove the effectiveness, performances and the robustness of the approach.

Keywords: Fault detection · Robot manipulator
High gain observer · Residuals · Nonlinear systems

1 Introduction

In recent years, robot reliability and its fault tolerance have received significant attentions, especially in robotic applications in remote and hazardous environments (the medicinal or biotechnological treatment of disease requiring patient robot interfacing).

In the literature, fault detection is used to indicate that something is wrong in the monitored system and fault detection is used to determine the location of the fault. There are many approaches based of fault detection, such as parameter estimation, parity relations and observer based approaches. In the observer based scheme, reconstruct some or all system outputs from accessible process variables. The fault indicators are generated by forming the differences between the estimated outputs and the actual outputs. In the absence of faults, the differences converge to zero as soon as a fault occurs, these differences are no longer equal to zero [1].

One of the most challenging topics in observer based in the FDI research and application areas is high-gain observers. They were first introduced by Gauthier et al. [2]. They are non-linear observers that take into account the nonstationarity, non-linearity of industrial processes and at the same time ensure a good estimate of the real state with easy adjustment of gain vector [3–5].

© Springer International Publishing AG, part of Springer Nature 2019
J. Mizera-Pietraszko et al. (Eds.): RTIS 2017, AISC 756, pp. 426–435, 2019.
https://doi.org/10.1007/978-3-319-91337-7_38

In [2], the authors gave a necessary and sufficient condition that characterizes the class of nonlinear affine systems mono-input, mono-output observable for any input. We show that this class of systems is diffeomorphic to the canonical form which consists of a fixed linear dynamic and a triangular nonlinear dynamic. Using this structure, the authors proposed a high gain observer synthesized under a Lipschitz hypothesis on nonlinear dynamics. The gain of the proposed observer is derived from the resolution of an algebraic Lyapunov equation which can be explicitly computed. Several generalizations of this result to multi-input systems are proposed in [6,7]. The main characteristic of the high gain observer proposed in [3], its simplicity of implementation since the observer is a copy of the dynamics of the system with a gain whose expression can be explicitly calculated. In addition, the observer setting can be completed by selecting a single synthesis parameter. The effectiveness of this type of observer is proved by several industrial applications [8,9]. In this paper, high gain observer will be applied to a robot manipulator in objective of fault detection and isolation.

The paper is organized as follows. The next section exposes the problem formulation. High-gain observers used for the design of residual generators are development in Sect. 3. Section 4 provides application for robot manipulator. The conclusion is given in Sect. 5.

2   Problem Formulation

Consider the MIMO nonlinear system with m inputs and p outputs defined by the following state representation [10]:

$$\begin{cases} \dot{x} = f(x, d, t) + \sum_{i=1}^{m} g(x, t)u(t) \\ y = Cx \end{cases} \tag{1}$$

where $x \in \mathbb{R}^n$ is the state variable vector, $u(t) \in \mathbb{R}^m$ is the input control vector and $y \in \mathbb{R}^p$ is the output vector, f(x, d, t) is the n-dimensional unknown nonlinear dynamics and d is the disturbance, $g(x, t)$ is the $(n \times m)$ nonlinear control dynamics matrix, C is the $(p \times n)$ output distribution matrix.

3   High Gain Observer

In general, an observer is a dynamic system that provides estimations of the current state of the system, by using the previous knowledge of the inputs and outputs of the system [6].

Consider the following class of affine nonlinear system:

$$\begin{cases} \dot{x} = f(x, d, t) + \sum_{i=1}^{m} g(x, t)u(t) \\ y = Cx \end{cases} \tag{2}$$

The system (2) has the input $u(t) \in \mathbb{R}^m$ in which has a set of admissible values of the input. It is also assumed that there exists a physical domain $\Omega \in \mathbb{R}^n$ (open, bounded) of evolution of the input and that is the domain of interest of the system.

Suppose that system (2) it is observable in the sense of rank and that $u = 0$ an input universal, then the jacobian $\{h_1, L_f h_1, \ldots, L_f^{n-1} h_1, h_2, L_f^{n-1} h_2, L_f^{n-1} h_p\}$ with respect to $x \in \mathbb{R}^n$ and x is of rank n. In the vicinity of a regular point one can select a subset of full rank:

$$\phi = \{z_1, \ldots, z_n\}\{h_1, L_f h_1, \ldots, L_f^{n-1} h_1, h_2, L_f^{n-1} h_2, L_f^{n-1} h_p\} \tag{3}$$

with $\sum_{k=1}^{p} \eta_k = n$.

The input $h_k(x)$ intervenes in the order η_k. This determines a local coordinate system in which the system (2) is written as:

$$\begin{cases} \dot{z} = Az + \tilde{\varphi}(z) + \bar{\varphi}(z)u \\ y = Cz \end{cases} \tag{4}$$

with $z \in \mathbb{R}^n, u \in \mathbb{R}^m, y \in \mathbb{R}^p$

$$A = \begin{bmatrix} A_1 & & \\ & \ddots & \\ & & A_p \end{bmatrix}; B = \begin{bmatrix} C_1 & & \\ & \ddots & \\ & & C_p \end{bmatrix}; \tilde{\varphi}(z) = \begin{bmatrix} \tilde{\varphi}_1(z) \\ \vdots \\ \tilde{\varphi}_p(z) \end{bmatrix}; \bar{\varphi}(z) = \begin{bmatrix} \bar{\varphi}_1(z) \\ \vdots \\ \bar{\varphi}_p(z) \end{bmatrix} \tag{5}$$

with

$$A_k = \begin{bmatrix} 0 & 1 & \ldots & 0 \\ \vdots & \vdots & \ddots & \vdots \\ 0 & 0 & \ldots & 1 \\ 0 & 0 & \ldots & 0 \end{bmatrix}; \tilde{\varphi}_k(z) = \begin{bmatrix} 0 \\ \vdots \\ 0 \\ \tilde{\varphi}_k(z) \end{bmatrix}; \bar{\varphi}_k(z) = \begin{bmatrix} \varphi_1 k(z) \\ \vdots \\ \varphi_{\eta_k} k(z) \end{bmatrix}; C_k = \begin{bmatrix} 1 & 0 & \ldots & 0 \end{bmatrix} \tag{6}$$

with $\tilde{\varphi}_k(z) = L_f^{\eta_k} h_k$, $\varphi_i k(z) = L_g L_f^{i-1} h_k$; $dim(A_k) = (\eta_k \times \eta_k)$, $dim(C_k) = (1 \times \eta_k)$ and $dim(\tilde{\varphi}_k(z)) = dim(\bar{\varphi}_k(z)) = (\eta_k \times 1)$ for $k = 1, \ldots, p$; $i = 1, \ldots, \eta_k$.

For the following theorem, is not used the linearity in u, the following system is then considered:

$$\begin{cases} \dot{z} = Az + \varphi(z, u) \\ y = Cz \end{cases} \tag{7}$$

Let K be a matrix $(n \times p, p)$ such that

$$K = \begin{bmatrix} K_1 & & \\ & \ddots & \\ & & K_p \end{bmatrix} \tag{8}$$

(with K_k of dimension $n \times 1$), such that for each block k, the matrix $A_k - K_k C$ have negative real parts. Then the system is uniformly locally observable, and it exists $T_0 > 0$, such that, for every T, such $0 < T < T_0$, the following system Constitutes an observer for the system (7):

$$\dot{z} = A\hat{z} + \varphi(\hat{z}, u) + \Lambda^{-1}(T, \delta)K(y - C\hat{Z}) \tag{9}$$

with $\hat{z}_\mu k = y_k$, $\hat{z}_j = \hat{z}_j \neq \mu_k$,

$$\Lambda(T, \delta) = \begin{bmatrix} T^{\delta_1}\Delta_1(T^{\delta_1}) & & \\ & \ddots & \\ & & T^{\delta_p}\Delta_p(T^{\delta_p}) \end{bmatrix} \quad with \quad \Delta_k(T) = \begin{bmatrix} T^{\delta_k} & & & \\ & \ddots & & \\ & & T^{2\delta_k} & \\ & & & \ddots \\ & & & & T^{\eta_k\delta_k} \end{bmatrix} \tag{10}$$

Moreover, the standard of the observation error is bounded by an exponential whose decay rate can be chosen arbitrarily large.

Remark 1. The system

$$\left\{ \dot{\hat{z}} = A\hat{z} + \varphi(\hat{z}, u) + \Lambda^{-1}(T, \delta)K(y - C\hat{Z}) \right. \tag{11}$$

Also an observer for system (7). If a change of variable $z = \phi(x)$ has been necessary, it is necessary to return to the old database by $(\hat{x}) = \phi^{-1}(\hat{z})$. By applying this change of coordinates to the previous system, we obtain the observer in the old coordinates:

$$\left\{ \dot{\hat{x}}(t) = f(\hat{x}(t)) + \sum_{i=1}^{m} g_i(\hat{x}(t))u_i(t) + [\frac{\partial\phi(x)}{\partial x}]_{\hat{x}}^{-1}\Lambda^{-1}K[y(t) - C\hat{x}(t)] \right. \tag{12}$$

Remark 2. The observer is written in the new coordinates as a system copy plus a non-linear correction. Implementation requires writing the observer into the original frame. Therefore the system becomes:

$$\begin{cases} \dot{\hat{x}}(t) = f(\hat{x}(t)) + \sum_{i=1}^{m} g_i(\hat{x}(t))u_i(t) + [\frac{\partial\phi(x)}{\partial x}]_{\hat{x}}^{-1}\Lambda^{-1}K[y(t) - C\hat{x}(t)] \\ y = Cx \end{cases} \tag{13}$$

The term of correction $[\frac{\partial\phi(x)}{\partial x}]_{\hat{x}}^{-1}\Lambda^{-1}K[y(t) - C\hat{x}(t)]$ then explicates as follows:

$$[\frac{\partial\phi(x)}{\partial x}]_{\hat{x}-1} = \begin{bmatrix} 1 & 0 & 0 & 0 & 0 \\ 0 & 1 & 0 & 0 & 0 \\ 0 & 0 & 1 & 0 & 0 \\ 0 & 0 & 0 & 1 & 0 \\ 0 & 0 & 0 & 0 & 1 \end{bmatrix}; \Lambda^{-1} = \begin{bmatrix} T^{-\delta_1} & 0 & 0 & 0 & 0 \\ 0 & T^{-2\delta_1} & 0 & 0 & 0 \\ 0 & 0 & T^{-\delta_2} & 0 & 0 \\ 0 & 0 & 0 & T^{-2\delta_2} & 0 \\ 0 & 0 & 0 & 0 & T^{-\delta_3} \end{bmatrix} \tag{14}$$

The gain T^{-1} must be selected as $0 < T \leq T_0 < 1$. T_0 is defined according to different parameters (η^2, the constant Lipschitz of the function $\varphi(z, u)$ defined by variable change,...). And the gain K is given by:

$$K = \begin{bmatrix} K_1 & & \\ & \ddots & \\ & & K_5 \end{bmatrix} \tag{15}$$

The pair (A,C) is observable, and it is an easy matter to assign eigenvalues to the matrix A-KC, that has the companion structure:

$$A - KC = \begin{bmatrix} -K_1 & 1 \ldots 0 \\ \vdots & \vdots \ddots \vdots \\ -K_{n-1} & 0 \ldots 1 \\ -K_n & 0 \ldots 0 \end{bmatrix} \tag{16}$$

If a n-pla $\lambda = (\lambda_1, \ldots, \lambda_n)$ of eigenvalues has to be assigned, the vector $K(\lambda)$ is the vector that contains the coefficients of the monic polynomial that has λ as roots. If the assigned eigenvalues are distinct, matrix A-KC can be diagonalized by a Vandermonde matrix:

$$V \equiv V(\lambda) \begin{bmatrix} \lambda_1{}^{n-1} & \ldots & \lambda_1 & 1 \\ \vdots & \ddots & \vdots & \vdots \\ -k_{n-1} & \ldots & \lambda_n & 1 \end{bmatrix} \tag{17}$$

So that

$$V(\lambda)(A - K(\lambda)C)V(\lambda)^{-1} = diag\{\lambda\} = \Lambda \tag{18}$$

Remark 3: Given a set λ of n eigenvalues to be assigned to A-KC, the gain $K(\lambda)$ is readily computed through the formula:

$$K(\lambda) = -V^{-1}(\lambda)[\lambda_1{}^n \ldots \lambda_n{}^n]^T \tag{19}$$

So, the term of correction of the system is:

$$[\frac{\partial \phi(x)}{\partial x}]_{\hat{x}}^{-1} \Lambda^{-1} K[y(t) - C\hat{x}(t)] = \begin{bmatrix} T^{-\delta_1} K_1(q_1 - \hat{q}_1) \\ T^{-2\delta_1} K_2(q_2 - \hat{q}_2) \\ T^{-\delta_2} K_3(q_3 - \hat{q}_3) \\ T^{-2\delta_2} K_4(q_4 - \hat{q}_4) \\ T^{-\delta_3} K_5(q_5 - \hat{q}_5) \end{bmatrix} \tag{20}$$

4   Application: Robot Manipulator

4.1   System Description and Modeling

To illustrate the effectiveness of the proposed approach, we consider Articulated Nimble Adaptable Trunk robot arm with 5° of freedom (DOF). The dynamic model is further specified by the well-known equation for rigid manipulators:

$$\ddot{q} = -M(q)^{-1} F(q, \dot{q}) + M(q)^{-1} \tau \tag{21}$$

where M is the inertia matrix, which is symmetric and positive definite. Thus, $M(q)^{-1}$ always exits. F is the centrifugal, coriolis, and gravity vector; q is the joint position vector; τ is the torque input vector of the manipulator. Let $x = [X_1^T, X_2^T]^T$ the state vector with $X_1 = [q_1, q_2, q_3, q_4, q_5]^T$ and $X_2 = [\dot{q}_1, \dot{q}_2, \dot{q}_3, \dot{q}_4, \dot{q}_5]^T$, and $y = X_1$ is the output vector. The description of the system can be given in state representation form as follows:

$$
\begin{cases}
\dot{x}_1 = x_6 \\
\dot{x}_2 = x_7 \\
\dot{x}_3 = x_8 \\
\dot{x}_4 = x_9 \\
\dot{x}_5 = x_{10} \\
\dot{x}_6 = f_1(x, d, t) + \sum_{i=1}^{5} g_{1i}(x, t) u_i(t) \\
\dot{x}_7 = f_2(x, d, t) + \sum_{i=1}^{5} g_{2i}(x, t) u_i(t) \\
\dot{x}_8 = f_3(x, d, t) + \sum_{i=1}^{5} g_{3i}(x, t) u_i(t) \\
\dot{x}_9 = f_4(x, d, t) + \sum_{i=1}^{5} g_{4i}(x, t) u_i(t) \\
\dot{x}_{10} = f_5(x, d, t) + \sum_{i=1}^{5} g_{5i}(x, t) u_i(t)
\end{cases}
\tag{22}
$$

where $g(x,t) = M(q)^{-1}$; $u_i = \tau_i$ for $i = 1:5$; $f(x, d, t) = -M(q)^{-1} F(q, \dot{q})$.

4.2 High Gain Observer Applied to Robot Manipulator

Using the system model (2), a high gain observer is developed as explained in Sect. 3 in order to improve the effectiveness of the proposed method in which is written as:

$$
\begin{cases}
\dot{\hat{x}}(t) = f(\hat{x}(t)) + \sum_{i=1}^{m} g_i(\hat{x}(t)) u_i(t) + [\frac{\partial \phi(x)}{\partial x}]_{\hat{x}}^{-1} \Lambda^{-1} K[y(t) - C\hat{x}(t)] \\
y = Cx
\end{cases}
\tag{23}
$$

The gain K and T, show the effectiveness of this observer. Thus, the choice of the gain of high-gain observer is based on a compromise between the speed of convergence of the observer and insensitivity to measurement noise. The gain K are chosen according to Remark 3, and T between 0 and 1.

4.3 Simulation Results

The state estimation error $r(t)$ can be calculated as:

$$
r(t) = y(t) - \hat{y}(t)
\tag{24}
$$

The residuals are supposed to differ from zero in case of faults $r(t) \neq 0$ and equal to zero when there are no faults on the sensors $r(t) = 0$.

So the residuals are evaluated as:

$$
r_1 = |q_1 - \hat{q}_1|; \quad r_2 = |q_2 - \hat{q}_2|; \quad r_3 = |q_3 - \hat{q}_3|; \quad r_4 = |q_4 - \hat{q}_4|; \quad r_5 = |q_5 - \hat{q}_5| \tag{25}
$$

Table 1 represents the fault signatures matrix for these residuals. Assuming that simultaneous faults cannot occur, we find that the signatures for each of the failures are quite different.

Table 1. Fault signatures matrix

d_i/r_i	r_1	r_2	r_3	r_4	r_5
d_1	1	0	0	0	0
d_2	0	1	0	0	0
d_3	0	0	1	0	0
d_4	0	0	0	1	0
d_5	0	0	0	0	1

We simulate the system during $T = 30\,\text{s}$ for residuals evolution. It is also noted that all faulty signals are additive.

– a fault d_1 is injected on the first joint (q_1) at the time $t = 15\,\text{s}$ and amplitude '1'. Figure 1 shows that the residual r_1 is different from zero during the time of fault existence, and the residuals $(r_2; r_3; r_4; r_5)$ are equal to zero.

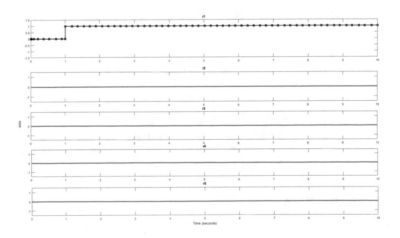

Fig. 1. Residuals evolution of the system with fault d_1

– a fault d_2 is injected on the second articulation (q_2) at the time $t = 15\,\text{s}$. Figure 2 shows that the residual r_2 is sensitive to this fault changed from zero during the failure time.

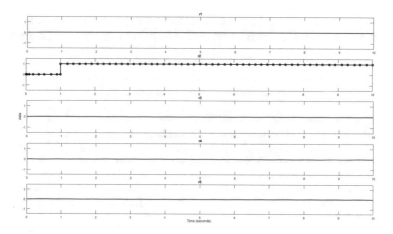

Fig. 2. Residuals evolution of the system with fault d_2

– a fault d_3 is injected on the joint number 3 (q_3) at the instant $t = 15$ s. Figure 3 shows the evolution of the different residuals. It can be seen that the residuals ($r_1; r_2; r_4; r_5$) are equal to zero and the residuals r_3 is sensitive to the fault.

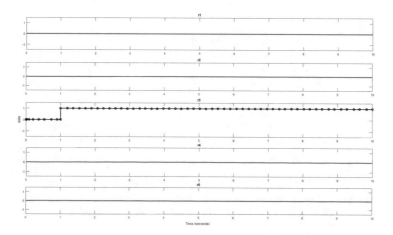

Fig. 3. Residuals evolution of the system with fault d_3

– a fault d_4 is injected on the fourth articulation (q_4) at the instant $t = 15\,\mathrm{s}$. Figure 4 shows that the residual r_4 differ from zero during the failure time.

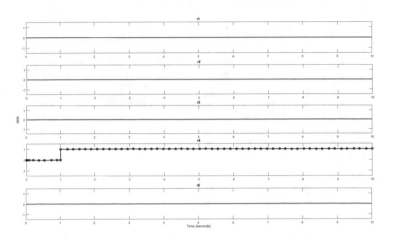

Fig. 4. Residuals evolution of the system with fault d_4

– a fault d_5 is injected on the articulation (q_5) at the time $t = 15\,\mathrm{s}$. Figure 5 shows that the residual r_5 is sensitive to this fault.

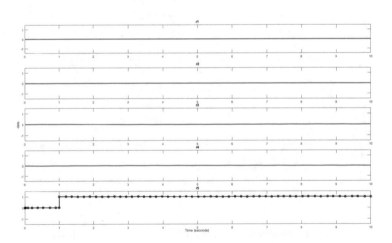

Fig. 5. Residuals evolution of the system with fault d_5

So, we can say that the proposed methods can give a good results and it can detect and isolate the sensor faults in a robot manipulator.

5 Conclusion

In this paper, an approach for fault detection and isolation based on high gain observer, for a class of affine nonlinear systems has been proposed. The approach is applied to a robot manipulator. Through the simulation results, it is shown that all faults defined in specifications are detected and isolated.

References

1. Filaretov, V.F., Vukobratovic, M.K., Zhirabok, A.N.: Parity relation approach to fault diagnosis in manipulation robots. Mechatronics **13** (2003). https://doi.org/10.1109/cca.1998.728495
2. Gauthier, J.P., Hammouri, H., Othman, S.: A simple observer for nonlinear systems - application to bioreactors. IEEE Trans. Autom. Control **37**, 875–880 (1992). https://doi.org/10.1016/j.automatica.2003.08.008
3. Farza, M., Msaad, M., Sekher, M.: A set of observers for a class of nonlinear systems. Int. Fed. Autom. Control-IFAC 2005 (2005). https://doi.org/10.1109/acc.2000.879247
4. Koobaa, Y., Farza, M., Msaad, M.: Obsevateur Adaptatif pour une Classe des Systmes Non Linaires. In: Cinquime Confrence Internationale des Sciences et Techniques de l'Automatique, STA 2004 (2004)
5. Nadri, M.: Observation et commande des systmes non linaires et application aux bioprocds. Thse de doctorat, Universite Claude Bernard Lyon-1 (2001)
6. Busawon, K., Farza, M., Hammouri, H.: Observer design for a special class of nonlinear systems. Int. J. Control **71**, 405–418 (1998). https://doi.org/10.1109/HPDC.2001.945188
7. Gauthier, J., Kupka, I.: Observability and observers for nonlinear systems. SIAM J. Control Optim. **32**, 975–994 (1994). https://doi.org/10.1137/S0363012991221791
8. Farza, M., Hammouri, H., Jallut, C., Lito, J.: State observation of a nonlinear system: application to (bio)chemical processes. AIChE J. **45**, 93–106 (1999). https://doi.org/10.1002/aic.690450109
9. Hammouri, H., Busawon, K., Yahoui, A., Grellet, G.: A Nonlinear Observer for Induction Motors. Eur. Phys. J. Appl. Phys. **15**, 181–188 (2001). https://doi.org/10.1051/epjap:2001181
10. Li, W., Slotine, J.J.: Applied Nonlinear Control. Printice-Hall International, Englewood Cliffs (1991)

Physiological Signals Based Automobile Drivers' Stress Levels Detection Using Shape and Texture Feature Descriptors: An Experimental Study

Abdultaofeek Abayomi[(✉)], Oludayo O. Olugbara, Delene Heukelman,
and Emmanuel Adetiba

ICT and Society Research Group, Durban University of Technology, Durban, South Africa
21451441@dut4life.ac.za, {oludayoo,deleneh,emmanuela1}@dut.ac.za

Abstract. Road traffic fatalities have been predicted to be the seventh leading cause of global deaths by the year 2030, if not quickly addressed. In proffering solutions, several methods have been applied in literature, yet road traffic deaths still remain high. Drivers' stress has been identified as capable of causing road traffic crashes. We present an experimental approach that utilizes digital image processing techniques to detect drivers' stress levels using physiological signals consisting of the electrocardiogram, electromyogram, galvanic skin response, heart rate and respiration data obtained from multiple drivers during real-life driving tasks, as contained in the physionet 'drivedb' corpus of stress recognition in automobile drivers. The histogram of oriented gradient and local binary patterns, which are shape and texture feature descriptor, were applied to the physiological data. The extracted features were used for supervised training of a multilayer perceptron artificial neural networks pattern classifier. This is to enable the detection of a driver's stress by correctly classifying the extracted features into three stress levels - low, medium and high. A classification accuracy of 98.2485% and mean square error of 0.0117 were obtained with the extracted local binary patterns feature. This promising result is better than what was obtained in the literature, using the same corpus and offers a good direction for automatic detection of stress in automobile drivers towards mitigating road traffic crashes and the resultant deaths.

Keywords: Artificial Neural Networks · Classification · Driving · Emotion
Histogram of Oriented Gradient · Physiological signals · Local binary patterns
Road · Stress

1 Introduction

The need for enhancing mobility and people commuting from one place to the other for diverse reasons including necessity, business and pleasure has compelled the use of various means of transportation such as automobiles, aircrafts, ships, bikes, trekking, and animal powered transportation, among others. As a result of its convenience, the automobile, which is a land transportation system and the most popular transportation means [1] in most parts of the world, need to be operated by driving. Land transportation

© Springer International Publishing AG, part of Springer Nature 2019
J. Mizera-Pietraszko et al. (Eds.): RTIS 2017, AISC 756, pp. 436–447, 2019.
https://doi.org/10.1007/978-3-319-91337-7_39

systems are considered to be more dangerous than the other means of transportation [2]. Each year since 2007, according to the World Health Organization (WHO), an estimated 1.3 million deaths occurred worldwide due to road traffic crashes while an additional 50 million people sustain various degrees of injuries [3]. In addition, pedestrians and (motor)cyclists are considered highly vulnerable on roads and constitute about half of these fatalities, as road traffic crashes have been predicted to be the seventh leading causes of global deaths by the year 2030 [3], if urgent action is not taken to stem this awful trend.

To proffer solutions, the United Nations (UN) in September 2015 aimed to reduce the total number of global deaths recorded and injuries sustained from road traffic crashes by 50% on or before the year 2020 [3]. Five risk factors, which include over speeding, drunk-driving, non-use of protective means such as helmet, child restraint, and seat belt [4] have been acknowledged as the fundamental causes of road traffic fatalities. It is noted that hitherto, virtually all these risk factors centered on visible and external factors about and around the drivers, passengers and other road users without considering the unseen internal factors such as the stress level as well as the emotional and cognitive states of the drivers behind the wheels. Despite the little success that has been achieved in reducing fatalities through the enforcement of safety strategies around the identified risk factors, global road traffic deaths continue to be unacceptably high [5] and need to be reduced. It is therefore imperative to consider research focusing on these internal factors as studies have shown that drivers experience and expresses emotions while driving cars and their emotional states [6] as well as their cognitive workload affect their driving.

Stress conditions are related to driving activity because driving is a complicated manual task. It involves mental efforts, visual activities, and requires drivers to manage and process information rapidly. This is aimed at making decisions relating to smooth transportation and a high level of safety for self, pedestrians, passengers and the environment. One major cause of road traffic crashes is the inability of drivers to manage their emotions while performing the driving task. Thus to ensure safety, drivers need to concentrate on the primary driving tasks, be aware of and manage both their negative and positive emotional states appropriately. Anger, anxiety, fear, frustration and stress have been identified as emotions that can negatively affect driving [6] and lead to road crashes if not well managed. However, there is more to human emotions than just the associated outward physical expressions. They consist of other factors including internal processes of feelings and thoughts [7] which the person experiencing them might not be privy to. These affective internal factors are closely associated with physiological signals [8] and the availability of wearable, non-intrusive physiological sensors have enabled seamless reading and monitoring of these physiological signals for efficient tracking of human affective and cognitive state.

As reported in literature, the physiological signals that have been measured during driving tasks include Electrocardiogram (ECG), Electromyogram (EMG), Electrodermal Activity (EDA)/Galvanic Skin Conductance (GSR), Electroencephalogram (EEG), electrooculogram, Heart Rate (HR), Heart Rate Variability (HRV), Respiration (RESP)/breathing and skin temperature [9, 10]. The HR, HRV and GSR physiological data have been experimentally shown to be highly correlated with elevated cognitive workload and driver's stress during driving task [10, 11].

The utilization of raw physiological data is not ideal for direct analysis and resultant computation for decision making, because they are often associated with artifacts and various errors. Discriminative features that represent the characteristics pattern in the raw physiological data need to be obtained to train a pattern classifier. Most of the papers reviewed for this study have extracted and utilized traditional statistical and spectral features from the physiological data in order to detect drivers' stress [8, 10, 11].

The extracted features are consequently fed into a pattern classifier for the purpose of training the classifier and building a model for stress detection. The Support Vector Machines (SVM), Decision Trees, Adaptive Neuro Fuzzy Inference System (ANFIS), Naïve Bayes, Linear Discriminant Analysis and Multilayer Perceptron Artificial Neural Networks (MLP-ANN) are some pattern classification algorithms that have been applied in the literature for the purpose of driver's stress detection [8, 10–12].

The aim of this study is to determine the suitability of the Histogram of Oriented Gradient (HOG) and Local Binary Patterns (LBP) to extract discriminatory features from the physionet 'drivedb' corpus of stress recognition in automobile drivers. The rationale behind choosing these descriptors is to swerve from the traditional statistical methods and apply these descriptors known to have recorded good performance in digital image processing and other fields in order to determine if results could be better. The Multilayer Perceptron Artificial Neural Networks (MLP-ANN) pattern classifier is then applied to classify the extracted features into three stress levels of low, medium, and high while comparing results with those obtained using other techniques. The choice of ANN is based on its efficient performance especially when data distribution is non-linear as obtained in emotionally laced physiological data.

2   Related Works

Although, research works using physiological signals to recognize stress levels or emotional states of automobile drivers while performing the driving task are relatively few, they are active and continuing. Some studies have used physiological data collected in simulated driving scenarios [8, 12] while others are during real-life driving tasks [10, 13]. Varied numbers of participants, driving conditions including routes, driving duration and weather, number of physiological signals measured and, features extracted as well as pattern classifiers have also been used [8, 10, 13] as all these affect classification results.

In the literature reviewed, only two databases of physiological data collected during real-life driving tasks are publicly available. These include the physionet 'drivedb' dataset produced by the MIT Media Lab [10] and the hciLab driving dataset [11]. We experimented with the physionet 'drivedb' dataset of stress recognition in automobile drivers in this study, because it has a higher number of physiological signals measured.

Katsis et al. [8] presented a wearable system methodology using physiological signals of ECG, EMG, RESP and EDA for recognizing four emotional states of disappointment, euphoria, high stress, and low stress of car-racing drivers. Statistical features

such as mean value, mean absolute first difference, root mean square, and mean amplitude extracted from the physiological data were trained using SVMs and ANFIS classifiers. Recognition accuracies of 79.3% and 76.7% respectively were obtained.

In another study [10], ECG, EMG, HR, RESP and GSR physiological data were
continuously recorded from drivers while performing a real-life driving task. These data
were analyzed using 5 min data intervals collected at rest, city driving, and highway
driving to recognize three different levels of a driver's stress – low (rest), medium
(highway) and high (city). A classification accuracy of 97.4% was recorded for multiple
drivers and different driving days using a linear discriminant classifier and a fisher
projection matrix.

Antonio et al. [14] reported an Autonomic Nervous System (ANS) fluctuation amidst
driving style modifications in response to stimulated incremental stress load during
simulated driving. The driving sessions involved steady motorway driving and two other
sessions consisting of added arithmetic questions and vehicle mechanical stimuli that
induce incremental stress load. The respiration activity, HRV and EDA response physiological data of 15 participants were measured. The physiological data in addition to
the vehicle's velocity, steering wheel angle and time responses recorded significant
changes in stress levels during the three driving sessions. A recognition accuracy of over
90% was obtained using a pattern classification algorithm.

In another research study, Ford - an automaker, in collaboration with Rheinisch-
Westfälische Technische Hochschule (RWTH) Aachen University has recently
launched a research project. The project involved using a seat sensor technology with
six sensors embedded at the driver's seat backrest to measure and monitor the heart rate
of a driver [15]. This is to determine his stress level amidst other heart related health
information such as detecting increased heart activity that could lead to heart attack and
notify the driver via a display board. Emergency services can also be contacted using
the device. It is also a boost to the in-car health and wellness solutions by providing
assistance to people with chronic heart illnesses or medical disorders while at the wheel.

3   Materials and Methods

3.1   Data Collection and Pre-processing

For the automatic detection of a driver's stress, which was realized by classification into
the appropriate stress level, the physionet 'drivedb' corpus was used. The corpus is
available on https://physionet.org/cgi-bin/atm/ATM [16]. It consists of physiological
data collected from 9 drivers in a real-world driving task for various days and time while
driving along a well-defined predetermined route in Boston, USA. The ECG, EMG,
GSR of the hand and foot, HR and RESP physiological signals which have been shown
in previous studies [17] to be capable of detecting stress were collected from the drivers.

The stress levels of low, medium and high have been carefully mapped to rest period,
highway driving, and city driving respectively as validated by drivers' self-reported
questionnaires. After our data pre-processing, in order to ensure data uniformity and
completeness only 10 out of 17 available driving records were utilized by us. These are
the only records that have all the six physiological signals collected in addition to having

clear markers indicating the driving duration of each driving segment of rest, city driving, and highway driving. An example of the Amplitude-Time plot of the physiological data collected, the marker and duration during drive5 and drive6 driving sessions are shown in Fig. 1 while Table 1 shows the driving time and data size of each drive that is utilized in this study. As shown in Fig. 1 (a-b), a visual inspection indicate that the HR and RESP data are both highly densed but HR has a higher amplitude as against the low amplitude of RESP data. This indicates that driving activity is strongly associated with these physiological signals.

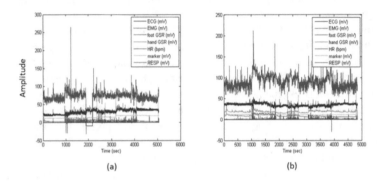

(a)   (b)

Fig. 1. (a-b) Amplitude-Time plots of the physiological signals collected during Drive5 and Drive6

Table 1. Duration and data size of the driving sessions

S/N	Drive Label	Drive Time (hr:min:s)	Physiological Data Size (including the marker)
1	Drive5	1:24:15.935	7x78367
2	Drive6	1:20:46.645	7x75123
3	Drive7	1:28:38.839	7x82442
4	Drive8	1:21:11.548	7x75509
5	Drive9	1:10:52.387	7x65912
6	Drive10	1:21:16.903	7x75592
7	Drive11	1:21:13.097	7x75533
8	Drive12	1:23:04.323	7x77257
9	Drive15	1:17:38.645	7x72209
10	Drive 16	1:04:57.935	7x60418
	Total Time	13:13:56.257	

A total of 13 h, 13 min and 56.3 s driving duration is recorded for all the 10 drives' records utilized as shown in Table 1. This duration is considered large enough for detection of stress level from changes in the pattern of physiological signals monitored. The physiological data collected, drive segment and duration are partitioned as shown in Table 2 to reflect the duration window of the physiological data captured and the driving segment involved.

Table 2. Drive sessions, drive durations (min) and the various drive segments within the driving task

Session	Rest1	City1	Highway1	Toll	Highway2	City2	Rest2	Drive Duration
Drive 5	15.1	16.0	7.8	6.7	7.6	15.1	15.9	84.2
Drive 6	15.5	14.5	7.3	7.1	7.8	12.6	15.8	80.6
Drive 7	16.0	16.8	11.7	9.9	7.6	11.4	15.2	88.6
Drive 8	15.0	12.3	7.2	9.5	7.6	13.4	16.1	81.1
Drive 9	15.7	18.2	8.5	5.7	7.1	7.2	8.4	70.8
Drive 10	15.5	15.8	8.7	5.9	7.5	12.3	15.5	81.2
Drive 11	15.0	15.8	7.6	7.5	7.4	12.7	15.2	81.2
Drive 12	15.9	14.4	8.6	7.5	9.1	11.7	15.8	83.0
Drive 15	15.5	12.5	7.7	6.5	7.3	12.6	15.5	77.6
Drive 16	15.0	16.1	7.1	5.1	6.8	12.9	1.9	64.9
Total Time	154.2	152.4	82.2	71.4	75.8	121.9	135.3	793.2

Each of the physiological datasets was framed in windows of 10 s as also done in similar work reported by other authors [8]. The 10 s period window has been selected and considered suitable because of the fact that there exists a time gap between the instance when a person experienced an increased mental workload resulting in an emotional state and the time a change is noticed in the selected bio-signal response.

3.2 Features Descriptors and Extraction

Feature extraction involves converting the raw physiological data into a series of feature vectors that bears characteristic information inherent in the raw data. In this study, the pre-processed physiological data are in 2-D matrix formats which is similar to a 2-D digital grayscale image representation as an M × N matrix. Consequently, we utilized appropriate MATLAB R2012a codes to read the pre-processed physiological data and applied the HOG and LBP digital image processing algorithms respectively in order to extract discriminatory features.

3.2.1 Histogram of Oriented Gradient (HOG)
The HOG is described as a texture, appearance, and shape extraction technique. It was developed by Dalal and Triggs [18] for human recognition and object detection purpose by considering that local object appearance as well as the shape of an image can be signified using the distribution of intensity gradients or edge orientations. Implementing the HOG algorithm involves dividing the image into cells and the histogram of gradient directions for the pixels within the cells are compiled. The aggregation of these histograms is the HOG features. By calculating a measure of intensity across a block of the image and using the calculated value to normalize all cells within the block, this is used to contrast normalized the local histogram. In order to compute the HOG, four basic steps are required which include masking, orientation binning, local normalization and block normalization. Further details in respect of this feature descriptor can be found in [18].

3.2.2 Local Binary Patterns (LBP)
The LBP descriptor was originally developed by Ojala et al. [19] by forming labels for the pixel block of an image. This is achieved by thresholding a 3 × 3 neighborhood of

each pixel with the center value. The result obtained is treated as a binary number. Since the neighborhoods to the center pixel consist of 8 pixels, the texture descriptor is derived from the histogram of the $2^8 = 256$ different labels. The LBP operator offer impressive performance in unsupervised texture segmentation when used together with a simple local contrast measure. The original LBP operator was later revised and extended to a more generic form to accommodate neighborhoods of various sizes including a circular neighborhood. In this study, the LBP algorithm was implemented in MATLAB R2012a on the pre-processed data and a $2^8 = 256$ elements feature vector was extracted from each of the 10 s windows of physiological data which was sampled at 15.5 Hz frequency. From the pre-processed data, the total number of samples as well as the total number of extracted HOG and LBP features from the samples for each of the driving segments is shown in Table 3. To avoid class imbalance, we trimmed the number of samples to within a reasonable spread such that Rest segment is 1137 (36.88%), City driving is 1046 (33.93%) and Highway driving is 900 (29.19%). This gives a total of 3083 training samples for the three driving segments.

Table 3. Total duration of each merged segment of the ten drives and number of HOG and LBP features extracted from the segments

Driving Segment	Duration (min)	Duration (s)	Number of samples (15.5Hz)	Number of Features extracted from samples (10s windows)
Rest	289.5	17,370	269,235	1,737
City	274.3	16,458	255,099	1,646
Highway	158.0	9,480	146,940	948
Total	721.8	43,308	671,274	4,331

3.3 Pattern Classifier

Pattern classification involves automatic categorization of features extracted from raw data samples into the appropriate classes based on the unique patterns inherent in the training samples. This is often done by using machine learning algorithms such as Support Vector Machines (SVM), Multilayer Perceptron Artificial Neural Networks (MLP-ANN) and Naïve Bayes among others. We decided to experiment with the MLP-ANN because of its popular usage and acceptable results in many pattern recognition and classification tasks. It also requires fewer formal statistical training, has several training algorithms and capable of implicitly detecting intricate non-linear relationships between independent and dependent variables [20].

3.3.1   Artificial Neural Networks

The Artificial Neural Networks (ANN) are pattern recognition algorithms that are capable of approximating a computable function to an arbitrary precision. They are inspired by the neuronal structure of the human brain and are made up of three layers [20]. These include the input layer, the hidden layer(s) and the output layer which are all highly interconnected. The number of hidden layers adopted is crucial to the performance and design of a neural network. One hidden layer network is sufficient to approximate a continuous function but the layer(s) could also be two or more, depending on the complexity of the problems at hand [21]. We utilized two hidden layers in our network design, supported by the fact that applying a higher number of neurons to more hidden layers can lead to fewer local minima.

Since we did not know the number of neurons to be used apriori, we decided to train the network with a varied number of neurons in the hidden layer to determine which number of neurons would give the best stress level classification accuracy.

The last layer is the output layer, whose activation function is determined by the nature of the problem being solved. For a multiclass problem, as is the case in this study, we opted for the hyperbolic tangent sigmoid activation function and configured the number of neurons in this layer to be three, which represents the number of stress levels to be classified. As a result of its simplicity, the most frequently used ANN is the Multi-layer Perceptron (MLP) [21].

The MLP-ANN training dataset consists of a finite set of patterns (x_p, t_p) where t_p indicates the target output vector for the p pattern classes. Each target output is encoded as a 3 element binary vector, because of the 3 stress levels required to be detected in our study. On the other hand, p indicates the number of patterns and x_p is the N-dimensional input vector having N as 256 for the LBP features and 81 for the HOG features that are used in this study. Thus, for the LBP descriptor, the MLP-ANN has in its input layer 256 neurons and 81 neurons for the HOG descriptor. On the other hand, the configured network has 3 neurons in its output layer for each of the two descriptors to represent the 3 stress levels to be detected. This MLP-ANN configuration was designed and implemented using MATLAB R2012a.

The performance evaluation metrics of classification accuracy and Mean Square Error (MSE) were used for the evaluation of the performance of the configured ANN.

4   Experimental Results

Using MATLAB R2012a, all the experimentations carried out in this study was conducted on an Intel (R) Core i5-3570, 3.40 GHz speed CPU personal computer running a 64-bit operating system with Windows 7 Enterprise and 4 GB of RAM. The configured MLP-ANN has two hidden layers and the number of neurons in the hidden layers was varied within the range of 10–100 in multiples of 10. The output layer has 3 neurons representing the low, medium and high stress levels to be detected by the network, while 3083 extracted feature samples were available for the training. The 81 elements each, HOG feature vector of the 3083 total sample representing the x_p were first fed into the configured network through the input layer. A supervised training of the network was carried out for a number of trials for the varied number of neurons in the hidden layer to enable stress level detection. The number of epoch was set to 1000 which indicate the number of training cycles on the training dataset, which we partitioned to be 70% for training, 15% for testing and 15% for validation. Once convergence is reached, the training is configured to stop. The performance goal was also set to 0 and validation checks to 500.

As shown in Table 4, it is noticeable that the results improve as more neurons are used. The improved result is at the expense of the execution time, which also increases as the number of neurons increases. Thus, as execution time increases, the detection accuracy increases except in some few instances. Though, the testing of the built model takes less than a second and the stress level classification result is obtained.

Table 4. Drivers' stress level detection results using HOG features and ANN

S/N	Number of Neurons in Hidden Layers	Accuracy (%)	Mean Square Error (MSE) %	Execution Time (s)
1	10	84.7551	0.0795	32
2	20	86.0525	0.0775	37
3	30	85.8579	0.0864	46
4	40	86.8310	0.0814	50
5	50	87.0905	0.0831	64
6	60	87.0581	0.0834	73
7	70	87.1554	0.0816	83
8	80	87.2851	0.0857	94
9	90	86.6688	0.0855	104
10	100	87.4473	0.0794	140

For the HOG feature, the best stress level detection result of 87.4473% and MSE of 0.0794 was reached at 100 neurons, after which the accuracy dwindles with any further increase of neurons in the hidden layers. The result is marginally better than the 87.2851% with MSE of 0.0820 obtained at 80 neurons. This best classification result though falls short of but compares with the 97.4% accuracy obtained by other authors [10] that utilized statistical features and a linear discriminant classifier on the same dataset. This could be as a result of the varied dimension of feature vectors utilised, number of training samples as well as number of target classes which can significantly affect the performance of a pattern classification algorithm [22] as well as its complexity.

In order to determine whether a better result could be achieved, we extended the experiment by utilizing the LBP feature extracted from the physiological data. The same configuration as earlier applied to the experimentation with the HOG feature was utilized, except for the fact that the number of input neurons are now 256. This represents the LBP feature vector elements for each of the 3083 samples to undergo supervised training. The results obtained are shown in Table 5. We observed, as contained in Table 5, that the LBP descriptor for all the varied number of neurons in the hidden layers, except for the 10 neurons, recorded stress level classification accuracy above the 97.4% which was obtained by the authors in [10]. Also, a progressive increase in the classification accuracy result is to an extent related to an increase in the number of neurons as well as execution time which directly impacts the decision. The best stress detection result of 98.2485% and MSE of 0.0117 is obtained with 60 neurons in the hidden layers and the accuracy dwindles subsequently.

Table 5. Drivers' stress level detection results using LBP features and ANN

S/N	Number of Neurons in Hidden Layers	Accuracy (%)	Mean Square Error (MSE) %	Execution Time (seconds)
1	10	97.1781	0.0178	43
2	20	97.6646	0.0138	49
3	30	97.9890	0.0134	58
4	40	97.5673	0.0180	65
5	50	97.7944	0.0130	78
6	**60**	**98.2485**	**0.0117**	**83**
7	70	98.2160	0.0108	99
8	80	98.0863	0.0130	108
9	90	97.5349	0.0151	118
10	100	98.2160	0.0111	129

The confusion matrix for this best result of the LBP descriptor is presented in Table 6. It is clearly observed in the Table for instance, that 1132 samples out of the 1137 samples of the rest period mapped to the low stress level was correctly classified.

Table 6. The confusion matrix of drivers' stress level detection result using LBP features and ANN

Classified as:	Low	Medium	High	Rate(%)
Low	1132	2	3	99.6
Medium	0	1016	30	97.1
High	0	19	881	97.9

The confusion matrix result is obviously a demonstration of a real-life driving scenario where expectedly, some portions of the city driving can be easier and smoother by being less congested for instance, hence demanding less concentration and a lower mental workload thereby mimicking highway driving. On the other hand, contingencies such as traffic accidents, road maintenances causing gridlock, weather conditions, excessive fidgeting with onboard electronic devices and using of mobile phones on some portions of highway driving can make it tiresome. The workload can impact the mental capability of the driver and make an otherwise smooth riding on a highway as cumbersome as busy city driving.

5   Conclusion

This research work was triggered by the threats posed by road traffic fatalities all over the world predicting this to be the seventh leading causes of death in 2030 as reported by the WHO. Research works using driver's physiological data to detect driver's stress level are active amidst the growing need to make automobiles more intelligent by using on-board devices. This is to enable proactive methods of preventing road crashes.

More synergies among international agencies, industries - automobile companies and research institutions are required. The heart-rate-monitor-seat research supported by Ford automaker, Faurecia's automaker Active Wellness seat which measures HR, RESP rate and detect stress levels as well as Volvo's corporation interests in the research work reported in [9] where high and low mental workload within highway and city drives were successfully classified using physiological data are instructive. The divergent experimental conditions of real-life and simulated, in addition to the palpable lack of many publicly available driver's physiological data corpus are notable challenges. These should be resolved to provide platforms on which different algorithms and methods can be tested and compared for real world deployments and applications.

In this current study, we considered drivers' stress level as one of the causes of road traffic crashes and swerved from the traditional statistical and spectral features applied in many related works, to experiment with the HOG and LBP features extracted from the physiological data collected from multiple drivers while performing a real-life driving task. The MLP-ANN pattern classifier trained with the extracted LBP features gave the best classification result of 98.2485% and MSE of 0.0117. This result is better than what was reported in the literature [8, 10, 14]. It also further indicates the capability of stress levels of drivers being correctly detected and suggestions could be offered in time to avert a road crash.

References

1. Most popular forms of transport. http://www.within-reach.org.uk/most-popular-forms-of-transport.html Accessed 22 Aug 2016
2. European Commission: Road transport - a change of gear. Publications Office of the European Union Luxembourg (2012)
3. World Health Organization (WHO): Global status report on road safety (2015)
4. World Health Organization (WHO): UN decade of action for road safety 2011–2020: saving millions of lives (2011)
5. Media Center, World Health Organization (WHO): Despite progress, road traffic deaths remain too high (2015). http://www.who.int/mediacentre/news/releases/2015/road-safety-report/en/. Accessed 22 Aug 2017
6. James, L., Diane, N.: Road Rage and Aggressive Driving Steering Clear of Highway Warfare. Prometheus Books, Amherst (2000)
7. Picard, R.W., Vyzas, E., Healey, J.: Toward machine emotional intelligence: analysis of affective physiological state. IEEE Trans. Pattern Anal. Mach. Intell. 23(10), 1175–1191 (2001)
8. Christos, D.K., Nikolaos, K., George, G., Dimitrios, I.F.: Toward emotion recognition in car-racing drivers: a biosignal processing approach. IEEE Trans. Syst. Man Cybern. –Part A: Syst. Hum. 38(3), 502–512 (2008)
9. Wiberg, H., Nilsson, E., Lindén, P., Svanberg, B., Poom, L.: Physiological responses related to moderate mental load during car driving in field conditions. Biol. Psychol. 108, 115–125 (2015)
10. Healey, J.A., Picard, R.W.: Detecting stress during real-world driving tasks using physiological sensors. IEEE Trans. Intell. Transp. Syst. 6(2), 156–166 (2005)

11. Schneegaas, S., Pfleging, B., Broy, N., Schmidt, A., Heinrich, F.: A data set of real world driving to assess driver workload. In: Proceedings of the 5th International Conference on Automotive User Interfaces and Interactive Vehicular Applications (Automotive UI 2013), New York, NY, USA, pp. 150–157 (2013). http://doi.acm.org/10.1145/2516540. 2516561.ACM
12. Miyaji, M., Danno, M., Kawanaka, H., Oguri, K.: Driver's cognitive distraction detection using AdaBoost on pattern recognition basis. In: ICVES, pp. 51–56 (2008)
13. Reimer, B., Mehler, B.: The impact of cognitive workload on physiological arousal in young adult drivers: a field study and simulation validation. Ergonomics **54**, 932–942 (2011)
14. Lanatà, A., et al.: How the autonomic nervous system and driving style change with incremental stressing conditions during simulated driving. IEEE Trans. Intell. Transp. Syst. **16**(3), 1505–1517 (2015)
15. The ford motor company. www.technology.fordmedia.eu/documents/Ford_DriverHeart Monitoring.doc. Accessed 26 Aug 2016
16. Goldberger, A. L., Amaral, L.A.N, Glass, L., Hausdorff, J.M., Ivanov, P.C.H., Mark, R.G., Mietus, J.E., Moody, G.B., Peng, C.K., Stanley, H.E.: PhysioBank, PhysioToolkit, and PhysioNet: Components of a New Research Resource for Complex Physiologic Signals. Circulation **101**(23), e215–e220 (2000). Circulation Electronic Pages; http:// circ.ahajournals.org/cgi/content/full/101/23/e215
17. Wilson, G.F.: An analysis of mental workload in pilots during flight using multiple psychophysiologic measures. Int. J. Aviat. Psychol. **12**(1), 3–18 (2001)
18. Dalal, N., Triggs, B.: Histograms of oriented gradients for human detection. In: Proceedings of the IEEE Computer Society Conference on Computer Vision and Pattern Recognition (CVPR 2005), San Diego, Calif, USA, pp. 886–893 (2005)
19. Ojala, T., Pietikainen, M., Harwood, D.: A comparative study of texture measures with classification based on feature distributions. Pattern Recogn. **29**(1), 51–59 (1996)
20. Abraham, A.: Artificial Neural Networks. Handbook of Measuring System Design. Wiley (2005). Edited by Sydenham, P.H., Thorn, R., ISBN 0-470-02143-8
21. Popescu, M.C., Balas, V.E., Perescu-Popescu, L., Mastorakis, N.: Multilayer perceptron and neural networks. WSEAS Trans. Circuits Syst. **8**(7), 579–588 (2009)
22. Nguyen, T.T., Nguyen, L.M., Shimazu, A.: Improving the accuracy of questions classification with machine learning. In: Proceedings of 2007 IEEE International Conference on Research, Innovation and Vision for the Future, pp. 234–241 (2007)

Two New Fast and Efficient Hard Decision Decoders Based on Hash Techniques for Real Time Communication Systems

M. Seddiq El Kasmi Alaoui[✉], Said Nouh, and Abdelaziz Marzak

TIM Lab, Faculty of Sciences Ben M'sik, Hassan II University, Casablanca, Morocco
sadikkasmi@gmail.com

Abstract. Error correcting codes are used to ensure as maximum as possible the correction of errors due to the noisy perturbation of data transmitted in communication channels or stored in digital supports. The decoding of linear codes is in general a NP-Hard problem and many decoders are developed to detect and correct errors. The evaluation of the quality of a decoder is based on its performances in terms of Bit Error Rate and its temporal complexity. The hash table was previously used for alleviating the temporal complexity of the syndrome decoding algorithm. In this paper, we present two new fast and efficient decoding algorithms to decode linear block codes on binary channels. The main idea in the first decoder HSDec is based on a new efficient hash function that permits to find the error pattern directly from the syndrome of the received word. The storage position of each corrigible error pattern is equal to the decimal value of its syndrome and therefore the run time complexity is much reduced comparing to known low complexity decoders. The main disadvantage of HSDec is the spatial complexity, because it requires to previously storing all corrigible error patterns in memory. For reminding this problem, we propose a second decoder HWDec based also on hash, but it requires storing only the weight of each corrigible error pattern instead of the error pattern itself. The temporal complexity of HWDec is more than that of HSDec but its spatial complexity is less than that of HSDec. The simulation results of the proposed decoders over Additive White Gaussian Channel show that they guarantee at 100% the correction of all corrigible error patterns for some Quadratic Residue, BCH, Double Circulant and Quadratic Double Circulant codes (QDC). The proposed decoders are simple and suitable for software implementations and practice use in particular for transmission of Big Data and real time communication systems which requires rapidity of decoding and prefer codes of high rates.

Keywords: Error correcting codes · Hard input hard output decoder · HWDec
HSDec · Hash function · Hash table · Syndrome decoding · Binary channel

1 Introduction

Frequently, data is exchanged using communication channels that are not fully reliable, which can cause the data to be altered by errors. The first solution to combat these errors

© Springer International Publishing AG, part of Springer Nature 2019
J. Mizera-Pietraszko et al. (Eds.): RTIS 2017, AISC 756, pp. 448–459, 2019.
https://doi.org/10.1007/978-3-319-91337-7_40

is the increase of the emission power; this solution is very expensive. The second solution is the addition of redundancy to the message to be transmitted or saved. Adding redundancy in the data to be protected is the principle of errors correcting codes which are increasingly used to detect and correct data transmission errors on computer networks, telecommunication and data storage systems.

In the case of transmission of Big Data and in real-time communication systems the rapidity of decoding is a first major criteria and the use of codes of height rates is generally preferred for reducing additional data of check in order to not complicate the problem of bulkiness and to reduce transmission time and cost.

The quality of a Coding-Decoding system is measured by the bit error rate (BER) which it can guarantee at a given Signal-to-Noise Ratio (SNR) and also by its run time complexity and the hardware resources it needs.

The decoding algorithms can be separated into two categories, the first one is hard decision and the second one is soft decision. Hard decision algorithms work on the binary form of the received information and it's compared with all possible codewords and the codeword which gives the minimum Hamming distance is selected, In contrast, soft decision algorithms work directly on the received symbols and generally they use the Euclidian distance as a metric to minimize the distance [1].

The remainder of this paper is structured as follows. In Sect. 2 we present some decoding algorithms as related works. In Sect. 3 we present the first proposed decoder HSDec, in the Sect. 4 the second proposed decoder will be cited, in the Sect. 5, we present the simulation of results of the proposed decoders and we make a comparison with other decoders. Finally, a conclusion and a possible future direction of this research are outlined in Sect. 6.

2   Related Works

In [2] an approach based on the link between syndromes and correctable errors pattern is developed by using hash techniques; in [3] the authors have presented a method called NESWDA to decode up to five errors in a binary systematic quadratic residue QR (47, 24, 11) code, this method is based on the weight of syndrome difference and proprieties of cyclic codes. The disadvantage of these two methods is that it's just applicable for Quadratic Residue (QR) code; By using the Bit Error Rate Term, the authors of [4] have released a performances study of BCH error correcting codes and they presented a comparative study between the BCH (15, 7, 2) and BCH (255, 231, 3) codes. In [5], the authors have discussed quadratic residue codes over a defined ring.

In [6] the authors have proved that the extended quadratic residue binary codes are the only nontrivial extended binary cyclic codes that are invariant under the projective special linear group; in [7] the authors proposed a performances study and a synthesis, of the new algorithms proposed in this domain, for Reed Solomon (RS), Bose Ray-Chaudhuri and Hocquenghem (BCH) and Low Density Parity Check Codes (LDPC). In [8] a deep learning method to improve belief propagation algorithm was proposed, by attribution of weights to the edges of the Tanner graph the authors generalized the standard belief propagation algorithm. In [9] the authors have presented an iterative hard decision decoding algorithm for binary linear block codes over a binary symmetric channel (BSC).

In [10] the authors have developed a Cyclic Weight (CW) algorithm decoding to decode the binary systematic (47, 24, 11) quadratic residue (QR) code; in addition to the properties of the cyclic codes, they based on the weights of syndromes; the same authors and in a previous paper [11] they have presented an algebraic decoding algorithm to correct all patterns of four or fewer errors in the binary QR (41, 21, 9) code. In order to decode up to five possible errors in a binary systematic QR (47, 24, 11) code, the authors of [12] have presented a table lookup decoding algorithm.

By using the Lagrange interpolation formula, the authors of [13] have calculated the needed primary unknown syndrome for the binary QR code and proposed hardware architecture to implement it, also by using the developed Berlekamp-Massey (BM) algorithm and Chien search they decoded the binary QR code. In [14] the authors have proposed a decoding of quadratic residue codes by using hashing search to determine error patterns.

In [15–18] several hard decoder based on genetic algorithms (GA) are developed, the first one is the HDGA (Hard decision Decoder based on Genetic Algorithms) [15], it used information sets to decode linear block codes; the second one is the Bit Flipping decoding algorithm (BF) [16, 17] developed initially for LDPC codes and generalized after on linear block codes, its principle is the verification of many orthogonal equations. In [18], an efficient decoder called ARDecGA (Artificial reliabilities based decoding algorithm) is presented; it uses a generalized parity check matrix to compute a vector of artificial reliabilities of the binary received word h and it exploited a genetic algorithm to find the maximum likelihood binary word to this vector. The algebraic hard decision decoder [19, 20] of Berlekamp-Massey is based on compute of syndromes and it has an efficient mechanism to localize all corrigible errors, it is applicable on BCH codes. Another version of this last decoder is adapted for Quadratic residue codes [21].

3   The First Proposed HSDec Decoder

The first proposed decoder HSDec (Hard decision decoder based on Hash and Syndrome decoding) works as follows:

Inputs:
   - b: the binary word to decode of length n.
   - The parity check matrix H or the generator polynomi-
   al h(x) of the dual code.
   - TH1: the hash table of 2^{n-k} rows.
   - POW2: the vector of n-k columns contains in each
   cell i the value of the 2^i
Outputs: the corrected word c.
Begin
   Compute the syndrome S of b.
   Position:=hash(S, POW2)
   c :=b⊕TH1[Position]
End
The function "hash" is defined as follows:
Function hash(S: binary vector of n-k digits, POW2): inte-
ger
Begin Function
   $Position := \sum_{i=0}^{n-k-1} S[i] * POW2[n - k - 1 - i]$

   Return Position
End Function

In the hash table TH1, each rows number m contains the error pattern e of weight less than or equal to the error correcting capability t of the code and having a syndrome of hash value equal to m = hash(syndrome(e)). In general, there are some syndromes which doesn't correspond to any corrigible error pattern; for these cases, the table TH1 contains in the corresponding rows the zero vector or only an indicator for this vector to reduce the used memory.

$$G = \begin{pmatrix} 1 0 0 0 1 1 0 \\ 0 1 0 0 0 1 1 \\ 0 0 1 0 1 1 1 \\ 0 0 0 1 1 0 1 \end{pmatrix}, \quad H^T = \begin{pmatrix} 1 1 0 \\ 0 1 1 \\ 1 1 1 \\ 1 0 1 \\ 1 0 0 \\ 0 1 0 \\ 0 0 1 \end{pmatrix} \quad and \quad TH1 = \begin{pmatrix} 0 0 0 0 0 0 0 \\ 0 0 0 0 0 0 1 \\ 0 0 0 0 0 1 0 \\ 0 1 0 0 0 0 0 \\ 0 0 0 0 1 0 0 \\ 0 0 0 1 0 0 0 \\ 1 0 0 0 0 0 0 \\ 0 0 1 0 0 0 0 \end{pmatrix} \begin{bmatrix} 0 \\ 1 \\ 2 \\ 3 \\ 4 \\ 5 \\ 6 \\ 7 \end{bmatrix}$$

4   The Second Proposed HWDec Decoder

The second proposed decoder HWDec decoder (Hard decision decoder based on decreasing of the error weigh) works as follows:

```
Inputs:
    - b: the binary word to decode of length n.
    - The parity check matrix H or the generator polynomi-
    al h(x) of the dual code.
    - TH2: the hash table of 2^{n-k} integers.
    - t: the error correcting capability
    - POW2: the vector of n-k columns contains in each
    cell i the value of the 2^i
Outputs: the corrected word c.
Begin
    e:=the zero vector of length n
    Compute the syndrome S of b.
    WE1:=WeightOfError(S,TH2,POW2)
    If (WE1≤t) then
        For i:=1 to n do
            copy b in b1;b1[i]:=1-b1[i]
            S:=Syndrome of b1
            WE2:=WeightOfError(S,TH2,POW2)
            If (WE2=WE1-1) then
                e[i]:=1
            End If
        End For
    End If
    c:=b⊕e
End
```

The HWDec uses the following function which computes the weight of the error associated with a given syndrome. The table TH2 associates for each syndrome S the weight of the unique corrigible error of syndrome equal to S; if the syndrome S doesn't correspond to any corrigible error then we choose to associate it with the weight n for indicating that it isn't corrigible.

```
Function WeightOfError(S: binary vector of n-k digits,
TH2, POW2): integer
Begin Function
    Position:=hash(S, POW2)
    w :=TH2[Position]
    Return w
End Function
```

5   Simulation Results and Comparison

To show the efficiency of HWDec and HSDec algorithms we give in this section its error correcting performances for some linear codes form many classes with a comparison with other decoding algorithms over the Additive White Gaussian Noise (AWGN) channel which can be viewed as a binary channel, with a BPSK (Binary Phase Shift Keying) modulation. All simulations are obtained by using the parameters given in Table 1. The error correcting performances will be represented in terms of Bit Error Rate (BER) in each Signal to Noise Ratio (SNR = Eb/N0).

Table 1. Default simulation parameters.

Simulation parameters	Value
Minimum number of residual bit in errors	200
Minimum number of transmitted blocks	100000

5.1   Simulation Results of HSDec and HWDec Algorithms and Comparison

In order to show the efficiency of the proposed decoders we made several simulations for plotting their error correcting performances for many linear block codes.

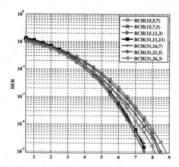

Fig. 1. Performances of HSDec for some BCH codes of length up to 31

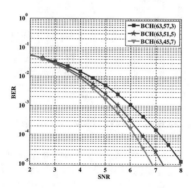

Fig. 2. Performances of HSDec for some BCH codes of length 63

The Figs. 1 and 2 represent the performances of HSDec for some BCH codes of length up to 63. Knowing that if the data are transmitted without coding in the sending step and without correction in the receiving step over AWGN channel then the BER reaches the value 10^{-5} at the SNR = 9.6 therefore the HSDec permits to obtain a gain of coding of more than 1 dB for BCH (15, 7, 5), about 2 dB for BCH (31, 21, 5) and about 2.7 dB for BCH (63, 45, 7). The Fig. 1 shows that for the same length 15, the performances are comparable even if the error correcting capabilities are different. These differences can be justified by the variation of the code rate which affects the error power and therefore the BER performances.

The Fig. 3 represents the performances of HWDec for some BCH codes of length up to 31. It shows that this decoder allows to obtain a gain of coding of more than 1 dB for BCH (15, 7, 5) and about 2 dB for BCH (31, 21, 5). We note that the gain for BCH (63, 45, 7) is about 2.7 dB.

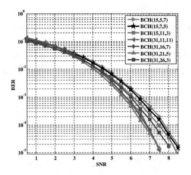

Fig. 3. Performances of HWDec for some BCH codes of length less than 31

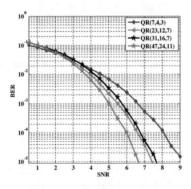

Fig. 4. Performances of HSDec for some QR codes

The Fig. 4 represents the performances of HSDec for some QR codes of length up to 47. It shows that this decoder allows to obtain a gain of coding of about 3 dB for QR (47, 24, 11) and about 2.2 dB for the QR (23, 12, 7) code.

Remark 1. QR (23, 12, 7) code is a perfect code that attains the Hamming bound and therefore all the syndromes correspond to corrigible errors.

The Fig. 5 represents the performances of HSDec for some QDC codes of length up to 40. It shows that this decoder allows to obtain a gain of coding of about 2 dB for QDC

(40, 20, 8). This figure shows that for the same rate and the same minimum distance the performances are better when the length decreases.

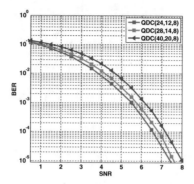

Fig. 5. Performances of HSDec for some QDC codes

The Fig. 6 represents a comparison of the BER performances for the HWDec and HSDec for the QR (47, 24, 11) code. This figure shows that these decoders have the same performances for the same code. We note that for all codes studied in this paper the simulation proves that our proposed decoders guaranty the correction of all corrigible errors.

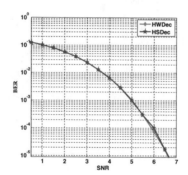

Fig. 6. Comparison of the performances of HWDec and HSDec for the QR (47, 24, 11) code

The Fig. 7 represents a comparison of the BER performances for the HSDec, HWDec and the BM decoder for the BCH (63, 51, 5) code. This figure shows that these decoders have the same performances.

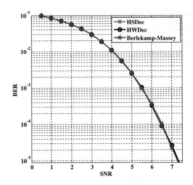

Fig. 7. Comparison of the performances of HSDec, HWDec and BM for BCH (63, 51, 5) code

The Fig. 8 represents a comparison of the BER performances for HSDec, HWDec, Deep Neural Decoder (DND) [8], HDGA [15] decoder for BCH (15, 7, 5) code. This figure shows that our decoders (HSDec, HWDec) have the same performances, pass absolutely the performances of DND and they have the same performances comparing to HDGA.

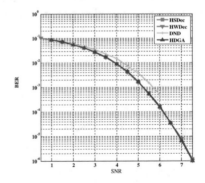

Fig. 8. Comparison of the performances of HSDec, HWDec, DND and HDGA for BCH (63, 45, 7) code

The Fig. 9 represents a Comparison of the BER of HSDec, ARDecGA [18] and BERT [4] decoder for the BCH (15, 7, 5) code. This figure shows that our decoder passes relatively the performances of BERT decoder and has the same performances comparing to ARDecGA.

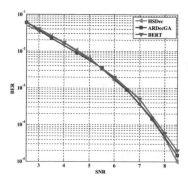

Fig. 9. Comparison of the performances of HSDec, ARDecGA and BERT decoders for the BCH (15, 7, 5) code

5.2 Study of Complexities

For a cyclic code in systematic form, the syndrome is simply the vector sum of the received parity digits and the parity-check digits recomputed from the received information digits [22]. The received vector r is treated in the form of a polynomial of degree $n - 1$ or less as it shows the formula (1).

$$r(X) = r_0 + r_1 X + r_2 X^2 + \dots + r_{n-1} X^{n-1} \tag{1}$$

Dividing r(X) by the generator polynomial g(X), we obtain (2). The remainder s(X) is a polynomial of degree $n - k - 1$ or less. The $n - k$ coefficients of s(X) form the syndrome s.

$$r(X) = a(X)g(X) + s(X). \tag{2}$$

Table 2. Complexity of ARDecGA, BM, HDGA, HSDec and HWDec algorithms.

Decoder	Complexity
ARDecGA (for linear codes)	$O(M.n + N_i * N_g[kn + \log(N_i)])$
Berlekamp-Massey (for cyclic BCH codes)	$O(n^2)$
HDGA (for linear codes)	$O(N_i * N_g[kn^2 + kn + \log(N_i)])$
HSDec (for cyclic codes)	$O(\log(n).\log(n - k) + 3n - 2k)$
HWDec (for cyclic codes)	$O(n * (\log(n).\log(n - k) + 3n - 2k))$
HSDec (for linear not cyclic codes)	$O(n^2)$
HWDec (for linear not cyclic codes)	$O(n^3)$

Table 2 presents the complexities of ARDecGA, BM, HDGA, HSDec and HWDec algorithms. This table shows that the complexity of the proposed two decoders is lowest than their competitors. The main advantages of HSDec is its low complexity comparing to all presented decoders and it's applicable on any linear block code, however it requires storing the syndromes of all corrigible errors in the memory. The main advantages of HWDec are its applicable on any linear block code and its reduced required space memory because it requires storing only weights of a corrigible error instead of the error itself; however its complexity is increased comparing to HSDec. So, HSDec requires more memory space and less execution time, while HWDec requires less memory space and more execution time.

6   Conclusion and Perspectives

In this paper, we have presented two new efficient hard decision decoding algorithms which use the hash techniques to reduce the run time of search. The simulation results show that the proposed decoders allows to correct all errors of weight less than or equal to t of the studied codes. The study of complexity shows that these decoders have low temporal complexity comparing to their competitors. Instead of many decoders that are specific for only some family of codes, the decoders proposed here are applicable on any linear code and exploit the polynomial form to reduce considerably the run time of decoding cyclic codes in addition to the gain of time that they offered by the proposed hash techniques. The success of these decoders will encourage us to continue on generalizing them for soft decision decoding case and for other families of error correcting codes.

References

1. Clarck, G.C., Cain, J.B.: Error-Correction Coding for Digital Communication. New York Plenum, New York (1981)
2. Huang, C.F., Cheng, W.R., Yu, C.: A novel approach to the quadratic residue code. In: Pan, J.S., Tsai, P.W., Huang, H.C. (eds.) Advances in Intelligent Information Hiding and Multimedia Signal Processing. Smart Innovation, Systems and Technologies, vol. 64. Springer, Cham (2017)
3. Yani, Z., Xiaomin, B., Zhihua, Y., Xusheng, W.: Decoding of the five-error-correcting binary quadratic residue codes. Am. J. Math. Comput. Model. 2(1), 6–12 (2017)
4. Elghayyaty, M., Hadjoudja, A., Mouhib, O., El Habti, A., Chakir, M.: Performance study of BCH error correcting codes using the bit error rate term BER. Int. J. Eng. Res. Appl. 7(2), 52–54 (2017)
5. Raka, M., Kathuria, L., Goyal, M.: $(1 - 2u^3)$-constacyclic codes and quadratic residue codes over $F_p[u]/< u^4 - u>$. Cryptogr. Commun. 9(4), 459–473 (2017). https://doi.org/10.1007/s12095-016-0184-7
6. Ding, C., Liu, H., Tonchev, D.T.: All binary linear codes that are invariant under PSL2(n). arXiv:1704.01199v1 [cs.IT] (2017)
7. El idrissi, A., El gouri, R., Lichioui, A., Laamari, H.: Performance study and synthesis of new Error Correcting Codes RS, BCH and LDPC Using the Bit Error Rate (BER) and Field-Programmable Gate Array (FPGA). Int. J. Comput. Sci. Netw. Secur. 16(5), 21 (2016)
8. Nachmani, E., Béery, Y., Burshtein, D.: Learning to decode linear codes using deep learning. In: IEEE 2016 Fifty-fourth Annual Allerton Conference (2016)

9. Esmaeili, M., Alampour, A., Gulliver, T.: Decoding binary linear block codes using local search. IEEE Trans. Commun. **61**(6), 2138–2145 (2013)
10. Lin, T., Lee, H., Chang, H., Truong, T.: A cyclic weight algorithm of decoding the (47, 24, 11) quadratic residue code. Inf. Sci. **197**, 215–222 (2012)
11. Lin, T., Truong, T., Lee, H., Chang, H.: Algebraic decoding of the (41, 21, 9) Quadratic Residue code. Inf. Sci. **179**, 3451–3459 (2009)
12. Lin, T., Lee, H., Chang, H., Chu, S., Truong, T.: High speed decoding of the binary (47, 24, 11) quadratic residue code. Inf. Sci. **180**, 4060–4068 (2010)
13. Jing, M., Chang, Y., Chen, J., Chen, Z., Chang, J.: A new decoder for binary quadratic residue code with irreducible generator polynomial. In: IEEE 2008 Asia Pacific Conference on Circuits and Systems APCCAS (2008)
14. Chen, Y., Huang, C., Chang, J.: Decoding of binary quadratic residue codes with hash table. IET Common. **10**(1), 122–130 (2016)
15. Azouaoui, A., Chana, I., Belkasmi, M.: Efficient information set decoding based on genetic algorithms. Int. J. Commun. Netw. Syst. Sci. **5**(7), 423–429 (2012)
16. Gallager, R.G.: Low-density parity-check codes. IRE Trans. Inf. Theor. **8**(1), 21–28 (1962)
17. Morelos-Zaragoza, R.H.: The Art of Error Correcting Coding, 2nd edn. Wiley, Hoboken (2006)
18. Nouh, S., El Khatabi, A., Belkasmi, M.: Majority voting procedure allowing soft decision decoding of linear block codes on binary channels. Int. J. Commun. Netw. Syst. Sci. **5**(9), 557–568 (2012)
19. Berlekamp, E.R.: Algebraic Coding Theory. Aegean Park Press, Laguna Hills (1984). rev. edn.
20. Massey, J.L.: Shift-register synthesis and BCH decoding. IEEE Trans. Inf. Theor. **15**(1), 122–127 (1969)
21. Chen, Y.H., Truong, T.K., Chang, Y., Lee, C.D., Chen, S.H.: Algebraic decoding of quadratic residue codes using Berlekamp-Massey algorithm. J. Inf. Sci. Eng. **23**(1), 127–145 (2007)
22. Lin, S., Costello, D.J.: Error Control Coding: Fundamentals and Applications. Prentice-Hall Inc., Upper Saddle River (1983)

Pattern Recognition

Content-Based Image Retrieval Using Convolutional Neural Networks

Ouhda Mohamed[✉], El Asnaoui Khalid, Ouanan Mohammed, and Aksasse Brahim

M2I Laboratory, ASIA Team, Department of Computer Science,
Faculty of Sciences and Technologies, Moulay Ismail University,
BP 509, Boutalamine, 52000 Errachidia, Morocco
ouhda.med@gmail.com, khalid.elasnaoui@gmail.com,
ouanan_mohammed@yahoo.fr, baksasse@yahoo.com

Abstract. Content-based image retrieval (CBIR) is a widely used technique for retrieval images from huge and unlabeled image databases. However, users are not satisfied with the traditional information retrieval techniques. Moreover, the emergence of web development and transmission networks and also the amount of images which are available to users continue to grow. Therefore, a permanent and considerable digital image production in many areas takes place. Hence, the rapid access to these huge collections of images and retrieve similar image of a given image (Query) from this large collection of images presents major challenges and requires efficient techniques. The performance of a content-based image retrieval system crucially depends on the feature representation and similarity measurement. For this reason, we present, on this paper, a simple but effective deep learning framework based on Convolutional Neural Networks (CNN) and Support Vector Machine (SVM) for fast image retrieval composed of feature extraction and classification. From several extensive of empirical studies for a variety of CBIR tasks using image database, we obtain some encouraging results which reveals several important insights for improving the CBIR performance.

Keywords: CNN · Feature · Classifier · Deep learning · CBIR · SVM

1 Introduction

Today's world is digital with the appearance of many devices that are used in image acquisition. Nowadays, it becomes easy to store huge amount of images by using image processing techniques. Since the collection of images and databases is fast and is increasing day by day, there is a need for new image retrieval techniques that should be efficient and fast.

The most common method for retrieving multimedia content from a huge collection of images consists of using meta-data associated to the images such as the timestamp, the geolocation, keywords, tags, labels or short descriptions, and performing the retrieval task through a Text-Based Image Retrieval (TBIR). Even though The TBIR requires expensive work and time, it often turns out not to be so effective. This is due to the subjectivity of the task compared to the meaning of its semantic content. Moreover, the use of a describing text for image labeling and tagging does not always reflect clearly

© Springer International Publishing AG, part of Springer Nature 2019
J. Mizera-Pietraszko et al. (Eds.): RTIS 2017, AISC 756, pp. 463–476, 2019.
https://doi.org/10.1007/978-3-319-91337-7_41

and definitely what an image represents because, for users, image often contains more meaning and is depicting realities in a direct and clear way better than long sentences that may turn to be confusing. Not only this, sometimes even the same word can have several meanings in different contexts.

Moreover, the content of an image can be described in two different levels: At the digital level, an image contains colored pixels from which color descriptors, textures, and shapes can be extracted. At the semantic level, an image can be interpreted and can have, at least, one meaning. Unfortunately, in today's information systems, images are described digitally while users are interested in their semantic content while it is currently difficult to find correspondences between the digital and semantic level.

CBIR is a system which uses visual contents to retrieve images from an image database. This system has now become indispensable because it can effectively overcome the problems written above. In CBIR, visual contents are extracted by several techniques: histogram, segmentation. Also are described by multidimensional feature vectors. The retrieval performance of content-based image retrieval (CBIR) system is mainly influenced by the feature vectors and similarity measures. Always a semantic difference exists between low-level image pixels captured by machines and the high-level semantics perceived by humans. The recent successes of deep learning techniques especially Convolutional Neural Networks (CNN) in solving the problem of computer vision applications has inspired us to tackle this issue so as to improve the performance of CBIR.

We should not neglect the fact that researches blooming in the past years were proposing the convolutional neural network in a primary position before feature extraction, classification and representation for CBIR.

Inspired by the advancement of CNN and the effectiveness of the Support Vector Machine (SVM) method in classifications, we ask whether we can profit more, and perfectly from the deep CNN to improve the feature extraction and, from the SVM for an efficient classification of CBIR. Instead of the use of the existing CBIR method, can we get a better classification the images in classes? When classifying images, the hybrid CNN-SVM highlights region in a given input image that provide evidence for or against a certain class. It surmount several gaps of previous methods and gives great additional insight into the decision making process of classifiers.

The rest of this paper is organized as follows. Section 2 presents the related work around both CNN, SVM and image retrieval, with Sect. 3 discussing proposed approach and then we tested this solution through experiments in Sect. 4. Finally comes Sect. 5 which presents the conclusion and the perspectives of the direction of future work.

2   Existing Work

According to several classic techniques developed in recent years having each one many disadvantages, for example the histogram; firstly, this representation leads to the loss of spatial information, which is important in order to correctly represent the content of the image. Secondly, the use of the histogram raises the quantification problem of characteristic spaces.

CNN specifically designed to deal with the variability of 2D shapes, are shown to outperform all other techniques. Recognition systems are composed of multiple modules including feature extraction, classification and paradigm learning. They are allowing such multimodal systems to be trained globally using gradient-based methods so as to optimize an overall performance measure.

In 1998, [1] firstly attained successful results on adopting supervised back-propagation networks for digit recognition. With the advance of machine learning and image processing retrieval and significant reduction on cost of computing hardware, such as Graphics Processing Unit (GPU), Researchers have taken advantage and have developed several models and techniques. [2, 3] have studied CNN. [2] they investigated a framework of deep learning with applications CBIR by testing a state of the art of deep learning method (Convolutional Neural Networks) for CBIR tasks under varied settings. [3] offers a survey, through the main information' fusion: ingredients that every recipe for the design of a CBIR system should include to meet the demanding needs of users.

[4] presented a CNN model parameterization work-flow developed on the cloud-computing platform of Microsoft Azure Machine Learning Studio (MAMLS), which is capable of learning from the feature maps and classifying multi-modal images with different variabilities using one common flow.

Unlike to previous methods, the binarization approaches [5, 6] require pairwise inputs for binary code learning the feature representation has the best performance provided by CNN, the extracted features generalization ability, the relationship between the dimensional reduction and the accuracy loss in CBIRs, the best distance measure technique in CBIRs and the benefit of the coding techniques in improving the efficiency of CBIRs, [5] proposed that feature binarization approach is presented for better efficiency of CBIRs. More precision, the binarization reduced 31/32 space usage of original data. [6] proposed a deep learning framework to generate binary hash codes for fast image retrieval. The idea is that when the data labels are available, binary codes can be learned by using a hidden layer for representing the latent concepts that dominate the class labels. The utilization of the CNN also enables image representations 'learning'. Their method passes hash codes and image representations in a point-wised manner, making it suitable for large-scale datasets.

[7] they have applied a support vector machine active learning algorithm for conducting effective relevance feedback for image retrieval. The proposed algorithm chooses the most informative images to query a user and quickly learns a boundary that separates the images that satisfy the user's query concept from the rest of the dataset.

In order to take advantage of the SVM classification methods the authors combined CNN with SVM [8–10]. [8] combined the CNN and support victor machine (SVM) applies in the CBIR, and uses (SVM) to train a hyperplane which can separate similar image pairs and dissimilar image pairs to a large degree. The input of the SVM is pair features which are assembled by pair of images: the query image and each test image in the image dataset. The test images then are ranked by the distance between the pair feature vectors and the trained hyperplane.

[9] proposed CNN-SVM model, CNN is feature extraction and SVM performs as a recognizer. While [10] combined CNN with linear SVM.

In contrast, we propose a simple and efficient deep learning approach to learn a set of effective which achieves more favorable results on the publicly available datasets. We will present the proposed method in the next section.

3   Proposed Method

Convolutional neural network (CNN) is a type of feed-forward artificial neural network where the individual neurons are tiled in such a way that they respond to over-lapping regions in the visual field. They are biologically-inspired invariant of Multi-layer Perceptrons (MLP) which are designed for the purpose of minimal preprocessing. These models are widely used in image and video recognition.

Compared to other feature extraction and classification algorithms, convolutional neural networks use relatively not much pre-processing. This means that the network is responsible for developing its own filters (unsupervised learning), which is not the case with other more traditional algorithms. The lack of initial parameterization and human intervention is a major advantage of CNN. The main objective of this work is to profit from the performance of CNN and SVM but with the minimum of material and time resources. For this reason we propose this architecture:

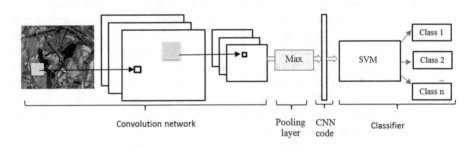

Fig. 1. Diagram of the proposed architecture

This network that we propose is a very small network that we can run on a CPU as well. Traditional neural networks that are very good at doing image classification have many more parameters and take a lot of time if trained on CPU.

The first part of a CNN is the convolutional phase. It works as an extractor of image features. An image is passed through a succession of filters, or convolution kernel, creating new images called convolution maps. Some intermediate filters reduce the resolution of the image by a local maximum operation. In the end, the convolution maps are flattened and concatenated into a feature vector, called a CNN code.

The SVM takes This CNN code at the outputs of the convolution phase as a new feature vector for training. When the SVM classifier has been well trained, it performs the recognition step and takes new decisions on testing images with such automatically extracted features.

3.1   Convolution Layer (Feature Extraction)

A feature is considered as an important part of an image and is used as a starting point for computer vision algorithms. An image can be described globally or also locally. Global models use the whole image to represent their aim while in the local approach; the selection of several regions or blocks of the image is utilized to extract the feature. In this case, there are sparse and dense representations. Sparse representation detects interest points or regions in the image. Then, this representation is extracted by a feature descriptor from each region. The characteristics of a good local feature are [10]:

- Should be highly distinctive: an efficient feature should be capable of correcting object identification with low probability of mismatch;
- Must be easy to extract;
- A good feature should be tolerant to image noise, to illumination' changes, to uniform scaling, rotation and to geometric forms;
- Should be practical and easy to match against a large database of local features.

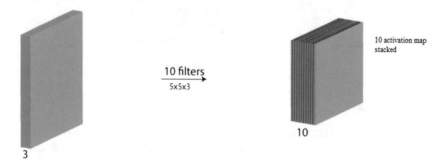

Fig. 2.   Input layer in CNN

Convolutional mapping function has the capacity of detecting some features Fig. 2; it is generally called a feature map. In practical image analysis, one feature is not enough to characterize the content of an image. Hence we use multiple features maps in a CNN. To extract the 3 feature maps from the convolutional input layer to the hidden layer. We uses 3 feature maps are defined by $\phi i(\cdot)(i = 1, 2, 3)$, respectively. We can say that each function learns a feature map on the input image, resulting in 3 feature maps of hidden units in the following layer, and we apply 10 filters of size 5*5*3 with valid padding.

$$x = \varphi\left(W * Y_{(ij)} + b\right) \tag{1}$$

Where:

Y denotes the input matrix
$\varphi(\cdot)$ the function is an active function.
b is bias term.
W weight matrix.

The objective of the training is to get the best possible values variable of the weights (W) and biases (b) or the parameters of the network which solve the problem reliably. We can find the best set of parameters using a process called Backward propagation, [8, 9] start with a random set of parameters and keep changing these weights such that for every training image we get the correct output. But this method is slow, There are many optimizer methods to change the weights that are mathematically quick in finding the correct weights. The networks are trained by stochastic Gradient Descent [10] is one such method, Backward propagation and optimizer methods to change the gradient is a complicated topic.

The visualized features should look like the following (Fig. 3):

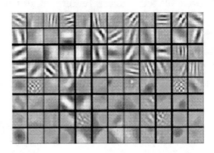

Fig. 3. The features learned by the CNN

3.2   Pooling Layer

After the convolution operation, the size of the hidden layer is very large. To reduce the computational complexity, a pooling layer or a sub-sampling layer is usually used immediately after a convolutional layer. Concretely, a pooling operation in a hidden layer summarizes the information in a local region and delivers the summarized statistic to the next hidden layer.

Let $Y = \{y_{i,j}\}$ denote the matrix in a pool, common used pooling functions include:

$$x = Max(Y) \tag{2}$$

- Max pooling: taking the maximal element in Y as the output.
- Average pooling: taking the average value of all the elements $\{y_{i,j}\}$ as the output.

$$X = \frac{1}{N} \sum_{i,j} y_{i,j} \tag{3}$$

Where N is the number of elements in Y.

We use (2) because the experiments show that max pooling is better than average pooling.

3.3 CNN Classifier

The process of feature extraction, as previously stated: an image is represented by a set of descriptors that structure the feature vectors. These feature vectors are considered as input variables and are introduced in a learning component. The outputs are classes of input image (Fig. 4).

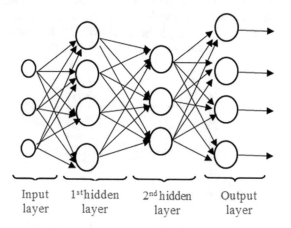

| Input layer | 1ˢᵗ hidden layer | 2ⁿᵈ hidden layer | Output layer |

Fig. 4. Multilayer perceptron architecture

When the data passes in one way, from the input layer to the output layer, as shown in the figure, MLP [2–4] is composed of four layers: the input layer, the output layer and the first and second hidden layers. The input layer receives the inputs features of the network. The first hidden layer contains the weighted inputs from the input layer and sends data from the previous layer to the next one. The use of additional layers makes the perceptron capable of solving nonlinear classification problems. The output layer contains the classification result. Several algorithms are used for the learning step of MLP. The supervised learning technique is called back propagation. It consists of four stages: initializing weights, feed forward, back propagation of errors and weight update.

3.4 CNN-SVM Model

Inspired by the statistical theory of learning by [16] SVM can change a nonlinear separable problem into a linear separable problem via projecting data into the feature space and then finding the separate hyperplane. This method was initially proposed to solve two-class problems (4).

$$
\begin{aligned}
& \min_{w,b,\xi} \frac{1}{2} w^T w + C \sum_{i=1}^{l} \zeta_i \\
& subject\ \ to\ y_i\left(w^T x_i + b\right) \geq 1 - \zeta_i \\
& \zeta_i \geq 0 \\
& i = 1, 2, \dots, l
\end{aligned}
\tag{4}
$$

Where w is an m-dimensional vector.
b is a scalar, x_i is the slack variables.
C is the penalty parameter.

Later a few strategies were suggested to solve efficiently perform a non-linear classification problem using what is called the kernel trick.

The proposed model of hybrid CNN-SVM [8, 14, 15] was designed by changing the latter output layer of the CNN model with an SVM classifier. The output values of the hidden layer is meaningless, but only makes sense to the CNN network itself, however, these values can be consider as features for SVM classifier (Fig. 5).

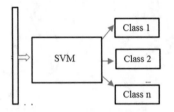

Fig. 5. Hybrid CNN-SVM architecture

Taking advantage of the efficiency and speed of the SVM method statistical theory of learning by [16] The SVM replaces the CNN output layer. The SVM takes the outputs from the hidden layer as a new feature vector for training. When the SVM classifier has been well trained, it performs the recognition step and takes new decisions on testing images with such automatically extracted features.

For implementation we use the LIBSVM [17] to build SVMs. LIBSVM is a popular and efficient open source library machine learning tool for classification problem. The one-against-one method is implemented for the multi-class SVMs in LIBSVM.

4   Experimental Result and Discussion

4.1   Dataset

- ImagNet database is publicly available at http://www.image-net.org. ImageNet uses the hierarchical structure of WordNet. Each meaningful concept in WordNet, can be described as "synonym set" or "synset" (Fig. 6).

Fig. 6. Example of images of dataset ImageNet

- Caltech256: holds 30607 images of objects, which were obtained from Google image search and from PicSearch.com. Images were classified to 257 categories and evaluated by humans in order to ensure image quality and relevance (Fig. 7).

Fig. 7. Example of classes in Caltech256 classes database

The learning of a CNN-SVM consists of optimizing the coefficients of the model and also of starting from a random initialization in order to minimize the classification error in output. One learns both the coefficients of the convolution kernel to extract relevant features, and the accurate combination of these characteristics (Fig. 8).

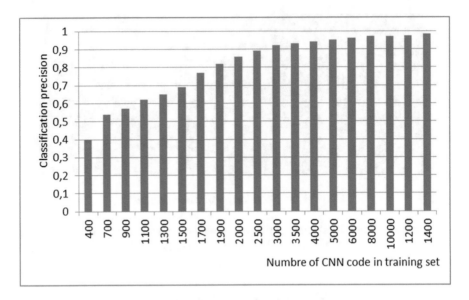

Fig. 8. Test: precision of 5000 images

In the following experiment, classifier SVM is tested by the CNN codes on learning samples of increasing size. It is observed that precision increases very rapidly: from 1000 images, or 100 per class, we obtain classification performances of 90%. 96% accuracy is achieved with 6000 images and 98.5% with the training data sets (minus those used for validation). It is therefore possible to train an efficient classifier with few images.

4.2   CNN-SVM vs CNN

We evaluated the complexity (computational time) of the SVM algorithm. It only depends on the number of inputs to be classified (d) and the number of training data (n) (5).

$$dn^2 \leq complexity \leq n \tag{5}$$

Obviously our model won't allow us to get such an impressive performance, but let's see what accuracy we can reach, 23% slower at test-time, because operations execute on the CPU (Table 1).

Table 1. Comparisons of the models in terms of accuracy time and test error on the test set

	Accuracy%	Time (min)	Test error%
CNN (GPU) [18]	99	6,57	11,9
CNN-SVM (CPU)	98,5	9,03	11

4.3   Performance Evaluation of CBIR

The most common measures to evaluate a system are response time and memory space used. A better system if the response time and the memory space are smaller this is the case of our system because we retrieve the query image in the appropriate class and not in the entire images database. But with CBIR systems, in addition to these two measures, we are interested in other measures. Usually in the CBIR system, the user is interested in the relevant responses of the system. So information retrieval systems require the evaluation of the accuracy of the response. This type of evaluation concerns the system research performance. The indexation system and Image searching is an information retrieval system.

In this section, we will describe the two most important measures: recall and precision. These measures are interrelated. Therefore this relationship is often described by a recall and precision curve. Then we present other measures that are also used to evaluate information retrieval systems.

The recall is defined as the ratio of the number of relevant images retrieved to the total number of relevant images in the database.

$$recall = \frac{Ra}{R} \tag{6}$$

where Ra is the relevant retrieved images, R is the total number of relevant images in the database.

Precision is the ratio of the number of relevant images retrieved to the total number of images retrieved.

$$precision = \frac{Ra}{A} \tag{7}$$

where A is the total number of images retrieved.

In practice, we use several queries. In these cases, to evaluate a system, we calculate the average precision for all the queries corresponding to each callback level. This value is given by:

$$P(r) = \sum_{i}^{N_q} \frac{P_i(r)}{N_q} \tag{8}$$

Where:

N_q is the number of queries.
$P_i(r)$ is the precision for recall r with query i.

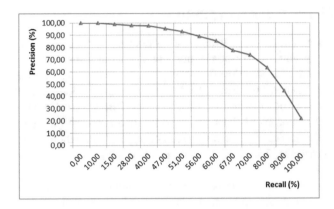

Fig. 9. Precision-recall result on the Caltech256 database

As shown in Fig. 9. We can consider that the retrieval performance can be important increased by adopting proposed scheme (Fig. 1). These optimistic results again validate the good generalization performance of the CNN-SVM model for learning effective features in a new field (Fig. 10).

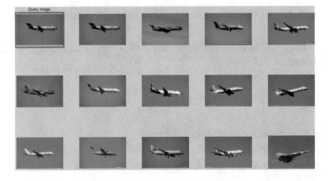

Fig. 10. Retrieval results of an airplane in the sky with the query image in the top-left corner

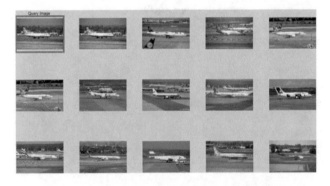

Fig. 11. Retrieval results of an airplane in the airport with the query image in the top-left corner

We notice that the proposed approach can detect the position of the airplane, due to the right and propitious choice of the feature maps and good training of SVM (Fig. 11).

5   Comment and Discussion

This method has many practical interests:

First, the image is transformed into a small dimension feature, which extracts feature that are very relevant. It reduces the size of the problem, it, thus, can replace standard methods of image processing, and, moreover, it is sophisticated. Note that this method works better and in its best way when working on a problem closed to the initial problem on which the model was driven. When the problem is very similar, for example for classifying objects on centered images, it is possible to further reduce the size of the problem by retaining the classification phase of the SVM.

Finally, since the extraction of features is only performed once per image, it can be performed quickly on the CPU. The machine learning libraries are generally sequential and also run on CPU. This method allows the exploitation of the power of CNNs without investing in GPUs.

6   Conclusion and Perspectives

By introducing the pooling intermediate layers of the model as feature representation, and SVM classifier, the CBIR task based on CNN can be improved to a full extent up-to-date. Especially, it can be saved with a high percent and can even achieve a higher speed computation.

The model CNN-SVM demonstrates their performance again. The use of pre-learning results in very good results: fast using them as extractors of feature image and classification, better by re-training them specifically (Fine Tuning). The extraction of characteristics is an excellent initial approach, with a very good compromise between performance and complexity.

The results obtained are promising, and open new perspectives; there are sometimes advantageous to pre-process the images, by cropping them or by normalizing their color histograms. The models can also be supplemented by conventional approaches for creating features, specific to the problem in question.

Acknowledgment. We want to thank all the researchers for their previous work. In the development of our algorithm, Libsvm [17] is utilized.

References

1. Lecun, Y., Bottou, L., Bengio, Y., et al.: Gradient-based learning applied to document recognition. Proc. IEEE **86**(11), 2278–2324 (1998)
2. Wan, J., Wang, D., Hoi, S., Steven, C.H., et al.: Deep learning for content-based image retrieval: a comprehensive study. In: Proceedings of the 22nd ACM International Conference on Multimedia, pp. 157–166. ACM (2014)

3. Piras, L., Giacinto, G.: Information fusion in content based image retrieval: a comprehensive overview. Inf. Fusion **37**, 50–60 (2017)
4. Ren, J.: Investigation of convolutional neural network architectures for image-based feature learning and classification. Thesis (2017)
5. Wang, H., Cai, Y., Zhang, Y., Pan, H., Lv, W., Han, H.: Deep learning for image retrieval: what works and what doesn't. In: IEEE International Conference on Data Mining Workshop (ICDMW), pp. 1576–1583 (2015)
6. Lin, K., Yang, H.F., Hsiao, J.H., Chen, C.S.: Deep learning of binary hash codes for fast image retrieval. In: Proceedings of the IEEE Conference on Computer Vision and Pattern Recognition Workshops, pp. 27–35 (2015)
7. Tong, S., Chang, E.: Support vector machine active learning for image retrieval. In: Proceedings of the Ninth ACM International Conference on Multimedia, pp. 107–118. ACM (2001)
8. Fu, R., Li, B., Gao, Y., Wang, P.: Content-based image retrieval based on CNN and SVM. In: 2nd IEEE International Conference on Computer and Communications (ICCC), pp. 638–642. IEEE (2016)
9. Niu, X.X., Suen, C.Y.: A novel hybrid CNN–SVM classifier for recognizing handwritten digits. Pattern Recognit. **45**(4), 1318–1325 (2012)
10. Tang, Y.: Deep learning using linear support vector machines. arXiv preprint arXiv: 1306.0239 (2013)
11. Paola, J.D., Schowengerdt, R.A.: Review and analysis of backpropagation neural networks for classification of remotely-sensed multi-spectral imagery. Int. J. Remote Sensing **16**(16), 3033–3058 (1995)
12. Krizhevsky, A., Sutskever, I., Hinton, G.E.: Imagenet classification with deep convolutional neural networks. In: Advances in Neural Information Processing Systems, pp. 1097–1105 (2012)
13. Bishop, C.M.: Neural Networks for Pattern Recognition. Oxford University Press, Oxford (1995)
14. Tang, Y.: Deep learning using linear support vector machines. arXiv preprint arXiv: 1306.0239 (2013)
15. Niu, X.X., Suen, C.Y.: A novel hybrid CNN–SVM classifier for recognizing handwritten digits. Pattern Recognit. **45**(4), 1318–1325 (2012)
16. Vapnik, V.: The Nature of Statistical Learning Theory. Springer, New York (1995)
17. Chang, C.C., Lin, C.J.: LIBSVM: a library for support vector machines (2001). http://www.csie.ntu.edu.tw/~cjlin/libsvm
18. Chen, Y., Jiang, H., Li, C., Jia, X., et al.: Deep feature extraction and classification of hyperspectral images based on convolutional neural networks. IEEE Trans. Geosci. Remote Sensing **54**(10), 6232–6251 (2016)
19. Hastie, T., Tibshirani, R.: Classification by pairwise coupling. In: Advances in Neural Information Processing Systems, pp. 507–513 (1998)
20. Tuytelaars, T., Mikolajczyk, K.: Local invariant feature detectors: a survey. Foundations and trends. Comput. Graph. Vis. **3**(3), 177–280 (2008)

A New Theoretical Pattern Based on a Methods Database for Dynamic Images Encryption

Faiq Gmira[1(✉)], Wafae Sabbar[2], Said Hraoui[3], and Abderrahmane Jarrar Ouilidi[3]

[1] Laboratory MAEG FSJESAS Hassan II University, Casablanca, Morocco
faiqgmira@hotmail.com
[2] Laboratory LIM@ FSTM Hassan II University, Casablanca, Morocco
swafae@gmail.com
[3] Laboratory LSO FS Sidi Mohamed Ben Abdellah University, Fez, Morocco

Abstract. In this article, we design optimally a new randomly-dynamic pattern suitable for encrypting image data. The proposed pattern splits the input image into S* sub-blocks and then encrypts them differently from their adjacent sub-blocks according to a set of techniques, a technique is composed of N* different Encryption Methods chosen with random queries from M*-DBMOE one Database constituted by M* Method Of Encryption. In this way, the proposed system introduces a random dynamic and strengthens key space, so image security in storage or transmission transactions is improved against the various cryptanalytic attacks.

Keywords: Dynamic image encryption
Random image encryption · Encryption by rechargeable database
Encryption by dictionaries

1 Introduction

Cryptography has very important stakes, the Image encryption applications fields in economics is vast and very varied, for example ensuring the security of commercial transactions, otherwise electronic commerce cannot be effectively promoted.

In a static encryption, a clear message is always coded in the same cryptogram [1]; in addition to that, the message by message processing is possible to carry out a cryptanalysis based on the exploitation of the general context of transactions [2, 3]. Therefore a certain number of attacks are possible especially differential cryptanalysis by searching frequency of occurrence of a message in whole or by parts [4]. This presents vulnerabilities, which are more acute to messages having repeating blocks as in the case of images that have intrinsic characteristics such as large volume, high redundancy and high correlation between pixels [4].

To protect image data against cryptanalytic attacks, it is therefore immediate to design evolutionary and robust cryptographic systems [5]. To fill the security vulnerabilities of static image encryption, we propose and design a randomly dynamic pattern wherein a same message has several possible cryptograms [6, 7].

© Springer International Publishing AG, part of Springer Nature 2019
J. Mizera-Pietraszko et al. (Eds.): RTIS 2017, AISC 756, pp. 477–484, 2019.
https://doi.org/10.1007/978-3-319-91337-7_42

This new pattern introduces dynamics, erases the correlations between the crypto-grams thus it reinforces the security against the most widespread attacks: Ciphertext-only, Known-plaintext, Chosen-plaintext, Adaptive chosen-plaintext and Related-key attack.

This article is organized as follows: after an introduction section, in the second section the proposed pattern is presented and explored step by step through its basic theory. The third section is reserved for perspective by proposing possible developments of the established pattern. Finally, we complete by a conclusion.

2   Proposed Pattern

Let "MOE" a Method of Encryption, M-DBMOE be a data base consisting of M Encryption Methods and let $\Omega_{M,N}$ the universal set of all possibilities of random N-MOE queries from The M-DBMOE.

2.1   Operating Mode Explanation

At each storage or transmission transaction, after having splitting the input image into several sub-blocks, the pattern elaborates the coordinates of each sub-block before encrypting it differently from its adjacent sub-blocks. Each sub-block is encrypted with a method so the clear image is encoded with an N-MOE of methods [4]. All of these N-MOE are randomly selected by requests from an Encryption Method Database M-DBMOE. The size M of the DBMOE and the number N of methods used at each execution are determined by solving optimization problems with constraints.

The proposed pattern is based on random uses of two dictionaries. The first defines the N-MOE used for each round of encryption while the second regulates the techniques of its application. Table 1 summarizes the protocols of the proposed pattern.

Table 1. The protocols of the proposed encryption pattern

Protocol 1	Protocol 2	Protocol 3	Protocol 4
Providing N-MOE methods: according to a random query from Methods Provider *Codebook-1*	**Splitting input images:** according to an optimal split into S sub-block and the determination the coordinates of each pixel sub-block	**Regulating techniques:** to manage the encryption of each Sub-block with the N-MOE provided by the *Codebook-1* a random query of the *Codebook-2* technique regulator is carried out	**Managing dynamics:** to recording configurations a backup copy of indexes associated with *Codebook-1* and *Codebook-2* is carried out in two key 256-pixels

The pattern is an abstract pattern based on a Database of Encryption Method. This Database is designed in such a way as to be rechargeable to set the usage and on the other hand, to be open on updates as needed with the various available Encryption Methods [4].

2.2 Determining M and N and Methods Provider "Codebook-1"

Determining M and N is amounts to realizing a maximum of encoding configurations indexed in two 256-pixel reserved a priori to the protocol Record/Restore of exchange of the parameters M and N. Accordingly the key parameters M and N are determined with an optimization problem under constraint (Table 2). The optimization problem is posed as follows:

Table 2. The Cardinality of the possible configurations as a function of M and N

M/N		N-MOE cardinality			
		3	4	5	6
M-DBMOE	1				
cardinality	2	not defined for M > N			
	3	6			
	4	24	24		
	5	60	120	120	
	6	120	360	720	720
	7	**210**	840	2520	5040

Maximize Card $\left(\Omega_{M,N}\right)$ under the constraints:

$$\begin{cases} N \leq M \\ \text{Card}\left(\Omega_{M,N}\right) \leq 255 \end{cases} \tag{1}$$

With the factorial notation:

$$\text{Card}\left(\Omega_{M,N}\right) = \begin{cases} A_N^M = \dfrac{M!}{(M-N)!} \text{ si } M \geq N \\ \text{Not defined elsewhere} \end{cases} \tag{2}$$

And so:

$$\text{Card (Codebook} - 1) = \text{Card}\left(\Omega_{M,N}\right). \tag{3}$$

M and N have a symmetric role, so we can set M and we vary N and vice versa. The scan performed for M = 1, 2, 3, 4, 5, 6 and N = 3, 4, 5, 6 is given below:

According to this table, which represents the cost function:

$$\left(M^*, N^*\right) : \begin{matrix} M^* = 7 \\ N^* = 3 \end{matrix} \tag{4}$$

is the unique optimal solution of the system (1) and therefore:

$$\text{Card (Codebook} - 1) = A_N^M == \frac{M!}{(M-N)!} = \frac{7!}{(7-3)!} = 210. \tag{5}$$

The calculation of the parameters gives rise to the development of the methods provider, "Codebook-1" is the ensuing result and is generated using the following pseudo code (Algorithm-1):

Algorithm 1. The generator of the "Codebook-1"

```
Codebook-1 ← {.,.,.,.}
index1 ← 0
for i,j,k: 1 ← M*=3
if ( MOEi!= MOEj and MOEi != MOEk and MOEj != MOEk)
  MOE:Draw1 ← MOE:i
  MOE:Draw2 ← MOE:j
  MOE:Draw3 ← MOE:k
Append at Codebook-1 ← {index1, MOE.Draw1, MOE.Draw2,
MOE.Draw3}
index1 ← index1+1
new record
end if
end for
```

After all, the methods provider "Codebook-1" (Table 3) is the indexed dictionary of the admissible assortments of the space $\Omega_{M,N}$:

Table 3. The Methods Provider "Codebook-1"

index $_1$	MOE$^{\text{Draw (1)}}$	MOE$^{\text{Draw (2)}}$	MOE$^{\text{Draw (3)}}$
000	MOE$_1$	MOE$_2$	MOE$_3$
001	MOE$_1$	MOE$_2$	MOE$_4$
002	MOE$_1$	MOE$_2$	MOE$_5$
...	...	...	...
208	MOE$_7$	MOE$_6$	MOE$_4$
209	MOE$_7$	MOE$_6$	MOE$_5$

In the implementation phase, this dictionary is used through its index which allows a fast search.

2.3 Input Image Splitter

Several splits are possible: rectangular, zigzag or circular; the following pseudo code (Algorithm 2) is used to determine the optimal split.

Algorithm 2. The calculator of the eligible encryption techniques

```
S: Sub-blocks number
ETN: encryption-technique number
varied S ← 1,2,3,4...
constraint 1: 'N=3 different MOEs are used'
constraint 2: 'the MOEs linked at two adjacent
 Sub-blocks are different'
ETN ← 0
varied Sub-block MOEs in{Map^Draw1, Map^Draw2,Map^Draw3}
if (constraint 1 and constraint 2 are true)
ETN ← ETN +1
end if
end varied
end varied
```

This pseudo code gives rise to this table (Table 4).

Table 4. The Encryption Techniques Number ETN

S	Rectangular	Triangular	Zigzag
7	186	126	186
8	378	252	378
9	762	510	762

The optimal split is the triangular splitting in 8* sub-blocks circularly positioned Fig. 1.

The $S^* = 8$ sub-blocks are defined by the following partitioning (Algorithm 3):

Algorithm 3. The Sub-blocks coordinates

```
Sub-block 1: {pixels p (i,j) / i>= L/2 et W+1-i>=j}
Sub-block 2: {pixels p (i,j) / i>= L/2 et W+1-i<j }
Sub-block 3: {pixels p (i,j) / i>= L/2 et i<= j }
Sub-block 4: {pixels p (i,j) / i>= L/2 et i>j }
Sub-block 5: {pixels p (i,j) / i< L/2  et W+1-I<=j }
Sub-block 6: {pixels p (i,j) / i< L/2  et W+1-i>j }
Sub-block 7: {pixels p (i,j) / i< L/2  et i>=j }
Sub-block 8: {pixels p (i,j) / i< L/2  et i<j }
```

L and W are respectively the length and the width of input image, i and j are the pixels coordinates.

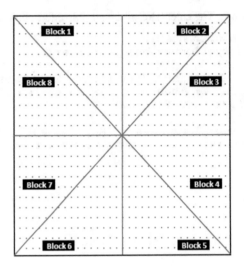

Fig. 1. The optimal split

2.4   Techniques Regulation

The encryption techniques regulator "Codebook-2" manages the mappings between the variants resulting from a 3*-MOE and the ways with witch the S* = 8 sub-blocks are encrypted.

In the execution step, to manage the encryption of sub-blocks with a given technique a randomized query from the "Codebook-2" is performed to provide a N-MOE. The Algorithm 4, shown below, generates the dictionary "Codebook-2".

Algorithm 4. The generator of Codebook-2

```
Codebook-2 ← { .,.,.,.,.,.,.,.,.,.}
index2 ← 0
for i, j, k, l, m, n, o, p : 1 ← N*=3
 if (the MOEs linked at two adjacent Sub-blocks are
 different)
 append at Codebook-1:
 Codebook-1←{index2, MOEi, MOEj, MOEk, MOEl, MOEm,
MOEn, MOEo, MOEp}
 index2 ← index2+1
 new record
 end if
end for
```

After the execution of this algorithm, the product dictionary "Codebook-2" is indexed in Table 5, where $\{MOE^{Drawi}, i\}$ are the draws related of the MOE and $\{Block_j \ 1 \le j \le S = 8\}$ are the S Sub-blocks of the split image.

Table 5. The Regulator of encryption "Codebook-2"

Randomization of the correspondence between methods and sub-block								
Index2	Block$_1$	Block$_2$	Block$_3$	Block$_4$	Block$_5$	Block$_6$	Block$_7$	Block$_8$
000	MOE^{Draw1}	MOE^{Draw2}	MOE^{Draw1}	MOE^{Draw3}	MOE^{Draw1}	MOE^{Draw2}	MOE^{Draw1}	MOE^{Draw3}
001	MOE^{Draw1}	MOE^{Draw2}	MOE^{Draw1}	MOE^{Draw2}	MOE^{Draw1}	MOE^{Draw2}	MOE^{Draw3}	MOE^{Draw2}
...	...	...	...	...	...	...	...	...
250	MOE^{Draw3}	MOE^{Draw2}	MOE^{Draw3}	MOE^{Draw2}	MOE^{Draw3}	MOE^{Draw2}	MOE^{Draw1}	MOE^{Draw2}
251	MOE^{Draw3}	MOE^{Draw2}	MOE^{Draw3}	MOE^{Draw2}	MOE^{Draw3}	MOE^{Draw2}	MOE^{Draw3}	MOE^{Draw1}

2.5 Management of Dynamic Transaction Parameters

The Dynamic functioning [4, 6, 7] of the two dictionaries provider/regulator Code-book-1/Codebook-2 takes place through two indexes associated with these two diction-aries. These indices of backup/restore are stored in two pixels reserved by the pattern.

The transmitting and receiving protocol appointee beforehand two 256-pixels to allow the recovery of the N-MOE used in the encryption process and also for each round the recovery of the technique with which N-MOE is applied to the $S^* = 8$ sub-blocks of the input image splitting.

Conversely in decryption by extracting the indexes stored in the two 256-pixels key, exploring the two dictionaries Codebook-1/Codebook-2 we can determine the 3^*-MOE used and also the Technique of its application to the encryption of the $S^* = 8$ sub-blocks of the input image.

3 Perspective

The designed pattern is evolutionary because it is based on an abstract data base of Encryption Methods; therefore we will consider, in order to have more security, to have it loaded by non-homogeneous methods [8–13]. We will also consider introducing this method in several rounds in an iterative way on smaller block sizes. This is convenient also for wavelet based compression [14] coupled with this approach.

4 Conclusion

Based on a database of methods and a random dynamic, we have optimally designed an abstract and reloadable pattern with the encryption methods available updates. This pattern eliminates the vulnerabilities of non-variational static systems and widens the potential space of the keys. In this way the time required to break such a pattern may be longer than its validity. Thus the security criteria are improved. This new proposed pattern is convenient for transmitting images in a secure technique, especially in the case where users transmitters/receivers do not have good flexibilities to change private keys.

References

1. Knudsen, L.R.: Dynamique encryption. J. Cyber Secur. **3**, 357–370 (2015)
2. Li, C., Álvarez, G., Chen, G., Nuñez, J.: On the security defects of an image encryption process, IACR's Cryptology (2007). http://eprint.iacr.org/2007/397
3. Li, S., Zheng, X.: Cryptanalysis of a chaotic image encryption method. In: Proceedings of IEEE International Symposium on Circuits and Systems, vol. 2, pp. 708–711 (2002)
4. Gmira, F., Hraoui, S., et al.: An optimized dynamically-random chaos based cryptosystem for secure images. Appl. Math. Sci. **8**(4), 173–191 (2014)
5. Gmira, F., Hraoui, S., Saaidi, A., Jarrar Oulidi, A., Satori, K.: Fast and secure image transfer with a mixed encoder based encryption combined with compression. Int. J. Imaging Robot. **15**(2), 124–133 (2015)
6. Yu, H., Zhang, Z., Gao, J., Zhu Z.: Cryptanalysis and improvement on a Block encryption Algorithme Based on dynamic sequences of Multiple Chaotics Systems. In: International Workshop on Chaos Fractals Theories and Applications, China (2010)
7. Gmira, F., Hraoui, S., Saaidi, A., Jarrar Oulidi, A., Satori, K.: Enhanced algorithm of Hill Cipher involving pseudo-randomized dynamic key approach for image encryption. Int. J. Tomogr. Simul. **29**(1), 86–98 (2015)
8. Gmira, F., Hraoui, S., Sabbar, W., Jarrar, O.A.: Image transaction encryption based on a dynamic upswing of hill cipher and JPEG compression. Int. J. Imaging Robot. **17**(3), 66–76 (2017)
9. Zhang, J., Tian, L., Khan, M.K.: Chaotic secure content-based hidden transmission of biometrics patterns. Chaos Solitons Fractals. **32**, 1749–1759 (2007)
10. Chai, X., Yang, K., Gan, Z.: A new chaos-based image encryption algorithm with dynamic key selection mechanisms. Multimed. Tools Appl. **76**(7), 9907–9927 (2017)
11. Federal Information Processing Standard (FIPS), Advanced Encryption Standard (AES), Publication 197, National Bureau of Standards, U.S. Department of Commerce, Washington D.C, November (2001)
12. Fan, H., Li, M.: Cryptanalysis and improvement of chaos-based image encryption scheme with circular inter-intra-pixels bit-level permutation. Math. Probl. Eng. **2017**(2017), 1–11 (2017)
13. Sun, S., Gerault, D., Lafourcade, P., Yang, Q., Todo, Y., Qiao, K., Hu, L.: Analysis of AES, SKINNY, And others with constraint programming. In: IACR Transactions on Symmetric Cryptology, 281–306 (2017)
14. Hraoui, S., Gmira, F., Saaidi, A., Jarrar, A.O., Satori, K.: Chaos based crypto-compression using SPIHT coding. Int. J. Imag. Robot. **15**(2), 67–78 (2015)

Convolutional Neural Networks for Human Activity Recognition in Time and Frequency-Domain

Lamyaa Sadouk[(✉)] [iD] and Taoufiq Gadi[(✉)]

Faculty of Science and Technology Settat, Settat, Morocco
lamyaa.sadouk@gmail.com, gtaoufiq@yahoo.fr

Abstract. Human activity recognition (HAR) is an important technology in pervasive computing because it can be applied to many real-life, human-centric problems such as eldercare and healthcare. Successful research has so far focused on recognizing activities using time-series sensor data. In this paper, we propose a multi-scale deep convolutional neural network (CNN) to perform efficient HAR recognition using smartphone sensors. Experiments show how a variation in the network parameters results in a better extraction of low and mid-level features. Also, an analysis of feature representations in the 1st layer gives us insights about the nature of physical movements. Our approach outperforms other datamining techniques in HAR for the UniMiB SHAR benchmark dataset, achieving an overall performance of 88.23% on the test set.

Keywords: Human activity recognition · Time series · Learning · Deep learning
Convolutional neural network

1 Introduction

Automatically recognizing human's physical activities has emerged as a key problem to ubiquitous computing, human-computer interaction and human behavior analysis. Human's activity is recognized using real time signals taken from multiple body-worn (or body-embedded) inertial sensors. For *human activity recognition* (HAR), signals acquired by on-body sensors are arguably favorable over the signals acquired by video cameras since on-body sensors alleviate the limitations of environment constraints, allow more accurate deployment of signal acquisition on human body and enjoy the merits on information privacy (signals by camera may contain information of nontarget subjects in the scene). In the past few years, body-worn based HAR made a significant advance in several domains including game consoles, personal fitness training, medication intake and health monitoring.

Typically, a human physical activity lasts a few seconds, and consists of few basic movements in each second. From the perspective of sensor signals, the basic continuous movements are more likely to correspond to the smooth signals, and the transitions among different basic continuous movements provokes variations in the signal. The goal of HAR is to perform a good feature extraction (i.e. find good representations of the *time series* collected from sensors) in order to identify basic continuous movements and the salience of the combination of basic movements.

© Springer International Publishing AG, part of Springer Nature 2019
J. Mizera-Pietraszko et al. (Eds.): RTIS 2017, AISC 756, pp. 485–496, 2019.
https://doi.org/10.1007/978-3-319-91337-7_43

Recently, *deep learning* has emerged as a family of learning models that aim to model high-level abstractions in data for image, speech and time-series recognition. *Convolutional neural networks* (CNN), in particular, made impressive results in object recognition [9] and audio classification [10].

In this paper, we advocate a CNN network designed for classifying human activities in the time series domain. In the sections below, we review related works of human activity recognition and deep neural networks (Sect. 2). We then describe the preprocessing and augmentation of input signal as well as the architecture of our CNN model (Sect. 3). Then, in Sect. 4, our experiments are carried out and our classification result is compared to state-of-art methods. Finally, Sect. 5 summarizes our work.

2   Related Work

Traditional hand-crafted features extraction methods have been investigated such as basis transform coding (i.e. signals with wavelet transform and Fourier transform) [2, 15], symbolic representation [16], statistics of raw signals (i.e. mean and covariance of time sequences) [14, 17], Hidden Markov Models (HMMs) [18], SVM classifier [19], k-nearest neighbor classifier [12], as well as logistic regression and artificial neural network classifiers [3].

Recently, automatic feature extraction methods, i.e. deep learning approaches, have been proposed. In [4], the selected network was CNNs while the input was a novel activity image constructed from acceleration signals taken from the gyroscope. The study in [5] employed a Deep Belief Networks (DBNs) for the HAR task by using spectrogram signal of the triaxial accelerometer data instead of raw acceleration data. It also combined deep learning with hidden Markov model (DL-HMM) for sequential activity recognition. The classification results of this study showed that deep models with more layers perform better than shallow ones. Hammerla et al. [6] applied the three types of deep networks for HAR: Deep feed-forward networks (DNN), convolutional networks (CNN), and recurrent networks (RNN). The work of [7] also used CNNs with input signals collected from the accelerometer and gyroscope triaxial sensor, using additional Fast Fourier Transform (FFT) information on the sensor data. Yang et al. [8] made use of multichannel time series data and came up with a novel approach based on temporal convolution and pooling that improved classification results. Their CNN outperformed the SVM with radial basis function kernel and deep belief network on Opportunity and Hand Gesture datasets.

In comparison to existing approaches which employ multiple varied sensor data (accelerometer, gyroscope, etc.) [4, 6–8], our method uses only the triaxial accelerometer data (x, y, and z component signals) collected from a single accelerometer sensor. Also, our contributions are the following: (i) train our CNN network on augmented transformed data, (ii) analyze different CNN architectures by tuning hyper parameters, (iii) study feature representations of the best architecture.

3   Methodology

In this section, we introduce the key components of our approach. First, we show how dataset signals were preprocessed and how our data is augmented (Sect. 3.1). Next, we discuss the network architecture that will be applied for the training phase (Sect. 3.2). Figure 1 illustrates our overall framework.

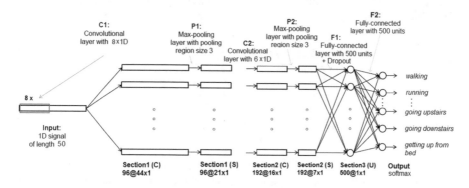

Fig. 1. Illustration of the overall framework. Symbols "C", "S", "U" in parentheses of the layer tags refer to convolution, subsampling, unification operations respectively. The numbers before and after "@" refer to the number of feature maps and the dimension of a feature map in this layer. Note that ReLU layers are not showed due to the limitation of space. ReLUs are right after each of the convolutional layers (C1, C2) and after the fully connected layer F1.

3.1   Signal Extraction: Extracting Input Signals

Before training our CNN, it is necessary to first extract input data signals. Experiments are conducted with two different datasets:

(i) dataset [1] which has already time-domain input signals extracted as a fixed time-length accelerometer data in the form of vector magnitude data of length 50. The resultant input is a $50 \times 1 \times D \times N$ matrix where D refers to the number of channels (number of sensors $\times$ 3 coordinates x, y and z) and N corresponds to the number of samples;

(ii) dataset [20] with 165,633 collected samples which has to be decomposed into input signals in both time and frequency domain.

Time-Domain Signals: We generate a 1 s time window (a frame of 8 reads/samples), with 125 ms (1/8 s) overlapping. We end up with input signals of length 8, which is not a matrix big enough to run CNNs. So, signals are resampled from 8 to 50. Resampling is done by applying an antialiasing FIR lowpass filter to the signals and compensating for the delay introduced by the filter. The resultant input matrix is a $50 \times 1 \times D \times N$ matrix.

Frequency-Domain Signals: The goal is to convert input signals in time-series into frequency signals of range 0–8 Hz. This frequency range is chosen based on the study of [20] that states that most body movements are contained within frequency components below 10 Hz. Instead of employing the Fast Fourier Transform that is restricted to a predefined fixed window length, Stockwell Transform (ST) is adopted since it adaptively captures spectral changes over time without windowing of data and therefore results in a better time-frequency resolution for non-stationary signals. It is computed using the following formula:

$$S(\tau,f) = \int_{-\infty}^{+\infty} h(t)w(\tau - t,f)e^{-2\pi i f t} dt \tag{1}$$

$$w_{GS}(\tau - t,f) = \frac{|f|}{\sqrt{2\pi}} e^{\left[\frac{-f^3(\tau - t)^2}{2}\right]} \tag{2}$$

where $S(\tau, f)$ is the ST computed for time sample τ and frequency f, $h(t)$ is the accelerometer signal, and $w_{GS}(t)$ is a Gaussian window with mean τ and variance $(1/f^2)$. We resample the input signal from 8 to 16 Hz. Then, for every input sample, we compute the ST to obtain the power of 50 frequencies in the range of 0–8 Hz, resulting in a feature vector of length 50. Thus, the extracted data consists of a vector of length 50 for each coordinate (x, y, and z) for each accelerometer. The resultant input matrix is a $50 \times 1 \times D \times N$ matrix.

3.2 Preprocessing Dataset Signals

Since data recorded is in the 3 directions (x, y and z axis), 3 data channels have to be processed separately. We normalize values of each axis/channel separately by subtracting the mean and dividing by the standard deviation:

$$x\_normalized = \frac{x - \bar{x}}{\sigma} \tag{3}$$

3.3 Data Augmentation

Next, the training dataset needs to be increased by undergoing transformations on them. Generally, deep neural networks perform very well on any recognition task provided that enough training examples are fed to them. Otherwise, with a small number of training examples, the network tends to overfit, providing wrong results during testing. To avoid overfitting, the training dataset should be large. One way to enlarge our dataset is to use data augmentation.

In our work, data augmentation consists of replicating acceleration signals with several transformations. In each epoch, we select half of the batch signals then perform the following transformations for each of the input signals. We randomly: (i) apply small noise or smoothing to the signal by adopting either a high frequency filter or a low frequency one (with multiple degrees of smoothness), (ii) shift the signal by a range of 1 to 10.

3.4 Network Architecture

Several architectures were conducted in our experiment, which will be detailed in the next section. However, in order to explain our methodology, the best architecture of the time-domain CNN is selected and summarized in Fig. 1. The first convolutional layer filters the 3-channel 50×1 input activity signal with 96 kernels of size 8×1, followed by 3×1 subsampling (with a stride of 2). The second convolutional layer takes the output of the first subsampling layer as input and filters it with 192 kernels of size 6×1, followed by 3×1 subsampling. The fully connected layer vectorizes the output of the second subsampling layer into a 500-dimensional feature vector.

Convolution ($C1, C2$): Each convolutional layer performs 1D convolution on its input maps. The output maps are generated for each convolution layer as:

$$c_i^{l,j} = \sigma\left(b_j^l + \sum_{m=1}^{M} w_m^{l,j} x_{i+m-1}^{l-1,j}\right) \tag{4}$$

where l is the layer index, σ is the activation function, b_j^l is the bias term for the jth feature map of the lth convolutional layer, M is the filter size and $w_m^{l,j}$ is the weight for the feature map j and filter index m of the lth layer.

Max-pooling ($P1, P2$): Max-pooling is adopted as the subsampling method. The output map p_j is achieved by computing maximum values over nearby inputs of the feature map:

$$p_i^{l,j} = max_{r \in R}\left(C_{i \times T + r}^{l,j}\right) \tag{5}$$

where R is the pooling size, and T is the pooling stride. Two convolutional and pooling layers are stacked on top of one another to form a deep CNN architecture.

Output ($F1, F2$): The output of these layers, which consists of a feature vector $p = [p_1, ..., p_I]$, is then fed into a fully connected layer ($F1$) and an activation layer to produce the following output:

$$p_i^l = \sum_j w_{ji}^{l-1} \times \left(\sigma\left(p_i^{l-1}\right) + b_i^{l-1}\right) \tag{6}$$

where σ is the activation function, w_{ji}^{l-1} is the weight connecting the ith node on layer l-1 and the jth node on layer l, and b_i^{l-1} is the bias term.

Then a dropout layer is added to prevent the neural network from overfitting. Figure 1 shows the fully connected neural network with dropout ($F1$). We set the percentage of the dropout to 0.7 in our evaluation experiment. Finally, this output is fed into a second fully connected layer (*F2*) and a softmax layer which infers the activity class:

$$P(c|p) = \frac{\exp\left(p_c^L\right)}{\sum_{k=1}^{N} \exp\left(p_k^L\right)} \tag{7}$$

where c is the activity class, L is the last layer index, and N is the total number of activity classes. This softmax function provides the posterior probability of the classification results. Then, an entropy cost function can be constituted based on the true labels of training instances and probabilistic outputs of softmax function.

4   Experiments and Results

4.1   Experimental Setup and Details

We perform our recognition evaluation on two datasets: the UniMibSHAR dataset [1] and the PUC dataset [20]. Our experiments include classification and retrieval.

UniMibSHAR Dataset: UniMibSHAR is a smartphone accelerometer dataset designed for activity recognition which includes 7,013 activities performed by 30 subjects. Activities are divided in 17 fine grained classes grouped in two coarse grained classes: 9 types of activities of daily living (ADL) and 8 types of falls. In our study, we are interested in the 9 ADL activities which are: walking, running, going upstairs, going downstairs, jumping, sitting down on a chair, standing up from the chair, lying down on a bed, getting up from the bed. Data is recorded with a sample frequency of 50 Hz along 3 Cartesian axes (x, y and z). Data instances consist of signal windows of 50 Hz width (50 samples/window) that are centered around a peak of magnitude higher than 1.5 g, with g being the gravitational acceleration.

PUC Dataset: Data was collected from 4 tri-axial ADXL335 accelerometers respectively positioned in the waist, left thigh, right ankle, and right arm, during 8 h of activities. The data recorded corresponds to 5 different activities: "sitting", "sitting down", "standing", "standing up", and "walking". With a sampling frequency of 8 Hz, a total 165,633 samples was collected. We train our CNN model using acceleration signals of the waist sensor as well as all 4 sensor acceleration signals. The purpose of such a training is to show how the number of sensors could affect the classification rate of HAR.

Simulation Details: Our model was implemented using the MatConvNet [13] toolbox of Matlab. Learning parameters are set according to Table 1.

4.2   Classification Results

Table 2 summarizes results of our experiments (CNN in time and frequency domain) as well as results of previous works [1, 20]. Examination of the performances on two datasets highlights the following remarks:

Table 1. Experimental Setup

Parameter	Value	Parameter	Value
The size of input vector	50	Learning rate	0.01
The number of input channels	3~12	Weight decay	0.0005
The number of feature maps	20~200	Momentum	0.9
Filter size	5x1 ~ 9x1	Dropout	0.5~0.8
Pooling size	3x1	Size of mini-batches	150

i. Results indicate that deep learning approaches allow us to recognize human activities in both time and frequency domain.

ii. In the UniMiB SHAR benchmark, our CNN model produces an accuracy of 88.23%, which is higher than other works including SVM [1], KNN [1]. Also, in the PUC benchmark, our CNN outperforms traditional methods with an accuracy of 99.9% against 99.4% (Decision tree & AdaBoost ensemble method. Thus, deep networks are able to capture a better feature representation from acceleration signals, and automatically learned feature representations are more efficient than engineered ones that are labor intensive and time consuming.

iii. Comparing results of time and frequency domain CNN demonstrates the efficiency of time over frequency in the human activity classification task.

iv. Training our model with only one sensor data (waist sensor data) performs less than the model trained on four sensors by 1.36% in time-domain and 0.05% in frequency-domain. This motivates the use of less sensors in the HAR task. Another reason for using less sensors is that we end up with less channels per input signal (i.e., 3 channels using one sensor versus 12 channels using four sensors in the PUC dataset), which is less consuming in terms of memory and training time.

Table 2. Comparison of HAR methods for the UniMiB SHAR and the PUC benchmarks.

Dataset	Reference	Number of Sensors	Method used	Accuracy(%)
PUC	[20]	4	Decision tree & AdaBoost	99.40
	Our solution	4	CNN in Time-domain	**99.90**
	Our solution	4	CNN in Frequency-domain	95.98
	Our solution	1 (Waist)	CNN in Time-domain	98.53
	Our solution	1 (Waist)	CNN in Frequency-domain	95.93
UniMiB SHAR	[1]	1	KNN	86.89
	[1]	1	SVM	86.47
	Our solution	1	CNN in Time-domain	**88.23**

4.3 Different Architectures by Parametrizing CNNs

We consider different time-domain CNN architectures applied on both datasets, with a varying number of hidden layers and feature maps as well as a varying convolutional filter size. The size of feature maps is varied between 10 and 100, 20 and 200 for the first 2 layers respectively. This variation is applied for two architectures: a 4 layer architecture (2 convolutional layers, 2 fully connected layers) and a 5 layer architecture (3 convolutional layers, 2 fully connected layers), as shown in Fig. 2. We notice that, as the number of feature maps increases, the network performs better in recognizing human activities. Indeed raising the feature map size from 20 and 40 in the 1st and 2nd layer respectively ("20–40" in Fig. 2) to 40 and 80 ("40–80" in Fig. 2) increases dramatically the classification accuracy from 52.77% to 85.50%. Also, it is worth mentioning that, with a small feature maps size like 20–40, the accuracy does not stabilize over epochs and keeps oscillating from 0.1 to 52.77% from one epoch to the other, meaning that even a small change in weights and biases of one epoch could make the network performance worse. Therefore, 20 feature maps in the 1st convolutional layer are not enough to represent and extract all low-level features present in the input acceleration signals.

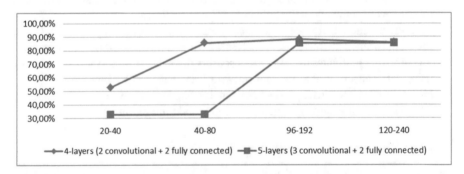

Fig. 2. HAR accuracies of the 4-layer and 5-layer architecture on testing data, with an increasing number of filters. "20–40" configuration has 20 feature maps in the 1st layer and 40 in the 2nd, "40–80" has 40 maps in the 1st layer and 80 in the 2nd, and so one.

Changing the Number of Feature Maps: Varying the feature map size from 40 and 80 in the 1st and 2nd layer respectively to 96 and 192 ("96–192" in Fig. 2) raises the accuracy by 2.73%. However, adding more feature maps seems to drop the performance of the network, suggesting that, beyond a certain feature map size, the complexity of the network is added without extracting any more useful low and mid- level features from input signals. Meanwhile, comparing the HAR accuracy of the 4 layer architecture to the 5 layer one shows that adding an extra layer/section (composed of a convolution, a ReLU activation and a Maxpooling layer) reduces the performance of the network – the accuracy goes down by 2.8% approximately for the "96–192" feature maps architecture as illustrated in Fig. 2. In fact, with this new section added, the gradient of the cost with respect to the weights/biases tends to get smaller as we move backward through the hidden layers, resulting in the vanishing gradient problem. This means that neurons in the earlier layers learn much more slowly than neurons in later layers. And, since all weights and biases of the network are randomly initialized, and the 1st layer weight and bias change very slowly, this 1st layer tends to throw away a lot of information about the input signal. Thus, for our case, it is more convenient to use simpler architectures with a small number of layers.

Changing the Filter Stride of the 1st Convolutional Layer: In addition to varying the feature map size and the number of layers, the filter size/stride of the 1st convolutional layer is varied between 5 and 10 across both architectures mentioned above (the 4 layer architecture and the 5 layer architecture), as shown in Fig. 3. An increase in the stride of the 1st convolutional layer from 5 ($\sim$ a time span of 0.12 s) to 8 ($\sim$ a time span of 0.17 s) results in a better classification, suggesting that a bigger stride seems to capture more low-level details of the input signal. On the other hand, applying a larger stride (9 and 10) diminishes the network performance. So, 0.17 is the best time span of the 1st convolutional layer that is able to retrieve the whole acceleration peaks and the best acceleration changes.

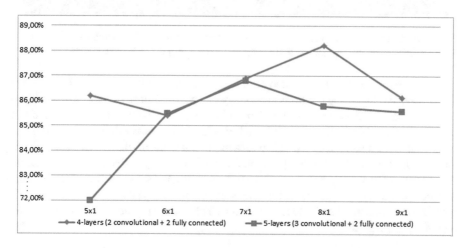

Fig. 3. HAR accuracies of the 4-layer and the 5-layer architecture on testing data, with an increasing size of the stride of the 1st convolutional layer. The range of the convolutional stride is between 5×1 and 9×1.

Therefore, it is clear that adjusting the number of feature maps and convolutional stride is necessary in order to retrieve more low and mid-level features to obtain better high-level features that provide us with better recognition results.

4.4 Feature Representations

We analyze the type of visual features learned by the units of the first convolutional layer by plotting some weights out of the 96 filters learned from the 4-layer architecture illustrated in Fig. 4. For a better visualization, weights of each 1st layer filters are converted from a 5×3 matrix (5 corresponding to the stride of the convolutional filter and 3 corresponding to the channels/axes x, y and z) into a 3×5 matrix. So, a filter has weights of the z-axis in the 1st row, weights of the y- axis in the 2nd row, and weights of x-axis in the 3rd row. White pixels correspond to weights of value 1 and black pixels to weights of value 0. We notice that some filters, have fluctuations in weights, especially in the z-axis (1st row). Thanks to the training phase, filters have learned such variations/ fluctuations in order to capture peaks and subtle changes in the input acceleration signals. However, we don't see as much variations in the learned weights on the y-axis (2nd row) and the x-axis (3rd row) as the ones in the z, suggesting that there is not as much sudden activity movements (i.e. change in acceleration) in the y and x-axis and that most of the sudden movements are in the z- axis. This is because most of dataset activities that our network has been trained on involve vertical up and down movements.

Fig. 4. Some weight filters out of the 96 feature maps/filters in the 1st convolutional layer. Each filter has three rows corresponding to the weights of the z, y and x axis respectively.

Furthermore, let's consider some sample feature maps of the 1st convolutional layer which have the highest activations across x, y and z axes. These samples are displayed in Fig. 5. Each of these samples has 3 plots, representing the x, y and z axis. We can clearly see from plots (a), (b) and (c) that weights in the z-axis show great variations compared to the x and y-axis weights with a minimum value of -0.5276 and a maximum value of 0.7053. We can see big peaks in the z-axis (either positive or negative) while the weights in the x and y-axis stagnate or vary slowly at a very small rate. This confirms our previous analysis which is that the network learned the peaks in the z-axis weights based on the sudden up and down movements. This motivates the use of z-axis signals only to perform HAR.

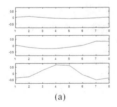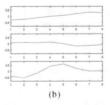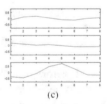

(a)   (b)   (c)

Fig. 5. The best 3 filters out of the 96 filters in the 1st convolutional layer, which are chosen based on the highest activations. Each of the plots has three subplots referring to weights of the x, y and z measurements (top to bottom). The x-axis represents filter weights whose size is the 1st convolutional layer stride, while the y-axis represents values of each weight.

5 Conclusion

In this paper, we presented a 1D CNN *human activity recognition* framework that uses triaxial accelerometer data collected by smartphones. Our method outperformed baseline methods in human activity classification, and exhibited better classification accuracy with data augmentation during the training phase. We found that the size of convolution stride, the number of layers and number of feature maps applied to the input vector could affect the activity recognition performance. Also, by analyzing feature representations of the 1st convolutional layer filters, the distribution of the weights in each x, y and z axis shows certain peaks and variations especially in the z-axis, which suggests that low-level features of our model capture up and down movements the most. For future works, we plan on performing transfer learning for HAR CNN across similar domains.

References

1. Micucci, D., Mobilio, M., Napoletano, P.: UniMiB SHAR: a New Dataset for Human Activity Recognition using Acceleration Data from Smartphones, SL (2016)
2. Bao, L., Intille, S.: Activity recognition from user-annotated acceleration data. Pervasive Computing, LNCS, vol. 3001, pp. 1–17 (2004)
3. Kwapisz, J., Weiss, G., Moore, S.: Activity recognition using cell phone accelerometers. SIGKDD Explor. **12**(2), 74–82 (2010)
4. Jiang, W., Yin, Z.: Human activity recognition using wearable sensors by deep convolutional neural networks. In: Proceedings of MM 2015. ACM (2015)
5. Alsheikh, M.A., Selim, A., Niyato, D., Doyle, L., Lin, S., Tan, H.-P.: Deep activity recognition models with triaxial accelerometers. In: Proceedings of the AAAI (2016)
6. Hammerla, N.Y., Halloran, S., Plotz, T.: Deep, convolutional, and recurrent models for human activity recognition using wearables. In: The IJCAI (2016)
7. Ronao, C.A., Cho, S.-B.: Human activity recognition with smartphone sensors using deep learning neural networks. Expert Syst. Appl. **59**, 235–244 (2016)
8. Yang, J.B., Nguyen, M.N., San, P.P., Li, X.L., Krishnaswamy, S.: Deep convolutional neural networks on multichannel time series for human activity recognition. In: IJCAI 2015 (2015)
9. Krizhevsky, A., Sutskever, I., Hinton, G.E.: Imagenet classification with deep convolutional neural networks. In: NIPS (2012)
10. Lee, H., Pham, P., Largman, Y., Ng, A.Y.: Unsupervised feature learning for audio classification using convolutional deep belief networks. In: NIPS (2009)
11. Bao L., Intille S.S.: Activity Recognition from User-Annotated Acceleration Data. In: PCC, pp. 1–17, May 2001
12. Wu, W., Dasgupta, S., Ramirez, E.E., Peterson, C., Norman, G.J.: Classification accuracies of physical activities using smartphone motion sensors. J. Med. Internet Res. **14**(5), e130 (2012)
13. Vedaldi, A., Lenc, K.: MatConvNet: CNNs for MATLAB (2014). http://www.vlfeat.org/matconvnet/
14. Huynh, T., Schiele, B.: Analyzing features for activity recognition. In: CSOAI (2005)
15. Lin, J., Keogh, E., Lonardi, S., Chiu, B.: A symbolic representation of time series, with implications for streaming algorithms. In: SIGMOD (2003)

16. Bulling, A., Blanke, U., Schiele, B.: A tutorial on human activity recognition using body-worn inertial sensors. ACM Comput. Surv. **46**, 33 (2014)
17. Ronao, C.A., Cho, S., Human activity recognition using smartphone sensors with two-stage continuous hidden markov models. In: The ICNC (2014)
18. Peterek, T.A., et al.: Comparison of classification algorithms for physical activity recognition. In: IBCA (2014)
19. Chatfield, K., Simonyan, K., Vedaldi, A., Zisserman, A.: Return of the devil in the details: delving deep into convolutional nets In: BMVC (2014)
20. Ugulino, W., Cardador, D., Vega, K., Velloso, E., Milidiú, R., Fuks, H.: Wearable computing: accelerometers' data classification of body postures and movements. In: BSAI 2012 (2012)

Analytical View of Augmenting Coarse Cloth Simulations with Wrinkles

Abderrazzak Ait Mouhou[1](✉), Abderrahim Saaidi[1],
Majid Ben Yakhlef[1], and Khalid Abbad[2]

[1] LSI, Department of Mathematics, Physics and Computer Science,
Poly-disciplinary Faculty of Taza, Sidi Mohamed Ben Abdellah University,
Fes, Morocco
{abderrazzak.aitmouh, abderrahim.saaidi,
majid.benyakhlef}@usmba.ac.ma
[2] ISA, Department of Computer Science, Faculty of Science and Technology
of Fez, Sidi Mohamed Ben Abdellah University, Fes, Morocco
khalid.abbad@usmba.ac.ma

Abstract. In this paper, we propose a comparative study between two categories of wrinkle augmentation approaches for clothing mesh simulation, namely, the geometric wrinkle augmentation approaches and data-driven approaches that use wrinkle samples stored in database. We aim to compare these categories of approaches in terms of the ability to generate and animate details on a virtual clothing mesh in motion. The comparison is carried out according to the following criteria: deformation detection, wrinkle shape parameters and run-time performance, as well as the possibility of implementation in GPUs.

Keywords: Clothing animation · Cloth simulation · Wrinkle augmentation
Example-based animation · Wrinkles

1 Introduction

Over the past few decades, the simulation of the behavior of clothes worn by the virtual characters has been an interesting subject for the community of graphic research [13]. Research methods recently carried out for garment wrinkles modeling and animation can be classified mainly in three categories: Methods based on the calculation of geometrical characteristics, methods based on physical forces and model analysis, and methods of wrinkles augmentation on the garment mesh.

First type: geometry based methods. In 1986, Weil's proposal [15] was the first pure geometric modeling method for virtual cloth simulation.

Second type: physical methods [4, 10, 11, 19, 20]. This type of methods, are the most used in computer animation production [10]. More specifically, for computing dynamic effects and collision handling [8].

Third type: methods of wrinkles augmentation on the cloth mesh. This type is the subject of our comparative study. It can be classified into two subcategories: geometric augmentation approaches [5, 7, 12] and methods driven by wrinkle samples stored in database [9, 16–18].

© Springer International Publishing AG, part of Springer Nature 2019
J. Mizera-Pietraszko et al. (Eds.): RTIS 2017, AISC 756, pp. 497–507, 2019.
https://doi.org/10.1007/978-3-319-91337-7_44

The cloth wrinkles represent a complex mechanical phenomenon. It is characterized by nonlinear or anisotropic behaviors, and by a high dynamic range. The physical methods, currently used in this context, force the artists to restart the simulations for many times with different parameters in order to approximate the desired visual results [2]. Therefore, time-machine investment and the requirement for artistic qualities for manual manipulation of parameters often make these methods difficult in the case of garment wrinkles [6]. Wrinkle augmentation research has attempted to bypass all the problems associated with physical methods by adding fine details to the cloth mesh after the initial simulation.

We propose, in this paper, a comparative study between the recently achieved works on the two subcategories of wrinkle augmentation approaches. This study aims to evaluate the ability of approaches to generate the details of the clothing worn by the virtual human character, as well as to highlight their advantages and disadvantages.

The rest of this article is composed as follows. In Sect. 2, We describe the approaches reviewed for the two subcategories of wrinkle augmentation. In Sect. 3, We present our results of the comparison according to the criteria of deformation detection, wrinkle shape parameters, run-time performance, and the possibility of implementation in GPU.

2   Augmentation Approaches Reviewed

wrinkle augmentation approaches offer an alternative to models with complex behavior, such as physical models, because their metaphor of modeling and animation of the wrinkles is based on the fact that the wrinkling operations are limited to mesh areas that undergo a significant deformation, which limits the number of computational operations.

In this section, we present the approaches recently realized under the two sub-categories of wrinkle augmentation approaches.

2.1   Geometric Approaches

Most of the geometric wrinkle augmentation approaches use a two-step-method: The first trace the wrinkle curves on the cloth mesh, and describe the approximate trend of the wrinkles. While the second generate the wrinkles in three-dimensional geometry, according to the curves already defined. The operating diagram of this category of approaches is summarized in Fig. 1.

The input data of the geometric approaches are the initial coarse mesh and its deformation mesh, while their output is the mesh of garment after enhancement with wrinkles. The process begins with tracing the wrinkle curves on the 2D mesh extracted from the initial mesh. Then, mapping these wrinkle curves to the three-dimensional deformation mesh. Finally, selecting and detailing the area of each wrinkle curve to create the geometry of the wrinkles.

Table 1, collects and presents the models and techniques of geometric augmentation examined in this work. The presentation is carried out according to the input parameters and the steps of modeling and animation of wrinkles for each approach.

2.2   Data-Driven Approaches

A wide range of data-driven wrinkle augmentation techniques add the wrinkles to the cloth mesh by utilizing wrinkle samples gained from offline physical simulations [16]. The operating diagram of this category of approaches is summarized in Fig. 2.

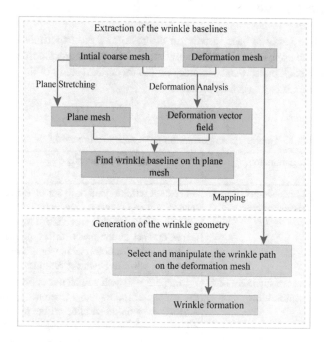

Fig. 1. Schematic diagram of geometric augmentation approaches.

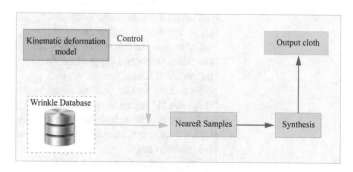

Fig. 2. Schematic diagram of data-driven approaches.

Researches on this approach are based on the observation that wrinkles in clothing behave in a predominantly kinematic way. They assume that each articulation affects only the two adjacent clothing segments. Therefore, most propositions segment the clothing mesh into influence regions according to a kinematic deformation model.

Table 2, collects and presents the techniques of data-driven augmentation examined in this work. The presentation is carried out according to the input parameters, the simulation techniques used to build the database and the steps of modeling and animation of wrinkles for each approach.

3   Comparative Study

Our comparative study is based firstly on the results of the experiments published by the authors of the approaches, and the simulation demonstrations recorded in the videos associated with each paper. And secondly on the theoretical study of the algorithms and the basic assumptions of each approach.

Table 1. Geometric augmentation approaches examined in this study

Approach (Author)	Input parameters	Modeling and animation steps
Cutler et al. [5]	Reference shape 10 to 12 of reference poses	– Create a wrinkle pattern (a set of wrinkle curves) manually using a custom interactive tool – Convert each wrinkle curve into a set of displacement forces that act on the mesh surface points – Create wrinkles deformations arbitrarily in some animation poses at character runtime
Rohmer et al. [12]	Reference shape Smooth garment motion	– Compute a smooth stretch tensor field [15] between the animation frames and a reference garment mesh – Trace the wrinkle baseline on the mesh, in areas of high compression, according to the results of the previous step – Generate the wrinkles geometry using a skeleton-based implicit deformer method by taking the wrinkle baselines as a skeleton
Gillette et al. [7]	—	– Construct a temporally local per-triangle reference shape for each animation frame – Compute a smooth stretch tensor field between each animation frame and the per-triangle reference mesh of the previous frame – Extract the motion patterns of the surface using a graph-cut [3] segmentation technique – Trace the wrinkle baseline on the mesh, in areas of high compression, according to the motion patterns in the previous step – Use programmable GPU tessellation to generate wrinkle geometry

We conducted our study according to five criteria. The first criterion is the deformation detection technique which concerns the mechanism used to track the cloth mesh motion. The second is the wrinkle shape parameters and the possibility of manipulation of these parameters by the artist. The third is the wrinkles persistence characteristics. The fourth is the run-time performance. The last criterion is the possibility of implementing the approach on GPUs.

Table 2. Approaches of data-driven wrinkle augmentation examined in this study

Approach (Author)	Input parameters	Simulation techniques	Modeling and animation steps
Wang et al. [16]	Human pose coarse cloth simulation.	Multiresolution simulation	– Take as input the joints of the human character – Find the wrinkle meshes corresponding to each joint in the pre-calculated database – Merge the joint meshes selected in one mesh – Transfer the detailed mesh to the coarse cloth simulation
Zurdo et al. [18]	Feature mesh Number of times of subdivision	Multiresolution simulation + Weighted Pose-Space Deformation	– Simulate the low-resolution mesh defined by the feature mesh – Subdivide the low-resolution mesh to obtain the smooth mesh – Apply the sample-based calculation to obtain the detailed mesh
Hahn et al. [9]	Rigid skeleton Undeformed surface mesh Animation sequences	Subspace Simulation	– Select from database the sample vectors close to the current pose – Select a set of vectors from the candidate vector base of step 1 – Perform the subspace simulation using the set of vectors selected
Xu et al. [17]	Body mesh Cloth mesh Query pose	Sensitivity Analysis simulation	– Find the nearest samples close to the input pose – Apply the sensitivity-optimized rigging scheme to the recovered samples – Merge the resulting deformations to synthesize the intermediate mesh – Resolve the penetrations then add dynamic effects and damping

3.1   Deformation Detection

In this section, we will compare the two categories of augmentation approaches in terms of their ability to simulate wrinkles, regardless of the source of cloth deformation.

Garment wrinkles are formed mostly because of the large number of contacts and frictions of the cloth and body, they are classified into two types, depending on the cause of mesh deformation (flexion or extension). The simulation of such a type of wrinkles depends mainly on the deformation model used. According to our study of the algorithms of each approach category and the associated demonstration videos, we found that:

- Geometric techniques are limited to wrinkles due to compression. Indeed, after calculating the stretch tensor field for each mesh triangle, the tracing of the wrinkle curves is performed following the vertices that have a maximum compression values.
- Data-driven techniques can detect both compression and stretching wrinkles, using a kinematic deformation model of the character (i.e. articulated skeleton), which represents a source of deformation of the garment mesh. This can be seen for example, in the cloth kinematic equation proposed by Hahn et al. [9]:

$$X_i(p) = \varphi_{LBS}(p, \overline{X})_i = \sum_j w_i^j T_j^T(p) \overline{X}_i \tag{1}$$

This equation is constructed by extending the skinning platform (LBS) of the character model to the cloth mesh. However, these types of kinematic deformation models have a disadvantage when they are applied to cloth with significant inertia effects, such as the deformations of loose-fitting garments, because the clothing deformation cannot be parameterized by the body pose, and the construction of the database of these types of deformation would require significant computational cost and storage exponential due to their number of degrees of freedom.

The summary of the comparison between the approaches, at the level of modeling the wrinkles caused by compression or stretching, is presented in Table 3.

Table 3. Comparison of techniques in terms of deformation detection model used, type of deformation detected and cloth categories supported.

Categories	Approach (Author)	Deformation detection technique	Deformation types	Garment kind
Geometric	Cutler [5]			Tight
	Rohmer [12]	Stretch tensor field	Compression	Tight Loose
	Gillette [7]	Stretch tensor field	Compression	Tight Loose
Data-driven	Wang [16]	Articulated skeleton	Compression	Tight
	Zurdo [18]	Weighted Pose Space Deformation (WPSD)	Compression Stretching	Tight Loose
	Hahn [9]	Linear Blend Skinning (LBS) rig	Compression Stretching	Tight
	Xu [17]	Articulated skeleton	Compression	Tight

3.2   Wrinkle Shape Parameters

In the real case, wrinkles shape is characterized by width and depth. That reflects both the amount of compression or stretching applied to the cloth surface and the fabric's thickness, or the internal structure. For example, the compression of silk fabric will generate many fine wrinkles, while the compression of a piece of leather will form much wider wrinkles.

These properties are modeled in geometric techniques, when creating the 3D geometry of the wrinkles, by the following formula:

$$R(u) = \left(\frac{1 - 2/\pi}{v(u)}\right) R_{min} \qquad (2)$$

With $R(u)$ the radius of the curve, R_{min} is the parameter that defines the cloth nature and $v(u)$ the compression ratio of the curve. This allows the generation of credible and close to reality wrinkles, and offers the artist a control of width and depth of wrinkles. Figure 3 illustrates a sectional view of a wrinkle modeled using a geometric approach proposed by Gillette et al. [8], as well as the wrinkles generated after the simulation.

However, data-driven methods do not have this mechanism to handle the wrinkles shape parameters. Their results depend on the samples of wrinkle stored in the database. Therefore, direct control of the artist seems impossible in this case; thus, the natural characteristics of depth and width depend on the simulation technique used in the stage of creating the wrinkles database. Figure 4 illustrates an example of this dependence in the proposition of Hahn et al. [10].

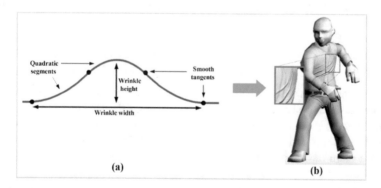

Fig. 3. (a) Sectional view of a wrinkle modeled using a geometric approach [7]. (b) Wrinkles generated after the simulation.

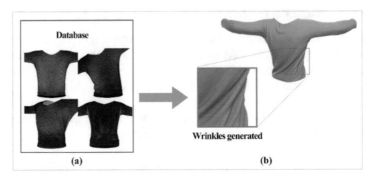

Fig. 4. (a) Visualization of four basis vectors extracted from four different sites in the database of Hahn method [9]. (b) Wrinkles generated after the simulation.

3.3 Temporal Persistence

In real-life, the clothing wrinkles are also persistent, changing their shape and position gradually over time. According to our study of the algorithms of each category of approach, we obtained the following results:

- Geometric methods detect the reaction of the cloth surface in each animation frame using a compute of stretch tensor field between the animation frames and a reference garment mesh, then use the results of this calculation in a process that ensures the persistence of the wrinkles.
- The data-driven methods are mainly based on a linear interpolation model, which takes into account the distribution of the joints weight on the cloth surface and the joints position. Therefore, the synthesis stage depends on the pose configuration and wrinkles examples, rather than the history of the global motion of the cloth. The following Fig. 5 takes as a case study the results of simulation with the method proposed by Wang et al. [16], during 80 animation frames. It represents the sum of mean curvature of the initial mesh, wrinkle samples and the final mesh.

This study shows that the curve of the mean curvature sum in the input high resolution mesh (blue) has stronger fluctuations than the detailed mesh (red). It can also be noted that the curve of the final result (red) has a shape similar to the curve of the wrinkle examples used in the synthesis stage (green). These results show that the final shape of a cloth mesh, after the use of data-driven methods, depends on the samples stored in the database. Indeed, for repetitive animation sequences, these methods generate simulations of similar wrinkle patterns. Therefore, the final mesh may have more wrinkles compared to the real case and these wrinkles appear to be more static.

Table 4, summarizes the advantages and disadvantages of the two sub-categories of wrinkle augmentation approaches at the level of wrinkles persistence.

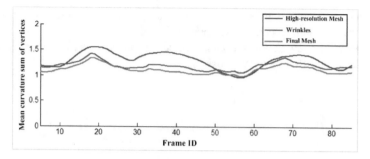

Fig. 5. Comparison of the mean curvature sum of vertices, over 80 frames [16], for input mesh (blue), wrinkles mesh (green) and final mesh (red).

Table 4. Advantages and disadvantages of the two sub-categories in the persistence of wrinkles

Categories	Advantages	Disadvantages
Geometric	Wrinkles synchronize with garment motion	– Wrinkles appear and disappear suddenly – The process used for wrinkle sliding is very sensitive to small changes in stretch tensor field – Absence of static wrinkles in cloth
Data-driven	Presence of static wrinkles in cloth	– Deformations caused by internal cloth friction are not detected – Deformations caused by friction between cloth and the human body are not detected – The quality of the results depends on the size of the database

3.4 Parallelization and Performance

The use of GPUs for simulation offers a massively parallel architecture that allows significant acceleration of a wide range of computational codes [1]. Our study of the operation processes of the two categories of wrinkles augmentation approach showed that:

- The calculations in the geometric methods are independent for each mesh vertex. therefore, it is an iterative algorithm that can be easily optimized in GPUs.
- One of the most interesting features of data-driven methods is that they can be massively implemented on GPUs. This can be seen, for example, in the Eq. (3), which calculates the wrinkles displacement Δx in the proposal of Zurdo et al. [18],

$$\Delta x = \sum_j w_j(u)b_j \tag{3}$$

With w_j the weight associated with a joint j, and b_j the corresponding wrinkle displacement vector in the database. This type of operations is very suitable for parallel implementation on GPUs.

The result of comparing the examined approaches, in terms of GPU implementation and average execution speed, is presented in Fig. 6.

These results show that data-driven methods are faster in performance, despite the static appearance of the generated wrinkles.

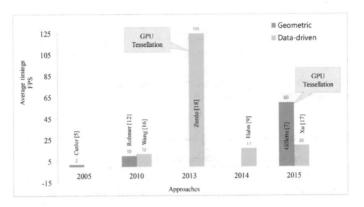

Fig. 6. Comparison of techniques in terms of average execution speed and use of GPU Tessellation.

4   Conclusion

The comparative study, that we have presented, is classified in the context of research work on modeling and animation of virtual garment wrinkles. Specifically, the category of approaches based on wrinkle augmentation on coarse cloth simulations. In this context, we have presented the approaches recently proposed. They are classified in two sub-categories (i.e. geometric methods and data-driven methods). These systems have been evaluated according to five criteria (i.e. applicability, wrinkle shape parameters, run-time performance and the possibility of GPUs implementation). The evaluation has shown that each of the sub-categories of approaches has some advantages compared to the other. Indeed, the geometric methods are operational, regardless of the nature of the garment; they allow a control of width and depth of wrinkles and they produce better results regarding persistence of wrinkles. Data-driven methods are faster in run-time performance; they allow the detection of both compression and stretching deformations [21]. Following this comparative study, we have concluded that the strength of geometric methods lies in their ability to manipulate the cloth mesh directly. And the strength of data-driven methods lies in the use of kinematic deformation model associated to character body, this captures the global shape of clothing in order to guide the recovery process of the wrinkle deformations from database.

References

1. Binotto, A.P., et al: Iterative sle solvers over a cpu-gpu platform. In: 12th IEEE International Conference on High Performance Computing and Communications (HPCC), pp. 305–313. IEEE (2010)
2. Bergou, M., Mathur, S., Wardetzky, M., Grinspun, E.: TRACKS: toward directable thin shells. ACM Trans. Graph. (TOG) **26**(3), 50 (2007)
3. Boykov, Y., Veksler, O., Zabih, R.: Fast approximate energy minimization via graph cuts. IEEE Trans. Pattern Anal. Mach. Intell. **23**(11), 1222–1239 (2001)
4. Baraff, D., Witkin, A.: Large steps in cloth simulation. In: Proceedings of the 25th Annual Conference on Computer Graphics and Interactive Techniques, pp. 43–54. ACM (1998)
5. Cutler, L.D., Gershbein, R., Wang, X.C., et al.: An art-directed wrinkle system for CG character clothing. In: Proceedings of the ACM SIGGRAPH/Eurographics Symposium on Computer Animation, pp. 117–125. ACM (2005)
6. Choi, K. J., Ko, H. S.: Stable but responsive cloth. In: ACM SIGGRAPH 2005 Courses, p 1. ACM (2005)
7. Gillette, R., Peters, C., Vining, N., Edwards, E., Sheffer, A.: Real-time dynamic wrinkling of coarse animated cloth. In: Proceedings of the 14th ACM SIGGRAPH/Eurographics Symposium on Computer Animation, pp. 17–26. ACM (2015)
8. Huh, S., Metaxas, D. N., Badler, N. I.: Collision resolutions in cloth simulation. In: The Fourteenth Conference on Computer Animation, 2001. Proceedings, pp. 122–127. IEEE (2001)
9. Hahn, F., et al.: Subspace clothing simulation using adaptive bases. ACM Trans. Graph. (TOG) **33**(4), 105 (2014)
10. Metaxas, D., Kakadiaris, I.A.: Elastically adaptive deformable models. IEEE Transact. Pattern Analy. Mach. Intell. **24**(10), 1310–1321 (2002)
11. Pabst, S., Thomaszewski, B., Strasser, W.: Anisotropic friction for deformable surfaces and solids. In: Proceedings of the 2009 ACM SIGGRAPH/Eurographics Symposium on Computer Animation, pp. 149–154. ACM (2009)
12. Rohmer, D., Popa, T., Cani, M.P., Hahmann, S., Sheffer, A.: Animation wrinkling: augmenting coarse cloth simulations with realistic-looking wrinkles. ACM Trans. Graph. (TOG) **29**(6), 157 (2010)
13. Seneshen, B., Prasso, L.: Clothing simulations in shrek. In: SIGGRAPH Conference Abstracts and Applications, p. 190 (2000)
14. Talpaert, Y.R.: Tensor Analysis and Continuum Mechanics. Springer, Heidelberg (2013)
15. Weil, J.: The synthesis of cloth objects. ACM SIGGRAPH Comput. Graph. **20**(4), 49–54 (1986)
16. Wang, H., Hecht, F., Ramamoorthi, R., O'Brien, J.F.: Example-based wrinkle synthesis for clothing animation. ACM Trans. Graph. (TOG) **29**(4), 107 (2010)
17. Xu, W., Umetani, N., Chao, Q., Mao, J., Jin, X., Tong, X.: Sensitivity-optimized rigging for example-based real-time clothing synthesis. ACM Trans. Graph. **33**(4), 1–11 (2014)
18. Zurdo, J.S., Brito, J.P., Otaduy, M.A.: Animating wrinkles by example on non-skinned cloth. IEEE Trans. Vis. Comput. Graph. **19**(1), 149–158 (2013)
19. Ning, J. Wenlong, L., Zhenglin, G., Ronald, P.F.: Inequality cloth. In: Proceedings of SCA 2017, Los Angeles, CA, USA, 28–30 July 2017 (2017)
20. Jianfeng, Z., Bing, H.: Garment fold enhancement simulation method based on energy field. In: Proceedings of the 2016 International Conference on Intelligent Information Processing, p. 4. ACM (2016)
21. Phuke, P.J., Rathod, S.B.: Learning influence probabilities in social networks. J. Netw. Technol. **8**(3), 91–99 (2017)

Is Stemming Beneficial for Learning Better Arabic Word Representations?

Ismail El Bazi$^{(\boxtimes)}$ and Nabil Laachfoubi

IR2M Laboratory, FST, Univ Hassan 1, Settat, Morocco
ismailelbazi@gmail.com

Abstract. In this paper, we investigate the effect of stemming on Arabic word representations (WR). We apply various stemmers on different word representations approaches, and conduct an extrinsic evaluation to assess the quality of these word vectors by evaluating their impact on the Named Entity Recognition (NER) task for Arabic, a highly agglutinative and inflectional language. Our findings suggest that the use of light stemming as preprocessing step before learning WR is beneficial and will produce better word representations, especially for morphologically rich languages similar to Arabic.

Keywords: Arabic · Named Entity Recognition · Natural Language Processing
Stemming · Word representations · Word embeddings

1 Introduction

Word representations are mathematical objects representing words in space, often as vectors [1]. Each dimension corresponds to a word feature and can eventually capture semantic and syntactic relationships between words. These representations have been proven to be very effective in numerous Natural Language Processing (NLP) applications.

Generally, we can classify word representations into three categories. The first one is *distributional representations* like Latent Dirichlet Allocation (LDA) [2] which are based upon a word co-occurrence matrix. The second type of WR is *Clustering-based word representations* which induce a clustering over words. A typical example of this category is Brown Clustering (BC) [3]. The third category is *Distributed representations*, commonly known as word embeddings. They are dense, low dimensional and continuous representations of words induced on large unlabeled corpora. Skip-gram Model [4] is an example of word embeddings.

While a lot of studies have focused on word representations for English, few attempts have been carried out to learn word representations for other languages.

In this paper, we focus on Arabic word representations. Arabic is a complex and morphologically rich language and the most important member of Semitic languages family. It has a high agglutinative and inflectional nature in which multiple affixes and clitics can be attached in different ways to words. This leads to a higher degree of sparsity with very large vocabularies and many rare words in comparison with languages such as English. Since each word in the vocabulary has a distinct vector without

© Springer International Publishing AG, part of Springer Nature 2019
J. Mizera-Pietraszko et al. (Eds.): RTIS 2017, AISC 756, pp. 508–517, 2019.
https://doi.org/10.1007/978-3-319-91337-7_45

taking into account the morphology of words, this makes it difficult to learn good word representations for a morphologically rich language as Arabic.

To overcome this limitation, we propose stemming as a way to reduce the data sparseness and enhance the quality of Arabic word representations.

In this study, we explore various stemming approaches and assess the quality of the learned word vectors extrinsically on the Arabic NER task.

Our main objective is to assess the impact of stemming on Arabic word representations to answer the question: Is using stemming as a preprocessing step beneficial for learning better Arabic word representations?

2   Related Work

Very few works have used word representations in NLP tasks for the Arabic language. One of the first ones was the study conducted by Zirikly and Diab [5] where they exploited Brown Clustering as features to enhance the performance of their NER system for Dialectal Arabic (DA). The proposed system delivers $\approx 16\%$ improvement in F-measure over state-of-the-art features when applied to an Egyptian Dialectal Arabic dataset manually annotated by the authors.

In [6], Zirikly and Diab studied the impact of WRs on Arabic NER system for social media data. They introduced gazetteers-free features which include Brown Clustering and Word2vec Cluster IDs and outperformed a state-of-the-art gazetteer based system with an F-measure of 72.68%.

Zahran et al. [7] compared different approaches to build word embeddings for Arabic and assessed the quality of the created word vectors via both intrinsic and extrinsic evaluations. The intrinsic evaluation was done by translating the English benchmark introduced in [4] and use it for Arabic. For the extrinsic evaluation, they employed the word embeddings in two NLP tasks, namely short answer grading and query expansion for information retrieval.

Dahou et al. [8] created a large multi-domain corpus crawled from the web in order to build neural word embeddings for Arabic and evaluated the quality of these word embeddings using sentiment classification. Finally, Elrazzaz et al. [9] introduced the first word analogy benchmark for the Arabic language which can be utilized to implement intrinsic evaluation of different word embeddings. They also evaluated the quality of available Arabic word embeddings using two extrinsic NLP tasks, namely NER and Document Classification.

To the best of our knowledge, no studies were yet carried out to assess the impact of stemming on the quality of word representations for the Arabic language.

3   Word Representations

3.1   Brown Clustering

Brown clustering [3] is an agglomerative hierarchical clustering algorithm which group similar word types into clusters using the mutual information computed according to a class-based bigram language model [10]. The algorithm takes an input sequence $w_1, \ldots, w_n$ of words and returns word clusters as a binary tree, where each leaf is an input word. Each leaf in the tree is tagged with a bit string marking the path from the root to that leaf, where 0 marks a left branch and 1 marks a right branch. Thus, we can uniquely identify each word by its path from the root.

3.2   Latent Dirichlet Allocation

Latent Dirichlet Allocation was initially suggested by Blei et al. [2] for topic modeling. The idea behind LDA is to find coherent topics shared among subsets of a collection of documents. LDA is a generative probabilistic model which learns a set of hidden topics, where each topic is described by a categorical distribution over words.

To induce LDA-based word representations, we use the probabilistic soft word-class algorithm proposed by Chrupala [11] and we add the 3 most probable cluster label of a word as features to our NER system.

3.3   Skip-Gram with Negative Sampling

Skip-Gram with Negative Sampling (SGNS) is a prediction-based model to learn word embeddings initially presented by Mikolov et al. [4]. It is a discriminative neural network model that aims to predict source context words from a pivot word trained using negative-sampling procedure [12]. The word2vec software of Tomas Mikolov and colleagues[1] provides an open-source implementation of SGNS neural network.

3.4   Hellinger Principal Component Analysis

Hellinger Principal Component Analysis (H-PCA) is a Count-based Model proposed by Lebret and Collobert [13] as an effective way of inducing word vector representations. It computes Hellinger PCA of the co-occurrence word statistics to generate word embedding. Lebret and Collobert have demonstrated that applying a well-known spectral method as PCA on the co-occurrence word statistics while optimizing the Hellinger similarity distance can efficiently generate word embeddings as good as the ones produced via complex deep learning architectures.

3.5   Global Vectors for Word Representation

Global Vectors for Word Representation (GloVe) is a method for learning word representations introduced by Pennington et al. [14]. This approach is a global weighted least squares regression model which trains only on the non-zero values in a global co-occurrence matrix. Glove combines the advantages of count-based methods and prediction-based ones to produce word embeddings with meaningful substructure that outperforms other models on NER and similarity tasks.

---

[1] https://code.google.com/archive/p/word2vec/.

3.6 FastText

FastText [15] is an extension of the continuous skip-gram model of Mikolov et al. [4]. FastText is very similar to skip-gram except that it enriches word vectors with subword information using character n-grams. The incorporation of this subword information allows the algorithm to outperform other approaches on rare words, morphologically rich languages and small training datasets [15].

4 Stemmers

4.1 KHOJA Stemmer

Khoja stemmer [16] is one of the first and well-known root-based Arabic stemmer [17, 18]. The Khoja stemmer algorithm follows these steps:

1. Remove diacritics of the input word.
2. Remove punctuation, non-letter characters and stop words.
3. Remove articles (بِالـ, كالـ, فالـ, لِولـ, الـ).
4. Remove conjunction (و).
5. Remove suffixes.
6. Remove prefixes.
7. Compare the remaining word to a set of verbal and noun predefined patterns. Once a match is found get the root.
8. Check the match with the roots dictionary. If no root is found, then leave the word un-stemmed.
9. Replace letters (ي , و , ا) by (و).
10. Replace (ذ, ئ) by (ا).
11. If the root consists of two letters, check if it should have a character in double. If yes, add the character to the extracted root.

4.2 ISRI Stemmer

The Information Science Research Institute's (ISRI) stemmer [19] is a root-based Arabic stemmer. ISRI has various similarities with Khoja stemmer [16]. Yet, his main advantage over Khoja is that ISRI does not use any type of dictionaries to linguistically validate the extracted roots. The ISRI stemming algorithm follows these steps:

1. Remove diacritics of the input word.
2. Normalize the Hamza to one form (ا).
3. Remove prefixes of three and two characters in that order.
4. Remove inseparable conjunction (و) if it comes before a word starting with (و).
5. Normalize all the forms of Hamza (آ, إ, أ) to (ا).
6. Return root if less than or equal to three characters.
7. Searches for any matches in a list of predefined patterns, if no match is found; it successively tries to remove affixes of single character and repeat the search.
8. Stop the stemming when it either a match is found and then it returns the root, or the word is too short to be stemmed.

4.3   Light10 Light Stemmer

Light10 is a light stemmer developed by Larkey et al. [20] for Arabic Information Retrieval. Light10 is a fast and straightforward algorithm. It follows this procedure:

1. Remove conjunction (و) if this leaves three or more characters.
2. Remove any of the prefixes that are found at the beginning of the word if this keeps at least two characters.
3. Remove all the suffixes if this keeps at least two characters.

The list of affixes to be removed by Light10 is reported in Table 1.

Table 1. Strings removed by Light10 [20]

Prefixes removed	Suffixes removed
ال،وال، بال، كا ل، فا ل، لل ، و	ها، ات، ان ، ون ، ين، ية ، يه ، ة ، ه ، ي

4.4   High Precision Stemmer

High Precision Stemmer[2] (HPS) [21] is an unsupervised multi-purpose stemmer that exploits both semantic and lexical information about words in the stemming process. Actually, HPS is suffix stripping stemmer which considers, that the longest prefix as the stem and the remaining as the suffix. The HPS consists of two main stages. In the first stage, we cluster words sharing the longest common prefix and lexically similar using Maximal Mutual Information clustering (MMI) [3]. The second stage applies a Maximum Entropy classifier with general features to strip suffixes and returns stems. We use the stemming candidates created from clustering at the first stage as training data for the classifier.

HPS was successfully tested in six languages, namely Czech, English, Hungarian, Polish, Slovak and Spanish [21].

The two main parameters to set for HPS are: The max length of suffix M and The minimum similarity between words with the same stem δ. In our study, we set M to 3 and δ to 0.7 according to the recommendation of the authors.

5   Experimental Setup

5.1   NER System

Our NER system is a Conditional Random Fields (CRF) sequence labeling classifier, which is considered as the state-of-the-art model for NER by many authors. We use CRFsuite a fast implementation of CRF provided by Naoaki Okazaki [22].

---

[2] http://liks.fav.zcu.cz/HPS/

All features are defined over a window of -1, ..., $+1$. We employ the following baseline feature set for our study:

Word – The word itself is used as a feature.

Affixes – Prefixes and suffixes of length from 1 to 4 are extracted from the current word.

Morphological Features – Aspect, Gender, Case, Number, NormWord and Part-of-speech tag. They were generated by the MADA toolkit and already available within AQMAR corpus. The detailed description of these features is presented in Table 2.

In addition to the aforementioned features, we use the word representations feature as follows:

- The dense continuous embeddings vector as features for all the word embeddings (SGNS, GloVe, FastText and H-PCA).
- The Brown Clustering IDs of variable prefix lengths (2, 3, 4, 5 and 6) as features for Brown Clustering.
- The top 3 most probable cluster label of a word as features for LDA.

Table 2. Morphological Features

Feature	Feature values
Part-of-speech	*Part of speech*: Nouns, Proper Nouns, Number Words, Verbs, Adverbs, Adjectives, Particles, Prepositions, Pronouns, Abbreviations, Digital Numbers, Conjunctions, Interjections, Punctuation, Foreign/Latin
Aspect	*Verb aspect*: Perfective, Imperfective, Command, Not applicable
Case	*Grammatical case*: Genitive, Accusative, Nominative, Not applicable, Undefined
Gender	*Nominal Gender*: Masculine, Feminine, Not applicable
Number	*Grammatical number*: Plural, Dual, Singular, Not applicable, Undefined
NormWord	Normalized spelling of the word form (romanized)

5.2 Evaluation Corpus

The Arabic Wikipedia Named Entity Corpus (AQMAR) is a small hand-annotated corpus of 28 Arabic Wikipedia articles for Arabic named entities [23]. Each article was annotated by 1 of 2 annotators with the traditional four entity classes: Person, Organization, Location, and generic Miscellaneous (MIS) following the BIO tagging format. The AQMAR corpus consists of 74000 tokens and 2687 sentences.

In this study, we used the test part as the training corpus. We divide the development part into half; one was used as development corpus and the other as the testing corpus.

5.3   Experimental Setting

We take the Arabic Wikipedia dump until December 2016 as our unlabeled data[3] and preprocess it using Gensim library[4] by removing all MediaWiki markups and tokenizing the texts. The resulting text represents the input dataset (ArWiki) used to learn all the normal word representations. For the stemmed version of the word representations, we first stem the words in ArWiki corpus with one of the stemmers used in our study to create a new stemmed corpus (S-ArWiki) which we exploit as input dataset to induce stemmed WRs. Table 3 summarizes the training details of the four word embeddings models (SGNS, FastText, H-PCA and GloVe).

For the Brown clustering features, we induce 100 brown clusters of words with the implantation provided by Percy Liang[5].

We use the software package provided by Chrupala[6] to generate the Soft Word-Class LDA features. The numbers of topics K is set to 100 and the hyper-parameters of the Dirichlet distribution α and β to $\alpha = 10/K$ and $\beta = 0.01$.

Table 3. Word embeddings training parameters.

	SGNS	FastText	H-PCA	GloVe
Vector size	50	50	50	50
Window size	2	2	2	2
Minimum word count	5	5	5	5
Sampling threshold	$1e-3$	$1e-3$	N/A	N/A
Hierarchical Softmax	No	No	N/A	N/A
Negatives samples	5	5	N/A	N/A
Max iterations	N/A	N/A	N/A	15
X_Max	N/A	N/A	N/A	10

Table 4. NER results for AQMAR dataset.

	SGNS	Glove	FastText	HPCA	BC	LDA
	F1	F1	F1	F1	F1	F1
NoStem	69,93	69,65	69,71	69,31	70,67	70,83
Light10	**70,07**	**69,78**	**69,82**	70,17	71,68	**72,52**
ISRI	69,1	69,28	**69,85**	69,67	72,02	71,36
Khoja	69,24	68,7	68,9	**69,43**	70,99	69,03
HPS	69,43	68,83	68,88	68,9	68,01	68,2

[3] https://dumps.wikimedia.org/arwiki/.
[4] https://radimrehurek.com/gensim/.
[5] https://github.com/percyliang/brown-cluster.
[6] https://bitbucket.org/gchrupala/lda-wordclass/.

6 Experiments and Results

Our experiments are quite straightforward. We learn word representations based on the techniques presented in Sect. 3 using each stemming approach and train a NER model with these WRs on the AQMAR training corpus. Then, we evaluate these models on the test corpus. For all our experiments, the features introduced in Sect. 5 are used. The evaluation is done using the CoNLL strict evaluation metric F-measure (F1). We consider the NER model without stemming as our baseline model. The results of our experiments are depicted in Tables 4. The values above the no stemming baseline model are in bold.

The best results were achieved using the Light10 stemmer with all its results above the baseline. The second best results were achieved with ISRI stemmer with four out of six results above the baseline. The Khoja stemmer is ranked third with just two results above the baseline. Surprisingly, we got the worst results using the HPS stemmer with all its results significantly lower than the no stemming baseline.

The good results reached by the Light10 stemmer can be explained by the fact that stemming copes with the agglutinative and inflectional nature of Arabic language by reducing the number of distinct word forms. Thus a reduced number of feature values leads to the construction of more reliable and accurate NER models, as it can be concluded from the results.

On the other hand, the lower values produced by Khoja stemmer are due to the use of roots dictionary to validate extracted roots. Since the number of roots in the dictionary is limited, a lot of words are not stemmed, specially named entities, which keep the number of word forms high and hence reduce the performance of the NER model. That is why the ISRI stemmer shows better results in comparison with Khoja since it did not rely on root dictionary to validate extracted roots and so it did not suffer from this weakness.

For the HPS stemmer, we were surprised that we got the worst results, even less than the baseline, using this stemmer. In fact, HPS was successfully used with highly inflectional languages as Czech and Hungarian [21], so normally it should be the same as Arabic, another highly inflectional Semitic language. These poor results can be explained by the approach used by HPS to stem words. Actually, HPS is a suffix stripping stemmer which assumes that each word is composed of a stem and a suffix. This is probably true for languages like Czech and Hungarian, but it is not the case of Arabic where the agglutinative morphology of the language allows having both prefixes and suffixes attached to the stems. Also, the unsupervised way adopted by HPS seems not dealing in the best way with the richness of Arabic by not correctly removing the right suffixes. Thus, the HPS stemmer cannot reduce the data sparseness sufficiently to enhance the performance of the NER model.

Generally, the empirical results show that the techniques based on the light stemming significantly outperform the techniques based on root extraction or unsupervised learning and are more likely to produce better Arabic word representations.

7 Conclusion and Future Work

We have tested four distinct stemming approaches in six word representations models and evaluated their impact on the Arabic NER task. The stemming approaches used include light, root extraction, and unsupervised stemming.

The results show that light stemmer significantly outperforms both the root extraction stemmers and the unsupervised stemmer and confirm that the use of light stemming can enhance the quality of the Arabic word representations.

Many paths can be followed to extend our current study, starting with the use of other stemming approaches and the assessment of the word vectors using both intrinsic and extrinsic evaluations. It will also be interesting if we can extend our study to other morphologically rich languages, especially within the Semitic family.

References

1. Turian, J., Ratinov, L., Bengio, Y.: Word representations: a simple and general method for semi-supervised learning. In: Proceedings of the 48th Annual Meeting of the Association for Computational Linguistics, pp. 384–394. Association for Computational Linguistics (2010)
2. Blei, D.M., Ng, A.Y., Jordan, M.I.: Latent dirichlet allocation. J. Mach. Learn. Res. **3**, 993–1022 (2003)
3. Brown, P.F., deSouza, P.V., Mercer, R.L., Pietra, V.J.D., Lai, J.C.: Class-based N-gram models of natural language. Comput. Linguist. **18**, 467–479 (1992)
4. Mikolov, T., Chen, K., Corrado, G., Dean, J.: Efficient estimation of word representations in vector space. In: Proceedings of Workshop at ICLR (2013)
5. Zirikly, A., Diab, M.: Named entity recognition for dialectal arabic. In: ANLP 2014, p. 78 (2014)
6. Zirikly, A., Diab, M.T.: Named entity recognition for arabic social media. In: VS@ HLT-NAACL, pp. 176–185 (2015)
7. Zahran, M.A., Magooda, A., Mahgoub, A.Y., Raafat, H.M., Rashwan, M., Atyia, A.: Word representations in vector space and their applications for arabic. In: Gelbukh, A. (ed.) Computational Linguistics and Intelligent Text Processing: 16th International Conference, CICLing 2015, Cairo, Egypt, 14–20 April, 2015, Proceedings, Part I, pp. 430–443. Springer, Cham (2015)
8. Dahou, A., Xiong, S., Zhou, J., Haddoud, M.H., Duan, P.: Word embeddings and convolutional neural network for arabic sentiment classification. In: COLING, pp. 2418–2427 (2016)
9. Elrazzaz, M., Elbassuoni, S., Shaban, K., Helwe, C.: Methodical evaluation of arabic word embeddings. In: Proceedings of the 55th Annual Meeting of the Association for Computational Linguistics (Volume 2: Short Papers), pp. 454–458. Association for Computational Linguistics, Vancouver (2017)
10. Liang, P.: Semi-supervised learning for natural language (2005)
11. Chrupala, G.: Efficient induction of probabilistic word classes with LDA (2011)
12. Mikolov, T., Sutskever, I., Chen, K., Corrado, G.S., Dean, J.: Distributed representations of words and phrases and their compositionality. In: Burges, C.J.C., Bottou, L., Welling, M., Ghahramani, Z., and Weinberger, K.Q. (eds.) Advances in Neural Information Processing Systems, vol. 26, pp. 3111–3119. Curran Associates, Inc. (2013)
13. Lebret, R., Collobert, R.: Word embeddings through hellinger PCA. In: EACL (2014)

14. Pennington, J., Socher, R., Manning, C.D.: Glove: global vectors for word representation. In: EMNLP, pp. 1532–1543 (2014)
15. Bojanowski, P., Grave, E., Joulin, A., Mikolov, T.: Enriching word vectors with subword information. arXiv preprint arXiv:1607.04606 (2016)
16. Khoja, S., Garside, R.: Stemming arabic text. Computing Department, Lancaster University, Lancaster, UK (1999)
17. Al-Sughaiyer, I.A., Al-Kharashi, I.A.: Arabic morphological analysis techniques: a comprehensive survey. J. Am. Soc. Inform. Sci. Technol. **55**, 189–213 (2004)
18. Larkey, L.S., Connell, M.E.: Arabic information retrieval at UMass in TREC-10. DTIC Document (2006)
19. Taghva, K., Elkhoury, R., Coombs, J.: Arabic stemming without a root dictionary. In: 2005 International Conference on Information Technology: Coding and Computing, ITCC 2005, vol. 1, pp. 152–157 (2005)
20. Larkey, L., Ballesteros, L., Connell, M.: Light stemming for arabic information retrieval. In: Soudi, A., Bosch, A. den Neumann, G. (eds.) Arabic Computational Morphology, pp. 221–243. Springer (2007)
21. Brychcín, T., Konopík, M.: HPS: High precision stemmer. Inf. Process. Manage. **51**, 68–91 (2015)
22. Okazaki, N.: CRFsuite: a fast implementation of Conditional Random Fields (CRFs) (2007). http://www.chokkan.org/software/crfsuite/
23. Mohit, B., Schneider, N., Bhowmick, R., Oflazer, K., Smith, N.A.: Recall-oriented learning of named entities in arabic wikipedia. In: Proceedings of the 13th Conference of the European Chapter of the Association for Computational Linguistics, pp. 162–173. Association for Computational Linguistics, Avignon (2012)

Arabic Sign Language Alphabet Recognition Methods Comparison, Combination and Implementation

Mohamed Youness Ftichi[(✉)], Abderrahim Benabbou, and Khalid Abbad

Department of Intelligent Systems and Applications, Faculty of Sciences and Technologies, Fes, Morocco
{Mohamedyouness.ftichi,abderrahim.benabbou,
khalid.abbad}@usmba.ac.ma

Abstract. Sign language can be defined as a combination of hand motion mainly used for communication purposes, especially for the deaf-mute community. More than 5% of the worldwide population (320 million) are concerned by the use of it. Through our work, we aim to provide a mean to automate the process of translation from Arabic sign language to written Arabic, in the static context. As a first step, we will produce a three-level process, allowing the recognition of static Arabic sign.

Keywords: Arabic sign language · Recognition · Skin color segmentation
Hull convex · Classification · Hand pose

1 Introduction

In literature, most static sign language recognition systems (based on images), and generally static hand gesture recognition systems, based on vision, are composed of three basic phases: segmentation, features extraction and recognition. Our work will be a combination of several methods, following the same schema to get a translation from Arabic sign language alphabet to written Arabic.

First step will be the skin color segmentation within a static image. This step will be crucial to get the correct segmented image for the rest of the process by generating a skin color model to work with.

The features extraction will allow us to get the characteristics best describing the sign (hand pose), and pass it on to the classifier. A number of classifiers will be tested and one will be chosen as the one with best results.

2 Skin Color Segmentation

Skin color segmentation is one of the widely used techniques to get a robust hand segmentation, it main aims to build a decision rule allowing as to separate skin from non-skin

pixels. A metric measuring the distance between the pixels values and skin tone is generally used, this metric is defined according to the method chosen to model the skin color.

2.1 Skin Modeling Method

Segmentation through a predefined set of rules is among the many efficient methods available to get skin clusters in images. Many researchers, such as [1–4], have tempted to use
this method in their works, essentially because of how easy it is to construct a fast skin
color classifier. Nevertheless, achieving a high recognition rate lies on the color space
choice and the tests to get the adequate set of rules.

Recently, there have been proposed a method that uses machine learning algorithms to
find both suitable color space and a simple decision rule that achieve high recognition rates
[5]. The authors start with a normalized RGB space and then apply a constructive induction algorithm to create a number of new sets of three attributes being a superposition of r,
g, b and a constant 1/3, constructed by basic arithmetic operations. A decision rule, is estimated for each set of attributes. The authors prohibit construction of too complex rules,
which helps avoiding data over-fitting, that is possible in case of lack of training set repre-
sentativeness. They have achieved results that outperform Bayes skin probability map
classifier in RGB space for their dataset.

In our work, and through experiment and comparison of the different available techniques, we have concluded that the following criteria, presented in the [1], is the best
suited for our usage, and is described as follow:

$$\text{If } \frac{3br^2}{(r+g+b)^3} > 0.1276$$

$$\text{And } \frac{r+g+b}{3r} + \frac{r-g}{r+g+b} \leq 0.9498 \tag{1}$$

$$\text{And } \frac{rb+g^2}{gb} \leq 2.7775$$

$$\text{Then Class} = \text{Skin}$$

2.2 Application

A standard hand gesture dataset, provided in [6–8], is used to test our process Fig. 1.

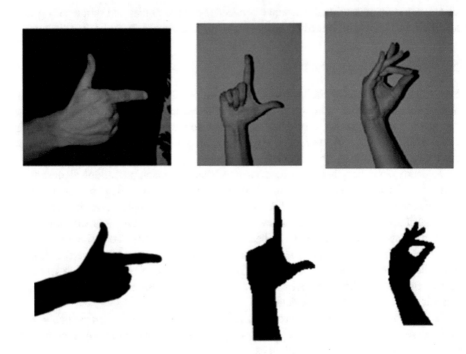

Fig. 1. Hand segmentation based on the chosen method, on top the original image, on the bottom the segmented one.

3   Features

3.1   In Literature

Features extraction is the major step toward the recognition of the Arabic sign language alphabet. This step is the mean to reduce the dimensionality of the problem, and give the minimum amount of data to the classifier (see Sect. 4), so we can get an identification of the hand pose correspondent in the sign language vocabulary.

In literature, many different methods have been used to characterize the main key elements of an image, in the hand pose recognition context. Among the various methods, SIFT [9], Hu Moments [10], dimensional Gabor Wavelets [11] and Fourier descriptors [12] have been used with more or less success to extract the features to use, efficiently, to recognize the sign.

3.2   Our Approach

For the purpose of work, we choose to use the hull convex [13] as a mean to detect the defection point that are the base on the features extracted for the recognition part (see Sect. 4).

Based on the results of the skin color segmentation, we apply the Hull Convex algorithm, following the extraction of the hand's contour to get the convexity defect points in those images as shown Fig. 2.

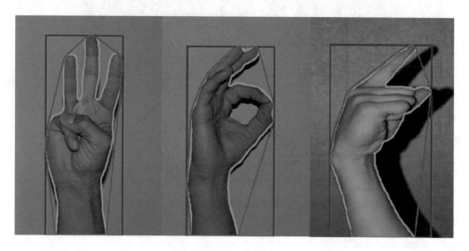

Fig. 2. Hull convex method applied to segmented images

The Hull convex step is followed by the drawing of the convexity defect points for each sign as shown Fig. 3.

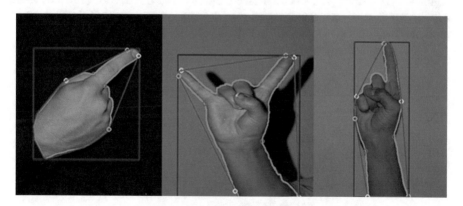

Fig. 3. Defect points identification

The points in yellow, blue and purple represent respectively the beginning of the defection (Bg), its end (En) and the defection point (Df). The next step is to calculate the following distances (Fig. 4):

- A = Distance from the defect point to the beginning of the defection;
- B = Distance from the defect point to the end of the defection;
- C = the angle $\left(Bg\widehat{Df}En \right)$

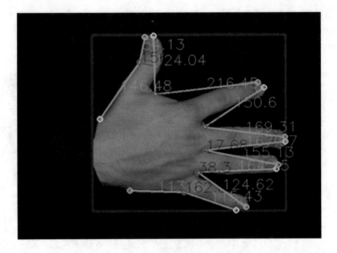

Fig. 4. Numbers in blue are the angles, the green ones are the distances.

3.3   Classification

Classification is the last, but most important step of the recognition process. In order to get the best result, we experimented different classification methods available in the literature, most specifically Stochastic Gradient Descent [14], Random Forest [15], Logistic Regression [16], KNN [17], Decision Tree [18], SVC and Linear SVC [19]. The vector used for classification contains the number of defect points, the angle C, the distances A and B, and the different locations of the defects point.

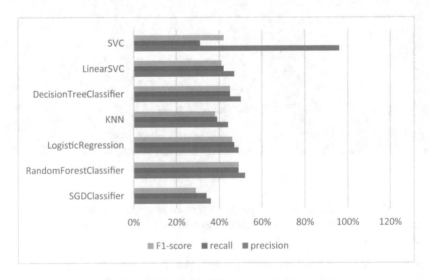

Fig. 5. Classification methods comparative results

We use the dataset provided by [6–8]. The learning process contains 10 images for each sign, for a total of 240 images. The next step is to apply those methods to the rest of the dataset which consists of a total of 899 images, approximately 33 images for each sign. The results of the recognition process are Fig. 5.

4   Conclusion

Through our work, we were able to address the matter of translation from Arabic sign language based on static images provided by the standard hand pose dataset on [6–8]. Test were realized on an intel core I7-4710MQ CPU 2.50 GHZ 2.50 GHZ computer and using Scikit image and OpenCV frameworks on python.

5   Future Work

We are currently working on the establishment of a translation process based on dynamic hand poses (videos), then, later, hand gestures in real time. Such objectives are realizable based on the work presented in this paper and are our main focus.

References

1. Peer, P., Kovac, J., Solina, F.: Human skin colour clustering for face detection. In: EUROCON 2003 – International Conference on Computer as a Tool (2003)
2. Ahlberg, J.: A system for face localization and facial feature extraction. Technical report LiTH-ISY-R-2172, Linkoping University (1999)
3. Fleck, M., Forsyth, D.A., Bregler, C.: Finding naked people. In: Proceedings of the ECCV, vol. 2, pp. 592–602 (1996)
4. Jordao, L., Perrone, M., Costeira, J., Santos-Victor, J.: Active face and feature tracking. In: Proceedings of the 10th International Conference on Image Analysis and Processing, pp. 572–577 (1999)
5. Gomez, G., Morales, E.: Automatic feature construction and a simple rule induction algorithm for skin detection. In: Proceedings of the ICML Workshop on Machine Learning in Computer Vision, pp. 31–38 (2002)
6. Kawulok, M., Kawulok, J., Nalepa, J., Smolka, B.: Self-adaptive algorithm for segmenting skin regions. EURASIP J. Adv. Signal Process. **2014**, 170 (2014)
7. Nalepa, J., Kawulok, M.: Fast and accurate hand shape classification. In: Kozielski, S., Mrozek, D., Kasprowski, P., Malysiak-Mrozek, B., Kostrzewa, D. (eds.) Beyond Databases, Architectures, and Structures. CCIS, vol. 424, pp. 364–373. Springer (2014)
8. Grzejszczak, T., Kawulok, M., Galuszka, A.: Hand landmarks detection and localization in color images. Multimed. Tools Appl. **75**(23), 16363–16387 (2016)
9. Carson, C., Belongie, S., Greenspan, H., Malik, J.: Blobworld: image segmentation using expectation-maximization and its application to image querying. IEEE Trans. PAMI **24**(8), 1026–1038 (2002)
10. Hu, M.: Visual pattern recognition by moment invariants. IRE Trans. Inf. Theory **8**, 179–187 (1962)

11. Chang, T., Kou, C.: Texture analysis and classification with tree-structured wavelet transform. IEEE Trans. Image Process. **2**(4), 429–441 (1993)
12. Wang, T., Liu, W., Sun, J., Zhang, H.: Using Fourier descriptors to recognize object's shape. J. Comput. Res. Develop. **39**(12), 1714–1719 (2002)
13. Youssef, M.: Hull Convexity Defect Features for Human Action Recognition. Dayton University, Ohio (2011)
14. Bottou, L.: Stochastic gradient descent tricks. In: Montavon, G., Orr, G.B., Müller, K.R. (eds.) Neural Networks: Tricks of the Trade. LNCS, vol. 7700. Springer, Heidelberg (2012)
15. Breiman, L.: Random forests. Mach. Learn. **45**, 5–32 (2001). https://doi.org/10.1023/A:1010933404324
16. Yu, H.F., Huang, F.L., Lin, C.J.: Dual coordinate descent methods for logistic regression and maximum entropy models. Mach. Learn. **85**, 41–75 (2011). https://doi.org/10.1007/s10994-010-5221-8
17. Andoni, A., Indyk, P.: Near-optimal hashing algorithms for approximate nearest neighbor in high dimensions. In: Proceedings of 47th Annual IEEE Symposium on Foundations of Computer Science (FOCS), October 2006, pp. 459–468 (2006)
18. Breiman, L., Friedman, J., Olshen, R., Stone, C.: Classification and Regression Trees. Wadsworth, Belmont (1984)
19. Cortes, C., Vapnik, V.: Support-vector networks. Mach. Learn. **20**, 273–297 (1995). https://doi.org/10.1007/BF00994018

Author Index

© Springer International Publishing AG, part of Springer Nature 2019
J. Mizera-Pietraszko et al. (Eds.): RTIS 2017, AISC 756, pp. 525–526, 2019.
https://doi.org/10.1007/978-3-319-91337-7

Printed in the United States
By Bookmasters